NUCLEAR STRUCTURE, REACTIONS AND SYMMETRIES

VOLUME 1

NUCLEAR STRUCTURE, REACTIONS AND SYMMETRIES

VOLUME 1

5-14 June 1986
Dubrovnik, Yugoslavia

Richard Adlin Meyer
Vladimir Paar

editors

World Scientific

Published by

World Scientific Publishing Co Pte Ltd.
P. O. Box 128, Farrer Road, Singapore 9128

Library of Congress Cataloging-in-Publication data is available.

NUCLEAR STRUCTURE, REACTIONS AND SYMMETRIES

Copyright © 1986 by World Scientific Publishing Co Pte Ltd.

All rights reserved. This book, or parts thereof, may not be reproduced in any form or by any means, electronic or mechanical, including photocopying, recording or any information storage and retrieval system now known or to be invented, without written permission from the Publisher.

ISBN 9971-50-141-4

Printed in Singapore by Kim Hup Lee Printing Co. Pte. Ltd.

THIS BOOK

IS

DEDICATED TO

THOSE WHO VOLUNTEERED

THEIR

TIME, TALENT AND HOMES

IN ORDER

TO

BRING NSRS

TO

FRUITION

PREFACE

This book is organized into three main parts. Part I concerns three issues in nuclear structure that are presently of considerable interest: Octupole and Dipole Modes in Nuclei (Sec. I), Bose-Fermi Symmetries and SUSYs (Sec.II), and Mixed Symmetry States in Nuclei (Sec. III). Part II contains mainly a survey of new experimental information and techniques. Sec. IV, the first chapter in this part, concerns experimental techniques and results from scattering reaction studies (note bene: the last part of Sec. III is related to Sec. IV in that it contains descriptions of scattering reaction studies aimed specifically at studying mixed symmetry states). Next is a chapter on Symmetries and Reactions (Sec. V). This chapter is juxtaposed to the previous one because it predominantly addresses the description of scattering reactions. The second part is completed with Sec. VI and VII: Experimental Techniques and Studies, and Systematic Properties, respectively. Part III contains material on the present status of several areas of nuclear theory: Microscopic Theory (Sec. VIII), Geometric and Algebraic Models (Sec. IX), and Recent Developments in Theory (Sec. X). The last of these, Sec. X, contains three subunits: A) The Symplectic Model, B) The IBM Model, and C) Various Models.

The chapters consist of manuscripts of invited lectures at the International Conference on Nuclear Structure, Reactions, and Symmetries (**NSRS**). The NSRS conference was organized as a forum for debating several issues and comparing various views of describing nuclear structure. To better focus on the issues and various models, leading experts were invited to present review lectures and requested to present bias towards revealing where the various views differed. These lectures, given in the evening, were followed by a discussion session that addressed not only the questions raised by each reviewer but also those raised during the day's invited presentations or instigated by the discussion session moderator. Because we believe that it is often the dynamics of these interchanges that can provide new physics, the discussion sessions were recorded, transcribed, and edited (by the participants) during the conference. Thus Sec. I, II, III, V, VIII, and IX start with a prepared version of the review talk and are accompanied by the discussion that followed the review lecture. Finally, because of the interest by the participants in another conference held at the same time as the NSRS, an invited lecture was given which summarized the proceedings of the "NATO Workshop on The Physics of Strong Fields". We are indebted to Prof. dr. Walter Greiner, University of Frankfurt am Main, F.R. Germany for arranging this survey of his workshop (see Sec. XI and the article by Dr. G. Soff).

The NSRS conference was held at the Inter-University Centre (**IUC**) in Dubrovnik, Yugoslavia from 5 to 14 June 1986. The total attendance of 251 scientists from approximately 50 countries certainly attests to the high and renewed interest in the area of nuclear structure that exists today. In fact, it was because of this interest that the original conference bifurcated into the NSRS conference and a parallel symposium on Symmetries and Nuclear Structure (SANS). The symposium contained 107 individual presentations of specific studies in nuclear structure and has its own published proceedings (Harwood Academic Publishers, Inc.).

A special effort was made to include graduate students who are presently engaged in thesis work in nuclear science. A separate session of student presentations took place and prizes were given for those who were voted to be most outstanding in content and presentation. The prizes were named in honor of Prof. Gaja Alaga of the University of Zagreb, who has made significant contributions to the field of nuclear structure as well as to science education in Yugoslavia (see Alaga Award announcement page at the end of this preface).

The conference and symposium benefitted greatly from being located at the IUC in Dubrovnik. Founded in 1970 by Prof. Ivan Supek, the IUC serves as a location for both conferences and small study groups, which can utilize its large auditorium and smaller independent seminar and course rooms. Mrs. Berta Dragicevic serves officially as executive secretary of the IUC, but in fact does the work of an executive director of the facility. Her efforts on behalf of the NSRS conference are greatly appreciated.

No conference can take place without sponsors and financial contributors. On a separate page we present a list of those organizations to which we are grateful for their support of the NSRS conference. We especially want to thank Prof. D.A. Bromley as well as Prof. B. Barrett, Prof. F. Iachello, Dr. G. Sher, and Prof. P. von Brentano for their guidance and assistance in obtaining conference support. The efforts of many others also made the NSRS conference a success.

Prof. Daeg S. Brenner of Clark University served as co-finance officer with Prof. Ruth P. Yaffe of San Jose State University. The conference is indebted to their efforts in obtaining support and administering the grants. The conference organization benefitted greatly from Prof. Franco Iachello of Yale University, who took time to attend organizational meetings throughout the year preceding the conference. The limited finances of the conference were greatly extended by the contributions of the "California Volunteers" (Joyce Plis, Ruth Randig Meyer, Martha Meyer, and Marlene Meyer), who provided all the secretarial labor and "office" space for both pre- and post-conference activities.

On-site, the conference was aided by the many individuals who volunteered their time and arranged for their own overseas travel support. On a separate page we give a list of these people. The conference benefitted from several graduate students who held conference-supported fellowships as discussion session monitors. Additionally, fellowships from the U.S. National Science Foundation provided sufficient funds for the attendance of several U.S. graduate students who also assisted as discussion session monitors.

We are indebted to Dr. Slobodan Brant of the University of Zagreb, who acted as on-site secretariat and contributed many hours every day assuring the smooth operation of the conference. We also want to thank Mrs. Ruth Randig Meyer, whose efforts in providing a positive social environment for both the participants and their companions led to the title of "conference house mother".

Last, but hardly least, we want to thank the administration of the Lawrence Livermore National Laboratory, particularly Drs. C. Gatrousis and G.L. Struble, for their support of activities in behalf of the conference, and to the Institut fuer Kernphysik II, KFA-Julich GmbH and its director Prof. dr. O.W.B. Schult for the hospitality and support received during part of the post-conference editing of these proceedings. We wish to express our gratitude to Jay Cherniak who provided both pre- and post-conference technical editing. In addition we are very indebted to our colleagues, Dr. E.A. Henry, R.W. Hoff, R.G. Lanier, and L.G. Mann who provided an understanding environment in which this conference was established.

Richard Adlin Meyer
Vladimir Paar

Livermore, California
25 August 1986

THE INTERNATIONAL CONFERENCE
ON
NUCLEAR STRUCTURE, REACTIONS, AND SYMMETRIES
AND
ITS PARTICIPANTS

are indebted to the following oganizations
for their role as:

Sponsors:

INTERNATIONAL UNION OF PURE AND APPLIED PHYSICS

LAWRENCE LIVERMORE NATIONAL LABORATORY

Providing fellowships for attendance
and
staff activities

U.S. NATIONAL SCIENCE FOUNDATION

U.S. COUNCIL FOR INTERNATIONAL EXCHANGE OF SCHOLARS

(FULBRIGHT FELLOWSHIPS)

BUNDESMINISTERIUM FUER FORSCHUNGS UND TECHNOLOGIE

KERNFORSCHUNGSANLAGE JULICH GmbH

PAN AMERICAN WORLD AIRWAYS, INC.

SPECIAL RECOGNITION IS GIVEN TO THE FOLLOWING INDIVIDUALS WHO ASSISTED THE NSRS AS:

On-site secritariat:

Dr. Slobodan Brant
Prirodoslovno-matematicki fakultet
Zavod za Teorijsku Fiziku
University of Zagreb
41000 Zagreb, Yugoslavia

On-site Conference Staff:

Ms. L. Reic, Institut Rudjer Boskovic, Zagreb, Yugoslavia
Ms. Joyce Plis*, LLNL, Livermore, California
Ms. Jackie Mooney*, BNL, Upton, New York
Mrs. Ruth Meyer, Livermore, California
Mr. C.B. Houser, Los Altos, California
Ms. M. Galekovic, University of Zagreb, Zagreb, Yugoslavia
Ms. V. Cicin, University of Zagreb, Zagreb, Yugoslavia

*We are indebted to Ms. Plis and Mooney for contributing their time and accepting a conference travel grant in order to perform on-site transcription of the recorded discussion sessions.

On-Site Operations & Assistance:

Prof. Ruth P. Yaffe (Co-Finance Officer)
Chemistry Dept, San Jose State Univ., San Jose, California
Dr. Nikita Kusnezov (Slavic-English Scientific aid)
Lockheed Research Corp., Palo Alto, California
Mrs. E. Kourkene-Kusnezov (Lingusitic aid)
PHOW Inc., Los Altos Hills, California

Conference-Sponsored Fellowship Holders:
(Graduate-student transcription & editorial aids)

Stephen Collins, University of Sussex, U.K.
Johnathan Copnell, University of Sussex, U.K.
Raoul Davie, University of Oxford, U.K.
Marie Harder, University of Sussex, U.K.
Dario Vretenar, University of Zagreb, Yu
Lauren Wood, University of Melbourne, Australia

California Volunteer Committee:
(Newsletters, Mailings, and Discussion Session Manuscript Preparation)

Ms. Joyce Plis
Ms. Martha Meyer
Ms. Marlene Meyer
Mrs. Ruth Randig Meyer

Program Notes:

Mrs Jean Wells Struble, San Francisco, California

GAJA ALAGA AWARD WINNERS

The International Conference on Nuclear Structure, Reactions, and Symmetries established the Alaga Award. This award was named in honor of Prof. Gaja Alaga of the University of Zagreb, who has contributed much to the field of nuclear structure and to science education in Yugoslavia. The award was given to the graduate students who had the most outstanding thesis content and presentation of their work at the conference. Because of the high quality of graduate student presentations and content, two main awards and two secondary awards were made. These were given to:

MARIE HARDER
(Advisor: Prof. D. Hamilton)
Physics Division, University of Sussex
Brighton BN1 9QH, U.K.

DENIS SUNKO
(Advisor: Prof. V. Paar)
Zavod za Teorijsku Fiziku, University of Zagreb
41000 Zagreb, Yugoslavia

LAUREN WOOD
(Advisor: Prof. I. Morrison)
School of Physics, University of Melbourne
Parkville, Victoria, 3052, Australia

MARK STOYER
(Advisor: Prof. J.O. Rasmussen)
Chemistry Department, University of California
Berkeley, California, U.S.A.

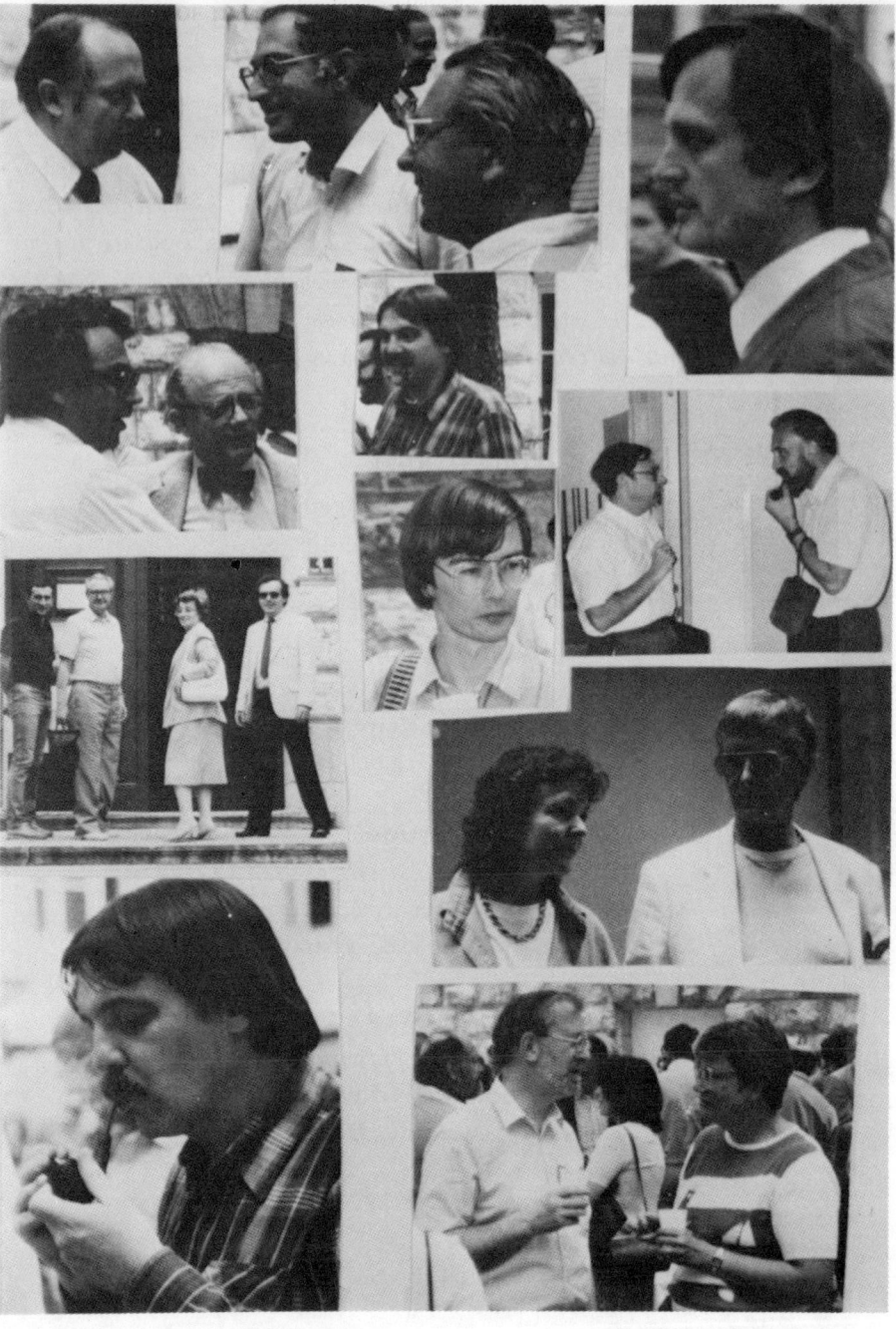

OPENING CEREMONIES

CONVENING LECTURE

by

PROF. I. TALMI

WELCOMING ADDRESS

by

PROF. B. GAGRO

CONVENING LECTURE

Igal Talmi
Weizmann Institute of Science, Rehovat, Israel

The title: "Nuclear Structure, Reactions, and Symmetries" actually covers the whole field of nuclear physics. However, there is a special character to this meeting. I do not even know how to define it exactly but I will try to describe it. First, we all start from the shell model. We do not ask why and how the shell model works, but we accept it as a reasonable starting point. At the same time we recognize the enormous complexity in trying to apply the shell model to any concrete problem in actual nuclei and therefore, all of us feel the need for models, for simplified versions of the shell model. The reason that we hope that this kind of approach will work is that nuclei, in spite of their complexity, exhibit experimentally, we know this only from experiment, many remarkably simple and regular features. As physicists we are used to believing that simple phenomena must have a simple description. Therefore, we believe that simple models or theories will be able to describe what we see in nuclei. This conference, in particular, concentrates on models for collective motion in nuclei, the relation of such models to the shell model, the interaction of collective modes with motion of individual nucleons, and so on.

Models in general pick out special aspects of the problem, they single out certain degrees of freedom which are believed to be relevant and important to the problem considered and treat them exactly. The other degrees of freedom are either ignored or hopefully contribute to various renormalization effects. Still, we should never forget that all these models can describe only part of reality and they cannot give a complete description of all the experimental data. Incidentally, in the conference the models are described in either as geometric or algebraic, and I pose a question to you. How would you describe the shell model? What kind of model is that? Maybe by the end of the conference we will find out. As far as models go there are two kinds of models. There are models that seem to be working even in cases where they have absolutely no reason to work but they work. The other kind of models are models that do not work even in the cases where you would expect them to work. Another aspect of models is their complexity. To paraphrase a well known saying, a model should be sufficiently detailed to be able to describe the phenomena and yet should be simple enough to be attractive. They should not be too simple or over simplified because then they will not be able to cope with the experimental data, but they should also not be too complicated because otherwise they would be completely useless.

The first and foremost requirement of any model is that it should agree quantitatively with experiment. But that is a very tricky question. How do you know whether a model really agrees with experiment? All models exhibit parameters --- they display them very prominently --- you see one, two, three, maybe five parameters. Many models, however, have what we can call "hidden parameters". They are buried somewhere in the pages that have no numbers in them, no figures, but words and words and you have to be very careful with those words. Some of the models make "reasonable assumptions" that may hide many parameters. Sometimes you have an expansion and the order of the expansion can be conveniently chosen as a hidden parameter of the model. You take as many terms as you need to explain the data and then you stop.

Suppose we overcome this difficulty and can say that a model agrees quantitatively with the experiment and it can be simple enough to be handled exactly. Are we then satisfied? Almost, but not yet. We would like to understand the physics that underlies this model. We want to know how it can be derived, which approximations are made in order to derive it from the more exact or complete theory. Incidently, the question: "What is a complete theory?" is also a difficult question. Do you stop at the shell model or do you go to the nuclear many-body problem or do you invoke quark structure of the nucleons or do you end up in superstring theory? In any case, we would like to understand how the model is related to a more precise theory. If the approximation from the exact theory to the model can be carried out in a simple fashion, then one should be able to calculate parameters of the model from the physical observable of the theory. But we know that this is far from being the actual situation, in the case of most models we do not know how to derive them exactly from the more detailed theory. Let us just look back at the shell model. It has been around for more than 35 years. There are hundreds, if not more, papers written on the subject: how to derive it from the nuclear many-body problem, how to derive the effective interactions from the interactions between free nucleons. Still, no reliable matrix elements of the effective interaction have been obtained nor has it been demonstrated that the shell model is a good approximation. In such cases, especially when the actual problem is very complicated, renormalization effects are prominent and very difficult to evaluate. Hence, we do not have to be too discouraged if we do not succeed in deriving those parameters right away.

Symmetries play a large and prominent role in this conference. If the problem has a certain symmetry it simply means that the Hamiltonian is invariant under a certain class of transformations which form a group. If we have such a Hamiltonian and such a group, then we or our colleagues can apply the elegant and powerful methods of group theory to derive very simple results about eigenvalues and eigenstates. It is amazing and really gratifying that such complex and complicated systems as nuclei do exhibit to a large extent such symmetries that can be handled by group theory. It is a minor miracle. We do not really understand the reasons for it, but it is very nice that such simple and elegant methods seem to be working exceedingly well for complex nuclei. Still we should always remember and realize that groups as well as all our models describe only idealized cases of the actual situation. Actual nuclei are much more complicated as we learn sometimes to our great disappointment from our experimental colleagues. It is very nice to have so many experimentalists among us even though many of us will not be able to listen to their talks. Still their presence here should keep us within the right proportions. We should know that whatever nice paper we write, someone is going to measure something and come back and tell you that you are all wrong. Let me quote the famous Latin American author Jorge Luis Borges who said that "Unfortunately life is real" and certainly nuclei are real. We have to realize that experimental facts or the truth are much more important than beauty in spite of the attractions of beautiful theories. For those of us who might forget it I have a mnumonic device to demonstrate my point taken from quark models. In the third generation of quarks the quantum number "<u>truth</u>" is associated with the top quark (which has not been discovered yet but so has truth...). The quantum number "beauty" is associated with the <u>bottom</u> quark. So also here truth comes on top of beauty. This is why physics is much more difficult than mathematics. This is why it is a very great pleasure when beautiful theories describes the truth, i.e. experimental data. That does not mean that one should not use groups and simple models. We certainly should use them. They give insight to the problem and probably offer good starting points from which to continue.

I want to remind some of you who have been at the Berkeley Conference that in his summary talk Herman Feshbach listed lots of new things which came out at that conference, new probes, new symmetries and so on. I do not know what he will do in his summary talk this time, but there is one thing that is new here: new people. It is a very very pleasant experience for us old timers to see so many young and new people. This is the guarantee of the future of our field. The first word I want to tell those young people is that there is still lots of work to do in the field of "classical" nuclear structure physics. Very much progress has been made but we have not solved everything, not everything is known. Nuclei exhibit lots of remarkable phenomena and we do not fully understand them even though we have spent many years on them. There is plenty of place for all of you to do both theoretical and, perhaps more important, experimental work. The presence of this large group gathered here, both of old and new people, is a clear demonstration that this field of "classic" nuclear physics is not dead. It is very much alive, temporarily in Dubrovnik.

Now I would like to join our chairman in the following message. All of us represent many different points of view. Not all are different, but there are many different points of view. All of us have our own models or our own way of applying these models or interpretation of those models. Everyone of us is, of course, sure that their model is the only correct model. Still, I think we should, in this coming conference, try to listen to what other people have to say. Believe me, it is very difficult, as I know it myself. It is very difficult to understand what somebody else is saying but sometimes he may be right or you may learn from him or maybe you may get a hint as to how you should modify your model. Instead of fighting fiercely and arguing fiercely to prove the truth of our model, we should try to be open minded and try to understand what other people are doing. Only if we do that, we will emerge from this meeting as a stronger group and perhaps a more influential group. We need this influence in the present world with severe budget cuts. Let us be better united in our efforts to find the truth. Then we will justify the ample and difficult task that has been carried out so well by the organizers. I would like to express the deep gratitude of all of us. Thank you.

WELCOMING ADDRESS TO THE INTERNATIONAL CONFERENCE ON NUCLEAR STRUCTURE, REACTIONS, AND SYMMETRIES

by

THE HONORABLE PROF. DR. BOZIDAR GAGRO

Minister of Education and Culture
of the
Socialist Republic of Croatia

It is my honor to greet you in the name of the Executive Council of the Parliament of the Socialist Republic of Croatia and to add my personal welcome as well. I am pleased to see that such a large number of you have gathered to discuss a topic of such great scientific importance. Especially since among you there is a large number of established experts and that you came from all over the world where many economic, political and ideological differences exist.

Your presence here points out at least two important facts: first, that contacts among scientists are possible in spite of all the barriers, and second: that these contacts are of the utmost necessity in the circumstances of today's life and destiny of modern civilization. A Meeting among scientists such as this one is important from the point of view of exchanging information about the results of their own research, but even more so if it serves the goal of putting science into the function of pursuit of the peaceful development of mankind.

It seems that the support of scientific research, in principle, does not pose a problem today. In all countries where the role of science and fast technological developments are recognized, and where the financial possibilities are, at least to some extent available, scientific research tends to be supported automatically. However, the scientific policy of each society must include even wider aims in the elaboration of which scientists have to participate.

In that sense, I appeal to your feelings of solidarity, your firm will to confront the abuses of science, and join your knowledge and abilities in the basic aims of the development of modern civilization.

Needless to say, I am glad that you are meeting here at our Inter-University Centre in Dubrovnik, This centre is more and more frequently becoming the meeting place of the intellectual elite of the world. One should add that, with its rich past as well as its present, Dubrovnik deserves your presence. This town has in its past given a number of internationally important people in the fields of literature, arts and especially science. Among the scientists, the most famous is the eighteenth century physicist, mathematician and philosopher Ruder Boskovic.

However, one could say that in the past of Dubrovnik which ranges over at least fourteen centuries, the most valuable is the town itself. It is the town as you see it today, with its medieval churches and monastaries, beautiful and magnificent fifteenth century walls, renaissance palaces and its urban fabric which took its definitive shape after the horrible earthquake of 1673. It is also the town as a notion, as a sense of oneness that wonderfully came in the statute of Dubrovnik from 1272, which is one of the oldest in Europe.

Dubrovnik, a small merchant republic on the Adriatic Coast has been fighting for centuries for its freedom. Its poets have written hymns on freedom and sayings have been encarved in the stone of its fortresses. It was not until 1906, that Napoleon's troops abolished the freedom of the old Republic. The tradition of freedom-loving, however, remained and lives today, just as it has, after all, in the entire country that is your host now. Upon the 10 July raising of the ceremonial flag of the Dubrovnik festival, an international cultural event known throughout the world, the word **LIBERTAS** can been seen inscribed on the flag. Its raising is accompanied by the sounds of the ceremonial song in honor of freedom by the seventeenth century Dubrovnik poet Ivan Gundulic'.

Thus, ladies and gentlemen one can learn a lot in Dubrovnik. With these words I again give you cordial greetings and wish you success in your conference.

TABLE OF CONTENTS

VOLUME I

Preface	vii
Convening Lecture *by Professor I. Talmi*	xvii
Welcoming Address *by Professor B. Gagro*	xxi

Part I: RECENT ISSUES

Section I: Octupole and Dipole Modes in Nuclei

Octupole correlations in the heavy elements R.R. Chasman	5
Opening statement J.A. Maruhn	30
Octupole and dipole modes Moderator: J.A. Maruhn	37
Description of the octupole degrees of freedom by shape variables S.G. Rohoziński & W. Greiner	50
An interacting boson version of reflection-asymmetric shape deformation J. Engel	73
A review of experimental evidence for octupole deformation J. Zylicz	79
Dipole collectivity and reflection asymmetric nuclear shapes M. Gai	92
In-beam and alpha decay studies of actinide nuclei P.A. Butler et al.	101

Anisotropic alpha emission of on line separated isotopes 109
   J. Wouters et al.

Nuclear moments and radii in the region(s) of octupole deformation 114
   R. Neugart

Collective properties of octupolly deformed nuclei 121
   R. Baranowski et al.

Parity doublets in ^{223}Ra and the octupole degree of freedom 126
   S.A. Eid et al.

Rotational alignment and stable octupole deformation 131
   Ch. Briançon & I.N. Mikhailov

Section II: Bose-Fermi Symmetries and SUSYs

Bose-Fermi symmetries and supersymmetry 148
   D.D. Warner

Bose-Fermi symmetries and SUSY in Nuclei 168
   Moderator: R.F. Casten

Experimental evidences for Boson-Fermion symmetries 175
and supersymmetries in nuclear physics
   J. Vervier

Experimental study of symmetries with transfer reactions 181
   J.A. Cizewski

Experimental tests of SUSYs using transfer reactions 187
   M. Vergnes

Overview of Bose-Fermi symmetries in odd-even nuclei 193
   R. Bijker

A new U(6/20) supersymmetry and its application to the 205
A = 130 mass region
   J. Jolie

The SO(7) Fermion dynamical symmetry and the Ru, Pd 215
transitional region
   R.F. Casten et al.

Dynamical symmetries for odd-odd nuclei 223
   A.B. Balantekin

Extension of supersymmetry to transitional and 231
odd-odd nuclei
   P. van Isacker

Approximate supersymmetry in odd-even and 240
odd-odd nuclei
    D.K. Sunko et al.

Extension of supersymmetry for the description of 246
odd-odd nuclei in the vibrational limit
    Z. Árvay

Theory of unified nuclei? 253
    T. Hübsch

Symmetries in odd-even nuclei: 259
$U^B(6) \otimes U^F(20)$ and $U^B(15) \otimes U^F(30)$
Boson-Fermion symmetry schemes
    V.K.B. Kota

Fermion-Boson model with isospin formalism 265
    S. Szpikowski et al.

Fermion dynamical symmetry and high spin physics 272
    M. Guidry et al.

U(12) multiplets 280
    R. Gilmore et al.

Section III: Mixed Symmetry States in Nuclei

Mixed-symmetry states in nuclei 288
    K. Heyde

Mixed-symmetry states in proton-neutron systems 306
    Moderator: I. Talmi

G-boson renormalizations and mixed symmetry states 315
    O. Scholten

Some remarks on mixed-symmetry states in IBA-2 322
    A.E.L. Dieperink

Magnetic properties of nuclei in the collective model 330
    M.S.M. Nour El-Din et al.

The nuclear spectroscopy of mixed symmetry states 338
    W.D. Hamilton

Investigation of mixed symmetry $J^\pi = 1^+$ and 2^+ states 350
with (e, e'), (γ, γ') and (p, p') techniques
    D. Bohle

Study of mixed symmetry states with photon scattering 362
    U. Kneissl

Strength distributions and mixed-symmetry states 368
   M. Pignanelli

Systematics of mixed-symmetry states 374
   H. Harter

PART II: CURRENT AREAS OF STUDY

Section IV: Scattering Reaction Studies

Experimental studies of the photon decays of giant resonances 385
   A.M. Nathan

Photon scattering in the region of giant dipole resonance 397
   S. Turrini et al.

Boson structure functions from inelastic electron scattering 404
   C.W. de Jager

Determination of boson densities by electron scattering 411
   D. Goutte

Experimental results on the electron scattering from nuclei 418
with the coincidence registration of charge secondary particles
   S.G. Popov

Determination of transition matrix elements and boson 424
effective charges in Pd isotopes from a coupled-channels
analysis of proton inelastic scattering
   V. Riech et al.

Fine structure of LEOR and GQR in ^{48}Ca and ^{208}Pb 431
   Y. Fujita et al.

Nuclear structure with protons and pions at LAMPF 438
   N.M. Hintz et al.

Section V: Symmetries in Reactions

A review of symmetries in nuclear reactions 455
   F. Iachello

Symmetries in reactions 469
   Moderator: L. Biedenharn

Medium energy scattering from "algebraic" targets 475
   R.D. Amado et al.

The interacting boson model and medium energy probes 483
  J.N. Ginocchio & G. Wenes

Group theory approach to scattering and its application 491
to heavy ion collisions
  Y. Alhassid

A survey of the group theoretical approach to scattering 507
  L.C. Biedenharn & A. Stahlhofen

Group deformations and the algebraic scheme for scattering 515
  A. Frank et al.

Molecular resonances and symmetries 520
  P. Kramer et al.

Physical boundary conditions for nuclear heavy ion scattering 529
in R-matrix theory
  R. Maass et al.

Description of two- and three-cluster systems of light nuclei 534
  R. Krivec & M.V. Mihailović

Nuclear structure in heavy-ion collisions 540
  K.V. Shitikova

PART I

RECENT ISSUES

SECTION I
OCTUPOLE AND DIPOLE MODES IN NUCLEI

SECTION II
BOSE-FERMI SYMMETRIES AND SUSYs

SECTION III
MIXED SYMMETRY STATES IN NUCLEI

SECTION I

OCTUPOLE AND DIPOLE MODES IN NUCLEI

R. CHASMAN

Reviewer

J. MARUHN

Discussion Moderator

Session Moderators

D. Burke
W.B. Walters

The submitted manuscript has been authored by a contractor of the U. S. Government under contract No. W-31-109-ENG-38. Accordingly, the U. S. Government retains a nonexclusive, royalty-free license to publish or reproduce the published form of this contribution, or allow others to do so, for U. S. Government purposes.

OCTUPOLE CORRELATIONS IN THE HEAVY ELEMENTS

R. R. Chasman

Physics Division, Argonne National Laboratory
Argonne, IL 60439-4843 USA

ABSTRACT

The effects of octupole correlations on the nuclear structure of the heavy elements are discussed. The cluster model description of the heavy elements is analyzed. The relevance of 2^6-pole deformation and fast E1 transitions to an octupole model is considered.

1. INTRODUCTION

In the past few years, the nuclear structure community has devoted considerable theoretical and experimental effort to the study of nuclides in the mass region $218 < A < 230$. The experimental studies have been carried out in spite of the fact that half-lives are short and radioactivity levels are high. The primary motivation for these studies is the strong octupole correlation effects, that are manifest only in this region of the periodic table. These octupole correlation effects will be the primary focus of my talk. I shall also discuss an alternate explanation that has been advanced for many of these same nuclear structure features: the cluster model. In addition, I shall discuss 2^6-pole (hexinda tessera-pole) deformation in this mass region and its relevance to octupole deformations. Also, I shall discuss high spin states and the E1 transitions seen in the nuclides of this mass region.

The nuclear structure features associated with octupole deformation (or any odd parity deformation mode) are strikingly different from those due to conventional quadrupole deformation (or any even parity mode). In a conventionally deformed even-even nucleus, there is a ground state rotational band with an $I^{\pi} = 0^+, 2^+, 4^+...$ sequence of states. In addition, we might see an $I^{\pi} = 1^-, 3^-, 5^-$ sequence of states at 2-quasiparticle state exicitation energies (~1 MeV in the heavy elements). Octupole deformation leads to a single rotational band[1] consisting of both the positive and negative parity states. In odd mass nuclides, an equally striking phenomenon is due to octupole deformation: the parity doublet. This doublet consists of a pair of states, almost degenerate in energy, having the same spins and opposite parities, and connected to each other through strong E3 transitions. See Fig. 1.

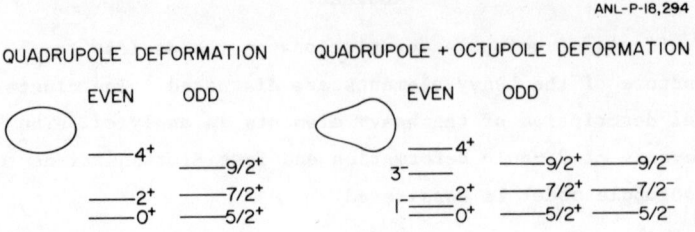

Fig. 1. Spectra of even and odd mass nuclides; (a) quadrupole deformation,(b) quadrupole plus octupole deformation

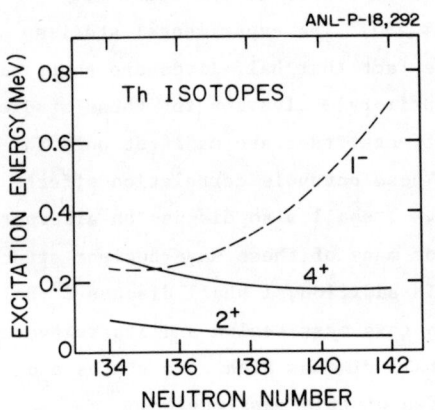

Fig. 2 Low lying states in the Th isotopes.

Are these simple signatures of octupole deformation found in nuclei? There are no known cases[2] of even-even nuclides in which the 1^- state occurs below the 2^+ state. However, there are several nuclides in the mass region $220 \leq A \leq 230$ in which the 1^- state is found at very low energy. In fact, the 1^- state at 216 kev in ^{224}Ra is the lowest non-rotational state in even-even nuclides. In Fig. 2, we plot the excitation energies of the 1^-, 2^+ and 4^+ states in the even isotopes of Th. The excitation energy of the 1^- state indicates that ^{232}Th is vibrational with respect to the octupole mode and that octupole correlations become increasingly important for the lighter isotopes. Although the energy of the 1⁻ states in the lighter isotopes clearly shows that none of them is deformed; it is equally clear that the correlations are much stronger than vibrational. We refer to this in between situation as incipient octupole deformation. Odd mass nuclides will be discussed below.

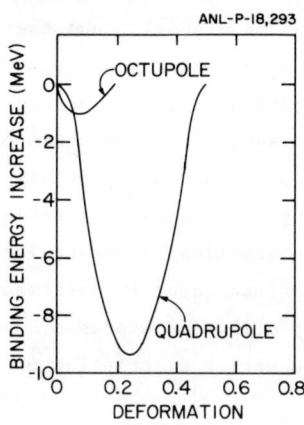

Fig. 3 Strutinsky method calculation of binding energy. The deformation axis denotes ε_2 for quadrupole case and ε_3 for octupole case.

This difference between incipient octupole deformation and quadrupole deformation can also be seen in a different way, making use of the Strutinsky[3] procedure, discussed in Sec. 3. In Fig. 3, we show the increase in binding energy, relative to a spherical state, for a deformed actinide as a function of quadrupole deformation; and, the energy gain for octupole deformation, relative to the energy minimum obtained for the even deformation modes. The octupole effects are substantially weaker.

2. A MANY-BODY TREATMENT OF OCTUPOLE CORRELATIONS

The important interactions for describing the low-lying states in the nuclides 220 ≤ A ≤ 230 are the pairing force, the quadrupole-quadrupole interaction that gives deformation and the octupole-octupole interaction that gives octupole correlations or deformation. The Hamiltonian we use[4] is

$$H = \sum_k \varepsilon_k a_k^+ a_k - G \sum_k a_k^+ a_{-k}^+ \sum_l a_{-l} a_l \quad \text{(pairing)}$$

$$-V_2 \sum_{i,j} (r^2 Y_2^0)_{i,j} a_i^+ a_j \sum_{k,l} (r^2 Y_2^0)_{k,l} a_k^+ a_l \quad \text{(quadrupole)} \qquad (1)$$

$$-(V_3)^2 \sum_{i,j} (r^3 Y_3^0)_{i,j} a_i^+ a_j \sum_{k,l} (r^3 Y_3^0)_{k,l} a_k^+ a_l \quad \text{(octupole)}$$

In the limit of strong residual interactions, the two-body pairing term can be replaced by a one-body term and a gap parameter (BCS approximation); the quadrupole interaction can be replaced by a quadrupole deformation parameter times a one-body term (Nilsson model); and the octupole interaction by an octupole deformation parameter and a one-body term. In deformed nuclides, the pairing interactions are not particularly strong and the BCS approximation is not adequate. The cause of this failure is that the correlations between pairs are not properly taken into account in the number non-conserving BCS theory. We get a much more accurate treatment of the pairing interaction by conserving particle number. Similarly, the introduction of an octupole deformation parameter leads to states of mixed parity. By utilizing states of good parity and good particle number, we obtain a superior description of octupole correlation effects.

The essential feature of the Hamiltonian of Eq. (1) is that all of the factored residual interaction terms conserve Ω, the projection of angular momentum on the nuclear symmetry axis; i.e. the interactions are of the form $(r^\lambda Y_\lambda^0) \cdot (r^\lambda Y_\lambda^0)$ where $\lambda = 0$ for the pairing interaction. This means that we can include the major effects of the correlations induced by all of the two-body interactions in a product type wave function. The pairing and the particle-hole interactions are placed on an equal footing. For the purpose of illustrating the structure of these wave functions, we consider a model system that consists of only one kind of nucleon and four doubly degenerate Nilsson levels. The levels have Ω^π values of $3/2^+$, $3/2^-$, $5/2^+$, and $5/2^-$. The product wave function is shown in Fig. 4. We project states of good parity and particle number from this wave function and then determine the amplitudes A_i and B_i. In this two level case, there are six configurations in each of the Ω subgroups. In the actual calculations, there are 5 doubly de-

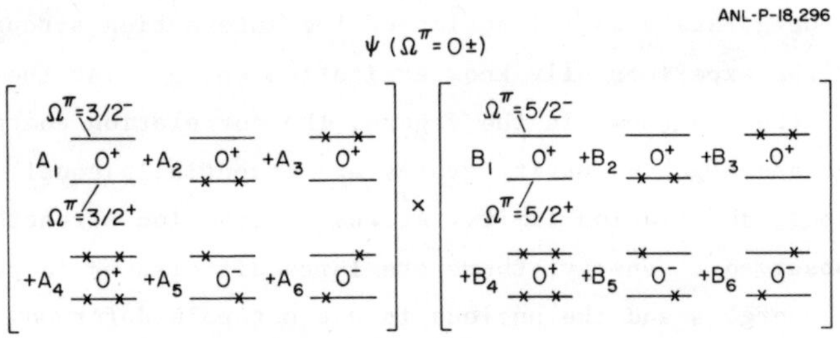

$$\psi(\Omega^\pi = 0\pm)$$

$$\begin{bmatrix} \Omega^\pi = 3/2^- \\ A_1 \underline{\quad 0^+ \quad} + A_2 \underline{\quad 0^+ \quad} + A_3 \underline{\quad 0^+ \quad} \\ \Omega^\pi = 3/2^+ \\ +A_4 \underline{\quad 0^+ \quad} + A_5 \underline{\quad 0^- \quad} + A_6 \underline{\quad 0^- \quad} \end{bmatrix} \times \begin{bmatrix} \Omega^\pi = 5/2^- \\ B_1 \underline{\quad 0^+ \quad} + B_2 \underline{\quad 0^+ \quad} + B_3 \underline{\quad 0^+ \quad} \\ \Omega^\pi = 5/2^+ \\ +B_4 \underline{\quad 0^+ \quad} + B_5 \underline{\quad 0^- \quad} + B_6 \underline{\quad 0^- \quad} \end{bmatrix}$$

Fig. 4 Representation of product wavefunction -x—— ≡ a_k^+ -x——x- ≡ $a_k^+ a_{-k}^+$, k^π is given for each configuration.

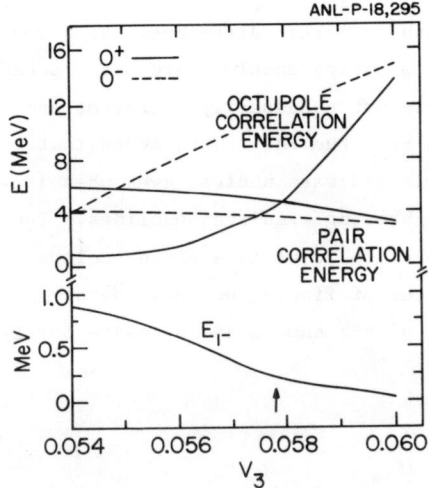

Fig. 5 Correlation energies calculated for $K^\pi = 0^+$ and $K^\pi = 0^-$ states in ^{226}Ra. The octupole correlation energy is the expectation value of the two-body octupole interaction in Eq. (1).

generate levels in each subgroup; and 252 different configurations in each subgroup rather that six. We project states of good proton number, neutron number, and do separate calculations for the positive and negative parity states. Actually, we do many such calculations for each positive and negative parity state. We vary the octupole interaction strength, the pairing interaction strength and the quadrupole interaction strength to generate wave functions that have rather different correlations in them. We then do a final diagonalization, taking the non-orthogonality of these wave functions fully into account. In Fig. 5, we present the results of such a calculation for the nuclide ^{226}Ra. The correlation energies are plotted as a function of the octupole interaction strength. In the lower part of the figure, we plot the calculated 1^-

excitation energy state as a function of the interaction strength with an arrow at the experimentally know excitation energy. At the largest interaction strength shown in the figure, the correlation energies of the positive and negative parity states are essentially equal, and this is the octupole deformation limit. At the interaction strength that gives the observed 1^- energy, there are large differences in the correlation energies and the nucleus is not octupole deformed. It is worth noting that the negative parity state is fairly close to the octupole deformation limit for the physically realistic value of the interaction strength. A major reason for the difference in the two states is that at least one pair is always broken in the 1^- state.

In odd mass nuclides, the situation is rather different. Both the positive and negative parity states of a parity doublet have one blocked level; i.e. the pairing correlations are not necessarily different and they are weaker than is the case in an even nuclide. This means that octupole deformation might be present in odd mass nuclei, even when it is not a proper description of the neighboring even-even nuclides. The octupole correlations may vary from level to level in a given nucleus depending on the value of Ω. The chances of finding octupole deformation are maximized when two orbitals of the same Ω and opposite parity

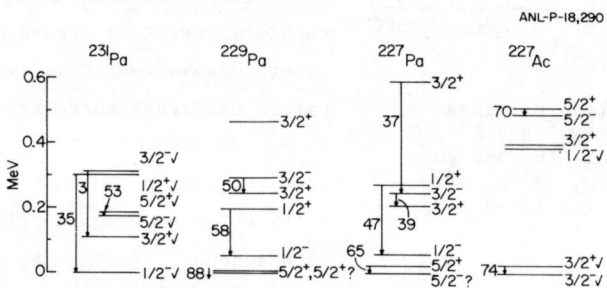

Fig. 6 Calculated[4] Spectra of odd proton nuclides using Hamiltonian of Eq. (1). The numbers beside the arrows are proportional to the value of $\langle i | r^3 Y_3^0 | f \rangle^2$. √ denote levels know at the time of the calculation.

have a large $(r^3 Y_3^0)$ matrix element and are on the same side of the fermi level. The latter feature is important because the matrix elements have

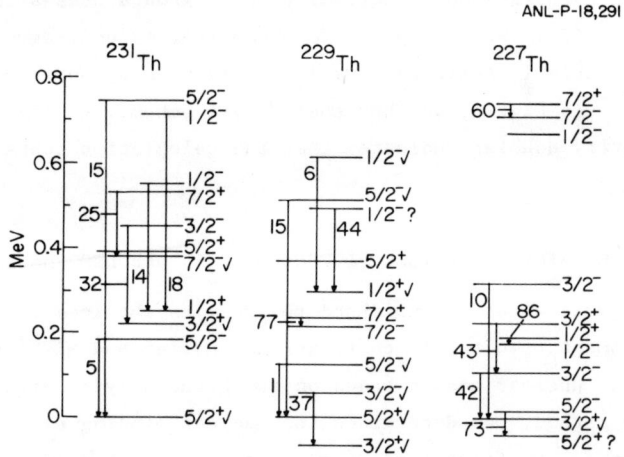

Fig. 7 Calculated[4]) Spectra of odd neutron nuclides.
See Fig. 6 caption.

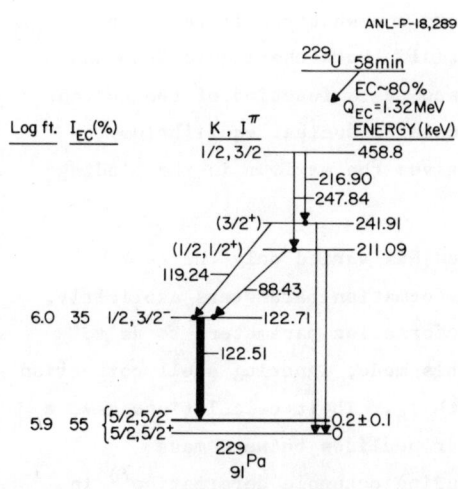

Fig. 8 Level scheme of ^{229}Pa.

a (uu'-vv') factor. With these thoughts in mind, we have carried out a series of calculations on odd proton and odd neutron nuclides and find several instances of parity doublets, with the positive and negative parity members being rather similar in structure. See Figs. 6 and 7. An important result of these calculations is the large changes in splittings of parity doublets and in octupole transition rates as a function of Ω in each nucleus.

Octupole deformed and undeformed states occur in the same nucleus. This is a signal of incipient octupole deformation. The most interesting result of these calculations is a prediction of ground states octupole deformation in ^{229}Pa. With very little prompting, my colleagues at Argonne particularly I. Ahmad and the late A. Friedman, undertook a study of ^{229}Pa. In Fig. 8, we show their level scheme.[5] Their finding of the $5/2^{\pm}$ parity doublet indicates that the calculation contains the essential ingredients.

3. THE DEFORMED SINGLE-PARTICLE PICTURE OF OCTUPOLE DEFORMATION

About one year after we published the results obtained from our many-body treatment of octupole correlations, Moller and Nix[6] published a calculation of nuclear masses based on the Strutinsky[3] method. In the Strutinsky method, one determines the nuclear binding energy as a sum of two terms. The first term is the energy of a charged liquid drop and the second is a shell correction. The shell correction gives increased binding when the nuclear single particle level spacings are large near the fermi levels for the protons and neutrons in the nuclide of interest. The shell corrections are somewhat sensitive to the details of the single-particle potential. Both the liquid drop and shell correction energies are calculated as a function of the nuclear deformation parameters, ε_i. The calculated nuclear equilibrium deformation is the deformation that gives the maximum in the binding energy.

In their calculations, Moller and Nix varied only the ε_2 (quadrupole) and ε_4 (hexadecapole) deformation parameters explicitly, and fixed the value of ε_6 (2^6-pole deformation parameter) so as to minimize the liquid drop energy of this mode, ignoring shell correction energies that might be associated with ε_6. Their calculations gave a region of substantial underbinding for nuclides between mass $220 \leq A \leq 230$. See Fig. 9. By including octupole deformation[7] in their potential they got a substantial increase in binding energy.

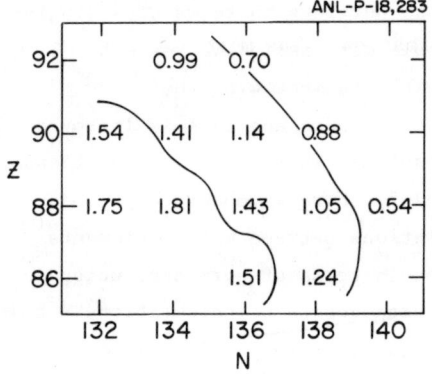
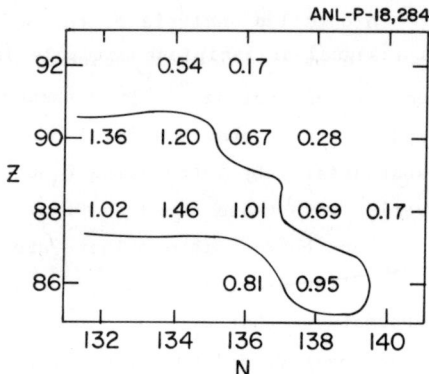

Fig. 9 Difference in calculated and measured binding energies. The calculations are those of Ref. 6 and consider only ε_2 and ε_4 deformation explicitly.

Fig. 10 Additional binding energy obtained with inclusion of ε_3 deformation. The calculations are those of Ref. 8.

These calculations were later extended[8] by Leander et al., who found an increase in nuclear binding energies of ~1.5 MeV in ^{222}Ra and ^{222}Th and a fairly rapid decrease with increasing neutron number. These changes, shown in Fig. 10, are relative to the values of Ref. 6. There is a substantial improvement in agreement between calculated and observed nuclear binding energies, however there remain several nuclides for which the discrepancies are still larger than 0.5 MeV.

Some aspects of this calculation are problematic vis-a-vis the experimental data. The first problem is that the energy of the 1^- state is lowest[2] for nuclides with 134, 136, and 138 nuclides; while the calculation suggests that these effects are most important for ^{222}Ra and ^{222}Th, which have 134 and 132 neutrons respectively. Secondly, structure studies of the odd-mass nuclides in this mass region do not indicate any great increase in octupole correlation effects as we go to the nuclides where the Strutinsky method gives the larger increases in binding energy.

A detailed analysis of the odd mass nuclides in terms of a single-particle potential that includes octupole deformation was made by Leander and Sheline.[9] They showed that agreement with the experimentally observed level orderings of these nuclides is improved substantially by introducing ε_3 deformation. Where these calculations overlap with those shown in Figs. 6 and 7, there is moderately good agreement between them. There are relations between matrix elements that follow directly from a description in terms of permanent octupole deformation, that are not a necessary consequence of calculations[4] that treat positive and negative parity states separately.

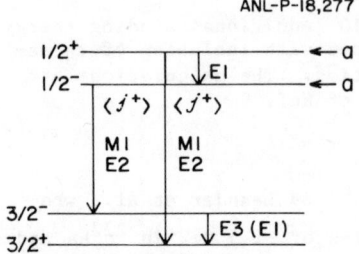

Fig. 11 Schematic $1/2^+$, $1/2^-$, $3/2^+$, $3/2^-$ bandheads and some of the matrix elements connecting them.

In the single particle model, the fact that both positive and negative parity states are generated from the same intrinsic states implies definite relations between matrix elements. In Fig. 11, we illustrate some of these relations for $3/2\pm$ parity doublet and $1/2\pm$ parity doublet band heads. The M1 transitions and the coriolis interaction matrix elements between the two bands should be the same for either parity and the decoupling parameters, a, for the positive and negative parity members of the $1/2\pm$ band should be equal in magnitude but opposite in sign. The E3 transition should be collective, and possibly the E1.

The observed shifts of matrix elements from the values calculated with a conventional deformed (even multipole) potential are in the direction given by the octupole deformed single particle calculations. In some cases, these shifts are extremely large. A success of this single particle approach is found[10] in the matrix elements of ^{225}Ac. The $\langle j^+ \rangle$ matrix elements are ~ 4 for the relevant single particle states when octupole deformation is not included in the potential. The inclusion of octupole deformation lowers the calculated value of $\langle j^+ \rangle$ to 1.05 in fairly good agreement with the experimental values of 0.9 for the positive-parity bands and 0.6 for the negative-parity bands. However, we note that the $\langle j^+ \rangle$ matrix elements differ by 50% experimentally. Also, the calculated intraband M1 transition probability in the 3/2+ band is brought into considerably improved agreement with experiment by introducing octupole deformation.

There is a similar trend with decoupling parameters. The experimental values of $|a|$ are better reproduced with the inclusion of octupole deformation in calculations. However, there are differences of ~100% in the positive and negative parity states. Cf. Table I. It is noteworthy that ^{223}Ra shows little change in this regard vis-a-vis ^{225}Ra since ^{222}Ra is expected to show the maximal octupole binding energy effects. Our values for ^{223}Ra come from a reanalysis of data in Ref. 13.

Table I Decoupling Parameters

Nuclide	$a^+_{1/2}$	$a^-_{1/2}$
^{227}Ac [11]	-2.0	4.5
^{225}Ra [12]	-2.2	1.3
^{223}Ra [13]	-2.1	1.2

Using the many-body method,[4] there is no such simple relation between the decoupling parameters of the positive and negative parity 1/2-bands. In Fig. 12, we display a calculation of the absolute values of the decoupling parameters for ^{227}Ac as a function of the interaction strength. At the interaction strength that gives the observed 1^- excitation energy in the neighboring even-even nuclides, we calculate decoupling parameters that are quite different for the positive and negative parity states and in good agreement with experiment. It is only for a somewhat larger interaction strength that we get the octupole deformation result, i.e. two decoupling parameters equal in magnitude.

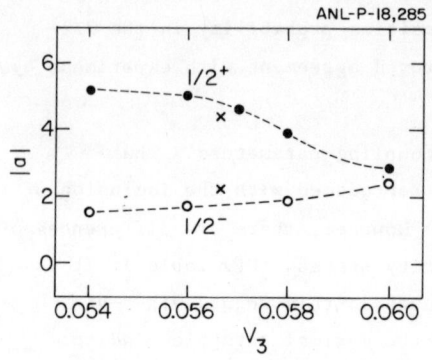

Fig. 12 Calculations of absolute value of decoupling parameters for $1/2^+$ and $1/2^-$ bands in ^{227}Ac, using many-body wave functions: calculated values given as a function of octupole interaction strength. The experimental values are denoted by x at the appropriate interaction strength.

4. 2^6-POLE (HEXINDA TESSERA POLE) DEFORMATION

The fact that there is no large increase in octupole correlation effects in ^{223}Ra vis-a-vis ^{229}Pa might mean that the Strutinsky approach is not particularly reliable for rather small changes (~1 MeV) in binding energy; there are uncertainties in the relative shell corrections of a few hundred keV and similar uncertainties in the relative liquid drop energies. Alternatively, there may be other deformation modes that should be considered. Typically, the calculations of shell corrections are done on a grid in deformation space. If we were to consider only 5 points in each dimension of a three dimensional deformation space, such as (ε_2, ε_3, ε_4) space, there are already 125 points at which we must calculate shell corrections for both protons and neutrons. Because the liquid drop energy increases sharply for the higher deformation modes, when the deformation departs from the liquid drop minimum, it is reasonable to fix the values of the higher deformation mode parameters such as ε_5 and ε_6 so as to minimize the liquid drop energy. The prescription[7,8] is

$$\varepsilon_5 = -0.9\varepsilon_2\varepsilon_3 \quad , \quad \varepsilon_6 = -\varepsilon_4(\varepsilon_2+0.1) \tag{2}$$

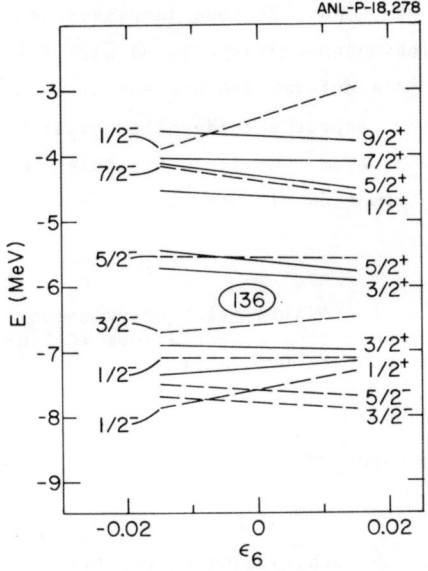

Fig. 13 Neutron single particle energy levels as a function of ε_6.

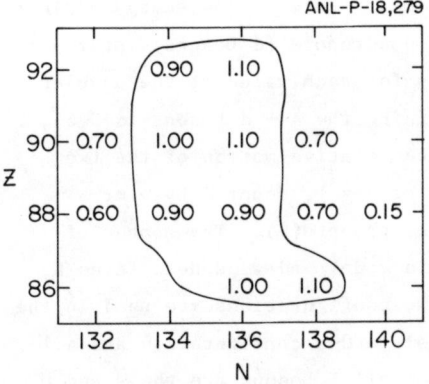

Fig. 14 Binding energy increase obtained from allowing ε_6 variation. Calculations are from Ref. 14.

In the light actinides, where $\varepsilon_2 \approx 0.15$ and $\varepsilon_4 \approx -.05$, we estimate $\varepsilon_6 \approx +.012$ using Eq. (2). However, in these nuclides the shell corrections are strongly affected by small changes in ε_6. In Fig. 13 we display a set of neutron single particle energies as a function of ε_6. The large decrease in the single particle level density near N = 136 with decreasing ε_6 is quite striking. Varying ε_6, and using a sharp surface liquid drop model, we find that the binding energies are increased substantially relative to the values calculated with ε_6 fixed by Eq. (2) and only ε_2 and ε_4 allowed to vary. There are ~1 MeV increases[14] in binding energy when ε_6 variations are included. See Fig. 14. This effect is centered at N = 136. There is, of course, additional binding energy from the ε_3 variation. These increases are given in Fig. 15. Comparing Figs. 10 and 15, there is a considerable decrease in the binding energy attributable to ε_3 in the nuclides near ^{222}Ra. This is qualitatively in agreement with the experimental observation of no marked increase in octupole deformation effects

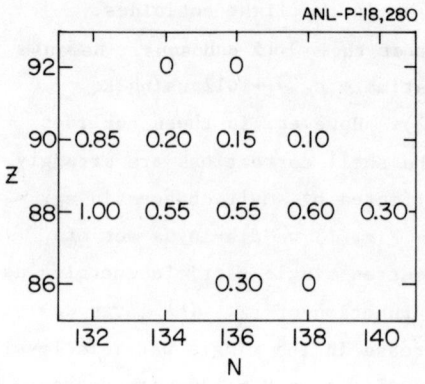

near ^{222}Ra. Because increases in binding energy associated with octupole deformation are small, we might expect considerable deviations from the simple octupole deformation picture.

Fig. 15 Additional binding energy increase from ε_3 variation. Calculations are from Ref. 14.

5. IBA TREATMENTS: CLUSTER MODEL AND F-BOSONS

5.1 Cluster Model

A very different model[15,16,17] for the description of negative parity states in the heavy elements is the α-cluster model of Iachello and coworkers. The model consists of a core cluster and an α-cluster, each described by s ($\ell=0$) and d ($\ell=2$) bosons. The s and d bosons represent the interactions that play a major role in low energy nuclear structure, i.e. pairing (s bosons) and quadrupole (d bosons) interactions. As there are ($2\ell+1$) substates for each value of the angular momentum, the most general group containing the s + d bosons is U(6), i.e. one s boson and five d bosons. The relative motion of the two clusters is allowed to be either $\ell = 0$ or $\ell = 1$, denoted by σ or π bosons, and accordingly described by the group U(4). The number of bosons in all cases is conserved in this model and we denote it as N, 1/2 the number of valence nucleons. Two configurations are used in the calculations done so far with this model, a 0-α configuration and a 1-α configuration. In the 0-α configuration all N bosons are the s and d bosons of the core. In the 1-α configuration, (N-2) bosons are s and d core bosons and the remaining 2 bosons are the (σ,π) bosons associated with the α-particle cluster relative motion. The α-cluster is inert. See Fig. 16.

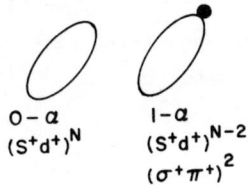

Fig. 16 0-α and 1-α configurations in the cluster model.

Actually, there is some ambiguity in the use of the σ and π bosons. Because the U(4) bosons describe the relative motion of the core and α-particle, rather than the internal structure of the α-particle, there is no reason why there should be exactly two-bosons involved in this non-particle degree of freedom. The relative motion bosons should not be involved in a boson number conservation law.

A more basic problem with the cluster model is the relation between the π bosons and effective nuclear interactions. The d-bosons are a representation of the strongly attractive quadrupole-quadrupole nuclear interaction. However, there is no such attractive dipole-dipole interaction. The dipole interaction is repulsive, and the collective dipole state is a high energy state.

There has been an argument advanced for enhanced α-cluster probabilities near ^{218}Ra based on the unusually short α-decay half-life (14 μ sec) reported by Valli et al.[18] When this half-life is converted to an α-decay radius parameter, via the Preston formula, it gives a very large value of Ro. However, this lifetime has been recently remeasured[19] by Rao et al. and found to be 25 μ sec. The new measurement removes the lifetime anomaly in ^{218}Ra, bringing it into line with the α-decay lifetimes of the neighboring nuclides.

The beauty of all IBA treatments is that the Hamiltonian is written in terms of Casimir invariants (constants of motion). By writing a Hamiltonian in terms of the constants of motion, we generate a spectrum of eigenvalues instantaneously. A very useful introduction to these methods is given in the lectures by Iachello.[20]

The cluster model Hamiltonian is

$$H = H_{sd} + H_{\sigma\pi} + H_{int} \tag{3}$$

with

$$H_{sd} = \underline{\varepsilon}_d N_d + \underline{K}_d Q_d \cdot Q_d + (\underline{K}' + 3/8\, K_d) L_d \cdot L_d \tag{4}$$

$$H_{\sigma\pi} = \underline{\varepsilon}_\pi N_\pi + \underline{\alpha}_\pi N_\pi (N_\pi + 2) + K'_\pi L_\pi \cdot L_\pi \tag{5}$$

$$H_{int} = \underline{K}\, Q_d \cdot Q_\pi + 2\, K'\, L_d \cdot L_\pi + \underline{Y}\, (s^+ s^+ \sigma\sigma + \sigma^+ \sigma^+ ss) . \tag{6}$$

$H_{\sigma\pi}$ is the U(3) version of the theory. We expand on this point below. In Eqs. (4), (5) and (6) the first mention of parameters of the theory are underlined. There is an additional parameter in the model

$$\underline{\Delta}_\alpha = E_{1\alpha} - E_{0\alpha} \tag{7}$$

used to adjust the energy differences of the 0α and 1α configurations. Additional parameters are underlined in Q_d^μ as well as in the E2 and E1 transition matrix elements shown below

$$Q_d^\mu = (d^+ s + s^+ d)_\mu^2 + \underline{\chi}\, (d^+ d)_\mu^2 ; \tag{8}$$

$$T_k^\mu (E2) = \underline{e}_d\, Q_d^\mu + \underline{e}_\pi(2)\, (P^+ P)_\mu^2 ; \tag{9a}$$

$$T_k^\mu (E1) = \underline{e}_\pi(1)\, (\pi^+ \sigma + \sigma^+ \pi)_\mu^1 \tag{9b}$$

Now there are two ways of constructing a chain of Casimir invariants from the group U(4). We follow these two sequences

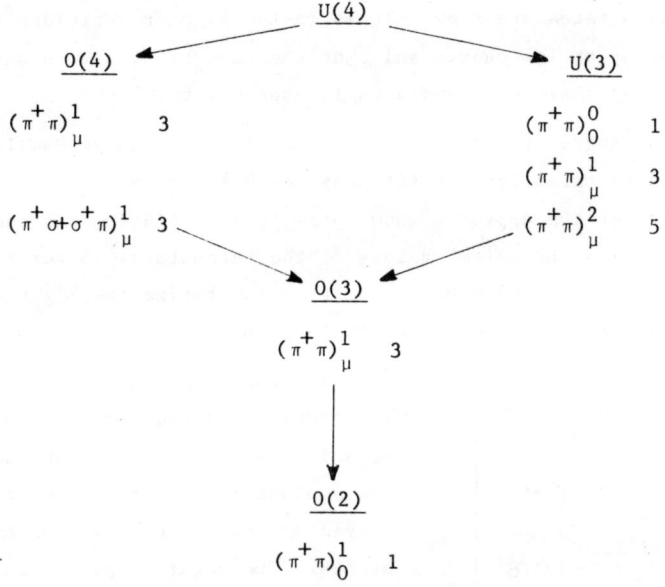

We have not written down the 16 elements of U(4). They are all products of the four boson creation operators with the 4 boson annihilation operators of U(4). The number of elements is shown to the right of each combination of boson operators. The point we emphasize here is that it is the O(4) group that contains the E1 transition matrix element $(\pi^+ \sigma + \sigma^+ \pi)$, of the model. <u>However</u>, in cluster model calculations, the Casimir invariants associated with the group chains

$$U(4) \supset U(3) \supset O(3) \supset O(2) \qquad (10)$$

are used. If one wants to describe the fast E1 transitions in the light actinides, the Casimir invariants of O(4) should be used. Recently arguments have been advanced[21] in favor of the U(3) chain of invariants. However, the arguments apply to the heavier actinides, where the

E1 transition rates are slow. It is in the lighter actinides that the 0⁻ bands occur at low energy and that the fast E1 transitions occur. Here, the O(4) chain of invariants is appropriate.

Because there are so many parameters in the cluster Hamiltonian, one can choose parameters to fit many features in any one nucleus. Inadequacies of the approach show up as large variations in these parameters or as unphysical values of the parameters. Several attempts[16,21,22,23] have been made to parameterize the light actinides. There are large differences between these parameterizations.

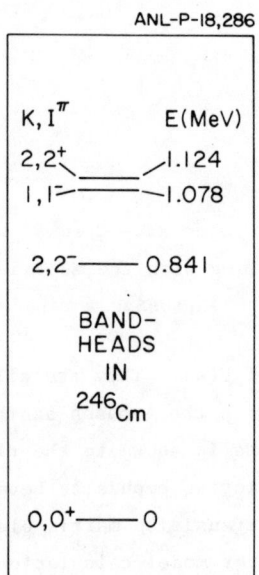

Fig. 17 Bandhead energies in ^{246}Cm. Levels are labelled by K,I^π. Data taken from Ref. 24.

A strong argument for octupole interactions as the explanation of negative parity states in the actinides and against the cluster model is the observed excitation energies of $K^\pi = 2^-$ states. The negative parity building block of the cluster model is the 1^- π boson and the only simple negative parity states are $K^\pi = 0^-$ or 1^-. In order to get a $K^\pi = 2^-$ band, one needs a two boson excitation of the ground states, e.g. $(d_2^2)^+(\pi_0^1)^+$ and $K^\pi = 2^-$ states should be higher in energy than the $K^\pi = 1^-$ states, i.e. roughly at an energy that is the sum of the 2^+ and 0^- band head energies. If the octupole description is correct, we expect to see $K^\pi = 0^-$, 1^-, 2^-, and 3^- states; with no systematic difference in excitation energy for the 2^- state relative to the $K^\pi = 1^-$ state. In ^{230}Th, the $K^\pi = 2^-$ band is[2] just 120 kev above the $K^\pi = 1^-$ band. In ^{232}U, it is[2] 130 kev below the $K^\pi = 1^-$ band. In both ^{246}Cm and ^{250}Cf, the $K^\pi = 2^-$ band is the lowest excited state band. See Fig. 17. This is not consistent with the cluster model picture.

We feel that the cluster model is not the correct description of the low-lying negative parity excitations in the even-even actinides. Octupole models provide a more natural, physically based, description of these states.

4.2 F-Bosons

Recently, Engel and Iachello[25] have introduced an extended IBA model to describe the negative parity states in the actinides by including p and f bosons in addition to the usual s and d bosons. One of their more interesting predictions is low-lying degenerate $K^\pi = 1^\pm$ bands. Such bands have not been seen in the light actinides. Further experimental searches for the $K^\pi = 1^+$ states would test the validity of this picture.

6. HIGH SPIN STATES IN THE LIGHT ACTINIDES

A triumph of the single particle octupole deformation picture is the treatment[26] of high spin states by Nazarewicz et al. Cranked HFB calculations dealing with the nuclides in the vicinity of ^{222}Th, using a single particle potential with only even parity deformation modes, lead to predictions of considerable back-bending in the vicinity of I=12.

Experimentally, there is no sign of this backbending. When octupole deformation of the potential is admitted, the calculations no longer show backbending, in agreement with experiment. The reasons for this are straightforward. In Fig. 18, we sketch the spherical single particle level scheme for heavy element neutrons. The N=7 $j_{15/2}$ neutron orbital is pushed down into the N=6 orbitals by

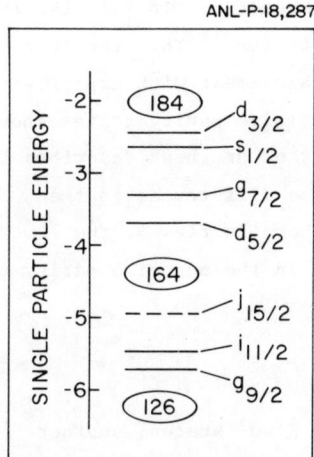

Fig. 18 Neutron single particle levels in the heavy elements.

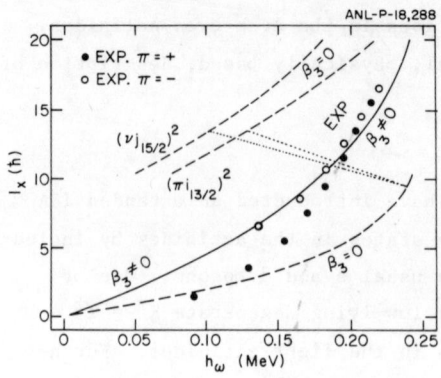

Fig. 19 Angular momentum, I_x vs. rotational frequency ω in ^{222}Th. This figure is taken from Ref. 26.

the spin orbit interaction, and it is the only negative parity state around. Quadrupole deformation does not mix this state with the N=6 states nor does the (j_x) cranking term of the CHFB. Increasing the cranking frequency gives a large spin align-ment in the lowest $j_{15/2}$ orbital. Therefore, putting a nucleon into this orbital gives a large increase in spin at little cost in energy, i.e. backbending. When octupole deformation is introduced, the $j_{15/2}$ level mixes level mixes strongly with the positive parity orbitals of lower angular momentum and the aligned angular momentum is substantially reduced. The calculation[26] now gives no backbending. In Fig. 19, the results are compared with the experimental data for ^{222}Th. The introduction of octupole deformation improves the agreement with experiment for the negative parity states and for the positive parity states above I=10. However, the lowest positive parity states are best described by a value of $\beta_3 \simeq 0$. This is consistent with the differences in the positive and negative parity wave functions shown in Fig. 5; the octupole correlations are substantialy larger in the negative parity state than in the positive parity state.

7. FAST E1 TRANSITIONS

Apart from the energies of the low-lying $K^\pi=0^-$ states, another extraordinary feature of the mass region 220 < A < 230 is the extremely fast E1 transitions. These transitions of $\sim 10^{-2}$ Weiskopf units are about two orders of magnitude faster than the E1 transitions in the mid-actinides. In Fig. 20, we display the E1 transition rates in several odd and even mass nuclides. There are extremely large fluctuations in E1 transition rates from one isotope to the next. The fluctuations are quite marked as we follow the sequence of ^{224}Ra - ^{225}Ra and ^{226}Ra or the sequence ^{225}Ac - ^{227}Ac.

Although we have been unable[10] to successfully calculate these E1 rates with many-body wave functions, our calculations give some insight into the problem. It is the collective E3 that is the signature of octupole correlation effects and we note that

$$B(E3) \propto \langle \Psi^+ | \sum_{\text{protons}} (r^3 Y_3^0)_{\alpha\beta} a_\alpha^+ a_\beta + \sum_{\text{neutrons}} (r^3 Y_3^0)_{\alpha\beta} a_\alpha^+ a_\beta | \Psi^- \rangle^2 \quad (11)$$

where Ψ^+ and Ψ^- are the relevant many-body states of opposite parity, i.e. the members of a parity doublet in odd-mass nuclides or the 0^+ and 1^- states in an even-even nuclide. Both sums are large and in phase with each other when octupole correlations are significant. Our calculations show that when $(r^3 Y_3^0)_{\alpha\beta}$ is large, the dipole matrix element $(rY_1^0)_{\alpha\beta}$ is also usually large, and very importantly the two are in phase. This means that the sums

$$\langle \Psi^+ | \sum_{\text{protons}} (rY_1^0)_{\alpha\beta} a_\alpha^+ a_\beta | \Psi^- \rangle \text{ and } \langle \Psi^+ | \sum_{\text{neutrons}} (rY_1^0)_{\alpha\beta} a_\alpha^+ a_\beta | \Psi^- \rangle$$

will also be large when octupole correlations are large and in phase. However, the E1 transition rate

$$B(E1) \propto \langle \Psi^+ | \sum_{\text{protons}} (rY_1^0)_{\alpha\beta} a_\alpha^+ a_\beta - (Z/N) \sum_{\text{neutrons}} (rY_1^0)_{\alpha\beta} a_\alpha^+ a_\beta | \Psi^- \rangle^2 \quad (12)$$

This means that it is possible to get large enhancements in the E1 transition rates only when the octupole correlations are substantial, because neither sum is coherent in the absence of octupole correlations. However, because the protons and neutron contributions to B(E1) are out of phase, we can get large fluctuations of B(E1) from nucleus to nucleus and in some cases almost complete cancellation between the two sums.

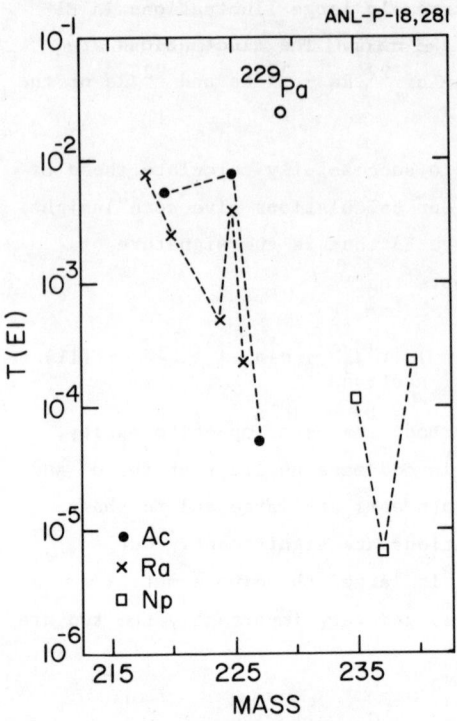

Fig. 20 E1 transition rates in Weiskopf units. Data taken from Refs. 5, 10, 22 and 27.

Large octupole correlations are necessary, but not sufficient, to ensure fast E1 transitions.

Leander has shown[28] that one gets large coherent sums of diagonal E1 matrix elements in octupole deformed single particle potential calculations. This indicates the possibility of obtaining large E1 transition rates from a single particle octupole deformation model. See Fig. 21. In order to make quantitative estimates, one must also include a liquid drop model contribution to the E1 moment, which is roughly comparable to the renormalized single particle contribution. There was some argument as to the sign of this liquid drop contribution, which has been resolved[28] in favor of the early arguments put forward by Strutinsky[29] that the protons preferentially concentrate in the sharply curved tip of an octupole deformed nucleus (lightening rod effect). Very recently, Dorso et al.[30] have examined liquid drop contributions to nuclear dipole moments, utilizing the droplet model. They find an additional contribution to the electric dipole moment from the neutron skin. This contribution is of the same magnitude as the charge redistribution effect[28] but opposite in sign. This gives an order of magnitude reduction of the liquid drop E1 moments which leaves the E1 calculations in even-even nuclides up in the air. No quantitative calculations have yet been performed for odd mass nuclei.

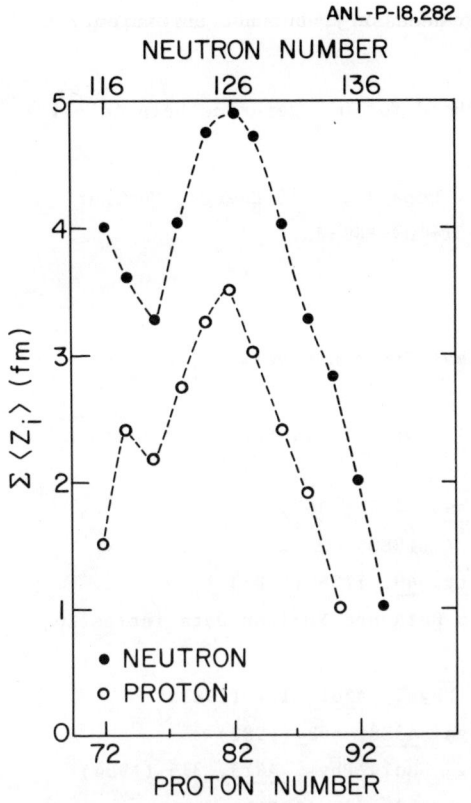

The F-boson model does not distinguish between protons and neutrons. Accordingly, it should not be used to discuss E1 transition rates. The cluster model calculations[22] fit the variations in B(E1) only by varying the E1 effective charge.

A quantitative explanation of the enhanced E1 transitions remains one of the open and exciting problems in nuclear structure calculations.

Fig. 21 Calculated values of $\sum_i \langle Z_i \rangle$ for protons and neutrons in octupole deformed potential. Results taken from Ref. 28.

8. SUMMARY

We find that the many-body treatment of octupole correlations provides a good description of the low energy structure of nuclides in the mass region $220 \leq A \leq 230$. Many of these features can also be described in terms of octupole deformation of a central potential. Deviations from the simple octupole deformation picture are due to the shallow minimum associated with octupole deformation, in the nuclear potential energy surface. We feel that the cluster model does not provide a natural description of these nuclides.

A detailed explanation of the E1 transition rates in the odd and even mass nuclides of this region remains an open and challenging problem.

It is a pleasure to thank I. Ahmad for his generous help in all aspects of this work.

This work supported by the U.S. Department of Energy, Nuclear Physics Division, under contract W-31-109-ENG-38.

9. REFERENCES

1. Bohr A. and Mottelson B., Nuclear Structurs Vol. II, pg. 19, (Benjamin, Reading, 1975)
2. Lederer C. M. and Shirley V. S., Table of Isotopes 7th Edition (Wiley, New York, 1978)
3. Strutinsky V. M., Nucl. Phys. A95, 420 (1967)
4. Chasman R. R. Phys. Lett. 96B, 7 (1980)
5. Ahmad I. et al., Phys. Rev. Lett. 49, 1758 (1982)
6. Möller P. and Nix J. R., Atomic Data and Nuclear Data Tables 26, 165 (1981)
7. Möller P. and Nix J. R., Nucl. Phys. A361, 117 (1981)
8. Leander G. A. et al., Nucl. Phys. A388, 452 (1982)
9. Leander G. A. and Sheline R. K., Nucl. Phys. A413, 375 (1984)
10. Ahmad I. et al., Phys. Rev. Lett. 52, 503 (1984)
11. Sheline R. K. and Leander G., Phys. Rev. Lett. 51, 359 (1983)
12. Piepenbring R., J. Physique Lett. 45, L-1023 (1984)
13. Sheline R., Phys. Lett. 166B, 269 (1986)
14. Chasman R. R., Phys. Lett. B (to be published)
15. Iachello F. and Jackson A., Phys. Lett. 108B, 151 (1982)
16. Daley H. and Iachello F., Phys. Lett. 131B, 281 (1983)
17. Daley H. and Iachello F., Ann. of Physics 167, 73 (1986)
18. Valli K. et al., Phys. Rev. C1, 2115 (1970)
19. Rao M. M. et al., BAPS 31, 875 (1986)

20. Iachello F., Lecture Notes in Physics Vol. 119, 140 (1979) (Springer Verlag, Berlin)
21. Daley H. and Barrett B. R., Nucl. Phys. A449, 256 (1986)
22. Daley H. and Gai, M., Phys. Lett. 149B, 13 (1984)
23. Shriner J. F. et al., Phys. Rev. C32, 1888 (1985)
24. Ahmad I. et al., Nucl. Phys. A258, 221 (1976)
25. Engel J. and Iachello F., Phys. Rev. lett. 54, 1126 (1985)
26. Nazarewicz et al., Phys. Rev. Lett. 52, 1272 (1984)
27. Fernandez-Niello J. et al., Nucl Phys. A391, 221 (1982)
28. Leander G., Proc. Fifth Int. Symposium on Capture γ-ray Spectroscopy, S. Raman ed. (1984)
29. Strutinsky V., Journal of Nucl. Energy 4, 523 (1957)
30. Dorso C. O. et al., Nucl. Phys. A451, 189 (1986)

OPENING STATEMENT

J. A. Maruhn[1]
Institut für Theoretische Physik
Universität Frankfurt
D6000 Frankfurt am Main, West Germany

ABSTRACT

It is emphasized that the recently discovered odd-multipole shape degrees of freedom in nuclei and the various types of asymmetric fission and cluster decay should be viewed in a unified framework.

1. INTRODUCTION

Recently there has been a lot of activity in the study of shape-asymmetric states of nuclei. On the one hand, unusually fast E1-transitions discovered in the Ra-Th region[1] seem to require a static octupole or even dipole deformation, and on the other hand the discovery of cluster radioactivity[2], probably not coincidentally in the same region of nuclides, points to an unexpectedly strong preformation of clusters such as ^{14}C or ^{24}Ne in heavy elements.

In this work we propose a collective description of these phenomena that allows a unified treatment of asymmetrically deformed ground states, cluster radioactivity, superasymmetric and "normal" fission. It will be based on the standard collective potential energy methods applied in fission theory[3], but we will indicate where new developments would have to be made to extend the description to more asymmetric shapes.

2. DEFINITION OF NUCLEAR SHAPES

Fig. 1 shows some of the shapes whose description is desirable for treating the processes discussed above. Models for asymmetric nuclear fission such as the two-centre shell-model[4] usually are suited only for describing less asymmetric shapes at large center separation, because the fragments are hemispheres (or hemiellipsoids) joined by a smooth neck. For highly asymmetric shapes allowance has to be made for a small nucleus "sticking out" of the large one. This requires only a trivial generalization of the two-center shape

---

[1]The results reported here were obtained in collaboration with K. Depta, R. Herrmann, W. Greiner, and A. Săndulescu.

parametrization, but has been responsible for the fact that fission models have been applied to superasymmetric fission[5] and not to cluster radioactivity.

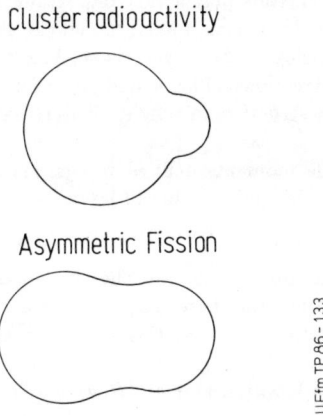

Figure 1: Typical shapes required for the description of asymmetric fission and cluster radioactivity. The extreme asymmetries required for cluster radioactivity make the use of conventional two-center models difficult.

The primary collective coordinates for these shapes will be the centre separation R and the asymmetry $\eta = (A_1 - A_2)/A$. It appears quite likely that an explicit octupole deformation will not be needed, as the shapes defined in the figure already contain strong octupole deformation. The octupole moment may be computed simply from

$$Q_{30}(R,\eta) = \int Y_{30}^*(\theta,\phi) r^3 \rho(\vec{r}; R, \eta) d^3 r \quad , \tag{1}$$

where coordinates are relative to the center-of-mass of the nucleus. Similarly the dipole moment would be given by

$$Q_{10}(R,\eta) = \sqrt{\frac{3}{4\pi}} \int z \rho(\vec{r}; R, \eta) d^3 r \quad . \tag{2}$$

Here $\rho(\vec{r}; R, \eta)$ is the charge density assumed in the model for given values of the two-center separation R and asymmetry η. For the octupole moment a uniform charge density for the intrinsic shapes might be sufficient but a dipole moment needs a more complicated internal state that allows for a difference between the centers of mass of protons and neutrons One possibility for including this effect is the use of a droplet model for determining the density distributions inside the shapes given. The dipole moments generated in this way have been investigated by Dorso et al.[6], although for quite different shapes. In any case this would lead to a non-uniform proton-to-neutron ratio and thus also to non-vanishing dipole moments. Alternatively, one might also introduce an additional charge asymmetry degree of freedom[7].

In the calculations of the probability of cluster radioactivity[8] it was found that the shell structures of the nascent clusters play a very important role in determining the most favorable decay processes. For this reason a shell model calculation should also be added, for example through an extension of the two-centre shell model to highly asymmetric shapes, to incorporate shell corrections. This would allow for a polarization of the nucleus by the presence of strong shell structure in a light cluster that favors an equal number of protons to neutrons.

Note that the multipole moments defined in eqs. (1) and (2) are not yet the observable ones. We will discuss this point in detail later.

3. COLLECTIVE POTENTIAL ENERGY SURFACE AND DYNAMICS

Let us assume that the nuclear shapes, the single-particle model, and the liquid drop or droplet model calculations for these shapes are available. If we denote the center separation by R and the asymmetry by η, this will yield a collective potential energy surface

$$V(R,\eta) = V_{drop}(R,\eta) + \delta U(R,\eta) + \delta P(R,\eta) \quad , \tag{3}$$

which contains the macroscopic contribution V_{drop}, the shell corrections δU, and the pairing correction δP. Associated with this surface are also the multipole moments $Q_{30}(R,\eta)$ and $Q_{10}(R,\eta)$ for each shape within the collective space.

In order to describe the dynamical behavior of the nucleus, mass parameters also have to be provided. This can be done either through the cranking model, which should be more appropriate for low excitation energies, or by using irrotational flow masses. Since none of these methods has been proven adequate, one might also consider a generator coordinate method or an ATDHF calculation in the long run. In any case we we will end up with a collective Schrödinger equation, which e. g. in the Pauli-Podolsky quantization method takes the form

$$i\hbar \frac{\partial}{\partial t} \Psi(R,\eta) = \left[-\frac{\hbar^2}{2m} \sum_{rs} \frac{1}{\sqrt{g}} \frac{\partial}{\partial u_r} \left(\sqrt{g}(g^{-1})_{rs} \frac{\partial}{\partial u_s} \right) + V(R,\eta) \right] \Psi(R,\eta) \quad , \tag{4}$$

where the coordinates R and η have been denoted by u_1 and u_2 and the metric tensor $(g_{rs}), r,s \in (1,2)$ is made up of the collective mass parameters.

Now the static properties of the nucleus concerning its dominant odd multipole moments will be determined by the eigenstates of the Hamiltonian in eq. (4), $\Psi_n(R,\eta)$. For example, the observable ground state octupole moment will be given by an average over the local moments:

$$Q_{30} = \int Q_{30}(R,\eta) \Psi_0^*(R,\eta) \Psi_0(R,\eta) g(R,\eta) dR d\eta \quad . \tag{5}$$

Here $g(R,\eta)$ is the volume element in the collective space obtained from the quantization procedure. *It is very important to realize that a non-vanishing effective octupole moment does not require an octupole-deformed equilibrium deformation: softness in the octupole direction is completely sufficient.* The same is true for the dipole moment.

4. STRUCTURE OF THE POTENTIAL ENERGY SURFACE

In fig. 2 we show a sample potential energy surface obtained in the Meyers-Swiatecki liquid drop model[9] with shell corrections added from the two-centre shell model. This is only a semi-realistic calculation, because, as mentioned before, the two-centre shell model does not yet describe very asymmetric shapes reasonably. However, it should be sufficient for pointing out some of the features expected in this collective approach.

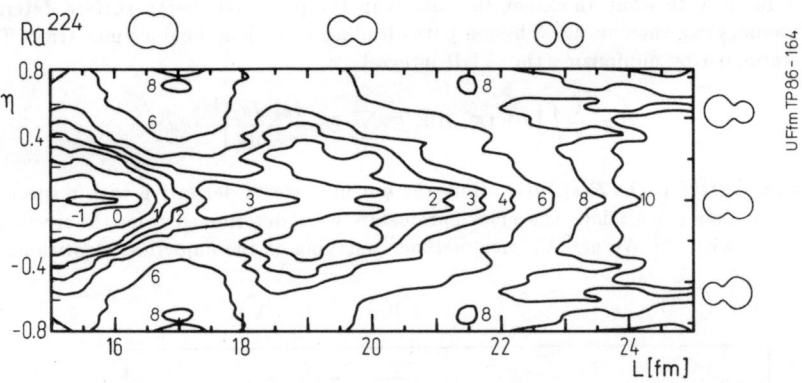

Figure 2: Potential energy surface of a typical heavy nucleus as a function of the two-center coordinates R and η. The ground state is located at $\eta = 0$ (symmetry) with the R-value indicating a prolate deformation, while for larger R-values the valleys crossing the fission barrier appear.

The major structures occurring in the potential energy surface are the fission barrier and the broad minimum near the ground state. Let us first examine the barrier in some more detail.

The fission barrier is crossed by a number of valleys that correspond to the shell-correction enhanced fragment configurations These should show up in the experimental data as the peaks in the mass distribution, ranging from the conventional asymmetric fission to the highly asymmetric process of cluster radioactivity.

On the other hand, the structures closer to the ground state determine the observable lowest odd multipole moments. As mentioned above, non-vanishing moments can be caused either by a minimum with η non-zero or by softness of the nucleus towards η-deformation.

Let $\Psi_0(R,\eta)$ be the ground state wave function, located in the vicinity of $R \approx R_0$ and spread out in the fragmentation degree of freedom. Obviously one might interpret the quantity

$$P(\eta) = \int dR \mid \Psi_0(R,\eta) \mid^2 \qquad (6)$$

as the preformation probability for the cluster configuration characterized by η.

Now the same valleys also have an influence on the fission probability: softness along a particular valley in the surface will lead to an enhanced probability density at that point where the valley meets the fission barrier and thus yield a larger "preformation probability" for that specific break-up. In this sense, we see a direct linkage between static multipole moments and highly asymmetric fission: both may be related to the same valleys in the (R, η)-plane.

To show to what an extent the valleys in the potential energy surface determine the dynamics, we show in fig. 3 fission paths leading to various final asymmetries. These are determined by minimizing the WKB integral

$$P = \frac{1}{\hbar} \int \left[2(V(R, \eta) - E) \sum_{kl} B_{kl} \frac{du_k}{ds} \frac{du_l}{ds} \right]^{1/2} ds \qquad (7)$$

over all collective paths $R(s)$, $\eta(s)$. Here the dot indicates a derivative with respect to s. For this sample calculation, the mass parameters were determined from the irrotational flow model with the Werner-Wheeler method[10] serving as the numerical algorithm.

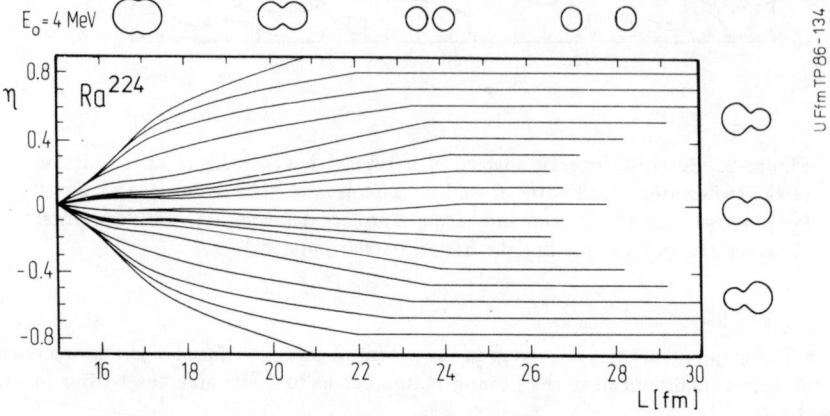

Figure 3: Family of fission paths in a surface like that of fig. 2 determined by a minimization of the penetration integral of eq. 7 with a fixed final asymmetry.

Apparently the valleys in the potential cause a strong concentration of dynamic paths and the fission properties will not be only given by the barrier height itself, but also by the structures in the surface closer to the ground state.

Unfortunately, the determination of the fission properties requires much more detailed theoretical investigation for highly asymmetric fission than for the conventional nearly symmetric fission. One reason is the much stronger shell structure in the light fragment, contrasted with the relatively smooth change in structure through the region of normal asymmetric fission. Also the correct treatment of the tunneling problem becomes much more crucial, because **the larger part of the barrier** is located in a region of separated

35

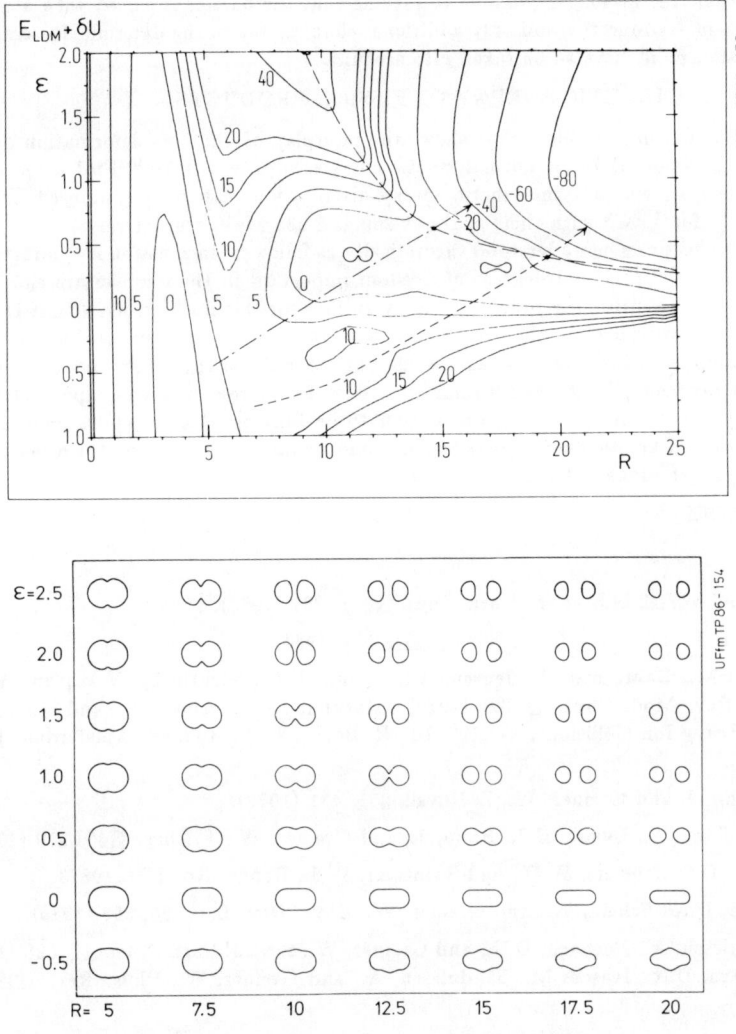

Figure 4: Upper part: Potential energy of ^{258}Fm as a function of the center distance R and the neck parameter ε in the two-center shell model. There are two independent fission paths indicated by dashed lines. Lower part: Nuclear shapes for the same shape parameters, to give an impression of the physical significance of the two fission paths.

nuclei, so that it is no longer possible to assume that the barrier is exited with a definite distribution of asymmetries and only additional changes due to the dynamic descent down to the scission point have to be taken into account.

5. ABNORMAL KINETIC ENERGY OF FISSION PRODUCTS

An additional problem that shows the interplay of different deformation degrees of freedom in fission is the abnormal kinetic energies observed near ^{258}Fm[11]; recently it has even been shown that the kinetic energy distributions can be decomposed into two Gaussians [12] for ^{260}Md with their peaks at 200 and 235 MeV, respectively.

This effect can be understood theoretically as follows: since fission is symmetric for these isotopes, the principal degrees of freedom important in these nuclei are elongation and necking. Computing the potential energy surface for this nucleus, fig. (4), it is found that there are two valleys in the surface, one with the lower barrier proceeding along the normal fission path, and one that takes a "detour" through shapes of less necking or even widened at the neck position and leading to scission for more elongated shapes. Although this explanation needs a dynamic calculation to be fully reliable, it shows qualitatively how the observed kinetic energy distribution could be produced and why the lower energy branch has lower probability.

REFERENCES

[1] Fernandez-Niello, J. et al., Nucl. Phys. A391, 221 (1982).

[2] Rose, H.J. and Jones, G.A., Nature 307, 245 (1984).

[3] Brack, M., Damgaard, J., Jensen, A.S., Pauli, H.C., Strutinsky, V.M., and Wong, C.Y., Rev. Mod. Phys. 44, 320 (1972).; Maruhn, J.A., Greiner, W., and Scheid, W., in: "Heavy Ion Collisions, Vol. 2", ed. R. Bock (North Holland, Amsterdam 1980), p. 399.

[4] Maruhn, J. and Greiner, W., Z. Physik 251, 431 (1972).

[5] Săndulescu, A., Lustig, H.J., Hahn, J., and Greiner, W., J. Phys. G4, L279 (1978).

[6] Dorso, D.O., Meyers, W.D., and Swiatecki, W.J., Report No. LBL-19873.

[7] Gupta, R.K., Scheid, W., and Greiner, W., Phys. Rev. Lett. 35, 353 (1975).

[8] Săndulescu, A., Poenaru, D.N., and Greiner, W., Sov. J. Part. Nucl. 11, 528 (1980); Poenaru, D.N., Ivaşcu, M., Săndulescu, A., and Greiner, W., Phys. Rev. C32, 572 (1985).

[9] Myers, W.D. and Swiatecki, W.J., Nucl Phys. 81, 1 (1966).

[10] Kelson, I., Phys. Rev. 136, B1667 (1964)

[11] Hoffmann, D.C., Phys. Rev. C21, 972 (1980).

[12] Schädel, M. and Sümmerer, K., GSI-Nachrichten.

OCTUPOLE AND DIPOLE MODES

Discussion Session
Following Review Lecture
by
R. Chaseman

Moderator

J. A. MARUHN
Institut fur Theoretische Physik
Universitat Frankfurt
D6000 Frankfurt am Main, F.R. Germany

Moderator: Apparently one of the main topics of this discussion will be the applicability the various models proposed. Phrased as a multiple choice question we have to answer: A) Dipole excitations,
B) Dipole deformation,
C) Octupole excitation,
D) Octupole deformation;
where Dr. Chaseman created a category between C and D. Now Franco Iachello has volunteered to comment on this: so let's give him the opportunity.

Franco Iachello, Yale: First of all I believe that the important point here is not that of pushing one particular model or the other. I'm not in favor of only the octupole or the clustering description. I think one should try to understand what is the physics involved in these nuclei and I dislike religious statements where one starts a talk by saying: "I believe that this is the truth." This is the wrong way to start understanding physics. So I will talk about neither politics nor religion. I would rather like to talk about some physics. It seems to me that the situation is rather complex. My illustration is not for a heavy nucleus but in the medium mass nucleus Nd 150. The situation that in my opinion is the actual one in Nd 150 is that at very low energy there are states corresponding in the classical picture to quadrupole shapes and this forms the sequence of levels with 0, 2, 4, 6, etc. together with their vibrations which for this shape are the beta and the gamma vibrations and possibly (if one can observe them) also the double vibrations. Then typically, in nuclei like this, there are octupole degrees of freedom. It is not clear whether these octupole degrees of freedom are of the vibrational type or of the deformed type. In this nucleus it appears that they are mostly the vibrational type and because of the coupling of the octupole degrees of freedom with the deformation it is split into two bands. In fact it is split into more than two bands, but here I wanted to point out that these two bands, one with K=0 and K=1, are known in this nucleus to occur at about 1 MeV. In these nuclei one expects to have clustering configurations at a little bit higher energy around 2 MeV in excitation, and again because of the alpha particle can either oscillate parallel to the symmetry axis or perpendicular to the symmetry axis, we have this mode split into two pieces. In my estimate, of the energy region in which the clustering occurs in this nucleus is between 2 and 3 MeV. It is interesting that some preliminary indications obtained with inelastic photon scattering (Kneissl will present the experimental results in a later session) appear to indicate that around 2.7 MeV there is a state connected by a large E1 transition to the ground state and it seems to me that there is no way to understand the large E1's of that order of magnitude going to the ground state for a state at 2.5 MeV. It's very hard to understand it in any other way. Then if you go even further there is the region of the giant resonances. Among these resonances there is certainly the giant dipole resonance which corresponds to oscillations of the protons versus the neutrons. So as far as E1 modes are concerned, the situation is certainly complicated. There are at least three possible explanations for large E1's. One coming from mixing of the giant dipole resonance into the low lying states. Two is coming from clustering and three is coming from octupole. I think that rather than quarreling about one or the other we should try to understand this point. At this stage my present feeling is that it's not completely clear to me which one of these three mechanisms is responsible for these large E1 transitions. The situation with the actinide nuclei is that perhaps in the actinide nuclei both the octupole and the clustering are moving down and they are coming to lower energy and perhaps the clustering is coming to about 1 MeV while the octupole is coming to about 200 or 300 kilovolts. So that's one comment. A second comment which seems to be misunderstood, is the fact that if we look at the clustering configuration, and expand the clustering configuration into multipoles, then because of the neck, you get to have all multipoles, 1,2,3,4,5,6. So, in fact, if from other reason like microscopic calculations one finds that one needs 2^6 multipoles, 2^5 multipoles, it seems to be an indication that perhaps clustering is playing an important role even in the so-called octupole states. That I think is an important point. Also I would like to remind some people of the fact that these nuclei do decay by emitting an alpha particle and therefore they must have some kind of preformation probability in the ground state. The estimates are that the preformation probability in the ground state is all about 5% or 10% but certainly they must have a fraction of the wave function which looks like that. One should not forget about this point. Also when one compares theory with experiment one should not select a set of data to compare with the theory but should compare everything. I give you an example. People have measured four particle transfer reactions in this region and they find very unusual properties. They find that both the ground state and of some excited states are excited. I doubt that there is any explanation of these data if one does not include four particle correlations. I do not see at all how one could explain the large excitation of 0+ excited states without some clustering correlations in the ground state. So, I repeat once more that my point of view is not that one model is to be preferred to the other. At this stage all the experimental information that I have seen, including that which was shown in the previous talk, could be explained in any of the four ways mentioned previously: octupole vibrations, octupole deformations, cluster vibrations, and cluster deformations. I think that I am not seeing any amount of data that could not be described by any of these four. I want to repeat another point which was in fact made by Chasman in his talk, that actually what really distinguishes whether one has one mode or the other mode is not the ground state band but are the excited bands. So certainly 2^+2 bands will tell us whether there is octupole or not. The side bands are the clue to the understanding of the structure.

Moderator: If you expand your cluster model into these multipoles wouldn't that imply a certain ratio between successive multipoles. Have you looked at that?

Iachello: I have not personally looked at that but I think that this is correct.

Moderator: Are there any direct comments to Franco's statement?

Tamura: About your point concerning alpha pickup, I just want to emphasize that a large alpha pickup cross section doesn't necessarily mean that the cluster has been formed. It may simply mean that the spectroscopic amplitude is large there.

Iachello: This is correct. There is a possibility that actually the pickup occurs not just by correlated alpha particle, but by two protons and two neutrons. In order to understand whether there is alpha clustering one should compare the Lithium-6 results with that of (p,t) for example, and if it would be available with two proton transfer and it is only by comparison that one could tell the difference. But if your point is correct, it does not necessarily mean there is an alpha cluster.

Frederick Riess, Munich: I think the question in deciding between a dipole model or an octupole model is the question if the dipole moment is the primary thing or if it is just an effect of an octupole deformation. From the shell model point of view the octupole deformation comes out much more naturally than the dipole deformation or dipole kind of excitations as was pointed by Dick Chasman looking at the single particle levels. I think that is an important point.

Iachello: It is important to comment on this point because I think it is an important point whether microscopically one could obtain octupole or dipole and so on. I would like to bring your attention to two recent calculations. One by Otsuka, who is here somewhere, who starts from the shell model but in a deformed basis, therefore similar to the calculations that have been presented before. Otsuka has tried to understand whether the dipole arises naturally. The microscopic calculation of the dipole is very difficult because you have to subtract the center-of-mass motion. Unless you subtract the centerofmass motion you cannot trust the calculation, but with some reasonable argument about the center-of-mass motion and how to deal with that, Otsuka finds that the dipole begins to appear as soon as you move from closed shells, and perhaps he would want to either comment or give you a copy of his paper. That's number one.

Number two, there is another calculation by Vitturi, Sambataro and others who are also here and they can discuss with you the recent calculation done in the laboratory frame, not in the intrinsic frame, but with a two-level system, and they find essentially the same conclusion --- that if you start from the shell model and if you want to have regular bands of both positive and negative parity you must put in the dipole. Without the dipole there is no way you can get out of it. And this goes back to the fact that the octupole is a dynamic model so even if you compute the potential energy surface and you find a very shallow minimum that does not guarantee automatically that you have an octupole deformation, because you should compute inertial parameters. What Kumar does for the quadrupole mode should be repeated here. It could be that the minimum is washed out completely by the dynamic effects. Thus, I think it is a difficult problem and it is not fair to present a statement saying that this is the truth when we know that the problem is very difficult and we have to try to understand it.

Moderator: Any more comments to this? Yes, Krishna?

Krishna Kumar: I would like to suggest a fifth box to the four boxes which have been mentioned: collective bands based on two quasiparticle states. There are two reasons for doing that: First, Dick Chasman showed a graph of E1 transitions. So far large E1's are not really so large, they are reduced by two to five orders of magnitude compared to Weisskopf units. They are large compared to some other things but they are not really strongly collective transitions. That's one point. A second point is that this is one of the worst features of the Copenhagen school that has been adopted by a number of people, which is to pull out a different mode for each new thing that we observe in nuclei.

Moderator: Anybody from Copenhagen here who wants to respond? I guess not.

Chasman: First I'd like to clear up a misconception. I am afraid I have associated Franco unfairly with the idea that all negative parity states in nuclides with $N \geq 140$ arise from the cluster model. If it were fair it wouldn't bother me. Franco, and I think it was clear from the statements, wants to disassociate himself completely from the position of associating all of the negative parity states in the even nuclides above $N = 140$ with a cluster model.

Daley, Daresbury: I would like to comment on the last comment. We have <u>not</u> said that all negative parity states must be associated with alpha clustering. Our philosophy was to keep things as simple as possible for as long as possible and then introduce the other degrees of freedom when we need them; when the data dictates. Let me now ask second question: Since we have been talking about octupole modes all morning and afternoon it would have been nice to see some experimental systematics of the E3's. We know that when we go from a spherical region to a deformed region with the E2 mode, the E2's increase from something on the order of 10 single particle units to over a hundred when you go from spherical to deformed. Do you expect a similar type of trend with octupole deformed situation, i.e. to have an increase over the octupole vibration region by a factor of 5 or 10 going to the deformation region?

Chasman: I thought I showed what I expected in the transparencies where I showed next to the those arrows connecting the parity doublets the $R^3 Y^0_3$ matrix element in the odd mass systems. In the even systems, in fact in the calculations they don't go up very dramatically in the many-body calculations that I did. I think in thorium230 I normalized that and that was the number that was like 20 and in the even nuclides I calculate in those same units whatever they are they never much went above 40 or 50. Only in the odd-mass nuclides did these matrix elements get quite a bit larger. They got close to a hundred. But in the even-even nuclides, the calculations that I did didn't seem to give any very large $R^3 Y^0_3$ matrix elements.

Daley: I meant the even-even cases. Is there octupole deformation in the ground state and low energy 0^- band?

Chasman: I thought I was trying to say that in the even systems we were sort of very much in an intermediate situation between deformation and vibration and that's just using words that aren't really descriptive of what's happening. The correlations were very different in the ground state and the excited state. It was really only in the odd mass systems that we had very large octupole correlation energies.

Daley: That's correct. The E3's are about the same for Ra-238 and the reported Ra-226 case where the 1^- energy is very low; near the minimum.

Chasman: That's in an IBA boson sort of picture. But I think if one does a detailed microscopic calculation I find that the one minus energy in fact, well I use that, and I find that the interaction strength that I used to fit the one minus energy didn't change very much in this region and that the B(E3)'s just don't get that large in the even nuclides. I'm trying to say that you can't use the simple words of deformed or vibrational.

Moderator: Maybe this is getting into too much detail.

Brentano: This is a question for Franco. The parity doublets are something very spectacular. Is there a simple explanation of the parity doublets also in terms of the boson model?

Iachello: Yes. The answer is yes. The parity doublet arise at any time when you break reflection symmetry. So either you break it with the dipole or you break it with the octupole, you have parity doublets. In fact we know the best example of parity doublet -- molecules. There the main degree of freedom is a dipole, just a vector, which is the distance between the two atoms. In fact all properties, i.e. parity doublets, equal magnetic moments, the A parameter being opposite, are properties of the breaking of reflection symmetry not a property of the octupole. It is a purely geometrical effect and therefore it comes in both cases.

Leviatan, Weizmann Institute: I want to comment that, as is well known, in usual IBA-1 there are no triaxial cases. However, the interaction of the dipole degree of freedom with the quadrupole may lead to a triaxial global minimum. I think that this is a very interesting case that should be considered. In such a case the ground state band would look much richer, there would be more states than in the usual axial case.

Iachello: Once you have complicated degrees of freedom, the topography of the energy surface becomes so complicated that I'm not sure you can not have minimum all over and I think it would require very careful investigation. I do not think that just the assumption that you have axial symmetry for the quadrupole, axial symmetry for the octupole, and for the dipole and they are all oriented along the same axis is in general a good assumption. We know from microscopic calculations of Kumar that this is not in general the case. In this case it would be even more complicated.

Leviatan: The interesting feature here is that one can obtain triaxial situations on the level of one and two body interactions. There is no need to introduce higher order terms in the Boson Hamiltonian.

Chasman: I'd like to comment on that because that's the part of my talk I didn't get to. I started doing some calculations in the last few months of introducing triaxial deformations and calculating the potential energy surface and what I seemed to find is that Z=86 and N=130 are best. Those numbers favor some sort of triaxial deformations and we've looked a bit at the experimental data in radium 225 and radium 223 and it appears that one has a gamma deformation. Just looking at interband E2 transitions indicate a gamma of about 15 degrees where we have the reflection asymmetric shapes. The other thing I didn't get to, which I guess I'm getting to now, is that some of the problems that we had with the decoupling parameters in Radium-225 go away to some extent when I introduced triaxial deformation into the single particle wave functions. This is a one body type of calculation.

Moderator: This seems to support the view that the coupling to the cluster structure could produce triaxial states or is your view completely different about the structure of those?

Chasman: It has nothing to do with it.

Moshe Gai: I would like to make a comment from the point of view of the experimentalist. First, comment is a general one. I feel that sometimes when I talk to people who are working in the actinide region, they really take theoretical models perhaps more seriously than the theoretician takes them. I really am amazed that in many cases, for example spin assignment has been put in a level scheme because they are predicted by the octupole deformed model without measurements. I think the question that is arising is at the moment the data is really consistent with all the four (or two) models and we really have to design an experiment that would actually show the difference between the two models and perhaps measure them. I just wanted to show you two data points which to me are not in clear support of stable octupole deformation in this region. The first one is a new data point from Princeton which I received permission from the author, Mike Lowry, to show you. The data, (p,t) on Ra-226 into Ra-224, presumably in the center of the region where octupole deformation should lie. In (p,t) we have of course kinematical factors, we see that the 0^+, 2^+, 4^+ cross sections come down. But if we now compare the cross section for the 1^- with the cross section for the 0^+ and 2^+ and that has been done by the authors to get spectroscopic factors, the 1^- compared to the 0^+ and 2^+, has a factor of 20 difference. Now if this state is somewhat the same state, rising from the same intrinsic state as the 0^+ ground state, I cannot understand this factor of 20 between the cross section (and the spectroscopic factor). To me this is an open question for a nucleus which is right in the center where presumably octupole deformation should be in the ground state including the 1^- and 0^+ and 2^+ but the cross section is again a factor of 20 different. You can do a lot of things to change the cross section for (p,t) by a factor of 3 to 4 from angular distribution and multistep processes but we're talking about a factor of 20. The second question that I have is on data from exactly the same nucleus concerning the alpha hindrance factor. The alpha hindrance factor, we are told for positive and negative parity should be the same if you have the same intrinsic state. The 1^- if I compare it to the 0^+ and 2^+ is a factor of eleven smaller. If I compare it to the 4^+ it is about the same. I have 0^+, 2^+, and 4^+ and a 1^- and 3^- alpha hindrance factor which are certainly not the same. So it's not clear to me that we don't have to shove a few data point underneath the rug to really not upset the stable octupole model. That was my comment.

Moderator: Any comment to the comment?

Butler, Liverpool: I think the answer must be that these nuclei are not reflection asymmetric in the ground state or in the low lying states and that's borne out by the B(E1)'s which are measured for these low lying transitions and also the systematics of the energy levels. But the measurements do not bear that out. And

I think that Nazarewicz, who arrives tomorrow, would agree. Octupole deformation is only stabilized by rotation.

Chasman: Let me repeat myself. One of the major points I was trying to make in my talk was that the octupole correlations are very different in the zero plus ground state and the one minus excited state and this seems to be quite true in general in the even systems and from that point of view I don't really see the point of what you're saying.

Gai: I was under the impression that if you have an octupole deformed state the 0^+ and the 1^- state arise from the same intrinsic state and the alpha hindrance factor should be the same for these nuclei.

Chasman: I am referring to the microscopic calculation I was discussing, and I guess I put everyone to sleep when I was discussing it if you missed that point that the octupole correlations are different than the ground state and the zero minus

Gai: I wasn't referring to your model. The model that I'm referring to is a model that predicts the hindrance factor for the negative parity state and the positive parity state to be the same. That was one of the evidences for octupole deformation because you're dealing with the same intrinsic state. It was not your statement I read but it was certainly one of the evidences quoted and I don't see that in the data.

Chasman: But, I think also it's misleading to say that the one and ten are very different because when you go to the mid-actinide region, and you look at the one minus state, there are factors of hundreds difference.

Gai: No I'm not talking about any nucleus, I'm talking about Ra-224.

Chasman: That's right.

Gai: But I was under the impression that the octupole deformation should be very well established in the ground state of Ra-224. Now I've been told that it's not very well established in the ground state. So I think we're in agreement, it's not in the ground state of Ra-224.

Chasman: I am also saying that one to ten is very close compared to what you find in other regions, so it's this intermediate situation that I've been trying to make a point about all night. As to what's going on; reality is very complicated and when you try to make a simple picture of octupole vibrations or octupole deformation, you can't do it.

Gai: It's very complicated to assign one model to the entire data. It's not a safe thing to do. I think on that I agree with you.

Chasman: I think Franco will second your last statement.

Butler: As Dick was saying, if you look at the systematics of the hindrance factors, as the neutron number changes from 140 to 132 the f values do reduce appreciably so it's clear that some correlation , alpha cluster correlations or octupole correlations, are setting in in that region. You cannot dispute that.

Gai: That is the reason why I was looking at Radium-224. I agree the alpha hindrance factor are very A (mass) dependent. The problem is they are close to unity where the calculations for the ground state do not show octupole deformation, as for example, in Radium-220. In Radium-224 where there is a developed minimum they are not equal to one. So there is a shift between the place where they are equal to unity and the place where the octupole correlation reduce a very well developed minimum in the ground state. So that's why I picked Radium-224. I agree that in Radium-220 the alpha hindrance factors are almost unity but there is no calculation which shows that radium 220 in the ground state has very well developed minimum for eta 3 not equal to zero.

Riess, Munich: I have a question that relates to the understanding where this deformation is coming from. We understand that quadrupole deformation comes from protons and neutrons being outside of closed shell and that can be very nicely shown. If you plot the $E(4^+)$ to $E(2^+)$ ratio as a function of N=126 (which is the closed neutron shell) you see that everything goes parallel independent of Z because there are enough protons outside of the closed shell anyway. If you plot the value of beta three which comes out from the binding energy calculations then you see that this is a function of A and that it has nothing to do with either Z nor N. My question: can that be understood?

Iachello: This is an interesting question about the origin and I don't want to say things which are not right. Certainly the experiences we know with octupole vibrations which is the part that we probably understand best, a crucial ingredient is the occurrence of the so-called intruder states those which come from the next oscillator shell. So for example the occurrence of h11/2 in the 50-82 shell is crucial to giving rise to octupole vibrations in the Gd region because that provides the possible shell model configurations which combine to three minus. Similarly, the next shell would be whatever comes after h11/2 and i 13/2 and the next one would be whatever comes next j 15/2 and so on. So the systematics has to do probably with the property of that state as far as the octupole. As far as the clustering, the systematics instead has to do with different features. First of all, in part, it has to do with the same feature because also in the clustering you need a negative parity state to be formed and so you require also these intruder configurations. But, you could also have particle or excitations which are very important. Probably the most important is how many particles are outside the doubly closed shells. For example a good case for cluster would be Lead-208 plus alpha which is Pollonium-212. Then Pollonium-212 plus alpha and so on. As as soon as you move away from the closed shell the presence of neutrons and protons block the possible formation of alpha clustering. Therefore if there is an alpha cluster it will tend to be destroyed by the presence of additional neutrons and protons. This is my feeling about the mechanism for these two things. I don't know whether other people agree with that but this is my feeling.

Engel, Yale: I would just like to raise a question and anybody can answer it. When a p boson or a dipole mode appears in the IBA, is it necessarily associated with clustering or could it be associated with a pair of nucleons as in Otsuka's calculation? Is that related to a cluster or is it something which is completely unrelated and is a p boson from another dynamical mode?

Iachello: Octupole degrees of freedom are probably more associated with pairs, like in the calculations of Otsuka while the p boson associated in the cluster is of completely different nature. It is probably associated with the relative variable between the alpha cluster and the core and therefore it probably had a very complicated description in terms of the shell model. This is a point that Chasman mentioned and he is right. The counting of the bosons for the relative motion should not be done the same way as it is done for the other bosons. So I think there are two different modes, and the question we have to understand is whether this modes are there, how important they are, and what is the interplay between these very complicated modes in nuclei. This is I think what we should do.

Raduta, Bucharest: I worked a long time ago on this business of octupole excitation in heavy nuclei but I still have some crucial problems I do not completely understand. Namely, in the case of the quadrupole picture one defines the deformation as an invariant against rotation. Taking only the octupole vibrations one should define a similar deformation to the quadrupole one, the octupole deformation, by means of the invariants of the SU(7) group. Now, considering simultaneously the SU(5) and SU(7) groups, one has to change the philosophy of defining the invariants. Then, what does it mean: the octupole deformation in the presence of quadrupole deformation? We can look at this problem from different point of view. Suppose we have a Hamiltonian which is not stable if one takes only the octupole degrees of freedom. Switching on the quadrupole degrees of freedom one may add terms which stabilize the shape of the energy function. One gets an equilibrium value for octupole deformation. Now, include additional degrees of freedom. How stable is that octupole deformation against adding some other degrees of freedom? I would like this question to be to clarified.

Iachello: Professor Raduta was absolutely right. This is another very difficult point about the octupole that nobody has been able to solve. The point is very simple. Think of the quadrupole, we have five degrees of freedom to start with. You can go to an intrinsic frame and define two degrees of freedom which you call beta and gamma and three Euler angles. Then if you rotate the system, the two intrinsic variables remain so because you have some invariants which in this case are beta square and beta cube cosine three gamma so you can define an intrinsic frame. But now think of the octupole. You have 7 degrees of freedom to start with, so you go to an intrinsic frame and you have three Euler angles and four intrinsic variables. How do you define the intrinsic frame? Because under rotations they will get all mixed up. So there must be a very complicated way and that complicated way is that you must go to higher order invariants in the 0(7) group. John Engel, I think, tried to compute those invariants with the result that some are of thirteenth order. So in order to be able to define an intrinsic frame you should go to a Hamiltonian, which is of the thirteenth order in the variables or the boson operators. Now usually all of these problems which are present with octupole are not discussed when people talk about the octupoles. It is just assumed that it is oriented around the quadrupole. As far as I know nobody has as yet started the question, what happens if you include other degrees of freedom whether the minimum disappears or not. In fact it could be that it isn't the minimum at all.

Engel: I guess in connection with that I'd just like to remark that in this IBA treatment of octupole degrees of freedom I talked about if you just include s and d bosons through this procedure of the coherent states you can associate a shape energy surface with every boson Hamiltonian. With just the f, d and s bosons I find it very difficult to find a Hamiltonian for which you actually have an axially symmetric minimum which is stable in all directions. However, when you add a p boson to that model it turns out that some of these dynamical symmetry Hamiltonians that I showed have the property that the corresponding surface minimum is stable in all directions; that is an instance where adding another degree of freedom to the model seems to provide a mechanism for stabilizing the deformation.

Zylicz: I would like to return to the question of Moshe Gai concerning the alpha decay to Radium-224. Who was saying that this nucleus has a stable octupole deformation? The 1^- state is close to the 4^+ state, not below the 2^+ state, of the ground-state band.. I was showing the results of simple model claculations by Adam Sobiczewski et al. for the 0.3 MeV octupole barrier. They got the 0^+ state above this barrier. It is a vibrational picture. The 0.3 MeV barrier is too low. One should assume somewhat higher barrier in this model to reproduce the data. But even then one will be perhaps closer to the vibrational picture than to the strong octupole deformation.

Gai: You say there is no large octupole deformation in the ground state?

Zylicz: No, there is only some minimum in the potential energy surface.

Moderator: So there is a minimum but it is a very shallow structure.

Zylicz: Yes, a shallow structure, and perhaps it is better to speak about an octupole instability than about the octupole deforamtion.

Gai: I have a question: wasn't there data about uneven staggering in the ground-state properties which was taken as evidence for octupole deformation in the ground state? What about evidence for effects of octupole deformation on the binding energies?

Moderator: But nobody was claiming that there was deformation in the extreme case.

Moderator: So one more question and maybe we should switch to another topic.

Chasman: One comment I'll say the same thing again. There is still a fair amount of octupole correlation even energy in the zero plus ground state not as much as in the one minus state but there is some there and in terms of any sort of mean square value there is something there in the calculations its not as much as in the one minus state but there is still something there.

Moderator: So one last question.

Butler, Liverpool: I just wanted to change the subject slightly. Dick Chasman in his talk referred to the calculations of the Berkeley school which said that the effect of the neutron skin was to reduce the liquid drop contribution to the E1's and this unfortunately had the effect of completely destroying the beautiful agreement of the octupole calculations of Leander, Nazarevicz, and coworkers with experiments. Now does this mean that we have throw away these calculations completely; or is it just an indication that E1's are not like M1's and notoriously difficult to reproduce and it is just another indication that one has great difficulty in reproducing these quantities?

Moderator: Well does anybody want to comment on that?

Chasman: Its the liquid drop contribution to the dipole moment that differs substantially and when I said it was up in the air at this time, but the thought I want to convey is that I think there an argument going on between people who are experts in the liquid-drop model which I am not. I think it's up to them to first clarify just what the liquid-drop contribution to the E1 moments is. Then we'll see how accurate the calculations are that were done of the dipole moments. They came out so well that my hope is that the old Strutinsky method maybe will turn out to be correct. That's just a hope, it's not a belief.

Butler: To carry on with this discussion, could it be perhaps that experimentalists should be concentrating on trying to measure B(E3)'s instead, especially at sufficiently high rotation frequencies where the octupole deformation is stabilized and one perhaps might expect to see to enhanced E3's as we have discussed before? Would you expect to see enhanced E3's at high spin states?

Chasman: I think the problem is the experimental one that the B(E1), rates even though they are very retarded, and B(E2) rates are so much faster than B(E3) rates that you are never going to observe them. As a theoretician would say, I think it would be nice to see them in the odd-mass nucleus if it's at all possible, perhaps an E3 excitation of the ground state with maybe a (d,d') reaction would be interesting. I have been trying to persuade my colleagues at Lawrence Livermore National Laboratory to do a (d,d') experiment on actinium 227 for the last four years now and they keep telling us they are about to get the target and I think this would be an interesting sort of thing to measure and maybe they will get the target this year.

Moderator: So maybe we should close this part of the discussion. From what I understood of the discussion it seems that the answer to the original question is that the cluster model and octupole model are able to describe the same physics if the cluster model is taken beyond the simple dipole approximation.

Moderator: I want to say a few words about the fission problem, namely there was this idea that maybe fission is also influenced by these things and if you look at the prediction of fission probability, or alpha decay you see that they also have two very different pictures. In alpha decay you try to describe the process as one of the preformed particle which then escapes from the nucleus and the potential that provides for the potential barrier is simply the potential of a point charge in the field of the nucleus; so that's the preformation probability times transmission essentially. Now if you go to the conventional symmetric fission or nearly symmetric fission everything is described in terms of surface modes, so instead of having a preformed particle escaping from the nucleus you have a surface mode which just happens to end up into a separate cluster forming. In this case there is no preformation probability; instead you have a zero point vibration and again the penetration probability through the barrier. So in some sense these are reminiscent of the cluster and the pure surface model and I wonder whether we can learn anything from fission about this problem? Finally, I would also like to say if you define the surface variables for a collective model of course in principle it is possible to have an asymmetry coordinate which goes over into the octupole for small separation and into something like the mass asymmetry for the fissioned nucleus; we've done a simple calculation of a potential energy surface which should show how this physics might look like. The numbers are no good probably because we didn't take a lot of degrees of freedom into account. Near R=0 you are describing octupoles corresponding to different asymmetric breakups and there is also rather shallow structure extending from the ground state out here where you would see the octupole deformed or maybe the octupole soft states, so in this sense if you observe the fission yield out here or alpha particles or carbon-14; can we in any sense measure the amplitude of the wave function on the other side of the barrier through this preformation probability? That is a topic I would also like to bring up in discussion. Does anybody have any ideas about how we could use the situation? Is it too far distant from the ground states to be measurably influenced by that? Any comments on this?

Gai: I would like to make a comment about a related subject which is the carbon-14 radioactivity. It has been suggested that perhaps carbon-14 clustering would be the main mode of the clustering and the main mode to form octupole deformation because carbon14 is very large. If I take the carbon-14 branching ratio and I divide by what I would call conventional penetrabilities, as was already done in the paper of Rose, I get the preformation for the carbon-14 which is 6 orders of magnitude below an alpha cluster. There is a paper by Swiatecki et al. in which they have chosen the potential and they got the carbon-14 to alpha branch ratio. This ratio translates to one in terms of the preformation probability. They go through a very elaborate potential, but if you look in the fine print, they took a radius parameter, which is about a 0.9 Fermi. In fact, the entire effect is the in radius parameter, not in the potential. I can take a square well potential and do Coulomb penetrability calculations, which in fact I did. Since we know that penetration factors are properties of the outside of the potential not to the inside. So the entire thing is given in the radius parameter. In fact, if you do take 0.9 fermi for the radius parameter you do get the preformation probability of carbon14 and alpha to be the same which would seem to say that the carbon14 clustering is as important as alpha. But then if you use the very same penetration factor to calculate the absolute width of the alpha particle and the carbon14 you are violating the Wigner limit sum rule by a factor of 20. My point is that if you do something to the radius, indeed you can get preformation probability of the carbon14 and the alpha to be the same, whereas if you take the conventional value for the radius it is 6 orders of magnitude below. So I think that before claiming that indeed carbon-14 clusterization is very important already in the ground state of Radium-220 or other actinides, one has to take a very serious look on this question of the penetratabilities.

Moderator: I also think it will be very hard to get penetratabilities precise to the level you need to extract much information about preformation. On the other hand, the effect should also be very large. If you think about the emerging carbon14 as a cluster inside the nucleus then you would accept 10 to the 6 as the ratio to the alpha, but if you think of the two in terms of surface modes you would probably not expect such a large factor or would anybody expect that.

Gai: No, I would not take the penetration factor to be indicative of maybe a factor of ten, but better. We are talking about 6 orders of magnitude difference and that is a big factor. I can only take away this big factor if I do a very drastic assumption on the radius parameter, which in fact violates the sum rule.

Moderator: Any other comments to the fission problem?

Zelevinsky: I would like to mention another interesting phenomenon which may have something to do with this question. I mean the parity nonconservation in the fission --- real parity nonconservation due to weak interaction. It was observed in different isotopes with polarized neutrons as an asymmetry of fission fragments with respect to the neutron polarization direction. It can be explained via the mechanism of strong mixing on the hot stage of this compound nucleus due to very high level density. The adjacent levels of opposite parity would mix very strongly at the hot stage, and after that due to the mass asymmetry this preliminary parity violation transfers to the motion of fragments. The observed asymmetry is enhanced by the factor of about 10^3 when compared to the expected one from the elementary particle level. It's difficult to understand how you can observe such an asymmetry in alpha cluster or quasi-molecular mechanism of fission.

Moderator: Any reactions to that statement? I hear the suggestion that it is too late anyway so if there are no more remarks we should close this session and thank all the contributors.

DESCRIPTION OF THE OCTUPOLE DEGREES OF FREEDOM BY SHAPE VARIABLES

Stanisław G. Rohoziński

Instytut Fizyki Teoretycznej Uniwersytetu Warszawskiego
Hoża 69, 00-681 Warszawa
POLAND

and

Walter Greiner

Institut für Theoretische Physik der J.W. Goethe Universität
Robert-Mayer-Str. 8-10, 6000 Frankfurt am Main
GERMANY

ABSTRACT

A general description is presented of the collective octupole excitations in terms of shape variables. The octupole degrees of freedom are, in general, coupled to the quadrupole core and thus both the quadrupole and the octupole mode are treated on an equal footing. The octupole-oscillator wavefunctions of a definite angular momentum are constructed. The problem of intrinsic frame of reference is discussed. Approximate solutions of the model for spheroidal and pear-shaped nuclei are presented. A role of the Coriolis and centrifugal forces is pointed out.

1. INTRODUCTION

The low-energy negative-parity collective excitations in even-
-even nuclei have been interpreted as an octupole motion of nuclear
surface for more than thirty years. Therefore, the title-problem of
our talk seems to be an old one. This is, however, only partly true.
Indeed, looking over the literature on the topic we see that it is not
very comprehensive and complete. Quite often the problem of octupole
shape vibrations has been presented deficiently and only on the analogy
of the quadrupole motion. Analogies between the two modes are, to our
mind, misleading and therefore often lead to an incorrect picture. It
seems to us that quite a few insinuations and false notions have been
accumulated on the problem and persist even today. It is not our
intension to criticize all that. Instead, we aim to present our point
of view and to recapitulate a general collective model of the nuclear
octupole motion, which we have formulated a few years ago[1].

Since early Bohr´s paper[2] on the collective motion in spherical
nuclei it has been generally accepted to introduce the concept of
nuclear surface multipole mode and to consider each mode separately.
However, such a treatment is not plausible a priori for deformed nuclei.
The octupole vibrations take place on the top of a deformed quadrupole
core and certainly feel it or interact with it. The idea, that the
octupole motion should be coupled to the quadrupole mode has already
been originated a long time ago[3-5]. Sharing such a point of view we
have formulated a general collective model of the quadrupole and octupole degrees of freedom treating both modes on an equal footing. This
model is presented in Section 2. The problem of individual octupole
haromnic vibrations is interesting not only by itself, because the
oscillator wavefunctions form a basis in the space of octupole variables
and can serve for solving more general problems. Methods of constructing that basis are reviewed in Section 3. The concept of intrinsic
frame of reference is crucial for deformed nuclei. It is discussed in
Section 4. Using this concept we present in Section 5 approximate
solutions of the model for spheroidal and pear-shaped nuclei.

2. THE GENERAL COLLECTIVE MODEL FOR COUPLED QUADRUPOLE AND OCTUPOLE MODES

In order to describe the low energetic nuclear excitations of both the positive and the negative parity within the collective model we introduce dynamical variables α_{2m} (m = 0, ±1, ±2) and α_{3m} (m = 0, ±1, ±2, ±3) forming a quadrupole and an octupole tensor, respectively. Usually, these variables appear in the model as parameters determining a deformation of the nuclear potential, density or shape. Just to have a simple geometrical interpretation we shall below refer to them as the parameters appearing in the equation for nuclear radius

$$R(\theta,\phi) = R_0 \{1 + \sum_{l=2}^{3} \sum_{m=-l}^{l} (\alpha_{lm} Y^*_{lm}(\theta,\phi) - \frac{1}{4\pi}\alpha_{lm}\alpha^*_{lm})$$

$$-\frac{3}{2}\sqrt{\frac{3}{4\pi}} \sum_{m=-1}^{1} [\alpha_2 \times \alpha_3]_{1m} Y^*_{1m}(\theta,\phi) \} \qquad (1)$$

where $[\ \times\ ]_{\lambda\mu}$ stands for the vector coupling to the multipolarity λ. However, only the tensor properties of α_{lm} (l = 2,3) and not a particular definition of them will be essential for our further discussion. In view of eq.(1) it is natural that α_{lm} are electrical tensors i.e. have the same transformation properties under space reflection (parity) and time reversal (complex conjugation) as those for the spherical harmonics Y_{lm}.

The most general classical collective Hamiltonian which is quadratic in velocities and invariant under time reversal and the space rotations and reflections has the following form

$$H = \frac{1}{2} \sum_{l,l'=2}^{3} \sum_{m,m'} B^*_{lml'm'}(\alpha_2,\alpha_3)\dot{\alpha}_{lm}\dot{\alpha}_{l'm'} + V(\alpha_2,\alpha_3) \qquad (2)$$

where the potential V is a scalar, real and isotropic function of the tensors α_{lm} and the inertial functions $B_{lml'm'}$ form a symmetric electric bitensor. The quantal counterpart of the Hamiltonian of eq.(2) up to a possible change of the potential[6] is

$$H = -\frac{\hbar^2}{2\sqrt{B}} \sum_{1,1'=2}^{3} \sum_{m,m'} \frac{\partial}{\partial \alpha_{1m}} \sqrt{B}\, B^{-1}_{1m1'm'} \frac{\partial}{\partial \alpha_{1'm'}} + V \qquad (3)$$

where

$$\sum_{1,m} B^*_{1m1'm'} B^{-1}_{1'm'1''m''} = \delta_{11''}\delta_{mm''}$$

and

$$B = \det(B_{1m1'm'})$$

In order to observe the structure of kinetic energy it is convenient to deal with the inertial tensors

$$A^{(11')}_{\lambda\mu}(\alpha_2,\alpha_3) = (-1)^{1+1'-\lambda} A^{(1'1)}_{\lambda\mu}(\alpha_2,\alpha_3) = (-)^{1+1'} A^{(11')}_{\lambda\mu}(\alpha_2,-\alpha_3) ,$$

$$= \sum_{m,m'} (1m\,1'm'|\lambda\mu) B^{-1}_{1m1'm'}(\alpha_2,\alpha_3) \qquad (4)$$

rather than with $B^{-1}_{1m1'm'}$ itself[7]. The Hamiltonian of eq.(3), when expressed in terms of the above tensors, looks

$$H = \frac{1}{2\sqrt{B}} \sum_{1,\lambda} \sum_{1'm} (-)^1\, \frac{2\lambda+1}{21'+1} [\pi_1 \times \sqrt{B} A^{(11')}_\lambda]_{1'm'} \pi^\dagger_{1'm'} + V$$

$$(5)$$

where

$$\pi_{1m} = -i\hbar \frac{\partial}{\partial \alpha^*_{1m}} \qquad (6)$$

is the momentum adjoint to α^*_{1m}.

We have much freedom in the Hamiltonian of eq.(3) or (5). It is given up to 78 independent inertial functions and the potential i.e. up to 79 functions of the twelve variables α_{1m}. All these functions ought to be either calculated within a microscopic theory or parametrized in terms of a number of parameters which are next to be fitted to the experimental data after the collective Schrödinger equation is solved. Only the pure quadrupole parts of the potential and the inertial bitensor are known from microscopic calculations for many nuclei. Also, methods of solving the collective eigenvalue problem are devoloped for the pure quadrupole mode. As far as the octupole degrees of freedom are concerned, calculations of the defomation energy surfaces were up to now performed only for axially symmetric shapes[8-13]. The knowledge of the octupole inertial functions is at the moment quite poor. Hitherto, calculations of the octupole inertial functions, again only for the motion conserving axial symmetry, have been performed aiming at a description of fission phenomena rather than octupole vibrations. The octupole part of the collective Hamiltonian, even in its very simplified form, can hardly be fitted in a unique way to the, scanty in most cases, experimental data.

The form of the angular momentum operator is established by the fact that α_{lm} (and hence π_{lm}) is a tensor of rank 1. It reads

$$L_{1\mu} = L_{1\mu}^{(2)} + L_{1\mu}^{(3)} \qquad (7)$$

where $L_{1\mu}^{(l)} = (-1)^l i \sqrt{(l(l+1)(2l+1)/3)} [\alpha_l \times \pi_l]_{1\mu}$ for $l = 2,3$.

The definite forms of other observables cannot be deduced only from symmetries and transformation properties and contain some arbitrary functions or parameters which should, like the parameters of Hamiltonian, be either calculated theoretically or fitted to the data. The $E\lambda$ transition operators $Q_{\lambda\mu}$ are, up to leading terms of the expansion in α_{lm}, equal to

$$Q_{2\mu} = q_2 \alpha_{2\mu} + q_2^{(22)} [\alpha_2 \times \alpha_2]_{2\mu} + q_2^{(33)} [\alpha_3 \times \alpha_3]_{2\mu} + \cdots\cdots ,$$

$$Q_{3\mu} = q_3 \alpha_{3\mu} + q_3^{(23)} [\alpha_2 \times \alpha_3]_{3\mu} + \cdots\cdots ,$$

$$Q_{1\mu} = q_1^{(23)} [\alpha_2 \times \alpha_3]_{1\mu} + \cdots\cdots \qquad (8)$$

Within the simplest version of the model treating the nucleus as a liquid drop of the shape given by eq.(1), uniformly charged, one has

$$q_\lambda = \frac{3ZR_0^\lambda}{4\pi}$$

$$q_\lambda^{(11')} = \frac{\lambda+2}{2} \sqrt{\frac{(2\mathfrak{l}+1)(2\mathfrak{l}'+1)}{4\pi(2\lambda+1)}} \; (\mathfrak{l}0\mathfrak{l}'0|\lambda 0) q_\lambda$$

for $\lambda = 2, 3$

$$q_1^{(23)} = 0 \qquad (9)$$

It is obvious that within the present model the E1 operator, which is a measure of displacement of the center of charge with respect of the center of mass, follows from an heterogeneity of the proton density. A few different phenomena are mentioned as possible sources of heterogeneous charge distribution inside the nucleus[15-19]. No matter which is the mechanism of the charge heterogeneity, the construction of the collective E1 operator requires some additional assumption and therefore an analysis of the E1 transitions is not a good probe of the model for octupole vibrations.

3. THE OCTUPOLE VIBRATIONS AROUND THE SPHERICAL SHAPE

The quadrupole and octupole vibrations around the spherical shape become decoupled in the harmonic approximation. Indeed, the potential, when expanded up to quadratic terms in both deformations, is

$$V = \frac{1}{2} \sum_{l=2}^{3} C_l \sum_m \alpha_{lm} \alpha_{lm}^* \qquad (10)$$

and does not contain any mixed term. Also, having no "material tensors" in our disposal, we can have only scalars out of the interial tensors of eq.(4), which do not vanish at zero deformation. Putting

$$A_{00}^{(22)} = \frac{\sqrt{5}}{B_2} \;,\quad A_{00}^{(33)} = -\frac{\sqrt{7}}{B_3} \qquad (11)$$

in eq(5) we obtain the Bohr Hamiltonian[2]

$$H = \frac{1}{2} \sum_{l=2}^{3} \sum_m (\frac{1}{B_l} \pi_{lm} \pi_{lm}^\dagger + C_l \alpha_{lm} \alpha_{lm}^*) \qquad (12)$$

being a sum of the quadrupole and the octupole harmonic oscillator. A quadrupole-octupole interaction is an anharmonic effect which leads to a splitting of the quadrupole-octupole phonon multiplets (Fig.1). A theory of such effects has been developed in early paper by Lipas[20].

A complete set of eigenfunctions with definite angular momenta of the Hamiltonian of eq.(12) is a basis in the quadrupole-octupole collective space and can be utilize to the diagonalization of an arbitrary quadrupole-octupole Hamiltonian of the form of eq.(3) or (5). These eigenfunctions are products of the quadrupole- and the octupole-oscillator wavefunctions coupled to definite angular momenta. A discussion of the quadrupole wavefunctions, which are, by the way, well-known, goes beyond the subject of the present talk. Hereafter we discuss briefly solutions for the octupole oscillator. The wavefunctions can be classified according to the following group chain :

$$SU(7) \supset SO(7) \supset G_2 \supset SO(3) \supset SO(2)$$
$$\quad N \qquad\quad \lambda \qquad\quad \lambda\, rqs \quad\; L \qquad\quad M$$

Under each group symbol the quantum number labelling its irreducible representations (IR's) is given. These are N , the octupole-oscillator number λ , the seniority, L and M, the angular momentum and its projection, respectively. We have no additional label for the IR's of G_2 since an IR of SO(7) contains only one IR of G_2 in our case. The quantum numbers r,q,s are internal labels in the chain SO(7) $\supset$ SO(3) . After Rakavy's paper[21] the reduction SU(7) $\supset$ SO(7) is done without any difficulty by introducing the "unit tensor" ε_{3m} such that

$$\alpha_{3m} = \beta_3 \varepsilon_{3m}, \quad \sum_m \varepsilon_{3m} \varepsilon_{3m}^* = 1 \tag{13}$$

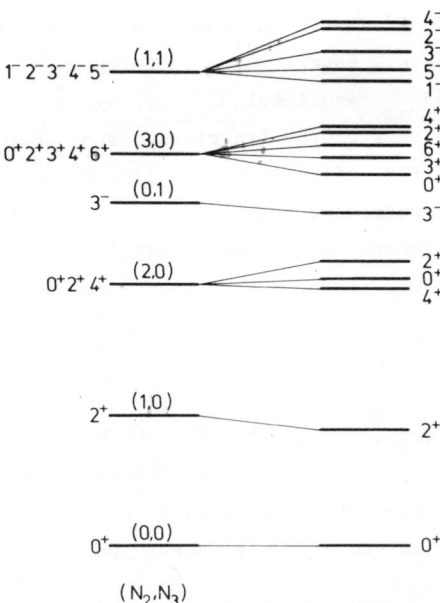

Fig. 1. The energy spectrum of the harmonic quadrupole-octupole vibrations (left). The multiplets are labelled by N_2 and N_3, quadrupole and the octupole oscillator numbers, respectively. Anharmonic corrections lead to a splitting of these multiplets (right). No importance should be attached to sequences of the levels within a multiplet.

which leads to the following factorization of the oscillator wavefunction:

$$\Psi_{NLM}^{\lambda rqs}(\alpha_3) = F_N^\lambda(\beta_3) T_{LM}^{\lambda rqs}(\varepsilon_3) \tag{14}$$

where

$$F_N^\lambda(\beta_3) \sim \beta_3^\lambda L_p^{(\lambda+5/2)}(\sqrt{B_3 C_3}\,\beta_3^2/\hbar) \exp(-\tfrac{1}{2}\sqrt{B_3 C_3}\,\beta_3^2/\hbar) \tag{15}$$

is expressible by the Laguerre polynomial $L_p^{(\alpha)}$ of degree $p=(N-\lambda)/2$. We refer to the functions $T_{LM}^{\lambda rqs}$ as the tensor spherical harmonics since these are spherical harmonics in the seven-dimensional space being at the same time irreducible tensors under the three-dimensional rotations. They have first been classified and constructed in terms of so called "traceless bosons"[22] in reference[23]. Different functions of the same λ and L are labelled there by q and s, the numbers of quartets and sextets of the traceless bosons coupled to spin zero, and r, the label for a " residual coupling". Because of strenuous commutations to perform, the traceless bosons are not very comfortable wo work with. Therefore, a much more useful, although only semi--analytical procedure is to expand the tensor spherical harmonics

$$T_{LM}^{\lambda n}(\varepsilon_3) = \sum_{\substack{\lambda_1 \lambda_2 \\ \mu_1 \mu_2 \mu_3}} \left(\begin{array}{c}\lambda\;\lambda_1\lambda_2 \\ \mu_1\mu_2\mu_3\end{array}\Bigg|\begin{array}{c}\lambda n \\ LM\end{array}\right) Y_{\mu_1\mu_2\mu_3}^{\lambda\;\lambda_1\lambda_2}(\varepsilon_3) \qquad (16)$$

in terms of the three-azimuthal spherical harmonics $Y_{\mu_1\mu_2\mu_3}^{\lambda\;\lambda_1\lambda_2}$ which are the bases for irreducible representations of the group chain[24]

$$\begin{array}{cccccc} SO(7) \supset & SO(6) \supset & SO(2) \times & SO(4) \supset & SO(2) \times & SO(2) \\ \lambda & \lambda_1 & \mu_1 & \lambda_2 & \mu_2 & \mu_3 \end{array}$$

The corresponding transformation brackets can easily be calculated numerically[25]. Then, however, the label n, which stands for the set of quantum numbers r,q,s, is nothing but the number of solution. Still other method of construction the tensor spherical harmonics has been proposed by Ogura[26]. It consists in expanding them in terms of the zonal spherical harmonics[27] or bases for IR˜s of the group chain

$$SO(7) \supset SO(6) \supset SO(5) \supset SO(4) \supset SO(3) \supset SO(2)$$

4. THE INTRINSIC FRAME OF REFERENCE

It is well-known that in the case of pure quadrupole shape

($\alpha_{3m} \equiv 0$) the system of principal axes of the tensor α_2 is a natural intrinsic frame of reference having two pleasant properties. Firstly, it is, at the same time, the system of principal axes of the ellipsoid of inertia which means that the corresponding tensor of inertia is diagonal. Secondly, there are no rotation-vibration terms in the kinetic energy.

The question is, how to define an intrinsic frame for the quadrupole-octupole shapes given by eq(1) . At first sight, again the system of principal axes of the ellipsoid of inertia would be the best one. Then, however, the rotation-vibration terms would always be present in the kinetic energy. Also, the definition of such a system would depend on a particular deformation dependence of the inertial functions and could appear to be quite incovenient for a parametrization of the vibrational energy.

Another idea, connected with the conception of an individual treatment of the quadrupole and the octupole motion, is to introduce two different intrinsic frames for the quadrupole and octupole subsystems separately. The quadrupole-octupole interaction would then be eventually replaced by a constraint condition stating that both frames move together. Here and there one comes across the definition of intrinsic frame for the octupole subsystem consisting in requirement, that the intrinsic components of the tensor α_3 are all real. However, such a frame does not exist at all. The Jacobion of the corresponding transformation from the laboratory to the intrinsic coordinates is just equal to zero. Moreover, there is no intrinsic frame of reference for the octupole degrees of freedom which rotates without coupling to the intrinsic motion.

The conclusion from the above discussion is that the best intrinsic frame for the quadrupole-octupole shapes seems to be still the system of principal axes of the quadrupole tensor α_2 . Below we discuss some consequences of this idea which are not always realized.

It is sometimes convenient to introduce variables a_{1m} and b_{1m} , the real and imaginary parts of the complex intrinsic components α'_{1k} of the tensors α_2 and α_3 :

$$\alpha'_{lk} = \sqrt{\frac{1+\delta_{k0}}{2}} \, (a_{lk} + ib_{lk}) \tag{17}$$

for $l = 2,3$ and $k \geq 0$. The intrinsic frame is defined by the conditions

$$a_{21} = b_{21} = b_{22} = 0 \tag{18}$$

It is well-known that eq(18) defines the Euler angels $\theta_1, \theta_2, \theta_3$ up to 48 transformations which change names and arrows of intrinsic axes and form the O_h group of symmetry of cube. For this reason it is sufficient to consider the quadrupole intrinsic variables a_{20} and a_{22} only within the range

$$a_{20} \geq 0, \quad \sqrt{3}\, a_{20} - a_{22} \geq 0 \tag{19}$$

These variables are, in turn, defined up to 8 transformations which change just arrows of axes and form the D_{2h} group of symmetry of cuboid. Since the variables a_{3k} and b_{3k} are still not invariant under D_{2h} it is sufficient to consider them within the range

$$a_{30} \geq 0, \quad \sqrt{3}\, a_{31} - \sqrt{5}\, a_{33} \geq 0 \quad \sqrt{3}\, b_{31} + \sqrt{5}\, b_{33} \geq 0$$

All other values are obtained from these obeying inequalities (20) by action of the group D_{2h} which does not change shape of nuclear surface but its orientation with respect to the intrinsic frame (Fig 2.).

4.1. The Collective Hamiltonian in the Intrinsic Variables

In order to transform the collective Hamiltonian into the intrinsic variables we can proceed in two ways. The one way is to transform the classical Hamiltonian of eq(2) and next quantize it according to the Pauli prescription (cf. ref.[28]). The other is to transform directly the quantal Hamiltonian of eq(3).

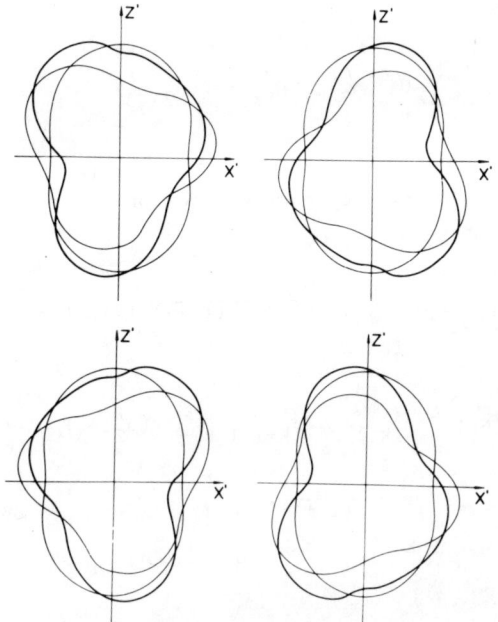

Fig 2. Four equivalent shapes of the nuclear surface (thick line) which differ from each other in the orientation of the octupole substructure (thin line) with respect to the intrinsic axes. The quadrupole substructure (thin line) has a fixed orientation.

The classical Hamiltonian, when transformed to the intrinsic variables, reads

$$H = \frac{1}{2} \sum_{1k1'k'} B^*_{1k1'k'}(\alpha'_2,\alpha'_3)\alpha'_{1k}\alpha'_{1'k'} +$$

$$+ \frac{1}{2} \sum_{\mu,\mu'} J^*_{1\mu 1\mu'}(\alpha'_2,\alpha'_3)\omega'_{1\mu}\omega'_{1\mu'} +$$

$$+ i \sum_{1,k,\mu} F^*_{1\mu 1k}(\alpha'_2,\alpha'_3)\omega'_{1\mu}\alpha'_{1k} + V(\alpha'_2,\alpha'_3) \qquad (21)$$

where $\omega'_{1\mu}$ are the intrinsic spherical components of the angular velocity vector and

$$J^*_{1\mu 1\mu'}(\alpha'_2,\alpha'_3) = - \sum_{1k1'k} \sqrt{1(1+1)1'(1'+1)}(1k1\mu|1k+\mu) \times$$

$$(1'k\ 1\mu|\ 1'k'+\mu')\alpha'_{1k}\alpha'_{1'k'} B^*_{1k+\mu\ 1'k'+\mu'}(\alpha'_2,\alpha'_3)$$

$$F^*_{1\mu 1k}(\alpha'_2,\alpha'_3) = \sum_{1'k'} \sqrt{1'(1'+1)}(1'k'1\mu|1'k'+\mu)\alpha'_{1'k'} B^*_{1k1'k'+\mu}(\alpha'_2,\alpha'_3)$$

$$(22)$$

We see that $J^*_{1\mu 1\mu'}$, the classical tensor of inertia depends explicitly on both the quadrupole and the octupole variables. The quantization of the Hamiltonian of eq(21) in the twelve-dimensional space of variables is a complicated task to perform. It seems that sometimes people have tried just to guess the result, knowing it for a pure quadrupole system. A much simpler and realible way to obtain this result is to transform the quantal Hamiltonian of eq(3) to the intrinsic variables. Then we have[1]

$$H = -\frac{\hbar^2}{2\sqrt{BD}} \sum_{1,1'} \sum_{k,k'} \frac{\partial}{\partial \alpha'_{1k}} \sqrt{BD} B^{-1}_{1k1'k'}(\alpha'_2,\alpha'_3) \frac{\partial}{\partial \alpha'_{1'k'}}$$

$$+ \frac{1}{2\sqrt{B}} \sum_{\zeta,\zeta'=x',y',z'} (L_\zeta - L_\zeta^{(3)}) \sqrt{B}\, A_{\zeta\zeta'}(\alpha'_2,\alpha'_3)(L_{\zeta'} - L_{\zeta'}^{(3)})$$

$$- \frac{1}{2\sqrt{BD}} i\hbar \sum_\zeta \sum_{1,k} ((L_\zeta - L_\zeta^{(3)})\sqrt{BD} M_{\zeta 1k}(\alpha'_2,\alpha'_3) \frac{\partial}{\partial \alpha'_{1k}}$$

$$+ \frac{\partial}{\partial \alpha'_{1k}} \sqrt{BD} M_{\zeta 1k}(\alpha'_2,\alpha'_3)(L_\zeta - L_\zeta^{(3)})) + V(\alpha'_2,\alpha'_3) \qquad (23)$$

where L_ζ are the intrinsic Cartesian coordinates of the total angular momentum, D is the Jacobian of transformation and

$$A_{\zeta\zeta'} = \frac{B^{-1}_{\zeta\zeta'}}{4 a_\zeta a_{\zeta'}}$$

$$M_{\zeta 1k} = \frac{B^{-1}_{\zeta 1k}}{2 a_\zeta} \qquad (24)$$

for $\zeta,\zeta' = x',y',z'$ with

$$a_{x'} = -\frac{1}{2}(\sqrt{3}\, a_{20} + a_{22})$$

$$a_{y'} = \frac{1}{2}(\sqrt{3}\, a_{20} - a_{22})$$

$$a_{z'} = a_{22}$$

and $B^{-1}_{\zeta\zeta'}$, $B^{-1}_{\zeta 1k}$ being some linear combinations[7] of $B^{-1}_{1k1'k'}(\alpha'_2,\alpha'_3)$.

Looking at eq(23) we see that, in accordance with our choice of the intrinsic frame, the object which rotates, is just the quadrupole subsystem (Fig 3.). It is by no means the rigid rotation of the system as a whole. Therefore, the rotational energy contains the Coriolis and the precession term. Also, $A_{\zeta\zeta'}$ is the inverse tensor of inertia of that subsystem and depends on the octupole variables only through the inertial functions. In particular it is not a tesor inverse to $J_{\zeta\zeta'}$, the classical Cartesian tensor of inertia, eq(22)

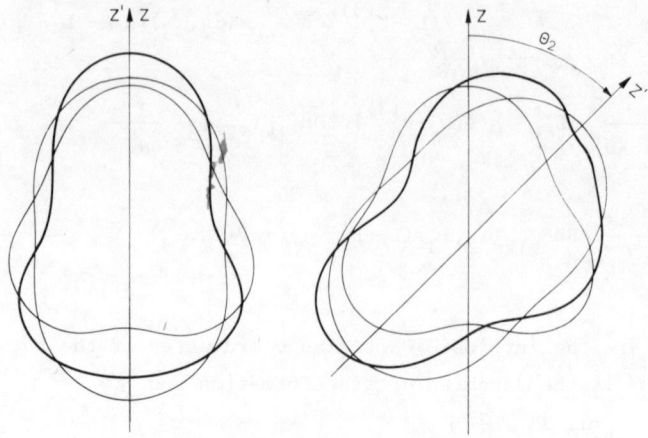

Fig 3. The quadrupole and octupole substructures (thin lines) of the nuclear surface (thick line). The intrinsic axes rotate together with the quadrupole substructure only, while the octupole substructure keeps the laboratory frame.

At the end, we notice that, since the octupole intrinsic variables are complex $(b_{3k} \neq 0)$, the octupole wavefunctions, when transformed to the intrinsic system, take complex values too. This is why a general form of the laboratory wavefunction of a definite parity π and angular momentum L reads

$$\Psi_{LM}^{\pi n}(\alpha_2, \alpha_3) = \frac{\sqrt{2L+1}}{4\pi} \sum_{K=0}^{L} (\phi_{LK}^{\pi n}(\alpha'_2, \alpha'_3) D_{MK}^{L}(\Theta_1, \Theta_2, \Theta_3)$$
$$+ \pi(-1)^{L+K}(\phi_{LK}^{\pi n}(\alpha'_2, \alpha'_3))^* D_{M-K}^{L}(\Theta_1, \Theta_2, \Theta_3)) \quad (25)$$

in distinction to the pure quadrupole case when the intrinsic wavefunctions $\phi_{LK}^{\pi n}$ can always be taken real.

5. THE OCTUPOLE MOTION IN DEFORMED NUCLEI

An approximate picture of the octupole motion in deformed nuclei is obtained when the potential is expanded around the equilibrium point up to the second order and the inertial functions are replaced by their equilibrium values. Below we briefly discuss such an approximation for the two cases of axially symmetric nuclei : spheroidal and pear-shaped ones.

5.1. Spheroidal Nuclei

When the static deformation is $a_{20} = \beta_2^{(0)}$, $a_{22} = a_{3k} = b_{3k} = 0$ the potential, when expanded around the equilibrium point, looks

$$V = \frac{1}{2} C_{20}(a_{20}-\beta_2^{(0)})^2 + \frac{1}{2} C_{22}a_{22}^2 + \frac{1}{2} C_{30}a_{30}^2 + \frac{1}{2} \sum_{k=1}^{3} C_{3k}(a_{3k}^2+b_{3k}^2) \tag{26}$$

Since the only intrinsic components of the inertial tensors of eq(4), which do not vanish at the equilibrium, are $A_{\lambda 0}^{(11)'}$ for $1 = 2,3$, and $\lambda = 0,2,4,6$ we have a similar decomposition for the vibrational kinetic energy. Furthermore, the rotation-vibration term in the Hamiltonian of eq(22) vanishes and the tensor of inertia $A_{\zeta\zeta'}$ is diagonal and axially symmetric. In consequence, we have seven octupole harmonic vibrations with respect to the intrinsic system. As the projection of the partial octupole angular momentum on the symmetry axis is

$$L_{z'}^{(3)} = - i\hbar \sum_{k=0}^{3} k(a_{3k} \frac{\partial}{\partial b_{3k}} - b_{3k} \frac{\partial}{\partial a_{3k}}) \tag{27}$$

the oscillators in variables a_{3k} and b_{3k} ($k = 1,2,3$) constitute the three two-dimensional vibrations carrying a definite eigenvalue $K_3 = k$ of $L_{z'}^{(3)}$ (Fig 4.). This is well-known picture which was, as mentioned in ref.[29,30], suggested by Christy thirty years ago. The octupole vibrations are coupled to the rotational motion through the Coriolis and centrifugal forces.

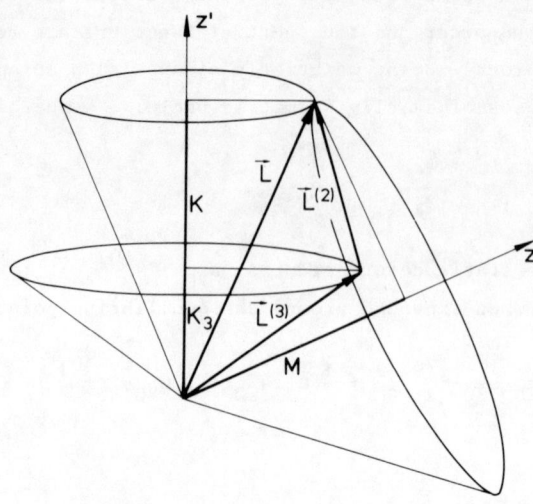

Fig 4. Precession of the total and the partial octupole angular momentum of a spheroidal nucleus. The projections M, K and K_3 are constants of motion.

5.2. Pear-shaped Nuclei

For pear-shaped nuclei the static deformation is assumed to be $a_{20} = \beta_2^{(0)}$, $a_{30} = \beta_3^{(0)}$, $a_{22} = a_{3k} = b_{3k} = 0$ for $k = 1,2,3$. Then, the potential takes in the harmonic approximation the following general form

$$V = \frac{1}{2} C_{20}(a_{20}-\beta_2^{(0)})^2 + \frac{1}{2} C_{30}(a_{30}-\beta_3^{(0)})^2 +$$
$$+ c_0 \sqrt{C_{20}C_{30}} \,(a_{20}-\beta_2^{(0)})(a_{30}-\beta_3^{(0)}) + \frac{1}{2} C_{22} a_{22}^2 + \frac{1}{2} C_{32}(a_{32}^2+b_{32}^2) +$$
$$+ c_2 \sqrt{C_{22}C_{32}} \, a_{22} a_{32} + \frac{1}{2} \sum_{k=1,3} C_{3k}(a_{3k}^2 + b_{3k}^2) \qquad (28)$$

for $a_{30} \geq 0$. In comparison to the potential of eq(26), we can additionally have a coupling between a_{20} and a_{30}, and a_{22} and a_{32}, respectively (Fig 5.)

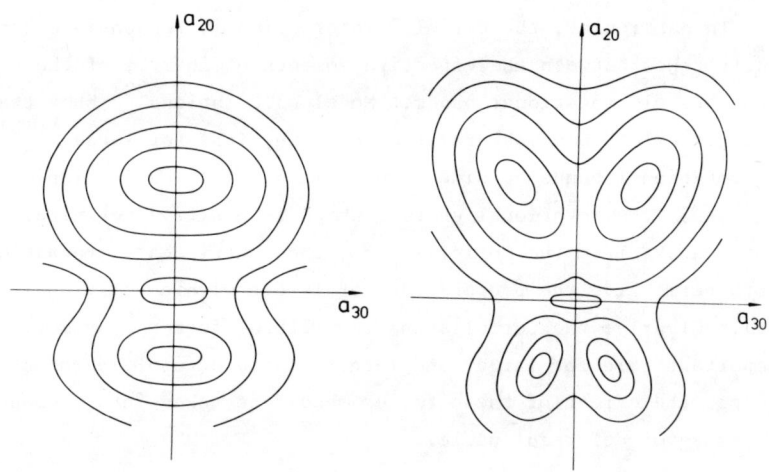

Fig 5. Schematic equipotential lines in the $a_{20} a_{30}$ plane for $\beta_3^{(0)} = 0$ (left) and $\beta_3^{(0)} \neq 0$ (right), showing a possible $a_{20} - a_{30}$ coupling in the latter case.

Couplings between these same degrees of freedom can appear also in the vibrational kinetic energy. Furthermore, the rotation-vibration term in the collective Hamiltonian, eq(23), need not vanish and can give an additional coupling of the variables a_{31} and b_{31} to the rotational motion.

To conclude, we see that apart from a widely discussed (cf. ref. [1]) anharmonicty of the $K = 0$ vibrations around pear-like shape, we can deal with a mixing of the quadrupole and octupole modes. Neglecting the quadrupole-octupole interation is, if possible at all, an additional dynamical assumption in the collective model for pear-shaped nuclei.

5.3. A Role of the Coriolis and Centrifugal Forces

In both cases of spheroidal and pear-shaped nuclei the rotational and the vibrational motion are coupled through the Coriolis and centrifugal forces. It has been known for a long time[31] that the Coriolis force affects considerably the octupole rotational bands in sheroidal nuclei. In particular, the Coriolis interaction is responsible for a large difference between the effective moments of inertia of the quadrupole and octupole rotational bands. Model calculations[32] show that the Coriolis effect becomes dramatic and unphysical for large $\beta_3^{(0)}$ turning rotational bands up side down (Fig 6.). It has a simple explanation : the rotational energy, which is positive definite, containts, apart from the Coriolis term, the centrifugal interaction. Only both terms give the proper order of levels within rotational bands (Fig 7.). Clearly, the Coriolis and centrifugal forces play more and more important role for larger and larger static octupole deformation. Concluding, the effect of these forces should be taken into account in every case of deformed nuclei.

6. CONCLUSIONS

We have presented a general collective model which claims to describing the lowest collective excitations of both, the positive and the negative parity in even-even nuclei. The model uses the quadrupole and octupole shape variables as degrees of freedom. Here, we have, in so far as possible, been interesed only in the octupole motion. Summarizing our discussion on the model we call to mind its main points.

The model, in its general form, is able to describe a variety of octupole motions in nuclei. In particular, it comprises all the models formulated so far for description of the octupole degrees of freedom in some special cases of nuclei. The construction of complete set of the tensor spherical harmonics in the space of octupole shape variables allows for solving the octupole motion in a general case.

The octupole mode is, as a rule, coupled to the dominant, quadrupole motion of the nuclear surface. From geometrical considerations one can learn something about the nature of this coupling in various

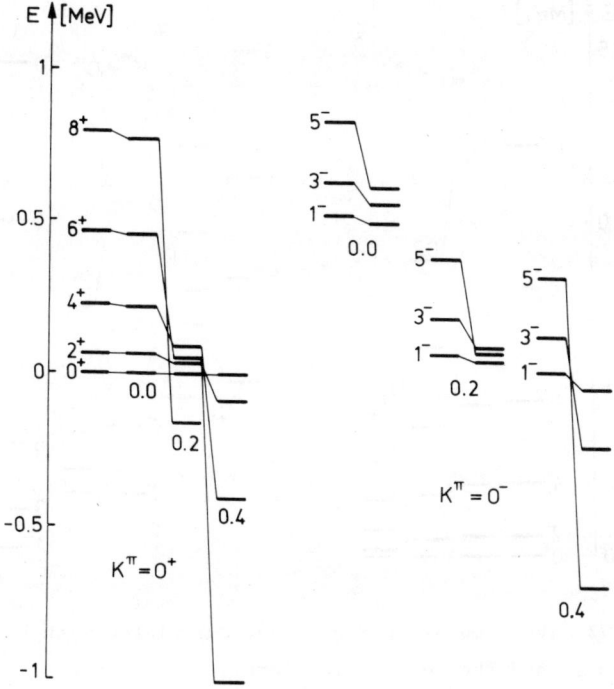

Fig 6. The lowest positive- and negative-parity K = 0 bands unaffected and affected with the Coriolis force for the octupole deformation $\beta_3^{(0)}$ = 0., 0.2, 0.4, indicated at the band . The unaffected levels are labelled with the spin and parity.

cases of nuclear deformation. Namely, the quadrupole-octupole interaction appears to be weak for spherical nuclei. It causes just a splitting, characteristic for the weak coupling scheme, of the quadrupole--octupole-phonon multiplets. The quadrupole and octupole degrees of freedom are also separated approximately for spheroidal nuclei. However, it is rather a strong coupling scheme, since an anisotropy of the octupole vibrations is due to a deformation of the quadrupole core.

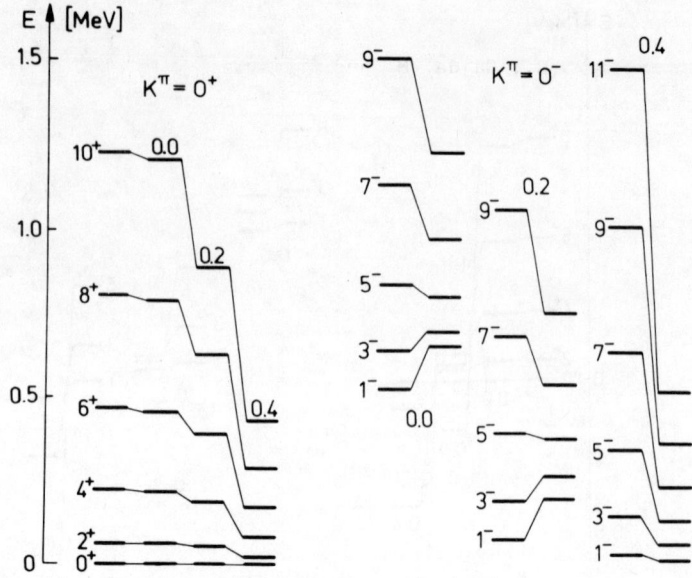

Fig 7. The same as in Fig 6. but calculated with both the Coriolis and the centrifugal force.

The separation of both modes need not be the case for pear-shaped nuclei, when a strong multipolarity mixing can take place.

Since only the quadrupole subsystem is fixed in the intrinsic frame of reference, the octupole vibrations are affected with the Coriolis and centrifugal forces. These forces should be taken into account in any case of deformed nuclei. Their effect becomes especially strong for a large pear-like deformation.

Last of all, let us notice that the interacting boson approach to the octupole motion[33] seems to be substantially different from this presented here, mainly because of introducing additional dipole degrees of freedom.

REFERENCES

1. Rohoziński, S.G., Gajda, M. and Greiner, W.
   J.Phys. G $\underline{8}$, 787 (1982).
2. Bohr, A. , K. Dan. Vidensk.Selsk., Math.-Fys. Medd. $\underline{26}$,
   no. 14 (1952).
3. Lipas, P.O. and Davidson, J.P., Nucl.Phys. $\underline{26}$, 80 (1961).
4. Leper, D.P., Nucl.Phys. $\underline{50}$, 234 (1964).
5. Donner, W. and Greiner, W., Z.Phys. $\underline{197}$, 440 (1966).
6. Hofmann, H., Z.Phys. $\underline{250}$, 14 (1972).
7. Rohoziński, S.G., "Coupling of the Nuclear Collective Modes",
   report IFT/18/84 (1984).
8. Vogel, P., Nucl.Phys. $\underline{A112}$, 583 (1968).
9. Möller, P., Nilsson, S.G. and Sheline, R.K., Phys.Lett. $\underline{40B}$,
   329 (1972).
10. Gyurkovich, A., Sobiczewski, A., Nerlo-Pomorska, B. and Pomorski,
    K., Phys.Lett. $\underline{105B}$, 95 (1981).
11. Leander, G.A., Sheline, R.K., Möller, P., Olanders, P., Ragnarsson,
    I. and Sierk, A.J., Nucl.Phys. $\underline{A388}$, 452 (1982).
12. Nazarewicz, W., Olanders, P., Ragnarsson, I., Dudek, J., Leander,
    G.A., Möller, P. and Ruchowska, E., Nucl.Phys. $\underline{A429}$, 269 (1984)
13. Bonche, P., Heenen, P.H., Flocard, H. and Vantherin, D., "Self-
    -consistent Calculation of the Quadrupole-Octupole Deformation
    Energy Surface of ^{222}Ra", report IPNO/TH 86-14 (1986).
14. Baran, A., Pomorski, K., Łukasiak, A. and Sobiczewski, A., Nucl.
    Phys. $\underline{A361}$, 83 (1981).
15. Strutinsky, V.M., Atomnaya Energiya $\underline{1}$, 150 (1956).
16. Bohr, A. and Mottelson, B.R., Nucl.Phys. $\underline{4}$, 529 (1957) ; $\underline{9}$,
    687 (1958).
17. Iachello, F. and Jackson, A.D., Phys. Lett. $\underline{108B}$, 151 (1982).
18. Dorso, C.O., Mayers, W.D. and Swiatecki, W.J., Nucl.Phys. (1986).
19. Leander, G.A., Nazarewicz, W., Bertsch, G.F. and Dudek, J., Nucl.
    Phys. $\underline{A453}$, 58 (1986).
20. Lipas, P.O., Nucl.Phys. $\underline{82}$, 91 (1966).
21. Rakavy, G., Nucl.Phys. $\underline{4}$, 289 (1957).

22. Chacon, E., Moshinsky, M. and Sharp, R.T., J.Math.Phys. $\underline{17}$, 668 (1976).
23. Rohoziński, S.G., J.Phys. G $\underline{4}$, 1057 (1978).
24. Raczka, R., Limić, N. and Niederle, J., J.Math.Phys. $\underline{10}$, 1861 (1966).
25. Rohoziński, S.G. and Greiner, W., J.Phys. G $\underline{6}$, 969 (1980).
26. Ogura, H., Progr.Theor.Phys. $\underline{63}$, 498 (1980).
27. Vilenkin, N.Y., Special Functions and Theory of Group Representations (AMS Translations, Providence, 1968).
28. Eisenberg, J.M. and Greiner, W., Nuclear Models (North-Holland, Amsterdam, 1970).
29. Alder, K., Bohr, A., Huus, T., Mottelson, B.R. and Winther, A., Rev.Mod.Phys. $\underline{28}$, 432 (1956).
30. Moszkowski, S.A., Handbuch der Physik $\underline{39}$, 411 (Springer, Berlin, 1957).
31. Neergard, K. and Vogel, P., Nucl.Phys. $\underline{A145}$, 33 (1970).
32. Rohoziński, S.G., Greiner, W., Phys.lett., $\underline{128B}$, 1 (1983).
33. Engel, J. and Jachello, F., Phys.Rev.Lett. $\underline{54}$, 1126 (1985).

AN INTERACTING BOSON VERSION
OF REFLECTION-ASYMMETRIC SHAPE DEFORMATION

Jonathan Engel
A.W. Wright Nuclear Structure Laboratory
Yale University, New Haven, Connecticut 06511
USA

ABSTRACT

In order to discuss octupole deformation, we present a model that treats f bosons on an equal footing with the s and d bosons of the IBA. A more flexible version incorporates p bosons as well and has two useful dyanamical symmetry limits. We discuss the application of these models to the light-actinide region of the periodic table, where stable reflection asymmetry is thought to be important.

1. TOWARDS AN IBA DESCRIPTION OF THE LIGHT ACTINIDES

In recent years, controversy has come to surround the interpretation of such structural features in the even-even light actinides as low-lying negative-parity states, enhanced cascading E1 transitions, and small alpha hindrance factors. Alpha clustering[1], octupole vibrations[2], and stable axially symmetric (reflection-asymmetric) octupole deformation[3] have all been put forth as candidate descriptions of the collective dynamics in these nuclei. The last interpretation, formulated most commonly in terms of the geometric collective model[4] or fermionic mean fields[3], has gained currency recently, in part because of its successful explanation of parity doubling, magnetic moments, and decoupling factors in certain odd nuclei. In even nuclei, the predictions of the models are somewhat more difficult to compare with experimental data. Those familiar with the Interacting Boson Approximation (IBA) and its application to collective states of quadrupole character in even-even nuclei may wonder whether the notion of octupole deformation can be successfully incorporated into a boson picture. If the answer is yes, there arises the additional question of whether the IBA can help pick out one of the above mentioned physical pictures as more faithful than the others. We will touch on both these issues in the course of this talk.

Octupole bosons have long been an ingredient in the IBA; however, they have always been considered in the context of one-f-phonon excitations built on ground states with no f bosons. While such a scheme is well suited to describe octupole vibrations, it will clearly not do as a model of stable octupole deformation, which ought to mix f bosons into the ground state. A more sensible approach is to treat the f bosons on an equal footing with the s's and d's. One may then proceed to examine the model for the existence of dynamical symmetries, or explore it numerically, to determine whether it is able to produce the sorts of spectra and transitions expected from octupole-deformed shapes.

Before considering the model in detail, let us first outline some of the properties of spectra and transitions associated with axially symmetric reflection-asymmetric (e.g. octupole) deformation. In the very rigid limit, rotational bands are expected to have degenerate partners of the opposite parity; the lowest-lying levels will consist of degenerate $K^\pi=0^+$ and $K^\pi=0^-$ bands, with L = 0, 2,... and L = 1, 3,... respectively. If the deformation is not completely rigid, the negative-parity band will be shifted upwards somewhat with respect to the other; this is what is observed in the light actinides. The doubled bands are expected to be connected to one another by strong E3 and E1 transitions; while E3's have not often been observed in the controversial nuclei, enhanced E1 transitions have been seen in several radium and thorium isotopes[5]. E1 transitions have always posed problems for collective models, and successfully accounting for them provides a crucial test for any theory claiming to describe the light actinides.

2. A U(13) s-d-f MODEL

When the "equal footing" idea is implemented it leads to the dynamical group (generated by all possible one-body operators) U(13). The first question we address is whether this model contains dynamical symmetries corresponding to stable axially symmetric octupole and quadrupole deformation. An examination of the way representations of various simple groups decompose under restriction to O(3) leads to the conclusion that the only possibilities for representing axially symmetric deformation are SU(3) and O(4); other groups do not allow the construction of K=0 bands. It is further apparent that U(13) contains neither of these (in a chain going through the physical rotation group O(3)). This can be deduced from the fact that there is no representation or combination of representations of either of the two groups that contains only the angular momenta 0, 2, and 3, the values associated with the one-boson states. While U(13) does have many dynamical symmetries, including some corresponding to octupole vibrations and unstable octupole deformation, there appears to be no hope of finding a chain that will produce the features we need.

This is not to say, however, that the U(13) model is not capable of generating these features via some Hamiltonian outside the restricted class of dynamical symmetries. While it is difficult to explore all possible model Hamiltonians, some likely candidates can be investigated. The most successful interaction we have found has the form

$$H=\varepsilon_d n_d + \kappa Q_{sd} \cdot Q_{sd} + \varepsilon_f n_f - \kappa_f [f^\dagger \tilde{f}]^2 \cdot Q_{sd} + \theta([f^\dagger f^\dagger]^2 \cdot [\tilde{d}\tilde{s}]^2 + h.c.) \quad (1)$$

where Q_{sd} is (for example) the SU(3) s-d quadrupole operator. The effects of the last two terms on the spectrum can be summarized as follows: the term involving Q_{sd} strongly couples the f bosons to the

SU(3) quadrupole-deformed "core" producing rotational bands, and the last term mixes components with all (even) numbers of f bosons into the ground state and causes the $K^\pi=0^+$ and $K^\pi=0^-$ bands to be degenerate. The mixing of n_f serves to correct another problem as well. E1 transition rates predicted by the usual one-body operator $[f^\dagger \tilde{d} + d^\dagger \tilde{f}]^1$ are strongly correlated with the number of f bosons in the initial state. For example, if the negative-parity levels in the light actinides are treated as one-f excitations around a zero-f ground band, the E1 transitions from the negative-parity to positive-parity bands are predicted to be an order of magnitude or so larger than those going in the opposite direction. The resulting oscillation of the E1 strength with J is not seen in the experimental data. The last term in equation (1) causes $\langle n_f \rangle$ to be roughly equal in the two bands and thus smooths out the E1's.

Despite the virtues of this kind of interaction, several facts deter us from applying it in a systematic treatment of light-actinide spectra. The first is that it is difficult to shift the negative-parity band upwards slightly (as it is shifted in reality) without reviving the E1 problem. Another is a lack of regularity in the predicted sidebands; in the data, what little has been seen of non-Yrast states is not reproduced by equation (1). Thus, while possessing many of the features we want, the interaction in (1) is not completely satisfactory for a description of radium and thorium. Other Hamiltonians may also be investigated. We have tested several, some of which include octupole-octupole and pairing terms. Unfortunately, none of them produce all the features we desire; it is difficult without the guiding hand of some dynamical symmetry to construct Hamiltonians that do just what you want them to. As a result, rather than spend a lot of time exploring the huge number of U(13) Hamiltonians, we have opted after testing the most likely interactions to alter the model so that it contains dynamical symmetries corresponding to stable reflection-asymmetric deformation.

3. A U(16) s-d-p-f MODEL

The addition of p bosons to the s's, d's, and f's results in a U(16) dynamical group. While p bosons appear at first sight to be unphysical in that they cannot be constructed from like-nucleon pairs in a single spherical valence shell, it has recently been argued[6] that p pairs arise naturally in a Nilsson scheme that includes octupole deformation. The inclusion of the additional p degree of freedom in our model creates two interesting dynamical symmetries:

I. $U(16) \supset U(6) \otimes U(10) \supset SU(3)_{sd} \otimes SU(3)_{pf} \supset SU(3) \supset O(3)$ (2)

II. $U(16) \supset U(4) \otimes U(4) \supset Sp(4) \otimes Sp(4) \supset SU(2) \otimes SU(2) \approx O(4) \supset O(3)$.

Chain I will be referred to as the SU(3) limit and chain II as the O(4) limit. Both chains produce spectral properties associated with reflection-asymmetric deformation, including parity doubling and smooth E1 and E3 transitions (though to obtain them in SU(3), we need a large number of bosons and somewhat ad hoc "signature-dependent" terms[7] in the Hamiltonian).

The SU(3) limit is discussed in detail in reference 7; here we shall briefly present some of the properties of O(4)[8]. The simplest rotational Hamiltonian in this chain has the form

$$H = \alpha\, C_{O(4)} + \beta\, C_{O(3)} \tag{3}$$

where the C's are Casimir operators. The lowest-lying band produced by (3) (if α is negative) is a (parity-doubled) $K^\pi = 0^\pm$ band belonging to the O(4) representation ($\sigma=3N, \tau=0$), where N is the total boson number. The first (parity-doubled) excited band has $K^\pi = 2^\pm$ and O(4) quantum numbers ($\sigma=3N-2, \tau=2$) & ($\sigma=3N-2, \tau=-2$). E1 transitions are obtained from the dipole O(4) generator and therefore do not connect different bands. O(4) tensor operators may be used to obtain E2 and E3 transitions that (along with the E1's) in the large-N limit go over to the Bohr-Mottelson expressions corresponding to rigid axially symmetric deformation. Finally, in contrast to what happens in the SU(3) limit, the parity doubling in O(4) emerges quite naturally, without resort to the signature-dependent terms discussed in reference 7. A drawback is that this degeneracy cannot be removed within the dynamical symmetry. The radium and thorium isotopes in fact do not correspond to either chain I or II, not only because the doubling is split, but also because the real nuclei are not free of such effects as rotation-vibration coupling, Coriolis interactions, etc. Nonetheless, the presence in U(16) of these two analytic limits provides us with a class of Hamiltonians that may serve as starting points in an attempt to describe the light actinides.

Through the use of coherent states, it is possible[9] to determine the ground-state shape associated with a given boson Hamiltonian. It turns out that in U(13) while we are able to find some Hamiltonians corresponding in this geometric large-N limt to axially symmetric quadrupole and octupole deformation, the minima always seem to be so shallow as to have no perceptible influence on the spectrum. In the U(16) model with the p bosons, both the SU(3) (with allowances made for the signature-dependent terms) and the O(4) limits can be shown to correspond to axially symmetric reflection-asymmetric ground-state shapes. We are led to conjecture that the p degree of freedom serves to stabilize the deformation in the quadrupole and octupole sectors of the model.

4. THE RADIUM AND THORIUM NUCLEI

In part because of the complexity of the U(16) model and the corresponding long computer times needed to construct and diagonalize Hamiltonians, we have not yet performed unrestricted, systematic fits to the radium and thorium isotopes. However, we are confident that given enough time or a well motivated truncation scheme, the task would be quite accomplishable. Shown below is a U(16) fit to the levels and B(E1)/B(E2) ratios in ^{218}Ra, which has five bosons. The quality of the fit is not very significant since discussion is restricted to one nucleus and the Hamiltonian contains several adjustable parameters. However, it does demonstrate the ability of the

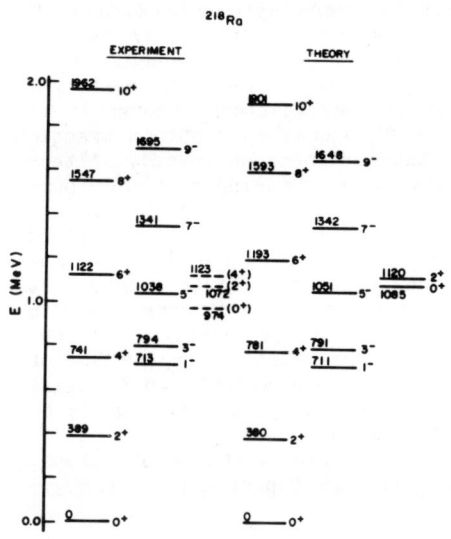

Fig. 1 (left): Fit to the energy levels in ^{218}Ra. The Hamiltonian contains terms from the SU(3) and O(4) limits described above plus single boson energies for the d, f, and p bosons and an additional quadrupole-quadrupole interaction in the s-d sector that deviates somewhat from SU(3).

Fig. 2 (below): Fit to the B(E1)/B(E2) ratios in ^{218}Ra. The wave functions come from the fit in Fig. 1 and the E1 operator is given by:

$$T^{E1} \alpha [p^\dagger \tilde{d} + d^\dagger \tilde{p}]^1$$

The E2 operator is the Q from the s-d interaction plus a small piece proportional to $[f^\dagger \tilde{f}]^2$.

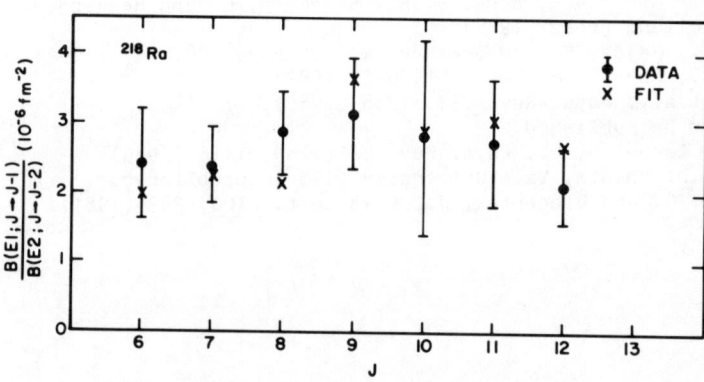

model to accurately reproduce the patterns seen in the light-actinide region.

Another reason we have not attempted more sytematic fits is a shortage of data that would discriminate among the different physical pictures put forth to describe the light actindes. Indeed, it is quite possible to fit existing data with an IBA scheme that treats the negative parity states as one-f vibrational excitations (though it is necessary to add a two-body term to the E1 operator to obtain transitions even remotely resembling those that have been observed). Alternatively, treating the negative parity sates as single "p-f vibrations" (mixtures of one-f and one-p configurations) one can obtain very nice systematic fits both to the spectra and measured transitions without resorting to a two-body E1 operator. In order that we may decide which scheme is the "correct" one, it is crucial that more data be obtained. More information about the locations and transitions to and from non-Yrast states would be particularly useful. In the meantime, we feel that the question of the relative importance of clustering, vibrations, and deformation in these nuclei, while still an open one, is not outside the realm of the IBA and may eventually be resolved with the help of the models described here.

I would like to thank F. Iachello for many useful discussions. This work was performed in part under the US Department of Energy, Contract No. DE-AC02-76ER03074.

5. REFERENCES

1. Iachello, F. and Jackson, A.D., Phys. Lett. 108B, 151 (1982); Daley, H. and Gai, M., Phys. Lett. 149B, 13 (1984).
2. Peker, L., Hamilton, J. and Rasmussen, J., Phys. Rev. C24, 1336 (1984).
3. For a review see Leander, G.A. in Niels Bohr Centennial Conferences 1985 -- Nuclear Structure, Broglia, R., Hagemann, G. and Heskind,B. eds. North Holland Press (1985).
4. Rohozinski, S., Gajda, M. and Greiner, W., J. Phys. G8, 787 (1982).
5. Gai, M. et al., Phys. Rev. Lett. 51, 646 (1983); Shriner, J. et al., Phys. Rev. C32, 1888 (1985).
6. Otsuka, T., to be published.
7. Engel, J. and Iachello, F., Phys. Rev. Lett. 54, 1126 (1985).
8. Engel, J., Ph.D. thesis, Yale University (1986) unpublished.
9. van Roosmalen, O. and Dieperink, J., Phys Lett. B100, 299 (1981).

A REVIEW OF EXPERIMENTAL EVIDENCE FOR OCTUPOLE DEFORMATION

Jan Zylicz

Gesellschaft für Schwerionenforschung mbH

6100 Darmstadt, F. R. Germany[*]

ABSTRACT

Experimental evidence for octupole correlations, which lead to octupole instability and octupole deformation of some nuclei, is illustrated through typical examples. Data are considered for both the 220 < A < 230 region and for a few medium mass nuclei.

1. INTRODUCTION

Experimental data on nuclei, which in the ground and/or excited states show features suggesting the presence of octupole correlations, are very rich [1,2]. In this paper a few typical examples are discussed in some detail. The rest of the data is presented very briefly and references are given to the original studies.

The manifestation of octupole correlations has been observed mostly in nuclei with the proton number Z from 87 to 91 (the Fr-Pa region) and the neutron number N around 134. This reflects primarily the strong octupole interaction between the $i_{13/2}$ and $f_{7/2}$ proton orbitals, as well as the $j_{15/2}$ and $g_{9/2}$ neutron orbitals, which in the nuclei of concern are close to the Fermi levels [3,4]. The $i_{13/2}$ and $f_{7/2}$ orbitals seem to

---

[*] On leave of absence from the Institute of Experimental Physics, University of Warsaw, 00-681 Warsaw, ul. Hoza 69, Poland

play a similar role in the N ≈ 90 nuclei[5]. Both groups of nuclei are considered in this review.

2. OCTUPOLE INSTABILITY AND OCTUPOLE DEFORMATION

Strong octupole correlations lead to octuple instability or, in the extreme case, to octupole deformation of nuclei. For a description of these effects it has been sufficient until now to assume axially symmetric shapes of nuclei given via the standard relation

$$R(\theta) = C(\beta) R_o \{1 + \beta_2 Y_{20} + \beta_3 Y_{30} + \beta_4 Y_{40} + \ldots\}$$

in which the coefficient $C(\beta)$ ensures a constant volume of the nucleus. For the non-zero values of the quadrupole and octupole deformation parameters, β_2 and β_3, one obtains a pear-like shape for the nuclear surface. By a change of the sign of β_3 this shape is transformed into its mirror reflection, fig. 1.

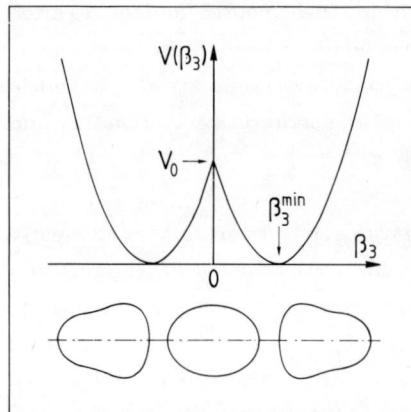

Fig. 1 Shape of a nucleus for a positive, zero and negative value of the octupole-deformation parameter β_3 (at $\beta_2 \neq 0$) and a rough double-oscillator approximation for the dependence of the potential energy of a nucleus upon β_3.

In the macroscopic-microscopic calculations of the nuclear potential-energy surface, e.g. ref. [6], usually a few deformation parameters are taken into account. However, in order to introduce the concept of octupole instability and octupole deformation, a one-dimensional approximation of fig. 1 is sufficient. For $\beta_3^{min} = 0$ in the ground-state one may speak about octupole deformation only in a dynamical sense related to the zero-point vibrations (the root mean square value

$(\langle\beta_3^2\rangle)^{1/2} \neq 0)$. For $\beta_3^{min} \neq 0$, the dynamical and static effects begin to interplay. The octupole instability corresponds to an incomplete separation of the two minima, when the energy of the lowest eigenstate exceeds the height V_o of the barrier or when the tunneling through this barrier is essential. An octupole deformation would correspond to a very high barrier and a negligible tunneling. In the latter case the intrinsic wave function, corresponding to the nucleus in one of the minima, has two opposite-parity components of equal amplitude. With the decrease of V_o, the degree of the parity violation gets smaller and disappears for $V_o = 0$. The total wave function must always be constructed in such way that parity is conserved in the laboratory system (if a very small contribution from weak interactions is neglected).

Sometimes the ε parametrization (e.g. in the modified oscillator model) is used in the description of the octupole effects. The relation $\beta_3 = -(4\pi/7)^{1/2}\varepsilon_3$ holds approximately for small ε_3 values[7].

3. OCTUPOLE INSTABILITY AND GROUND-STATE PROPERTIES

The calculations of the potential energy surface are obviously related to theoretical predictions of atomic masses. For radium isotopes around $N \simeq 134$ a comparison of masses calculated by Möller and Nix[8] with the experimental ones shows a large discrepancy if only ε_2 and ε_4 are used as free parameters. The discrepancy is greatly reduced when additionally ε_3 is taken into account. This argument in favour of the octupole instability of ^{222,224}Ra (and neighbouring nuclei[9]), however, is weakened by the fact that the octupole deformation correction becomes much smaller when ε_6 is also included in the calculations[3] (in the case of the β parametrization the role of β_6 is not essential[7]).

In the single-particle models, a change of the octupole deformation parameter leads to a rearrangement of the single particle levels. It has been noticed that this rearrangement is necessary to explain the ground-state spins of odd-A nuclei in both regions of octupole instability[10-14,29]. Furthermore, for the same nuclei the agreement between the calculated and experimental magnetic moments of ground states improves when $\varepsilon_3 \neq 0$ is taken into account (see also section 5). Finally, the octupole instability seems to be reflected in the changes of the mean square nuclear charge radii. This indirect evidence for octupole instability is discussed at this conference by Neugart[15] with reference to the results of laser measurements.

4. OCTUPOLE INSTABILITY AND LOW-ENERGY EXCITATIONS OF EVEN NUCLEI

To discuss the main patterns of low-energy states of even nuclei having an octupole instability one may refer to the eigenstates of the double harmonic oscillator displayed in fig. 2 for a large interval of β_3^{min} values. For $\beta_3^{min} = 0$, $\beta_2^{min} \neq 0$, one expects the K=0 band built on the ground state and on the octupole-vibrational states. Due to the symmetry conditions[17], for positive- and negative-parity bands one has only the states with even and odd spins, respectively. The energy of the 0_2^+ two-phonon state is twice that of the (virtual) 0_1^- one-phonon state*). For very large values of β_3^{min} one reaches the conditions for a stable octupole deformation which should manifest itself via the alternating parity states forming a regular rotational band. The pattern of the ^{224}Ra levels is somewhere in between the above two extremes.

Let us recall that ^{224}Ra represents the group of a few even nuclei with the 1^- states at an unusually low excitation energy, table 1. Some of these states were already observed by Stephens et al.[19] about 30

---

*) According to ref.[18], such harmonic-oscillator picture does not apply to a well deformed nucleus; the two-phonon state must be shifted upwards due to the Pauli principle. It is not clear whether this is true also in the case of transitional nuclei considered here.

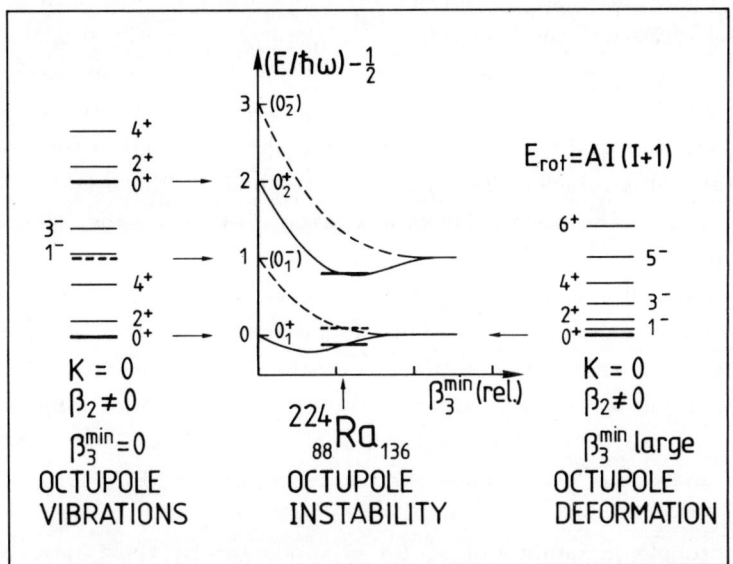

Fig. 2   Energies of intrinsic states of a nucleus vs the value of the octupole deformation parameter corresponding to the minimum of the potential energy approximated by a double oscillator (see fig. 1 and ref.[16]) and rotational bands in the two extreme cases (with the moment-of-intertia parameter A chosen arbitrarily). The position of the ^{224}Ra levels is defined by the experimental ratio $E(0^+)/E(0^-) = 4.6$ (the $E(0^-)$ value being deduced from the energies of the 1^- and 3^- states).

Table 1   Selected data for the $^{220-228}_{88}$Ra[a] and $^{146}_{56}$Ba nuclei[19,21-28]

A	220	222	224	226	228	146
N	132	134	136	138	140	90
$E(1^-)$ keV	412	242	216	254	474	739
$E(0^+)/E(1^-)$	?	3.77	4.24	3.25	1.5(2.2)	1.42
$E(2^+)/E(1^-)$[b]	0.43	0.46	0.39	0.27	0.13	0.24

[a] Properties of the Th isotopes are similar to those of their Ra isotones.

[b] The value of 3 is expected for a stable octupole deformation.

years ago. Soon after that observation its possible relevance to a nuclear octupole deformation was recognized[20]. However, the vibrational origin of the 1^- states could not be ruled out a priori. Under the assumption of near harmonic octupole vibrations one had to expect the two-phonon 0^+ states at the energy $E(0^+) \simeq 2\, E(1^-)$. They should be fed in the α-decay of relevant mother nuclei. Bjornholm[21] searched for these states with a negative result. The same problem was taken up later by the Mainz-Warsaw collaboration[22]. The existence of the harmonic two-phonon states was exluded with a high accuracy and, instead, the 0^+ states were found at much higher energy. This finding, together with the fact of low $E(1^-)$ values, is interpreted as a signature of the octupole instability of the N = 134-138 nuclei. At the same time the $E(2^+)/E(1^-)$ ratios indicate how far one is here from the stable-octupole-deformation limit. A more advanced analysis of these questions has been performed in refs.[30,31].

The octupole instability of ^{146}Ba is suggested by the relatively low $E(1^-)$ value [28]. The 0^+ state observed in this nucleus is not likely to be related to the octupole mode.

5. PARITY DOUBLETS AT LOW ENERGY IN ODD-A NUCLEI

In odd-A nuclei with a stable octupole deformation one expects [10,11] rotational bands composed of degenerate parity doublets (PD), that is, pairs of equal energy states with the same quantum numbers I and K but opposite parity. The mixed-parity (see section 2) intrinsic wave functions of PD states should be identical which means, in particular, that both states should have the same values of magnetic moment μ. The decoupling parameter a for octupole-correlated $K^\pi = 1/2^\pm$ bands are predicted to have the same absolute value but opposite sign.

The octupole instability should manifest itself via the $K^\pm$ bands being somewhat shifted against each other, as for instance observed in ^{223}Ra, fig. 3. However, in heavy nuclei the density of the Nilsson levels is so high that a small energy spacing between the K, $I^\pm$ states may be accidental. A better test is provided by the μ and a values which are sensitive to the structure of the intrinsic wave function. For ^{223}Ra, the μ values of the $I^{\pi} = 3/2^{\pm}$ states are fairly close, while the absolute values of a determined for the $K^{\pi} = 1/2^{\pm}$ band differ by a factor of 2. As discussed by Sheline[35], the $\varepsilon_3 = 0.1$ model predictions are closer to the experiment (fig. 3) than those for $\varepsilon_3 = 0$. An account for the Coriolis coupling may further improve the agreement, making the assignment of octupole instability to ^{223}Ra rather justified.

Similar evidence for octuple instability is available for ^{225}Ra (ref.[36]), ^{227}Ra (ref.[37]), ^{225}Ac (ref.[38]), ^{227}Ac (ref.[39]) and perhaps ^{145}Ba (ref.[40]). In the case of ^{229}Pa a $5/2^{\pm}$ PD has been observed[41] with the energy splitting of only 0.2 keV. This could be an indication of a stable ground-state octupole deformation but further studies are necessary before a definite conclusion can be made. Some deviation from the reflection symmetric shape is likely also in the third potential well of the ^{231,233}Th fission barrier[42].

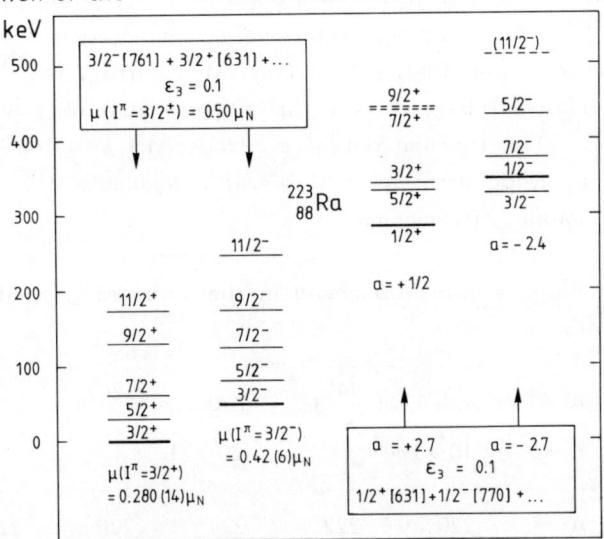

Fig. 3   The $K^{\pi} = 3/2^{\pm}$ and $K^{\pi} = 1/2^{\pm}$ bands of ^{223}Ra (refs.[34,35]). Experimental values of magnetic moment μ and decoupling parameter a are compared to predictions of a model assuming octupole instability of this nucleus[11,35].

6. OCTUPOLE DEFORMATION AT HIGH ROTATIONAL FREQUENCY

For several nuclei in the Ra-Th region (table 2) data is available on medium-high spin states. One observes the alternating-parity level sequences expected for nuclei with octupole instability or deformation[2,13]. The deformation seems to stabilize with increasing rotational frequency, as illustrated by the example of ^{220}Ra, fig. 4. At low exctitation energy this nucleus exhibits a rather vibrational character (in agreement with the theory predictions[6]) of a spherical ground state). However, with increasing spin the states of positive and negative parity tend to form approximately one rotational band. The ω^-/ω^+ ratio of rotational frequencies derived from the bands of negative and positive parity approaches one. The differences δE between the observed energies of I^- states and those calculated (using the simple rotational formula) from the position of the $(I+1)^+$ and $(I-1)^+$ levels vary from the large positive values at lower spins to relatively small negative values at high spin (zero would correspond to the ideal octupole-deformation case). The pattern of the ^{226}Th levels at high rotational frequency gets much closer to the stable-octupole-deformation limit (fig. 4). Contrary to ^{220}Ra, this nucleus shows features characteristic for octupole instability already at the ground state. The thorium data are discussed by Butler[43]. Some evidence for octupole deformation at high spin is available also for ^{150}Sm and for lighter isotopes of samarium[48,49].

Table 2. The highest spins observed in the alternating-parity bands of Ra and Th isotopes

$^A_{88}$Ra	A	218 (ref.[44])		220 (refs.[26])		
	I^π	$16^+/17^-$		$20^+/21^-$		
$^A_{90}$Th	A	220	222	224	226	228
(ref.[45])	I^π	$10^+/15^-$	$14^+/15^-$	$16^+/15^-$	$18^+/19^-$	$14^+/13^-$

For ^{219}Ra (ref.[46]) and ^{221}Th (ref.[47]) spins were observed up to $(I+22)^+/(I+20)^-$ and $(I+12)^+/(I+13)^-$, respectively.

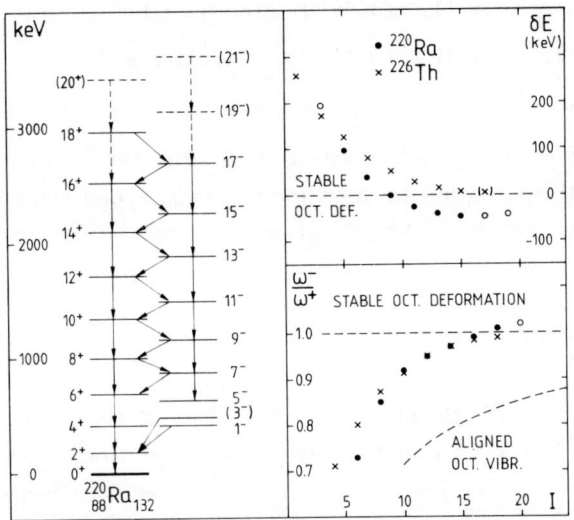

Fig. 4   Illustration of the establishment of an octupole deformation with increasing spin for the nuclei ^{220}Ra (ref.[26]) and ^{226}Th (refs.[43,45]). For definitions of symbols and for comments, see text.

7. THE ENHANCED E1 TRANSITIONS

The reduced probabilities of E1 transitions, B(E1), can be expressed in Weisskopf units (W.u.) of $6.45 \times 10^{-2} \times A^{2/3}$. For heavy elements, experimental B(E1) values are typically $10^{-4} - 10^{-6}$ W.u. However, in the Ra-Pa region the E1 transitions between high-spin states are enhanced, the B(E1) values being $10^{-3} - 10^{-2}$ W.u. In ref.[50] this enhancement is explained as being due to an intrinsic nuclear electric dipole moment Q_1 induced by the β_3 deformation of the nucleus. One expects that in a pear-shape nucleus the protons have an increased density at the pointed end, where the curvature of the nuclear surface is the largest. The centre of mass of protons is shifted with respect to that of the nucleus as a whole which means $Q_1 \neq 0$. This simple picture based on the droplet model is to be corrected for the shell effects[50].

For the E1 and E2 transitions within a rotational band composed of alternating parity states one has (in standard notation):

$$B(E1, I \to I-1) = (3/4\pi)Q_1^2 <IK10|I-1\ K>^2,$$
$$B(E2, I \to I-2) = (5/16\pi)Q_2^2 <IK20|I-2\ K>^2,$$

where the moments Q_1 and Q_2 are expressed in fm and fm^2, respectively[50]. For transitions starting from the same initial state of spin I, one gets the following relation for the ratio R of E1 and E2 transition probabilities:

$$R = B(E1)/B(E2) = 1.6[(2I-1)/(I-1)]\ (Q_1/Q_2)^2.$$

This expression is used to determine Q_1 if Q_2 value is known from the experiment or systematics.

For ^{220}Ra, e.g., one obtains $R \simeq 1 \times 10^{-6}$ fm^{-2} from the intensity ratios determined[26] for E1 and E2 transitions deexciting the high-spin states. From $Q_2 = 718(12)$ fm^2 of ^{226}Ra, ref.[51], and the semiempirical dependence[52] $Q_2 \times [E_\gamma(2^+ \to 0^+)]^{1/2} \simeq$ constant, one estimates $Q_2 \simeq 450$ fm^2 which leads to $B(E1) \simeq 4 \times 10^{-3}$ W.u. and $Q_1 \simeq 0.26$ fm. The latter value is not far from the theoretical estimate[50] of $Q_1^{th} = 0.13 - 0.19$ fm at $\beta_3 = 0.08 - 0.10$ ($\beta_2 = 0.11$, $\beta_4 = 0.65$). In the case of ^{226}Th, the experimental value $Q_1 = 0.30(2)$ fm is in very good agreement with the predicted value of 0.29 fm (for more extensive presentation of the Th data, see refs.[43,45]). The agreement between experimental and calculated Q_1 values could be used as an argument in favour of the octupole deformation of Ra-Th (and Sm, refs.[48,49]) nuclei at high spin. However, Dorso et al. have shown[53] that inclusion of droplet model effects associated with the presence of the neutron skin, neglected in ref.[50], brings a reduction of the droplet model contribution to Q_1^{th} to small or even negative values. Further theoretical studies are clearly needed in this context.

In some odd-A nuclei of Ra, Ac and Pa an enhancement of E1 transitions has been observed even between low-energy, low-spin states[54]. However, as emphasized by Ahmad et al.[55], there is no simple relation between the E1 transition rates and the degree of octupole instability of nuclei.

8. CONCLUSIONS

The data on the medium-high spin states (the alternating-parity bands) in a few isotopes of Th, Ra and Sm are consistent with an assumption of nuclear octupole deformation. This deformation seems to stabilize with the increase of rotational frequency; it would be important (see also ref.[56]) to obtain data on states with even higher spins (I>20). The data on ground and low-energy states of several nuclides in the Fr-Pa (and Ba) region can be interpreted in terms of octupole instabilty. There are alternative theoretical approaches[32,33] (not presented in this lecture) which explain the same nuclear-spectroscopy data without taking into account the octupole instability. However, the octupole-instability approach is favoured by the calculations of the potential energy of nuclei. Let us hope that further experimental studies (e.g. on side bands and E3 transition probabilities), combined with relevant theoretical predictions, will lead to more definite conclusions.

The author would like to thank GSI for hospitality and excellent working conditions.

REFERENCES

1. Leander, G.A., in "Nuclear Structure 1985", ed. by R.Broglia, G.Hagemann and B.Herskind, North-Holland, Amsterdam 1985, p. 249
2. Nazarewicz, W., ibid., p. 263
3. Chasman, R.R., this Conference
4. Nazarewicz, W., this Conference
5. Leander, G.A. et al., Phys. Lett. 152B, 284 (1985)
6. Nazarewicz, W. et al., Nucl. Phys. A429, 269 (1984)
7. Leander, G.A. and Nazarewicz, W., private communication (1986)
8. Möller, P. and Nix, J.R., Nucl. Phys. A361, 117 (1981)
9. Leander, G.A. et al., Nucl. Phys. A388, 452 (1982)
10. Ragnarsson, I., Phys. Lett. 130B, 353 (1983)

11. Leander, G.A. and Sheline, R.K., Nucl. Phys. A413, 375 (1984)
12. Ahmad, S.A. et al., Proc. AMCO-7, ed. by O. Klepper, TH Darmstadt 1984, p. 361
13. Nazarewicz, W. and Olanders, P., Nucl. Phys. A441, 420 (1985)
14. Coc, A. et al., Phys. Lett. 163B, 66 (1985)
15. Neugart, R., this Conference
16. Merzbacher, E., Quantum Mechanics, Wiley, New York 1961
17. Bohr, A. and Mottelson, B.R., Nuclear Structure, Vol. I, Benjamin, New York 1975
18. Soloviev, V.G. and Shirikova, N.Y., Z. Phys. A301, 263 (1981)
19. Stephens, F. et al., Phys. Rev. 96, 1568 (1954) and 100, 1543 (1955)
20. Alder, K. et al., Rev. Mod. Phys. 28, 432 (1956)
21. Bjørnholm, S., Thesis, Munksgaard, Copenhagen 1965
22. Kurcewicz, W. et al., Nucl. Phys. A270, 175 (1976), A289, 1 (1977) and A304, 77 (1978)
23. Kurcewicz, W. et al., Nucl. Phys. A356, 15 (1981)
24. Ruchowska, E. et al., Nucl. Phys. A383, 1 (1982)
25. Ruiz, C.P., Thesis (1961), UCRL-9511
26. Burrows, J.D. et al., J. Phys. G10, 1449 (1984); Celler, A. et al., Nucl. Phys. A432, 421 (1985); Cottle, P.D. et al., Phys. Rev. C30, 1768 (1984), C33, 1855 (1986)
27. Van den Berg, A.M. et al., Nucl. Phys. A422, 45 (1984)
28. Scott, S.M. et al., J. Phys. G6, 1291 (1980)
29. Rozmej, P. et al., Proc. of the XXIV Int. Winter Meeting on Nuclear Physics, ed. by I. Iori, RSEP, Milano 1986, p. 567
30. Rohozinski, G. and Greiner, W. Phys. Lett. 128B, 1 (1983)
31. Böning, K. et al., Phys. Lett. 161B, 231 (1985); Baranowski, R. et al., this Conference
32. Piepenbring, R., Z. Phys. A323, 341 (1986) and earlier papers quoted there-in
33. Daley, H. and Iachello, F., Phys. Lett. 131B, 281 (1983)
34. Briançon, Ch. et al., CSNSM Orsay, Annual Report 1983-84, p. 27 and Eid, S.A. et al., this Conference
35. Sheline, R.K., Phys. Lett. 166B, 269 (1986)

36. Nybo, K. et al., Nucl. Phys. A408, 127 (1983); Sheline, R.K. et al., Phys. Lett., 133B, 13 (1983); Lovhoiden, G. et al., Nucl. Phys. A452, 30 (1986)
37. V. Egidy, T. et al., Nucl. Phys. A356, 26 (1981)
38. Auger, P. et al., Nucl. Phys. A202, 37 (1973); Ahmad, I. et al., Phys. Rev. Lett. 52, 503 (1984)
39. Teoh, W. et al., Nucl. Phys. A319, 122 (1979); Anicin, I. et al., J. Phys. G8, 369 (1982); Sheline, R.K. and Leander, G.A., Phys. Rev. Lett. 51, 359 (1983)
40. Robertson, J.D., this Conference
41. Ahmad, I. et al., Phys. Rev. Lett. 49, 1758 (1982)
42. Blons, J. et al., Nucl. Phys. A414, 1 (1984)
43. Butler, P., this Conference
44. Fernandez-Niello, J. et al., Nucl. Phys. A391, 221 (1982); Gai, M., Phys. Rev. Lett. 51, 646 (1983)
45. Schüler, P. et al., Phys. Lett. 174B, 241 (1986) and earlier papers quoted there-in
46. Cottle, P.D. et al., Phys. Rev. C33, 1855 (1986)
47. Dahlinger, M. et al., Z. Phys. A321, 535 (1985)
48. Sujkowski, Z. et al., Nucl. Phys. A291, 365 (1977)
49. Urban, W., private communication (1986)
50. Leander, G.A. et al., Nucl. Phys. A453, 58 (1986) and earlier papers quoted there-in
51. Goldhaber, A.S. and Scharff-Goldhaber, G., Phys. Rev. C17, 1171 (1978)
52. Grodzins, L., Phys. Lett. 2, 88 (1962)
53. Dorso, C.O. et al., Nucl. Phys. A451, 189 (1986)
54. Reich, C.W. et al., Phys. Lett. 169B, 148 (1986) and earlier papers quoted there-in
55. Ahmad, I. et al., Phys. Rev. Lett. 52, 503 (1984)
56. Briançon, Ch. and Mikhailov, I.N., this Conference

DIPOLE COLLECTIVITY AND REFLECTION ASYMMETRIC NUCLEAR SHAPES*

Moshe Gai
A.W. Wright Nuclear Structure Laboratory
Yale University, New Haven, CT 06511

ABSTRACT

Collective dipole bands observed in light and heavy nuclei, are characterized experimentally by states of alternating parities and enhanced E1 decays. Such states arise from an **intrinsic state** which is **reflection asymmetric** and, therefore, is not invariant under the parity operation. This state may correspond to any configuration that is reflection asymmetric (e.g. cluster state, or stable octupole deformed state). Dipole collective states have been observed in ^{18}O and the Ra-Th region. The experimental signatures for dipole collectivity in these nuclei are discussed, and similarity and differences between the cluster model and stable octupole shape model are discussed. Search for such states in other regions of the chart of nuclei are proposed.

The study of dipole collective bands in light and heavy nuclei is a very active area of research in nuclear structure. Several experimental[1-9] and theoretical[10-19] pioneering works are now being extended and studied in further details, as reflected in the program of this conference.

Dipole collective bands are characterized by an intrinsic state which is reflection asymmetric, and which, in most cases, acquires a nonvanishing intrinsic dipole moment; such an intrinsic state gives rise to bands of states of alternating parities and enhanced intraband E1 decay matrix elements. They are therefore identified experimentally by the presence of bands linked by enhanced E1 gamma transitions, in contrast to the usual enhanced E2 deexcitations.

The reflection asymmetric intrinsic state can arise from any such asymmetric shape. Two have been proposed: first, a state that acquires a stable octupole shape[14-19], and second, a cluster state[10-13]. In light nuclei it appears that cluster states dominate the structure and give rise to the observed enhanced E1s[1,11,20]. In heavy nuclei both the cluster model and the stable octupole shape model appear to give a good description of the data and it is the point of this talk to discuss the similarities and differences of these two approaches to the description of heavy nuclei. In addition both models include the limit at which the deformation is dynamic

*Work supported by U.S.D.O.E. Contract No. DE-AC02-76ER03074.

rather then static; these are the well recognized octupole vibrational limit and (cluster) dipole vibrations.

Both models predict enhanced E1 decay modes. In the stable octupole shape model, while enhanced E1 are less natural, they appear as a perturbation of the pear shape that polarizes the nucleus and yields a nonvanishing dipole moment[15,16] as given by:

$$D = + 0.000687 \, AZ \, \beta_2 \, \beta_3 \quad (efm) \qquad (1)$$

and:

$$B(E1:0^+ \to 1^-) = \frac{9}{4\pi} \langle D^2 \rangle \qquad (2)$$

In the case that $\langle \beta_2^2 \rangle \approx \langle \beta_3^2 \rangle \approx 0$ (e.g. ^{218}Ra) it has been suggested that shell corrections[21] can give rise to such enhanced B(E1)s. In that context it will be very interesting to further study these shell corrections, appearing in the vicinity of shell closures, as a possible source of the occurrence of cluster states in the vicinity of shell closures.

In the cluster model[10,11,20] enhanced E1 appear naturally, and are given as a fraction of the sum rule[11] given by:

$$S_1(E1:\alpha+A_2) = \frac{9}{4\pi} \frac{(N-Z)^2}{A(A-4)} \frac{\hbar^2 e^2}{2m}; \qquad (3)$$

such cluster states indeed have nonvanishing dipole moment given by:

$$D(\alpha+A_2) = 2 \frac{N-Z}{A} e \, d_0 \qquad (4)$$

where d_0 is the separation distance. Since cluster states are not pure, we normally expect the E1 decays to be smaller than would be given by the above dipole moment equation (4). In addition, antisymmetrization of the cluster wave function yields a reduction in the B(E1) strength by a factor of 3-4[22]. For cluster states, we may define a renormalized **single nucleon** E1 Weisskopf unit given by:

$$mol. \, W.u. = \left(\frac{R_{cc} - R_{cm}}{1.2 A^{1/3}}\right)^2 \times W.u. \qquad (5)$$

It is immediately clear that

$$B(E1: \text{pure cluster}) \approx \frac{16}{3} Z^2 \, mol. \, W.u. \qquad (6)$$

and the two estimates provide an upper and lower limits on the expected E1 strength.

It should be emphasized that both estimates--the classical estimate for pure cluster states given by Eq. (4) and the single nucleon (proton or neutron) estimate given by Eq. (5)--are derived in Ref. 11. The nomenclature "molecular Weisskopf unit" is appropriate for Eq. (5) [and not for Eq. (4)] since Weisskopf units usually represent the strength of a single nucleon. Collective dipole states should yield E1s larger than such a normalized single particle strength.

The nucleus ^{18}O: We have carried out two different experiments[1] to study ^{18}O with high accuracy, including detection of very weak, previously unknown, electromagnetic decays. **Extremely selective E1 transitions** were found to connect only cluster collective states, as shown in Fig. 1 and Fig. 2. The B(E1) found are indeed very large B(E1)⩾10^{-3} W.u. The experimental data appears to be intermediate between the two limits of the vibron cluster model[10], the O(4) deformed limit and U(3) vibrational dipole limit, as shown in Fig. 1. The E1 radiative width correspond to 13% of the E1 molecular sum rule[11] and the alpha widths are known to be about 20% of the Wigner limit, reflecting the fact that these states are not pure cluster states. It is interesting to note that all the low spin low-lying states of ^{18}O, such as the first excited 0^+ state (at 3.63 MeV) and the two 1^- states (at 4.45 and 6.30 MeV) are associated with selective enhanced E1 decays to collective states, suggesting that

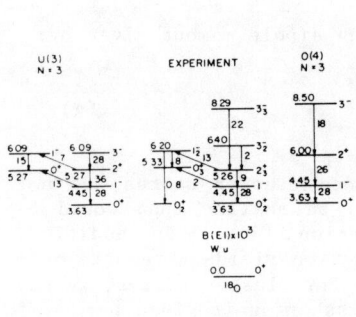
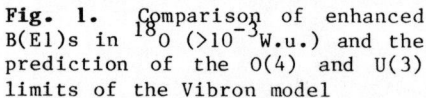
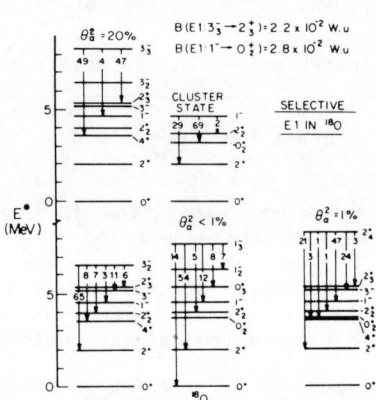

Fig. 1. Comparison of enhanced B(E1)s in ^{18}O (>10^{-3} W.u.) and the prediction of the O(4) and U(3) limits of the Vibron model

Fig. 2. Selective and fast E1 decays in ^{18}O, from cluster states.

the dipole cluster molecular degree of freedom plays a major role in the description of ^{18}O, even though the states in general are not pure cluster states.

The nuclei ^{218,220}Ra and the Ra-Th region: We also have studied the ^{218}Ra and ^{220}Ra isotopes[3,7) and the corresponding data are presented in Fig. 3 and show part of the systematic behavior in this region.

The best evidence for dipole collectivity in the Ra-Th region is the smooth behavior of B(E1)/B(E2) ratios as functions of A and J as shown in Figs. 3,4,5. Clearly the effect of two quasiparticle 11^- states on the negative parity states in ^{218}Ra shown in Fig. 5 and discussed in details in Ref. 7, reflects the collective nature of these E1 transitions. Indeed a Yale-GSI-Munich experiment[23) yielded enhanced B(E1)s in ^{218}Ra, B(E1) ≈ 6 x 10^{-3} W.u. which correspond to 8% of the molecular sum rule for E1, and 200 E1 molecular Weisskopf units.

Rn nuclei: In Fig. 6 we show the systematic behavior of the Rn isotopes[24,25), while the data on the Rn isotopes are more fragmented than those for Ra isotopes, they are suggestive. In the Rn isotopes we find the 1^- and 3^- states at higher energies than in the Ra isotopes, and in addition, while throughout the $^{220-226}$Ra isotopes we find the 1^- state to be consistently ~ 70 keV below the 3^- states, in the Rn isotopes these differences are A dependent, and in ^{218}Rn the 1^- and 3^- states are inverted as opposed to ^{220}Ra[7). The alpha particle hindrance factors for the 1^- and 3^- states in the Rn isotopes are constant while in the Ra-Th region they have a strong A dependence, as shown in Fig. 6. All these suggest that while the Rn isotopes behave like usual octupole vibrational nuclei, with the lowering of the 1^- state as a deformation effect, in the Ra-Th region these states are of a different collective nature arising from reflection asymmetric states.

We conclude that in the Ra-Th nuclei low-lying dipole collective states arising from reflection asymmetric intrinsic state were found, but it appears that the data available thus far can not establish whether these intrinsic states are cluster states or octupole deformed ones. We further note that octupole states give rise to a dipole term[14) and that cluster states acquire nonvanishing octupole moment[11) suggesting that the two models are related.

A suggestion in favor of the cluster interpretation for ^{218,220}Ra nuclei comes from the alpha width and alpha particle hindrance factors for these nuclei, shown in Figs. 6 and 7. The alpha particle hindrance factors appear to be small and the ground state alpha particle reduced widths appear to be large in those nuclei where we find enhanced E1 decay matrix elements.

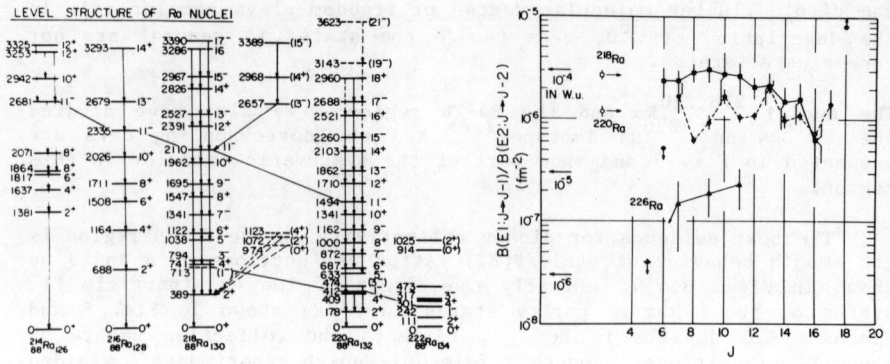

Fig. 3. Level structure of Ra isotopes and B(E1)/B(E2) ratios.

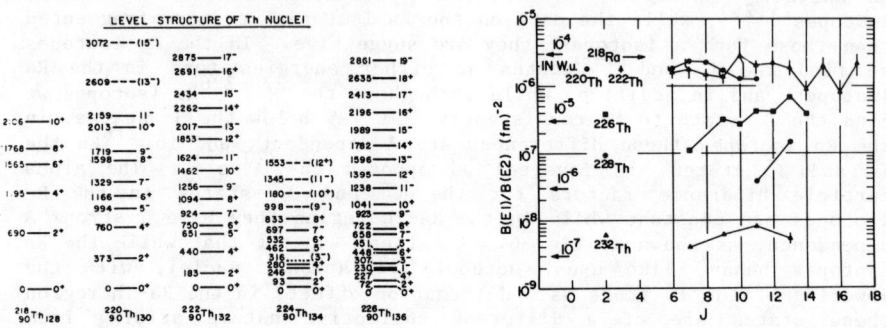

Fig. 4. Level structure of Th isotopes and B(E1)/B(E2) ratios.

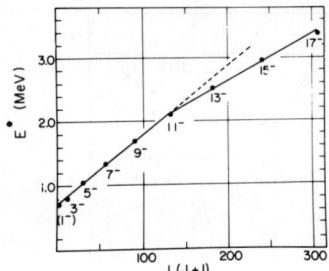 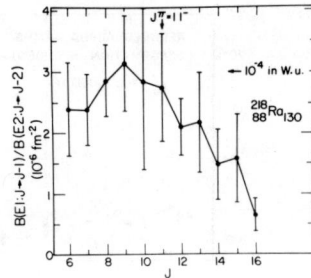

Fig. 5. Moment of inertia and B(E1)/B(E2) for states in ^{218}Ra.

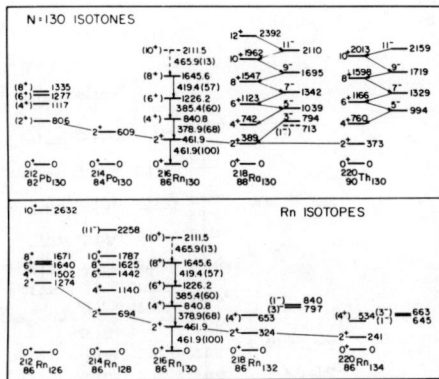 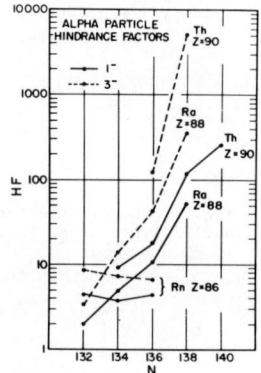

Fig. 6. Level structure of Rn isotopes and N=130 isotones, and the alpha particle hindrance factor for Rn- Ra-Th isotopes.

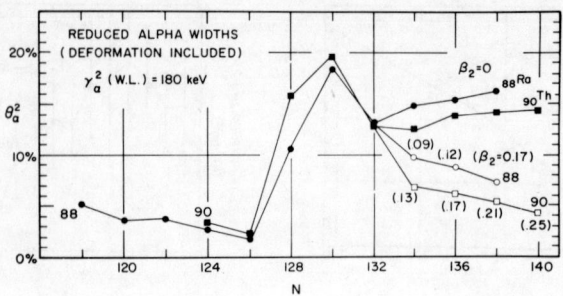

Fig. 7. Alpha reduced widths for ground states of Ra-Th isotopes, as a fraction of the Wigner limit. These calculations include the effect of deformation (in open symbol). The value for β_2 is given in parenthesis.

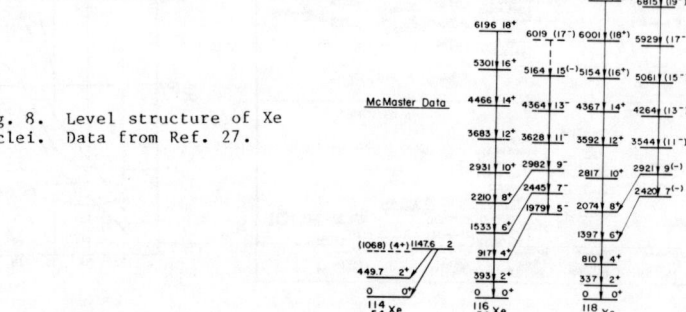

Fig. 8. Level structure of Xe nuclei. Data from Ref. 27.

Alpha Particle Widths: The reduced alpha particle widths are calculated using the W.K.B. penetration factor:

$$P = \exp[-2\int_{R_i}^{R_o} k\, dr] \qquad (7)$$

$$\hbar^2 k^2 = 2\mu\, [V(r) + \frac{2Ze^2}{r} + \frac{\hbar^2}{2\mu}\frac{\ell(\ell+1)}{r^2} - E]$$

It was already noted by Rasmussen[26] that in deformed nuclei, because of the variation of the radius (in angle), these penetration factors may not be entirely physical. Therefore, in deformed nuclei, where we have a non central potential which is dependent on both r and θ, $V(r,\theta)$, we suggest that the one dimensional radial integral should be evaluated over two dimensions involving both the radial and angle variables, where the angular dependence of the potential is given through:

$$r = R_o\, [1+\beta_2 Y_{20}] \qquad (8)$$

with β_2 evaluated from the energy of the 2^+ state $E(2^+) = 1225/(A^{7/3}\beta_2^2)$. Results obtained in this way are shown in Fig. 7, including the deduced values for β_2, and they suggest large reduced alpha particle widths in ^{218,220}Ra nuclei.

Several experiments can be suggested to distinguish between the cluster model and the stable octupole shape model; theses are now underway at Yale and will be discussed in my talk. In addition, it is quite clear that a large body of data is required to map out dipole collective states, especially in medium mass nuclei. Indeed, the heavy Xe nuclei,[27] shown in Fig. 8, show behavior resembling that of the heavy Ra-Th nuclei and it will be interesting to extend these studies to lighter Xe isotopes to investigate whether phenomena similar to those observed in the light Ra isotopes exist in the light Xe isotopes. Such studies can be carried out with high energies tandems, such as the Daresbury tandem and the new Yale ESTU tandem.

We conclude that low-lying collective dipole states are experimentally established both in light and heavy nuclei. In light nuclei they appear to correspond to cluster states while in heavy nuclei both the stable octupole shape model and cluster model can account for the available data.

(1) M. Gai, M. Ruscev, A.C. Hayes, J.F. Ennis, R. Keddy, E.C. Schloemer, S.M. Sterbenz and D.A. Bromley; Phys. Rev. Lett. **50** (1983) 239.
(2) J. Fernandez-Niello, H. Puchta, F. Riess, and W. Trautmann, Nucl. Phys. **A391** (1982) 221.
(3) M. Gai, J.F. Ennis, M. Ruscev, E.C. Schloemer, B. Shivakumar, S.M. Sterbenz, N. Tsoupas and D.A. Bromley; Phys. Rev. Lett.

(4) W. Bonin, M. Dahlinger, S. Glienke, E. Kankeleit, M. Kramer, D. Habs, B. Schwartz and H. Backe; Z. Phys. **A310** (1983) 249; ibid Z. Phys. **A322** (1985) 59.
(5) D. Ward, G.D. Dracoulis, J.R. Leigh, R.J. Charity, D.J. Hinde, and J.O. Newton, Nucl. Phys. **A406** (1983) 591.
(6) J.D. Burrows, P.A. Butler, K.A. Connell, A.N. James, G.D. Jones, A.M. El-Lawindy, T.P. Morrison, J. Simpson, and R. Wadsworth; J. Phys. **G10** (1984) 1449.
(7) J.F. Shriner, Jr., P.D. Cottle, J.F. Ennis, M. Gai, D.A. Bromley, J.W. Olness, E.K. Warburton, L. Hildingsson, M.A. Quader, and D.B. Fossan; Phys. Rev. **C32** (1985) 1888.
(8) A. Celler, Ch. Briancom, J.S. Dionisio, A. Lefebvre, Ch. Vien, J. Zylicz, R. Kulessa, C. Mittag, J. Fernandez-Niello, Ch. Lauterbach, H. Puchta, and F. Reiss; Nucl. Phys. **A432** (1985) 421.
(9) W. Kurcewitz, E. Ruchowska, N, Kaffrell, T. Bjornstad, and G. Nyman, Nucl. Phys. **A356** (1981) 15.
(10) F. Iachello, and A.D. Jackson; Phys. Lett. **108B** (1982) 151.
(11) Y. Alhassid, M. Gai, and G.F. Bertsch; Phys. Rev. Lett. **49** (1982) 1482.
(12) H. Daley, and F. Iachello; Phys. Lett. **131B** (1983) 281.
(13) H.J. Daley, and M. Gai; Phys. Lett. **149B** (1984) 13.
(14) J. Engel, and F. Iachello; Phys. Rev. Lett. **54** (1985) 1126.
(15) V.M. Strutinski, At. Energ. **1** (1956) 150 [Sov. J. At. Energy **4** (1957) 523].
(16) A. Bohr, and B.R. Mottelson; Nucl. Phys. **4** (1957) 529, **9** (1958) 687.
(17) R.R. Chasman; Phys. Lett. **96B** (1980) 7.
(18) G.A. Leander, R.K. Sheline, P. Moller, P. Olanders, I. Ragnarsson, and A.J. Sierk; Nucl. Phys. **A388** (1982) 452.
(19) W. Nazarewicz, P. Olanders, I. Ragnarsson, J. Dudek, and G.A. Leander; Phys. Rev. Lett. 52 (1984) 1272, Nucl Phys. **A429** (1984) 269.
(20) B. Buck, and A.A. Pilt; Nucl. Phys. **A280** (1977) 133.
(21) G. Leander, Niels Bohr Centennial Conference on Nuclear Structure, North Holland, 1985, p.249, and W. Nazarwicz, ibid. p.263.
(22) H.J. Assenbaum, K. Langanke, and A. Weiguny; Z. Phys. **A318** (1984) 35.
(23) J.F. Ennis, M. Gai, D.A. Bromley, F. Azgui, H. Emling, E. Grosse, G. Seiler-Clark, H.J. Wollersheim, C.M. Mittag, F. Riess; Bull. Amer. Phys. Soc. **29** (1984) 1050.
(24) L.K. Peker, J.H. Hamilton, J.D. Rasmussen; Phys. Rev. **C24** (1981) 1336.
(25) P.D. Cottle, M. Gai, J.F. Ennis, J.F. Shriner, Jr., S.M. Sterbenz, D.A. Bromley, J.W. Olness, E.K. Warburton, L. Hildingsson, M.A. Quader, and D.B. Fossan; Bull. Amer. Phys. Soc. **31** (1986) 875.
(26) John O. Rasmussen; Phys. Rev. **113** (1959) 1593.
(27) V.P. Janzen, J.C. Waddington, J.A. Cameron, D. Popescu, and D. Rajnanth; Report McMaster, 1985.

IN-BEAM AND ALPHA DECAY STUDIES OF ACTINIDE NUCLEI

P. A. Butler[1], Y. K. Agarwal[2], K. P. Blume[2], J. D. Burrows[1], K. A. Connell[3], J. R. Cresswell[1], J. de Boer[4], A. M. Y. El-Lawindy[1], K. Euler[2], G. D. Jones[1], Ch. Lauterbach[4], M. Loiselet[5], Ch. Fleischmann[4], C. Günther[2], E. Hauber[4], V. Holliday[1], H. J. Maier[4], M. Marten-Tölle[2], Ch. Schandera[4], P. Schüler[2], R. S. Simon[6], J. Simpson[3], R. Tanner[1], R. Tölle[2], J. Vervier[5], R. Wadsworth[7], D. L. Watson[7] and P. Zeyen[2].

[1] Oliver Lodge Laboratory, University of Liverpool U.K. [2] Institut für Strahlen - und Kernphysik, Universität Bonn, Federal Republic of Germany. [3] Science and Engineering Research Council, Daresbury Laboratory, U.K. [4] Sektion Physik, Universität München, Federal Republic of Germany. [5] Institute de Physique, Universite Catholique de Louvain, Belgium. [6] Gesellschaft für Schwerionenforschung, Darmstadt, Federal Republic of Germany. [7] Department of Physics, University of Bradford, U.K.

Abstract

The properties of $^{224, 226, 228}$Th and ^{223}Th from in-beam spectroscopy, and ^{219}Ra, ^{223}Th from α decay measurements, will be discussed in the framework of the octupole model.

1. Introduction

Nuclei with $Z \approx 88 - 90$, $N \approx 132 - 136$ are of interest because they are transitional between spherical and deformed regions where effects from octupole coupling between orbitals near the Fermi surface will be important in determining the intrinsic shape[1]. In this mass region the presence of low lying bands with spin sequence $1^-, 3^-, 5^- ...$ in even-even nuclei is seen as evidence for reflection asymmetry about the x-y plane (where the z axis is the symmetry axis) arising from octupole deformation. It has also been pointed out these low lying negative parity states could arise from molecular states formed by alpha-clustering in these nuclei[2,3,4].

We have attempted to obtain more systematic information on the nature of the low lying negative parity states by measuring the property of the yrast bands in even-even (see section 2) and even-odd nuclei (section 3) up to moderate spin values, where the octupole deformation appears to be well stabilised. We have also measured the alphas decay hindrance factors for the sequence $^{227}U \to ^{223}Th \to ^{219}Ra$ (see section 4) where both daughter nuclei exhibit reflection asymmetry at rotational frequencies > 150 keV. We have preliminary results on the decay for an odd-odd system ($^{224}Pa \to ^{220}Ac$).

2. In-Beam Studies of $^{224,226,228}Th$

We have studied the nuclei $^{224,226,228}Th$ with (α, xn) reactions on a ^{226}Ra target, which have much higher cross sections than previously used reactions such as $^{208}Pb(^{18}O,2n)^{224}Th$ [5], $^{230}Th(\alpha, \alpha', 2n)^{228}Th$ [6]. Targets of 200 μ g/cm^2 $^{226}RaBr_2$ were bombarded with 33, 42 and 55 MeV α -particles at the Bonn cyclotron. Conversion electron spectra, e^--e^-- and e^--γ co-incidence spectra were measured with a double orange spectrometer and a 20% HPGe detector. We are able to assign E1 and E2 multipolarities to all reasonably strong lines in ^{224}Th and ^{226}Th from the measured conversion electron intensities. The level schemes for ^{224}Th, ^{226}Th and ^{228}Th are shown in fig. 1. For ^{224}Th the level scheme given by ref 5) up to the 10^+ state is confirmed. The remaining assignments in ^{224}Th and the scheme of ^{226}Th are based on coincidence relations and energy sums. The parity assignments of the levels up to I $\approx$ 12 are confirmed by the multipolarities determined from the conversion electron spectra, as discussed above.

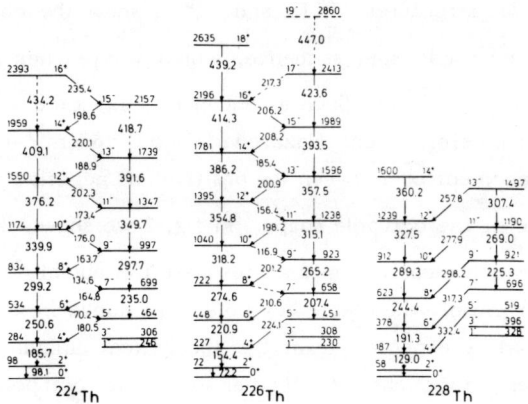

Fig. 1. Level schemes of ^{224,226,228}Th as observed in the ^{226}Ra(α, xn) reaction.

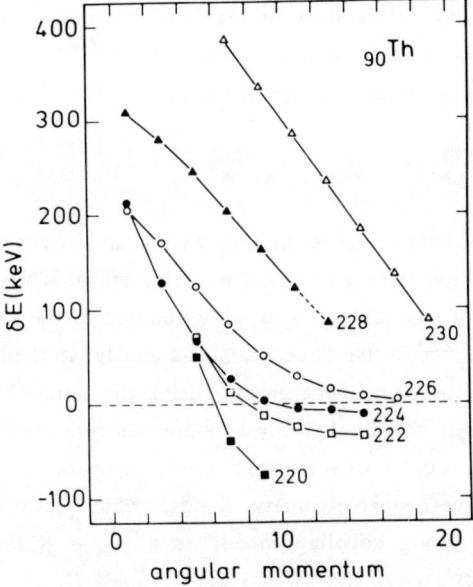

Fig. 2. Displacement δE of the positive and negative parity bands in the Th isotopes.

The level structures ^{224}Th and ^{226}Th show the characteristic interleaving even-odd spin sequence, which was earlier observed in ^{218,220}Ra and ^{220,222}Th. Such a sequence is expected in nuclei with stable octupole deformation. Nazarewicz and Olanders[7] introduce the energy displacement δE between the positive and negative parity bands as a measure of the octupole shape. In fig. 2 we show δE vs. I for $^{220-230}$Th, which clearly demonstrates that the nuclei ^{222,224,226}Th follow rather closely the stable octupole limit above I $\approx$ 10. This is precisely what is expected from the theoretical calculations of the potential energy functions[1,7]. Further support for octupole instability comes from the en-hanced stretched E1 transitions connecting the opposite parity states. The resulting induced intrinsic dipole moment has the form $D_0 = c\,A\,Z\,B_2\,B_3$, where c is a model dependent constant. For ^{220,222}Th the B(E1)/B(E2) ratios determined from the measured γ-branching ratios show no significant variation with spin for I $\geqslant$ 6. We also obtain this spin-independence for ^{224,226,228}Th from the intensities of the transitions shown in fig. 1.

3. In-Beam Study of ^{223}Th

In-beam studies of odd nuclei in this region have revealed single bands of alternate negative and positive parity states which can be interpreted as the odd particle being weakly coupled to the even-even core. The nucleus ^{223}Th has been simultaneously studied by the Liverpool group and the Heidelberg group[5] using the same ^{208}Pb(^{18}O, 3n)^{223}Th reaction, but the final decay scheme has remained elusive. Both the groups agree, however, that there appears to be <u>two</u> sequences of positive-negative parity bands, which can best be interpreted by the strong coupling model as a single K band with combined signature $\sigma = +1/2$ and $\sigma = -1/2$.

4. Alpha Decay Studies
4.1 ^{223}Th → ^{219}Ra

Recent experiments using the ^{208}Pb (^{14}C, 3n) reaction[8,9] have revealed the presence of positive and negative parity bands in ^{219}Ra which are connected by strong E1 transitions. However, the low lying structure has not been experimentally studied in detail and it is not known whether the ground state deformation is well developed or exhibits reflection assymmetry as indicated at higher rotational frequency. We have measured the properties of low lying states in ^{219}Ra (and ^{215}Rn) populated by the alpha decay of ^{223}Th.

The parent nucleus ^{223}Th was populated using the ^{208}Pb(^{18}O, 3n)^{223}Th reaction at a bombarding energy of 83 MeV. In this experiment, the strongest transitions in the daughter nuclei were identified by collecting alpha spectra and alpha-gamma coincidence data. The beam, supplied by the Oxford Nuclear Physics Laboratory Folded Tandem, irradiated Pb foils enriched to 99% in ^{208}Pb, of various thicknesses in the range 1-2 mg/cm^2. The target was mechanically moved from the in-beam position to a position adjacent to a 75mm^2 surface barier Si detector and a 23% efficient Ge detector. All charged particle data and coincidence particle-γ events were collected synchronously with the target being out-of-beam and the beam pulsed off. The in-beam and out-of-beam periods were both 0.75s.

Internal conversion measurements were carried out using the electron spectrometer at the Daresbury Laboratory. Fig. 3. shows the out-of-beam γ-ray spectrum and electron spectrum in coincidence with ^{223}Th alpha particles.

The alpha branching ratios are taken from the intensities assuming the decay scheme given in fig. 4. for ^{219}Ra which show the deduced hindrance factors. We observe that the alpha transitions with hindrance factors < 6 populate two levels at 140 and 178 keV which

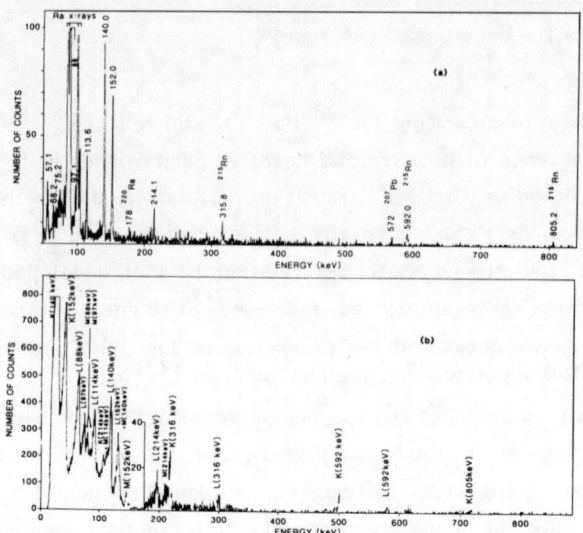

Fig. 3. Out-of-beam spectra from the reaction ^{208}Pb$(^{18}O, 3n)^{223}$Th: a) γ-ray spectrum b) electron spectrum.

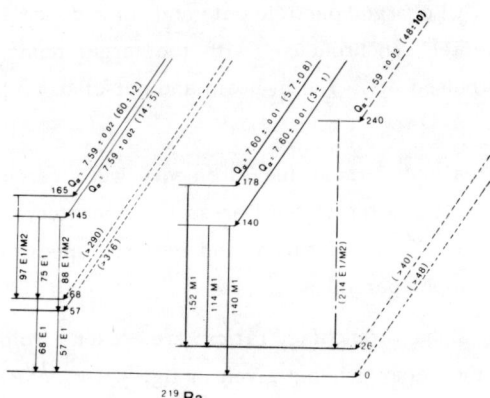

Fig. 4. Level scheme of ^{219}Ra deduced from this work. The alpha transitions are labelled with ground state Q-value and (in brackets) hindrance factors.

decay via M1 or M1/E2 transitions to the ground state and an excited state at 26 keV. The occurence of so many M1 transitions in ^{219}Ra can be accounted for if the low lying structure arises from (ν g9/2)^n near spherical configurations. These states would then have structure similar to that observed in N = 87 nuclei[10] in which the f7/2 - h9/2 shells above the N = 82 neutron gap play a similar role to the g9/2 shell above the N = 126 gap.

Low lying negative parity states are observed in this work, such as the 57 and 68 keV levels in ^{219}Ra. The alpha hindrance factors to these states are > 290, which suggests that they have a different intrinsic structure to the ground states of ^{223}Th and ^{219}Ra. They cannot easily be accounted for by invoking pure configurations. For example a prolate deformation of ε > 0.1 would be required to bring the low K members of the j15/2 orbital sufficiently down in energy. It is more likely that these states arise from the mixed (g9/2-j15/2) configurations which can give rise to stable octupole deformation in this region.

In the present work the non-observation of γ -transitions in ^{219}Ra which were observed in the (^{14}C, 3n) reaction [8,9] such as the 234 keV transition which was placed as the lowest transition in these decay schemes, implies that these states have either high spin (J > 9/2) or that the observed band does not have as its bandhead the ground state, but is connected by a hitherto unobserved transition to the levels observed here.

4.2 ^{227}U → ^{223}Th

In order to obtain complementary information on ^{223}Th to the in-beam studies (see section 3). We have studied that α decay of ^{227}U ($T_{1/2}$ = 1.1 min) prepared by the ^{208}Pb (^{22}Ne, 3n) reaction. The 110 MeV ^{22}Ne beam was supplied by the CYCLONE facility of the University Catholique de Louvain. In this experiment two targets were continuously rotated in and out-of-beam for 30 seconds each. In each of the two out-of-beam positions the target irradiated a silicon α-detector, subtending about 25% of 4π, and a Ge detector, with peak efficiency of about 2% at 300 keV. Alpha-γ coincidences were recorded for a 56 hour irradiation with a 5 pnA beam. Transitions in nuclei in the ^{227}U decay chain were identified on the basis of energy and time of the event following irradiation. We are able to estimate the relative hindrance factors to the lowest observed negative parity states in ^{223}Th from the observed intensity of the 129 keV transition, which feeds the $7/2^+$ member of the $K^\pi = 5/2^+$ ground state band of this nucleus[11]. This intensity is measured to be 6 ± 2% of the intensity of the 316 keV tansition in ^{215}Rn, which implies an absolute branching of 4 ± 2%. From the total observed intensity of known ^{223}Th γ-rays and K X-rays, we can estimate that the branching to the lowest members of the $K^\pi = 5/2^+$ band in ^{223}Th is at least 20% of the total decay. The deduced hindrance factors are respectively < 4 and 5 ± 2 to different parity members of the $\sigma = 1/2$ $K = 5/2$ band.

References

1) Leander G. and Sheline R.K., Nucl. Phys. A413 375 (1984)
2) Iachello F, and Jackson A. D., Phys. Lett. 108B 151 (1982)
3) Gai M., Proc. Inst. Symp. In-Beam Nucl. Spectr., Debrecen (Akademiai Kiado) 331 (1984)
4) Daley H. J. and Barrett B. R., Nucl. Phys. A449 256 (1986)
5) Dahlinger M. et. al., Jahresbericht MPI - Heidelberg 100 (1984)
6) Hardt K. et. al., Nucl. Phys. A419 34 (1984)
7) Nazarewicz W. and Olanders P., Nucl. Phys. A441 420 (1985)
8) Mittag C. et. al., Jahresbericht Munich 40 (1984)
9) Cottle P. D. et. al., Yale preprint 3074-860 (1985) (to be published)
10) Kleinheinz P. et. al., Nucl. Phys. A282 189 (1977)
11) D. Habs, private communication and Dahlinger M. et. al., Jahresbericht MPI-Heidelberg (1985).

ANISOTROPIC ALPHA EMISSION OF ON LINE SEPARATED ISOTOPES

J. WOUTERS, D. VANDEPLASSCHE, E. VAN WALLE, N. SEVERIJNS and
L. VANNESTE
Instituut voor Kern- en Stralingsfysika, Leuven University,
B-3030 Leuven, Belgium

Abstract

A systematic on-line nuclear orientation study of heavy isotopes by use of anisotropic α-emission is reported. The anisotropies recorded for ^{199,201,203,205}At are remarkably pronounced and strongly varying. This may be interpreted in terms of the high sensitivity of the α-emission probability to changes in the nuclear shape.

Our recent technical development of particle detection at low temperatures (4K) at the on line nuclear orientation facility KOOL has made it possible to study for the first time anisotropic α-, β- and conversion electron emission besides the γ-anisotropy measurements.[1] Results have been obtained by use of a combination of production by heavy-ion reactions, on-line separation, on-line implantation and orientation at low temperatures.

We recently started a systematic study of shortlived Po, At and Fr isotopes and their Bi and Pb daughters (from nuclear mass 195 to 209 and with shortest lifetime of oriented nucleus $t_{1/2}$ = 3.9 s). Apart from the large amount of information on decay schemes and magnetic moments, very intriguing first results on anisotropic α-decay have been obtained.

The immense attraction of α-detection in connection with oriented nuclei lies in the fact that the directionality of the α-emission and therefore the anisotropy of the α-radiation, is strongly connected with the shape of the emitting and daughter nuclei.[2] A comparison of the data on the α-decays of the type $\pi h9/2^- \rightarrow h9/2^-$ in the odd ^{199,201,203,205}At isotopes shows a change in sign of the anisotropy (fig. 1): the α-particle emission shifts from preferentially parallel to preferentially perpendicular to the nuclear spin direction as one goes further away from the N=126 closed shell structure.[3] The same trend can also be seen in the ^{205,207,209}Fr α-decays.

A first analysis reveales the L≠0 contribution in the α-decay to be very small although they - the mixing ratios δ of emitted α-particles

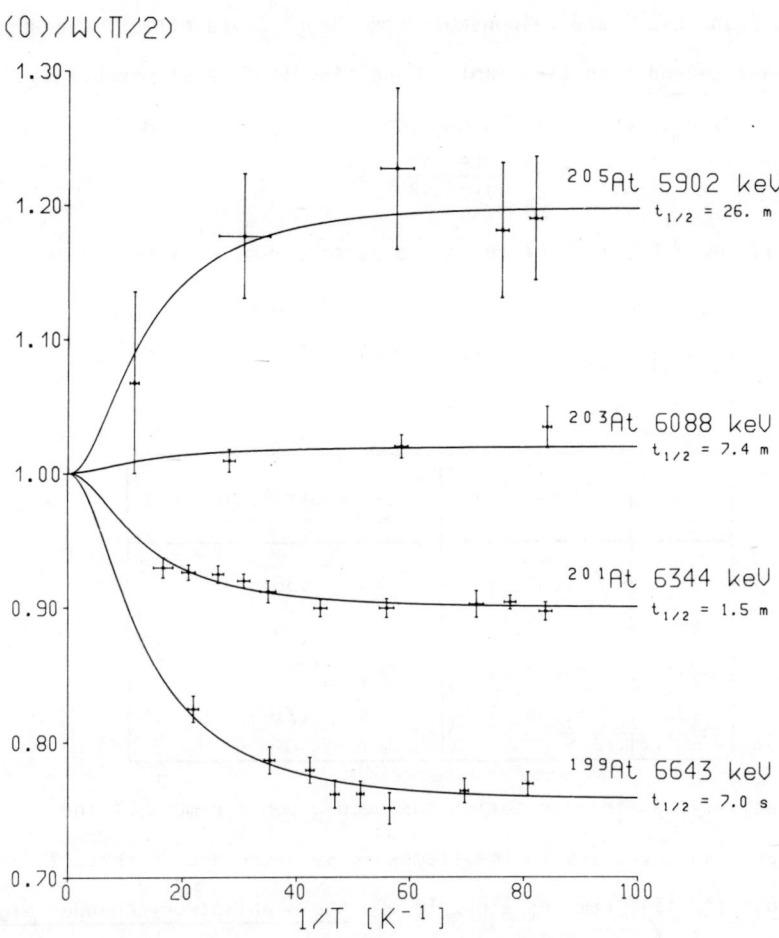

Fig. 1. Alpha-particle emission anisotropies as a function of 1/T of the $\pi h_{9/2}^- \to \pi h_{9/2}^-$ α-decays of ^{199,201,203,205}At.

with L=2 and L=0 - are responsable for the observed effect: a change of magnitude <u>and</u> sign (see table 1 and fig. 1). The anisotropy is $W(0)/W(\Pi/2)$ with $W(\Theta) \simeq 1 + A_2B_2P_2(\cos\Theta)$ and $A_2 \simeq 2\delta$ and
$$\delta = \frac{<||L=2||>}{<||L=0||>} \cos(\sigma_2-\sigma_0).$$

<u>Table 1.</u> The product $f\delta$ of the solid state parameter f (which is the same for all 4 isotopes) and the mixing ratio's $\delta = \frac{<||L=2||>}{<||L=0||>} \cos(\sigma_2-\sigma_0)$ of the L=2 α-wave contributions in the decays of ^{199,201,203,205}At

A	fδ	L=2 contribution (%)
199	-.059(2)	.35
201	-.023(1)	.05
203	+.004(2)	/
205	+.046(7)	.21

Qualitatively, an interpretation has been given for most of the observed α-anisotropies in the framework as described in refs. 1, 2 and 3 but it still remains a puzzle <u>why the α-anisotropy changes sign</u> in the $\Pi h9/2^- \rightarrow \Pi h9/2^-$ α decays! Specific calculations of the L=2 α-wave contribution in these decays are being performed now.

In summary, we have measured α-emission anisotropies of short lived Po, At and Fr isotopes. Alpha anisotropies have been measured with separated beams of 10^3 ions/s which clearly shows the sensitivity of the technique and the feasibility in connection with on-line systems. These experiments provide the first systematic study of anisotropic α-decay and bring to light some very interesting aspects of α-decay and may contribute to a better understanding of the α-decay process. Nevertheless, a quantitative explanation for the observed effects is not possible yet, and a lot of analysis has to be done still. Furthermore, due to the high sensitivity of this technique and with the right source preparation, exciting new experiments become possible now: measuring the α-particle angular distributions of actinides (octupole deformations, α-clustering?) and nuclei in the transuranium region to characterize the nuclear structures.

More generally, the technical realization of particle detection of on-line oriented isotopes opens also new avenues for systematic studies of weak interaction (β-decay of mirror nuclei, e.g. ^{17}F as reported at this conference [4]) and nuclear penetration effects (conversion electron angular distributions).

REFERENCES

1. J. Wouters, D. Vandeplassche, E. van Walle, N. Severijns and L. Vanneste in Proceedings of the Int. Symp. on Nuclear Orientation and Nuclei far from Stability, Leuven, Belgium, eds. B.I. Deutch and L. Vanneste, Hyp. Int. 22, 527 (1985)
2. J.O. Rasmussen in Alpha-, beta- and gamma-ray spectroscopy, ed. K. Siegbahn, Vol. 1 (North-Holland, Amsterdam, 1965) p. 701
3. J. Wouters, D. Vandeplassche, E. van Walle, N. Severijns and L. Vanneste, Phys. Rev. Lett. 56, 1901 (1986)
4. L. Vanneste, N. Severijns, J. Wouters, D. Vandeplassche, E. van Walle, this conference.

NUCLEAR MOMENTS AND RADII IN THE REGION(S) OF OCTUPOLE DEFORMATION

R. Neugart
Institut für Physik, Universität Mainz,
D-6500 Mainz, F.R. Germany

Features of strong ocutpole correlation have been observed in the spectra of nuclei in the Ra-Th region with neutron numbers around N = 132 - 138 and it has been suggested that these should be attributed to a well-developed octupole-deformed nuclear shape (see ref. 1, 2 and references therein). From the ground-state properties we can certainly not expect a straightforward proof of this hypothesis: Unlike the quadrupole deformation a possible octupole deformation will not be manifested in a spectroscopic observable such as the quadrupole moment of nuclei with spins $I \geq 1$. For well-developed spheroidal shapes this spectroscopic quadrupole moment can be related to the intrinsic one by the simple
strong-coupling projection formula

$$Q_s = \frac{I(2I-1)}{(I+1)(2I+3)} Q_o. \qquad (1)$$

We shall discuss which indirect evidence for octupole deformation can be drawn from the nuclear ground-state properties.

A first indication of octupole effects in the ground state came from the masses around ^{222}Ra. The systematic deviations of calculated masses[3] from the experimental values were removed by the inclusion of the octupole degree of freedom into deformed shell-model calculations[1] which then resulted in reflection-asymmetric equilibrium shapes for the nuclei with proton numbers between Z = 86 and 92 (i.e. Rn and U) and neutron numbers between N = 130 and 138.

The most detailed information about the nuclear ground states emerges from laser spectroscopy experiments yielding the spins, the magnetic dipole and electric quadrupole moments, and the variation of the mean square nuclear charge radii from the measurement of atomic hyperfine structures and isotope shifts (see e.g. ref. 4). Such experiments were performed recently on $^{208-214,220-232}$Ra[5,6,7] and

$^{202-212,219-222}$Rn[8]) by collinear laser fast beam spectroscopy, and on $^{207-213,220-228}$Fr[9]) by optical pumping combined with magnetic state selection of a thermal atomic beam. All three experiments used the ISOLDE isotope separator which provides $10^4 - 10^9$ atoms per second of these radioactive nuclides from 600 MeV proton induced spallation in uranium or thorium carbide targets. The measurements on radium and francium cover the whole region of interest in connection with octupole effects, while the radon data obtained so far for N = 133 - 136 are centered on the maximum of the predicted octupole instability.

We start with a discussion of the single-particle states of the odd-A isotopes. In comparison with the usual Nilsson diagrams the calculations of Leander and Sheline[2]) for a reflection-asymmetric potential with ε_3 = 0.08 - 0.10 lead to a complete re-arrangement of the level ordering for the respective neutron and proton orbitals. This is due to the mixing of parities of the intrinsic single-particle wave functions. Such an effect should be reflected in the isotopic sequence of the ground-state spins. In fact, the spins of 5/2, 3/2, 1/2, 3/5, 5/2 for the sequential filling of neutron orbitals in the odd-A isotopes $^{221-229}$Ra (N = 133 - 141) are perfectly reproduced by the calculations for ε_3 = 0.08 (Fig. 1), if one assumes a gradual increase of ε_2 from about 0.14 to 0.20 as suggested by B(E2) values or the experimental quadrupole moments. The first two members of the corresponding sequence ^{219,221}Rn (N = 133, 135) have the spins 5/2 and 7/2 of which the latter has no obvious explanantion. The reason may be a strong octupole correlation on top of a small ε_2 which would bring the Ω = 7/2 component from $g_{9/2}$ close to the Fermi surface. In $^{221-227}$Fr the spins 5/2, 3/2, 3/2, 1/2, arising from the 87th proton, again fit qualitatively into the octupole-deformed scheme under the assumption of increasing ε_2 and decreasing ε_3 from N = 134 to 140.

The magnetic moments test rather sensitively the parity mixing of single-particle states in the octupole-deformed potential, in particular if pairs of spin-flipped orbitals are involved. In fact, the calculations by Ragnarsson[10]) and by Leander and Sheline[2]), with ε_3 = 0.1 for A = 221 - 225 decreasing towards A = 229, reproduce much

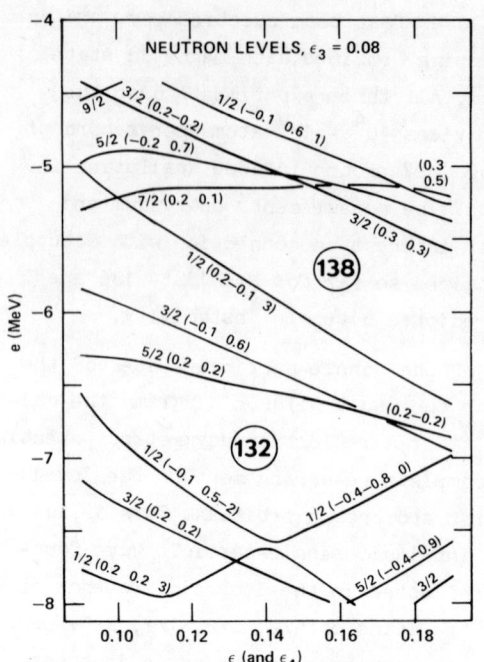

Fig. 1:
Single-neutron orbitals in a reflection asymmetric potential (from ref. 2).

better the experimental moments for the whole sequence of $^{221-229}$Ra than the conventional particle-rotor calculations[5] including only ϵ_2 and ϵ_4. In addition, the latter have to assign ground-state orbitals which are expected rather far from the Fermi surface. A particularly interesting nucleus is the $1/2^+$ ^{225}Ra of which the magnetic moment μ = -0.75 μ_N fits sensitively to the same ϵ_3 = 0.1 that has been successful in reproducing the decoupling parameter of the ground-state band and its parity splitting[2,10]. For ^{219,221}Rn a detailed analysis of the magnetic moments is still missing, but from the data μ^{219} = -0.43 μ_N (I = 5/2); μ^{221} = 0.02 μ_N (I = 7/2) it seems clear that the situation is different from radium, although inconsistent with vanishing octupole effects. The almost zero magnetic moment for I = 7/2 in ^{221}Rn probably requires a considerable modification of the present theoretical approaches.

For the odd-proton ground states of ^{223,225}Fr one finds the spin I = 3/2 with magnetic moments very similar to the largely discussed ^{227}Ac[11,2]. Here the magnetic moment, together with the M1+E2 mixing in the intraband transitions gave the same gyromagnetic ratios for both members of the $3/2^{\pm}$ parity doublet, as expected for identical intrinsic states. Calculations by Nazarewicz[12] reproduce the ^{223,225}Fr moments with ε_3 = 0.09 and indicate a vanishing octupole deformation for the I = 1/2 ground state of ^{227}Fr. On the other hand, it has been pointed out that the I = 3/2 moments of ^{223,225}Fr and ^{227}Ac can as well be associated with the Nilsson orbital [532 3/2] in a reflection symmetric picture[13]. This shows the importance of measuring the magnetic properties not only in the ground state, but also in its presumable parity-doubled partner.

As expected, the quadrupole moments are insensitive to octupole correlations. However, in this context they can serve as a guide for assigning the levels, and provide information about the quadrupole deformation and the strength of its coupling to the single-particle motion.

The isotope shifts, or the changes in the mean square nuclear charge radii, contain the total collective motion of the nuclei, as expressed by

$$\delta<r^2> = \delta<r^2>_{sph} + \frac{5}{4\pi} <r^2>_{sph} (\delta<\beta_2^2> + \delta<\beta_3^2> + ...), \qquad (2)$$

where $<r^2>_{sph}$ and $\delta<r^2>_{sph}$ represent spherical nuclei of the given volume. (The deformation parameters β_ν are chosen here because of the expansion in spherical harmonics.) These volume terms are rather well described by the droplet model[14], and changes in deformation $\delta<\beta_2^2>$ or $\delta<\beta_3^2>$ give pronounced deviations from the regular increase of the radii with neutron number. This is plotted in Fig. 2 for the example of radium. The upper curve represents the slope of $<r^2>$, i.e. $\delta<r^2>^{A-2,A}$ for the addition of neutron pairs to the doubly-even isotopes. The droplet model slope (represented by a horizontal line) is close to the values of the light isotopes below N = 126. The large $\delta<r^2>$ values of the heavier isotopes mainly reflect the gradual increase of quadrupole deformation up to A ≈ 230. On top of this,

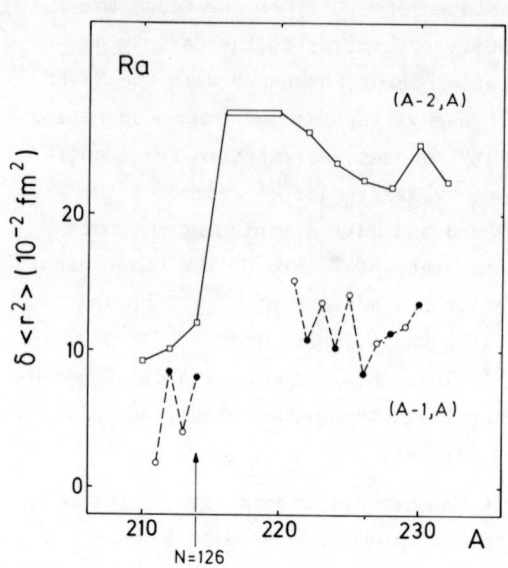

Fig. 2:
Variations of mean square charge radii and odd-even staggering. Full dots (open circles) represent steps from odd to even (even to odd) neutron number.

there is a slight reduction between ^{222}Ra and ^{228}Ra which may be associated with the dissipation of the octupole mode. A quantitative comparison with eq. (2) is difficult, because the small $\langle \beta_3^2 \rangle$ contributions require a precise knowledge of the dominating $\delta \langle r^2 \rangle_{sph}$ and $\delta \langle \beta_2^2 \rangle$ terms. Still, such an analysis suggests quite appreciable contributions from higher-order deformations, e.g. β_3 [6].

A very interesting feature is observed for the odd-neutron isotopes in the region $^{220-227}$Ra as well as $^{220-226}$Fr, and for the few measured isotopes $^{219-222}$Rn. This is shown in the lower curve of Fig. 2 which gives $\delta \langle r^2 \rangle^{A-1,A}$, i.e. the differences between odd and even or even and odd isotopes. Below N = 126 the pronounced odd-even effect follows the normal behaviour that is observed all over the chart of nuclides, namely a reduction of $\langle r^2 \rangle$ for the odd isotopes. This corresponds to the steeper collective potential which is reduced in the doubly-even isotopes by the effect of pairing.

For the neutron numbers between N = 132 and 138 we find the opposite behaviour, which suggests a relationship to the occurence of octupole deformation. It has indeed been shown that the soft octupole potential should be more stabilized in the odd than in the neighbouring even nuclei, due to polarization by the parity-mixed single-particle state.[2] This should result in an additional contribution to $\langle \beta_3^2 \rangle$. Reinhard[15] has pointed out that the odd-even effect of octupole deformation on top of a well-developed quadrupole deformation should be opposite to the normal odd-even staggering due to pure quadrupole deformation: The one-dimensional dynamics in β_3 is sensitive to the inner potential barrier at $\beta_3 = 0$, while the three-dimensional dynamics in β_2 overweights the outer potential barrier with the volume element $\beta_2^2 \, \delta\beta_2$. If we compare the observed inverted with the "normal" odd-even staggering of 0.04 fm², the "octupole" effect of 0.07 fm² corresponds to about half the value of β_3 which is at least the right order of magnitude. Of course, the quantitative significance of the inverted staggering has to be clarified by detailed calculations including also the coupling between quadrupole and octupole modes and their influence on $\langle r^2 \rangle$.

In general, it seems that the coherence in the interpretation of the ground-state properties favours a description in terms of a stable octupole deformation rather than dynamic excitation modes.

From the experimental point of view it is less obvious whether the neutron-rich Ba-La isotopes follow the same trends. The spectroscopic information in this region is rather poor, but the spin and magnetic moment of ^{145}Ba[16] - unexplained in the reflection symmetric particle-rotor scheme - fits well into the calculations assuming octupole deformation[17]. Also the odd-even staggering in the radii of the heavy barium isotopes disappears gradually in the sequence $^{139-145}$Ba. It will be a task of future experiments and theoretical analyses to establish all significant features for octupole-deformed states in this region.

References

1) G.A. Leander et al., Nucl. Phys. A388 (1982) 452
2) G.A. Leander and R.K. Sheline, Nucl. Phys. A413 (1984) 375
3) P. Möller and J.R. Nix, Nucl. Phys. A361 (1981) 117
4) E.W. Otten, Nucl. Phys. A354 (1981) 471c
5) S.A. Ahmad et al., Phys. Lett. 133B (1983) 47
6) S.A. Ahmad et al., Proc. 7th Int. Conf. on Atomic Masses and Fundamental Constants, Ed. O. Klepper (Darmstadt, 1984) p. 361
7) S.A. Ahmad, W. Klempt, R. Neugart, E.W. Otten, G. Ulm and K. Wendt, submitted to Nucl. Phys.
8) W. Borchers, H.T. Duong, R. Neugart, E.W. Otten, G. Ulm and K. Wendt, to be published
9) A. Coc et al., Phys. Lett. 163B (1985) 66
10) I. Ragnarsson, Phys. Lett. 130B (1983) 353
11) R.K. Sheline and G.A. Leander, Phys. Rev. Lett. 51 (1983) 359
12) W. Nazarewicz, private communications to the authors of ref. 9)
13) C. Ekström, L. Robertsson and A. Rosen, submitted to Physica Scripta
14) W.D. Myers and K.H. Schmidt, Nucl. Phys. A410 (1983) 61
15) P.G. Reinhard, private communication
16) A.C. Mueller et al., Nucl. Phys. A403 (1983) 234
17) G.A. Leander et al., Phys. Lett. 152B (1985) 284

COLLECTIVE PROPERTIES OF OCTUPOLLY DEFORMED NUCLEI [*]

R. Baranowski, K. Böning and A. Sobiczewski

Institute for Nuclear Studies, Hoza 69, PL-00-681 Warszawa, Poland

Last years, much attention is given to the problem of the nature of the lowest collective states of nuclei around radium. Specific properties of these nuclei are interpreted as being connected with the octupole deformation.

In the present paper, we concetrate on the properties of even-even nuclei. An anharmonic character of the spectra of the nuclei[1] and a small value of the energy of the lowest negative-parity states of them (down to about 200 keV) seem to be connected with a strong anharmonicity of the collective potential of a nucleus, treated as a function of the octupole degree of freedom. The strong anharmonicity may be manifested by a presence of a minimum of the potential at the octupole deformation different from zero, $\varepsilon_3^o \neq 0$ (and not at the zero deformation, $\varepsilon_3^o = 0$), making the shape of the potential closer to that of the double harmonic oscillator[2] than to the shape of the single oscillator. Such minimum is usually obtained now in the calculations of the potential energy (e.g. refs.[3-5]).

In the present contribution, we limit ourselves to a discussion of the energy spectrum of the lowest collective states and of the reduced transition probabilities B(E2) and B(E3) between them. These quantities have already been calculated in refs.[6,7] The states have been treated as large-amplitude quadrupole and octupole vibrations, coupled with each other by the potential energy. The quadrupole- and octupole-deformation parameters, ε_2

---

[*] Supported in part by the Polish-US Maria Sklodowska-Curie Fund, Grant no. P-F7F037P.

and ε_3, respectively, have been used as the dynamical variables. A large sensitivity of the spectrum and of the probabilities $B(E\lambda)$ to the shape of the potential has been found. The potentials used for the discussion of this sensitivity have been based on the Nilsson single-particle model with the standard "A=225" parameters[8] obtained by the interpolation (for the radium, A ≈ 225, region) between the values proper for the rare-earth and actinides regions.

The scope of the present analysis is to look at the effect of using improved single-particle parameters, introduced in ref.[9] They allow to better reproduce the ground-state properties of odd radium izotopes and lead to better quadrupole moments of even-even isotopes. The values of the parameters are: κ_p =0.0580, μ_p =0.630 for protons and κ_n =0.0526, μ_n =0.457 for neutrons.

The collective potential energy $V(\varepsilon_2, \varepsilon_3)$ obtained with the new parameters is shown in fig.1 for ^{224}Ra. It is calculated in two variants. One, (Y), is the macroscopic-microscopic energy with the Yukawa-plus-exponential model (with parameters of ref.[10]) taken for the macroscopic part, and the Strutinski shell correction used for the microscopic part. The other, (BS), is the sum of the single-particle energies smoothed by the pairing interaction. A comparison of the present V with that obtained with the "A=225" parameters shows a smaller effect of the octupole degree of freedom on V obtained presently. In the case (BS), the depth of the octupole minimum (i.e. the gain in V due to the ε_3 degree of freedom) is presently around 0.5 MeV, while it was about 1.6 MeV previously. In the case (Y), there is presently no gain in V due to ε_3, while it was about 0.2 MeV previously. It seems that the smaller effect of ε_3 on V is connected with a larger quadrupole deformation ε_2 at the equilibrium point, which makes the octupole interaction less effective.

The quadrupole moment found for ^{224}Ra experimentally[11] is (6.34 ± 0.04) b. The theoretical value obtained by us is 7.55 b for the (Y) case and 3.21 b for the (BS) case. It is obtained in the dynamical way, i.e. with an account for the structure of the

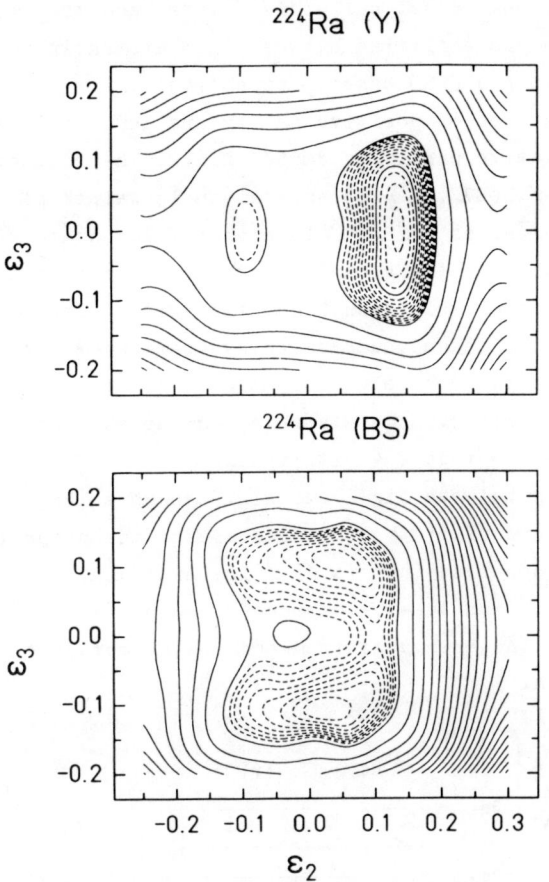

Fig.1. Contour maps of the potential energy $V(\varepsilon_2, \varepsilon_3)$ calculated for ^{224}Ra in two variants (Y) and (BS) described in text. The difference in energy between the neighbouring solid lines is 1 MeV and between dashed lines 0.1 MeV.

ground-state wave function. Thus, it is too large in the (Y) case and too small in the (BS) approach.

Spectrum of the lowest collective states and the B(E2) and B(E3) transition probabilities between them are shown in figs. 2 and 3 for the (Y) and (BS) cases, respectively. Six excited states are shown. The states of positive parity are denoted by solid lines and those of negative parity by dashed lines. The reduced transition probabilities $B(E\lambda)$, $\lambda = 2,3$, are given in Weisskopf units (W.u.). Only the largest values of $B(E\lambda)$ are shown. The state with a large B(E2) transition to the ground state (g.s.) is considered as the first quadrupole excitation (1q), that with a large B(E2) to the 1q state as the second quadrupole excitation (2q) and so on. The same concerns the octupole excitations 1o, 2o,..., with B(E2) being replaced by B(E3). Some states appear as superpositions of the quadrupole and octupole excitations.

A comparison between fig.2 and fig.3 shows a large sensitivity of the spectrum and of $B(E\lambda)$ to the shape of the potential, similarly as it was observed in refs.[6,7] In particular, the B(E3) value for the transition 1o→g.s. strongly depends on the octupole position ε_3^o of the energy minimum and on the octupole contribution to its depth.

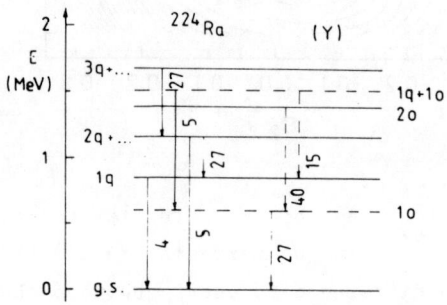

Fig.2. Energy spectrum of the lowest 7 states and the largest B(E2) and B(E3) transition probabilities between them, given in Weisskopf units (W.u.), obtained with the variant (Y) of the potential V.

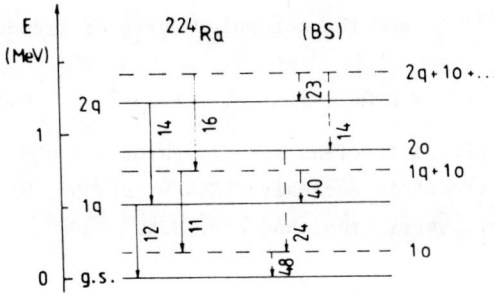

Fig.3. Same as in fig.2, but for the variant (BS) of the potential V.

A noticeable thing is a large value of the B(E3) probability for the transition from 1o state to the ground state. Even in the (Y) case, for which the octupole deformation is of purely dynamical character ($\varepsilon_3^o = 0$), the B(E3) is quite large.

REFERENCES

1) Kurcewicz, W. et al., Nucl.Phys. A356, 15(1981).
2) Merzbacher, E., "Quantum Mechanics", Wiley, New York 1961, ch.5.
3) Gyurkovich, A. et al., Phys.Lett. 105B, 95(1981).
4) Leander, G.A. et al., Nucl.Phys. A388, 452(1982).
5) Nazarewicz, W. et al., Nucl.Phys. A429, 269(1984).
6) Böning, K. et al., Phys.Lett. 161B, 231(1985).
7) Böning, K. et al., Proc. 21st Winter School on Nuclear Physics, Zakopane 1986, in press
8) Nilsson, S.G. et al., Nucl.Phys. A131, 1(1969).
9) Rozmej, P. et al., Proc. 24th Int. Winter Meeting on Nuclear Physics, Bormio 1986, in press
10) Möller, P. and Nix, J.R., Nucl.Phys. A361, 117(1981).
11) Goldhaber, A.S. and Scharff-Goldhaber, G., Phys.Rev. 17, 1171 (1978).

Parity doublets in ^{223}Ra and the octupole degree of freedom

S A Eid*, Ch Briançon[+], W D Hamilton*, C F Liang[+], R J Walen[+] and V Green[†]

*Physics Division, University of Sussex, Brighton BN1 9QH, U.K.
+Laboratoire de Spectrometrie Nucleaire, CSNSM, 91406, Orsay, France
†Daresbury Laboratory, Warrington, WA4 4AD, U.K.

It is recognized that the light actinide nuclei with $N \sim 134\text{-}136$ have particular shape properties and appear soft or slightly unstable with respect to octupole deformation. We have thoroughly studied the level structure of ^{223}Ra which is of special interest since it has 135 neutrons and lies at the centre of this region.

Despite experimental work extending over three decades, there remained inconsistencies in the level structure of this nucleus and inadequacies in the descriptions used to explain this structure. In this paper we report on a set of measurements of the ^{227}Th α-decay which combined α-γ and γ-γ directional correlation experiments, γ-ray directional distribution measurements from oriented ^{227}Th, with γ-γ coincidence measurements and internal conversion electron data [BRI71].

We show that the spin-parity assignment of the ground state is $3/2^+$ which is in disagreement with the previously accepted value $1/2^+$, but in agreement with the recent laser spectroscopy measurements [AHM83].

In figure 1 is illustrated the analysis of our γ-γ angular correlation data for the 286.2 keV [236.0 keV (E1)] 50.2 keV [50. keV (E1)] 0 keV cascade, for which the correlation coefficient is

found to be:

$$a_2 = -0.187 \pm 0.006, \quad a_4 = 0$$

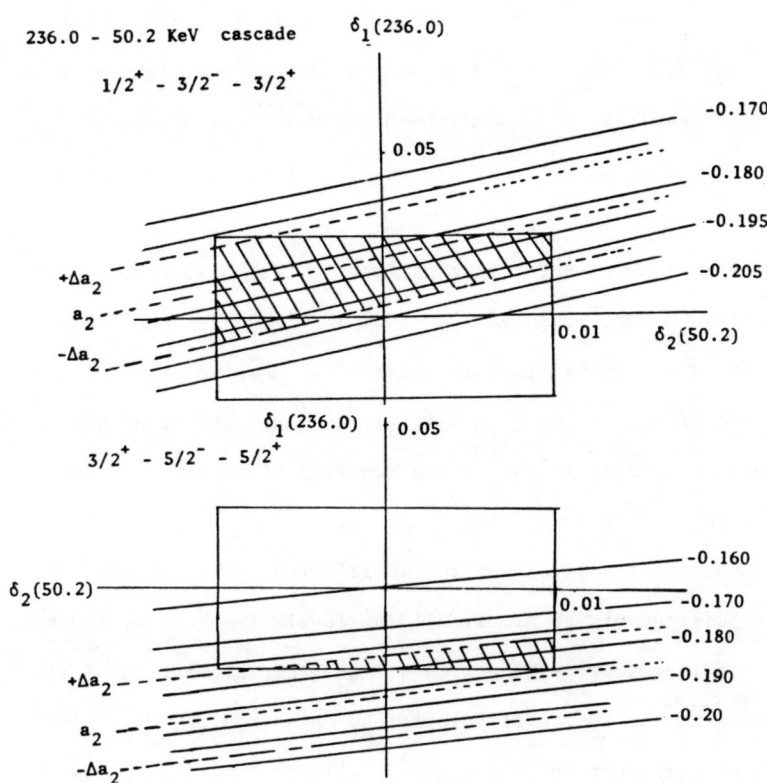

Figure 1

The analysis of the γ-γ angular correlation data for the 236.0 keV - 50.2 keV cascade. From the upper limit of the mixing ratios of the M2 components (resp. δ_1(236.0 keV), δ_2(50.2 keV) and the experimental value of the correlation coefficient $a_2 = -0.187 \pm 0.006$, one may deduce the compatible spin assignments.

Taking into account the upper limit of the mixing ratio of the M2 component in these essentially pure electric dipole transitions, as deduced from the conversion electron measurements (resp. δ_1(236.0 keV) $\leq$ 0.025 , δ_2(50.20 keV) $\leq$ 0.010), one may deduce that the ground state is $3/2^+$. Moreover a $5/2^+$ assignment is ruled out by the α-γ directional correlation data for the 79.8 keV transition as shown in figure 2.

In the nuclear orientation experiments, the anisotropies of 35 γ-ray transitions were measured. An example of the data analysis is presented for a few transitions in figure 2 which gives the total correlation coefficient $a_2 = A_2(\alpha)A_2(\gamma)$ as function of $\Delta'(\alpha) = \Delta(\alpha)/(1 + \Delta^2(\alpha))$ where $\Delta(\alpha)$ is the mixing coefficient of the α-waves with a ground state $3/2^+$ for ^{227}Th.

The consistent analysis of the whole set of data provides unique spin-parity assignments to almost all the levels lying below 450 keV.

The detailed level scheme with all the characteristics of the transitions (intensities, multipolarities, anisotropies ...) and their assignment to different rotational bands will be published elsewhere. In figure 3 is shown a partial level scheme with rotational bands based on two parity doublets as proposed in ref [BRI85] and which is in fair agreement with the recent work of ref [SHE86].

The splitting in energy of these parity doublets as well as the values of the decoupling parameters of the $1/2^{\pm}$ doublet bands reflects the softness of this nucleus with respect to the octupole deformation but it is difficult to conclude that the left-right asymmetry is well stabilized in this nucleus.

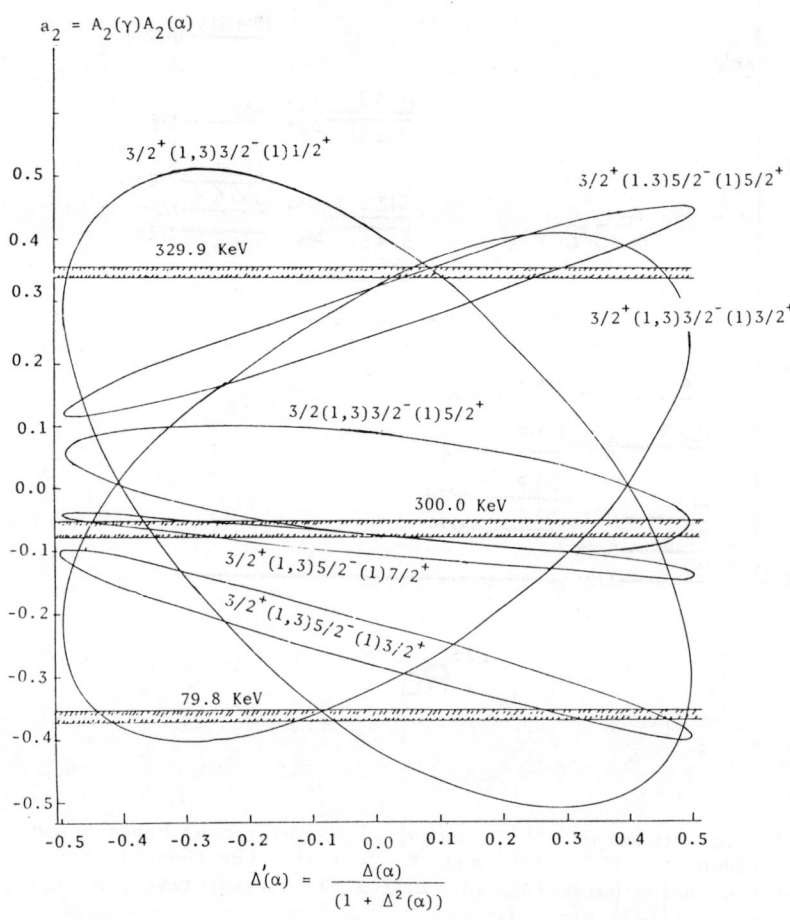

Figure 2

The total correlation coefficient $a_2 = A_2(\alpha)A_2(\gamma)$ as function of the mixing coefficient of the α-waves with a spin $3/2^+$ for ^{227}Th. The experimental values are given as an example for the 79.8, 300.0 and 329.9 keV γ-ray transitions and compared with the calculated ellipses taking into account the $\ell_\alpha = 1, 3$ α-waves and the pure E1 multipolarity of these transitions.

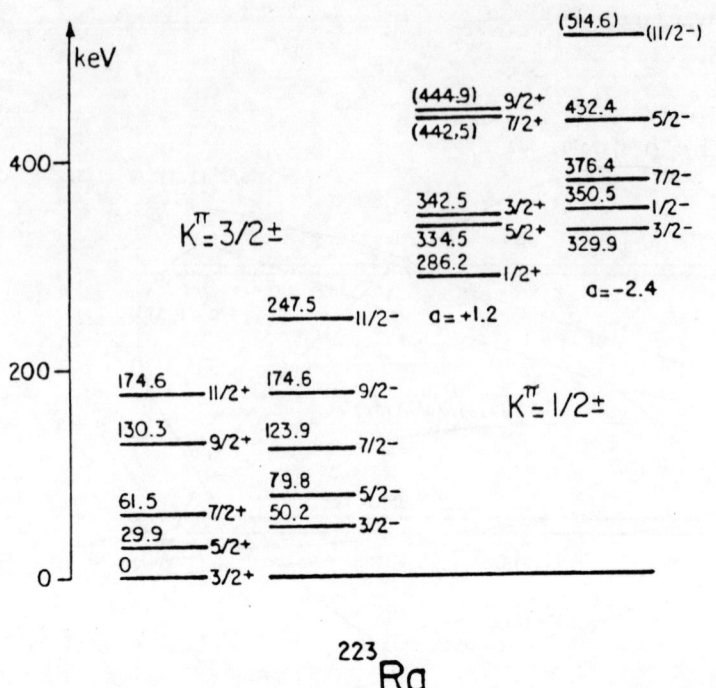

Figure 3

A partial level scheme of ^{223}Ra showing the rotational bands based on the parity-doublets $K^\pi = 3/2^\pm$ and $K^\pi = 1/2^\pm$. For these latter bands the decoupling parameters are indicated as respectively a $(1/2^+)$ = + 1.2 and a $(1/2^-)$ = - 2.4.

References

[BRI71]   Briançon Ch. et al, J. Phys. 32, 373 (1971) and 32, 381 (1971)

[AHM83]   Ahmad S. A. et al., Phys. Lett. 133B, 47 (1983)

[BRI85]   Briançon Ch. et al., CSNSM Activity Report (1983-1984)

[SHE86]   Sheline R. K., Phys. Lett. 166B, 269 (1986)

ROTATIONAL ALIGNMENT AND STABLE OCTUPOLE DEFORMATION

Ch. Briançon and I.N. Mikhailov
C.S.N.S.M. Orsay and J.I.N.R. Dubna

Abstract

The properties of transitional and well-deformed nuclei in the actinide region are discussed in relation with the existence of octupole deformation of the mean field.

The Coriolis effects leading to the alignment of the vibrational angular momentum turn out to be important for the appearance of the octupolly deformed shapes at high spin. As follows from the model developed here, such shapes have neither reflection nor axial symmetry. The onset of such a deformation is shown to occur at moderate spin in nuclei which present already in their ground state a softness with respect to the axial octupole deformation.

The parity and axial symmetry violating type deformation is predicted as a general property of nuclear high spin states.

Introduction

The properties of negative parity states in heavy nuclei at high angular momentum attracted much attention during the last years. This is because some of these nuclei are exceptionally soft with respect to the octupole deformation.

Two important factors determine the softness of nuclei with respect to the deformation of the surface : the bulk properties of nuclei, viewed as those of liquid droplets, and the shell-effects. The first factor determines the general tendencies while the second is responsible for the individuality of each nucleus.

The interplay of these two factors is clearly seen in fig. 1 where the lowest negative parity states in nuclei are systematized. This is essentially the same figure as in ref. [1] to which a few experimental points (see for example ref. [2]) are added. The theoretical estimations of ref. [3] based on the Fermi-liquid droplet model for the excitation energy of octupole phonon states in spherical nuclei on the β-stability line are represented in this figure by the solid curve. One sees that the softness of nuclei indeed increases with increasing A as it must be in the liquid droplets with increasing size and increasing fissioning parameter $X = 49 \, Z^2/A$:

$$\varepsilon_{3-} = 31.4 \sqrt{\frac{7}{4} - X} \, / \, A^{1/2} \tag{1}$$

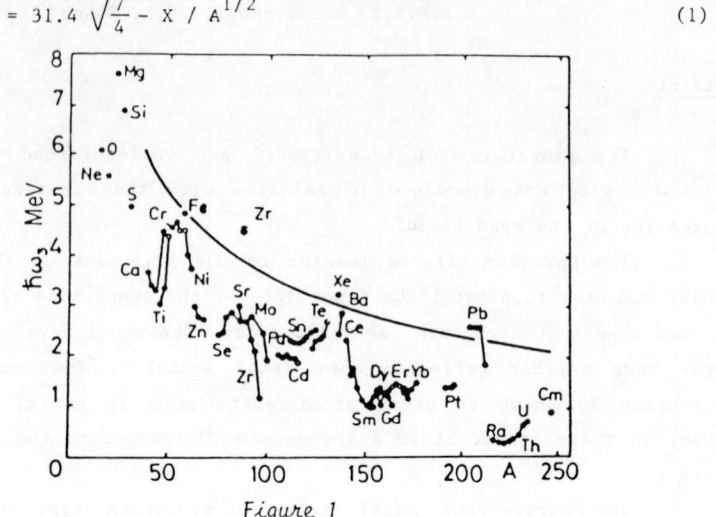

Figure 1

The energy of the lowest negative parity states as function of A. The solid curve shows the theoretical estimations of ref.[3], based on the Fermi-liquid drop model.

One also notices that there are regions of nuclei where the excitation energy becomes exceptionnaly small due to the shell-effects.

One of these regions includes the light Ra, Th isotopes in which the excitation energy of the lowest negative parity state is not much different from the excitation energy of the 2^+ state of the ground band. This feature could give a reason to believe that the ground state of these nuclei is octupolly deformed, as one of the fingerprints for the octupole deformation is the presence of well developped rotational bands including the parity doublets : ...I^+, $(I+1)^{(-)}$, $(I+2)^+$, $(I+3)^{(-)}$,... The experimental situation is shown in Fig. 2 with an example of two nuclei, one of which is soft and the other is one of the most rigid inside this group of nuclei. One sees that the hybridization of parity doublets takes place already at spins of the order of 10 in ^{220}Ra while in the other nuclei (^{232}Th) this part of the spectrum resembles the spectrum of a quadrupolly deformed axial nucleus including the $K^\pi = 0^+$ (ground state) and $K^\pi = 0^-$ (octupole-vibration) rotational bands well separated in energy.

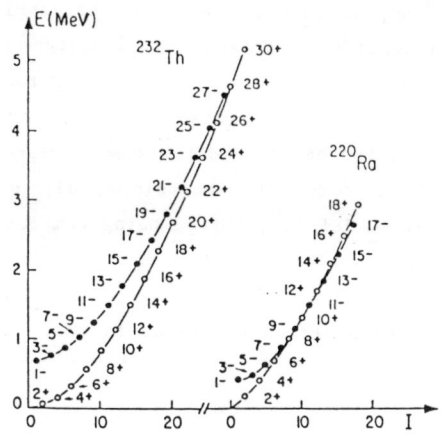

<u>Figure 2</u>
Energy of the nuclear states as function of the spin for ^{232}Th (ref.[4]) and ^{220}Ra (ref.[5]) nuclei.

The recent calculations of the potential energy surfaces [6-8] which have been performed to determine the stability of the ground state with respect to the octupole deformation predict that the even-even nuclei with N ~ 134-136 are indeed soft or slightly unstable with respect to the octupole deformation, but it is difficult to say that the left-right asymmetry is well stabilized in the ground state. There remain still some problems in understanding the effects of the zero point octupole motion to agree completely with the conclusions based on such calculations. These show that the octupole deformation is the feature of a very small island of nuclei, but the experimental data in this region do not fit completely the criteria for the existence of a stable octupole deformation in the ground state, while at high spin the properties of these nuclei present the characteristics (energies and transition probabilities) of such a deformation.

This feature brings about the idea that the rotation may induce and stabilize the octupole deformation as pointed out in ref. [9]. To analyse this possibility one must answer the following questions :

i) What kind of octupolly-deformed shapes could be expected as resulting from the combined effects of internal forces of the nucleon-nucleon interaction and of the centrifugal and Coriolis forces appearing due to the rotation ?

ii) Which criteria may be used to distinguish between the stable deformation in the ground state and the deformation generated by the rotation ?

Both questions demand some modeling of the phenomenon. We may remark that the investigation of stability of rotating bodies with respect to the octupole deformation plays an important role in the theory of gravitating masses. Reaching some critical frequencies of rotation such systems become unstable in respect to the pear-shaped distortions [10]. The same is found for the axially symmetric rotating charged liquid droplets with surface tension but here the unstability in respect to the octupole deformation is preceeded by the unstability in respect to the fission. The analysis of all possible stationary rotating shapes of unaxial nuclei is far from being complete even in the framework of the liquid-droplet model and thus the possibility of stationary octupolly deformed rotating shapes remains open. For the time being, only the simple phenomenological model as described in ref. [9] allow to come to some answer.

Energy spectra.

Consider the Hamiltonian :
$$H = H_{rot}(\vec{R}^2) + H_{intr} \qquad (2)$$
where
$$H_{rot} = E_{core}((\vec{I} - \vec{j})^2) \qquad (3)$$
represents the rotational part of the energy depending on the difference between the total angular momentum $\vec{I}$ and the angular momentum of octupole vibrations $\vec{j}$.

The intrinsic part H_{intr} of the Hamiltonian may be expressed as :
$$H_{intr} = \sum_{K=-3}^{+3} \omega_{|K|} b_K^+ b_K + h \left[\sum_{K=-3}^{+3} b_K^+ b_K\right]^2 \qquad (4)$$
The energies and the corresponding wave-functions of the states may be obtained by solving the equation :
$$(H_{intr} - \omega_{rot} j_x)|\psi> = \varepsilon |\psi> \qquad (5)$$
where
$$\omega_{rot} = \frac{dE_{core}}{d\sqrt{I(I+1)}} \qquad (6)$$
and j_x is the aligned angular momentum defined as
$$j_x = <\psi| \hat{j}_x |\psi> \qquad (7)$$
The energy of the one phonon states may be thus expressed as :
$$E(I) = E_{core}(I) + \varepsilon(I) \qquad (8)$$
where $\varepsilon(I)$ is the effective excitation energy :
$$\varepsilon(I) = <\psi| \sum_K \omega_{|K|} b_K^+ b_K |\psi> - \omega_{rot} j_x + h \qquad (9)$$

When ω_{rot} becomes large, one or several ε solutions of eq. (5) may become negative, and the family of vacuum states is shifted upwards with respect to the same states containing phonon excitations. Which precisely phonon configuration becomes the lowest one depends on the anharmonic properties of the octupole vibrations.

Let us consider as a simple example the case of a degenerate octupole excitation ($\omega_K = \omega$), $(j_x)_{KK'} = (j_x/3) K \delta_{KK'}$, with the anharmonicity determined by the second term of eq.(2). With this choice of the anharmonic part of the Hamiltonian, the effective excitation energy for the multiphonon states is expressed as :

$$\varepsilon = \sum_K n_K (\omega - (j_x/3) K \omega_{rot}) - h (\sum_K n_K)^2 \qquad (10)$$

where $n_K = 0, 1, 2, \ldots$ is the number of phonons with different values of K.

The number of phonons in the two lowest configurations increases stepwise with ω_{rot} and the parities of states corresponding to the two configurations are opposite. At $\omega_{rot} > (\omega + h)/j_x$ the states of the two configurations may be regarded as those of a single rotational band hybridized in parity with the positive and negative partners shifted in respect with each other but remaining close with increasing ω_{rot}.

The onset of such a picture is determined by the vanishing of the effective excitation energy ε of the most aligned phonon. Figure 3 shows such effective excitation energies for some nuclei calculated on the basis of experimental data in the framework of a model similar to the one discussed here and described in ref.(9,11-14). It may be seen that in the well deformed

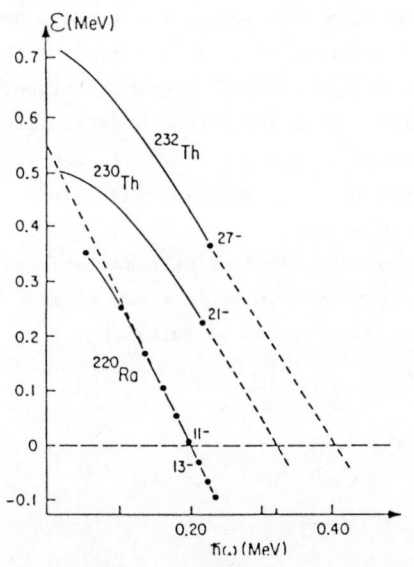

Figure 3

The effective excitation energy as deduced from the experiment for ^{232}Th, ^{230}Th and ^{220}Ra nuclei.

actinides, the effective excitation energy vanishes at spins higher than those presently accessible by the experiment, while in ^{220}Ra this situation occurs already for moderate spin values.

The hybridization of the positive and negative parity bands becomes more clear when expressing the ratio of the energy intervals in the form :
$$R = 2\left[E(I+1)^- - E(I)^+\right] / \left[E(I+2)^+ - E(I)^+\right]$$
In the case of a perfectly hybridized band, this ratio should be = 1. The experimental results shown in fig. 4 indicate however that this ratio is quite large at low spin and tends to 1 at very high rotational frequency in ^{232}Th, while in light radium and thorium nuclei R becomes lower than one and even show a tendency to increase after reaching a minimum.

In fig. 5 are shown together the experimental and calculated ratios R of the energy intervals in ^{220}Ra. The oscillating behaviour of the calculated R around unity is closely related to the change in the number of phonon excitations and may be considered as a mark distinguishing the stable octupole deformation in the ground state from the situation in which the hybridization of parity doublets is related to the changes of intrinsic structure as function of the spin.

The tendency to such oscillations is clearly visible in the experimental curve, although it would be very exciting to get more experimental information at higher spin. This oscillating behaviour of the experimental curve shows clearly that some changes in the intrinsic wave function take place at some critical rotational frequency $\sim (\omega + h)/j_x$.

If one extends the calculations to very high spins, one finds indeed that the alignment of the vibrational angular momentum can really produce the effects on the spectra which are expected in the case of a stable octupole deformation (see fig. 6). One may thus expect that the octupole deformation of the nuclear shape is indeed a rather general property of nuclear high-spin states. Only in some exceptional cases, due to shell-effects, such a deformation is present at already small spin values.

In order to determine precisely to which kind of deformation leads the alignment of the vibrational angular momentum, one needs a new element : a hypothesis concerning the microscopic structure of the operators b_K, b_K^+.

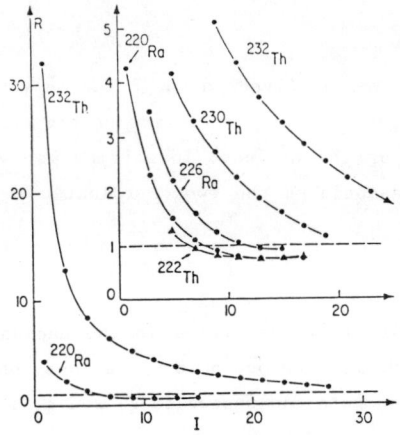

Figure 4

The experimental ratio

$$R = 2 \times \frac{E(I+1)^- - E(I)^+}{E(I+2)^+ - E(I)^+}$$

as function of the spin I

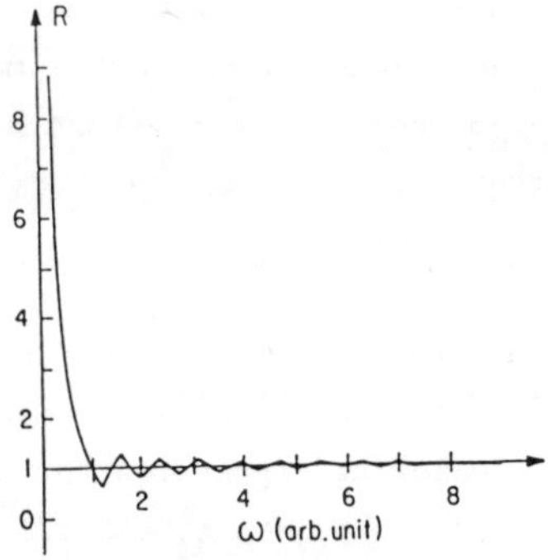

Figure 6

The same ratio R as in fig. 5, calculated for high spin values (0x unit = 0.18 MeV/$\hbar$).

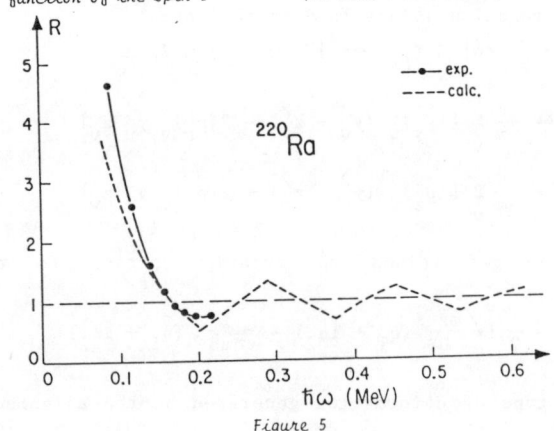

Figure 5

The calculated and experimental ratio $R = 2 \times \dfrac{E(I+1)^- - E(I)^+}{E(I+2)^+ - E(I)^+}$

- The deformation

For the collective coordinates involved in the octupole vibrations of the surface, one may take with a good measure of confidence the components of the octupole moment of the mass of the nucleus. In particular, with the aligned phonon operators there are associated the two coordinates q_1 and q_2 such that :

$$Q_{3,3} = q_1 + iq_2 = \sum_\nu (y_\nu^3 - 3y_\nu z_\nu^2) + i \sum_\nu (3z_\nu y_\nu^2 - y_\nu^3) \tag{11}$$

Assuming that the role of the velocity dependent terms in the nuclear many-body Hamiltonian is not very important for the octupole vibrations, one comes to the following expression for the momenta conjugated to q_1 and q_2 :

$$p_i = B \dot{q}_i = B \, i/\hbar \, [T, q_i] \tag{12}$$

with $T = \sum \vec{p}^2/2m$ being the operator of the kinetic energy of the nucleus. The mass parameter B is given by the following relation :

$$B = \hbar^2 / \langle 0 | [q_i, [T, q_i]] | 0 \rangle \tag{13}$$

where $|0\rangle$ is the many-body wave function of the ground state. Straightforward calculations lead to the formula :

$$B = \frac{m}{9} \langle 0 | \sum_\nu (y_\nu^2 + z_\nu^2)^2 | 0 \rangle = 35m/72 \, AR^4 \tag{14}$$

$$P_1 = \frac{3B}{m} \sum_\nu \{(p_y)_\nu (y_\nu^2 - z_\nu^2) - 2(p_z)_\nu y_\nu z_\nu\}$$

$$P_2 = \frac{3B}{m} \sum_\nu \{(p_z)_\nu (y_\nu^2 - z_\nu^2) + 2(p_y)_\nu y_\nu z_\nu\} \tag{15}$$

With the above definitions, the creation operator for the aligned phonon becomes :

$$b^+ = \frac{1}{2} \{\sqrt{\frac{\omega B}{\hbar}} (q_1 + iq_2) - \frac{i}{\sqrt{\hbar\omega B}} (P_1 + iP_2)\} \tag{16}$$

To find the type of deformation generated by the alignment we consider the coherent state : $b|\lambda\rangle = \lambda |\lambda\rangle$,
with $|\lambda\rangle$ related to the ground state of the nucleus by :

$$|\lambda\rangle = e^{\lambda(b^+ - b)} |0\rangle = e^{i\lambda S} |0\rangle \tag{17}$$

and with the numerical parameter λ chosen so as to minimize the energy, i.e. to obtain the best approximation within the family of coherent states to the true energy of the lowest nuclear state with a given value of the momentum.
Within the simple model defined above :

$$|\lambda|^2 = \begin{cases} 0 & \text{if} \quad \omega_{rot} j_x < \omega + h \\ \\ (\omega_{rot} j_x - \omega - h)/2h & \text{if} \quad \omega_{rot} j_x > \omega + h \end{cases} \tag{18}$$

The phase of λ remains undetermined and we assume it to be a real number. The corresponding effective excitation energy in the coherent state with λ determined by eq.(18) is :

$$\varepsilon(\lambda) = \begin{cases} 0 & \text{if } \omega_{rot} j_x < \omega + h \\ -(\omega + h - \omega_{rot} j_x)^2/4h & \text{if } \omega_{rot} j_x > \omega + h \end{cases} \quad (19)$$

The approximation of the "true" energy so obtained is quite good. Thus it seems reasonable to associate the deformation properties of the coherent states with those of the "intrinsic" states of the system. To define the deformation we recall that the probability to find a nucleon at the point

$$\vec{X}' = e^{i\lambda S} \vec{X} e^{-i\lambda S} \quad (20)$$

in the state $|\lambda\rangle$ is the same as the probability to find a nucleon at the point $\vec{X}$ in the state $|0\rangle$. Accordingly $\vec{X}'$ may be regarded as the position occupied by the same element of the nuclear matter in the state $|\lambda\rangle$ which is located at the point $\vec{X}$ in the ground state $|0\rangle$. If the nuclear surface in the state $|0\rangle$ is determined by the equation $f_o(\vec{X}_s) = 0$, then in the state $|\lambda\rangle$ the surface is given by the equation $f_\lambda(\vec{X}_s) = f_o(\vec{X}'_s(x)) = 0$.

The relation between the coordinates $\vec{X}$ and $\vec{X}'$ is :

$$x' = x$$
$$y' = y - \frac{3\hbar}{m}\sqrt{\frac{B}{\hbar\omega}} \lambda (y^2 - z^2) \quad (21)$$
$$z' = z + \frac{6\hbar}{m}\sqrt{\frac{B}{\hbar\omega}} \lambda y z$$

To the axial deformation of the surface in the ground state with z-symmetry axis, there corresponds the following surface equation describing the nuclear shape in the coherent state $|\lambda\rangle$:

$$R(\lambda;\vec{n}) = R_o \{1 + \beta_2 Y_{2,0}(\vec{n}) + \beta_3 Y_{3,0}(\vec{n}) + \beta_4 Y_{4,0}(\vec{n}) + \ldots$$
$$+ \frac{\tilde{\beta}_3}{\sqrt{2}} \{(Y_{3,3}(\vec{n}) + Y_{3,-3}(\vec{n})\} + 0(\beta_3^2, \beta_3, \beta_2 \ldots)\} \quad (22)$$

In eq.(22), $R(\lambda;\vec{n})$ means as usual the distance from the origin to the surface in the direction defined by the unit vector $\vec{n}$. The deformation parameters β_2 β_3 $\beta_4 \ldots$ describe the ground state configuration, while $\tilde{\beta}_3$ caracterizes the octupole deformation induced by the alignment.

$$\tilde{\beta}_3 = \sqrt{2\pi} \frac{\hbar}{A^{1/2} R} \sqrt{\frac{1}{m\hbar\omega}} \lambda = \frac{13.4}{A^{5/6}} \cdot \frac{\lambda}{\sqrt{\hbar\omega \text{ (MeV)}}} \quad (23)$$

The spherical harmonics $Y(\vec{n})$ appearing as factors of the deformation parameters β_2, β_4 and $Y(\tilde{\vec{n}})$ appearing as factors of $\tilde{\beta}_3$ depend on different angles.

$$Y_{lm}(\vec{n}) = Y_{l,m}(\theta,\phi) \quad \begin{cases} x = R \sin\theta \cos\phi \\ y = R \sin\theta \sin\phi \\ z = R \cos\theta \end{cases} \quad (24)$$

$$Y_{lm}(\tilde{\vec{n}}) = Y_{lm}(\vartheta,\varphi) \quad \begin{cases} x = R \cos\vartheta \\ y = R \sin\vartheta \cos\varphi \\ z = R \sin\vartheta \sin\varphi \end{cases} \quad (25)$$

One sees that the alignment of the vibrational angular momentum creates the octupole deformation of a new type (see fig. 7) which ruins the axial symmetry of the surface.

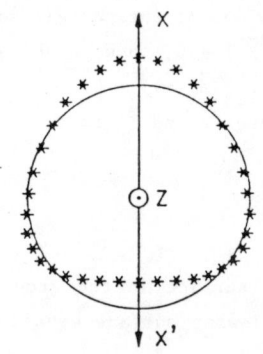

Figure 7
The type of deformation induced by the rotation in ^{220}Ra

The estimate of $\tilde{\beta}_3$ as function of ω_{rot} may be obtained from eq.(18) and (22):

$$\tilde{\beta}_3(\omega_{rot}) = \begin{cases} 0 & \text{if } \omega_{rot} j_x < \omega + h \\ \dfrac{13.4}{\sqrt{2}\, A^{5/6}\, h^{1/2}} (\hbar\omega_{rot} j_x - \hbar\omega - h)^{1/2}, & \text{if } \omega_{rot} j_x > \omega + h \end{cases} \quad (26)$$

One may remark that in the microscopic calculations [7,8], the dependence of the octupole deformation on the frequency of rotation of the mean field is not found as a general feature of the nuclei. But one must notice that this remark applies only to the axial octupole deformation which is not expected to appear as a result of alignment of the intrinsic angular momentum. It would be of interest to analyse microscopically the octupole deformation in rotating nuclei using a parametrization similar to that of eq.(22).

- <u>Electric-dipole transition properties</u>

The stricking difference in the spectra of the two nuclei considered here is accompanied by different electric properties : the reduced probabilities of electric-dipole transitions from the negative-parity states to the states of the ground band are much smaller than one Weisskopf unit, but the retardation of E1 transitions in ^{220}Ra is about two orders of magnitude smaller than in ^{232}Th. This evidence is usually taken as a confirmation of the presence of octupole deformation.

The absolute values of B(E1) factors drew much attention in recent studies. Much less often other properties are discussed : the branching ratio $\rho = B(E1, I^- \to (I-1)^+)/B(E1, I^- \to (I+1)^+)$ for example.

If the K number is conserved (with or without octupole deformation) and if the intrinsic states do not change with I one should have for ρ :

$$\rho_{adiabatic} = \frac{I}{I+1} \sim 1.$$

The experimental data for ρ are shown in fig. 8 for ^{232}Th. The solid line gives the theoretical estimate for ρ based on the calculations of the alignment of angular momentum.

The unadiabatic effects manifest themselves clearly in the case of ^{232}Th while in ^{220}Ra only the transitions $I^- \to (I-1)^+$ are allowed energetically, however the ρ ratio may be estimated assuming that the quadrupole deformation of the core is the same for the states with adjacent values of spin with positive and negative parities. If these conditions are satisfied, then, for $I \gg 1$:

$$\rho \simeq \frac{B(E1, I^- \to (I-1)^+)/B(E2, I^- \to (I-2)^-)}{B(E1, (I+1)^+ \to I^-)/B(E2, (I+1)^+ \to (I-1)^+)}$$

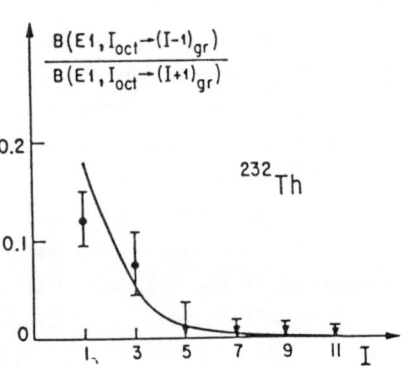

Figure 8

The branching ratio
$$\rho = \frac{B(E1, I^- \to (I-1)^+)}{B(E1, I^- \to (I+1)^+)}$$
of the electric dipole transitions from the octupole band to the ground band in ^{232}Th.

The appropriate experimental data on ^{220}Ra from$^{(5)}$ are summarized in table 1.

Table 1

The branching ratio ρ as function of the spin in ^{220}Ra ($\rho = \frac{a/b}{c/d}$)

I	a/b	c/d	ρ
9	1.2 (2)	-	-
11	1.6 (3)	1.2 (2)	1.3 (4)
13	1.9 (4)	0.8 (2)	2.1 (5)
15	1.3 (3)		
17	2.1 (5)		

These data are too fragmentary to give a clear picture of the changes in the nuclear states as function of the spin, but still show the presence of some changes.

The softness of the actinide nuclei in respect to the octupole deformation favours the γ-decay between negative and positive parity states with the emission of electric dipole γ-quanta. This happens because the octupole deformation in the mass distribution of the nucleus is accompanied by the generation of an electric dipole moment.

In the case of a stable octupole deformation conserving the axial symmetry of the field, the dipole electric moment depends on the deformation parameters in the following way :

$$\langle \eta(E1,0) \rangle = (C_2 \beta_2 + C_4 \beta_4) \beta_3 \qquad (27)$$

The square of $\langle \eta(E1,0) \rangle$ multiplied by the relevant geometrical factor gives the reduced probability of the transitions in the band with the alternating parities. When the K-number is not conserved, the reduced probabilities depend on the decomposition of the intrinsic states into the basic states with definite K-quantum numbers and on the matrix elements of the two operators $\eta(E1,0)$ and $\eta(E1,1)$. If the number of phonons is conserved the two matrix elements :

$\langle K^\pi = 0^- | \eta(E1,0) | gr \rangle$ and $\langle K^\pi = 1^- | \eta(E1,1) | gr \rangle$, with $|K^\pi\rangle$ being the one-phonon states, determine the strength and the branching ratios of the dipole transitions $^{(24)}$.

Estimating the mass parameter for the axial symmetry conserving octupole vibrations in the same way as it was done for the aligned phonons one comes to the following expression for the first of these matrix elements:

$$\langle K^\pi = 0^- | \eta(E1,0) | gr \rangle = \sqrt{2\pi} \frac{\hbar}{A^{1/2} R} \sqrt{\frac{1}{m\hbar\omega}} (C_2 \beta_2 + C_4 \beta_4)$$

$$= \frac{6.7}{A^{5/6}} \cdot \frac{1}{\sqrt{\hbar\omega}} (C_2 \beta_2 + C_4 \beta_4) \qquad (28)$$

with the same coefficients C_2 and C_4 as in eq.(27). The coefficient before the bracket in the last expression determines the fluctuations of β_3 in the vacuum state for phonons. Its numerical value is close to 0.1 for the nucleus ^{220}Ra which in turn is close to the estimate of the octupole deformation parameter of the ground state of this nucleus based on the microscopic calculations. This means that the data on the strength of dipole transitions in ^{220}Ra cannot be taken as a strong argument for the presence of the stable octupole deformation in the ground state. It resembles more a very soft nucleus in respect to the octupole deformation in which the admixtures of states with different numbers of phonons and the changes with spin due to the alignment play equally important roles already at moderate spins.

In fact, the physical processes which are responsible for the appearance of the dipole moment are of different nature, mainly :

i) the volume polarization effect due to the Coulomb forces, which has been discussed as the "lightning-rod" effect [15] which tends to push the protons toward the surface regions of maximum curvature.

ii) the shell-effect originated by the quantal details in the probability to find a nucleon occupying a particular level at a given point in the nucleus. The protons and the neutrons occupy different levels and this brings about a difference in the center of mass of the protons and the neutrons.

The precise balance of these two ways to generate a dipole moment is responsible for the observed variation of the strength of the dipole transitions among different nuclei (see ref.[16]).

CONCLUSION

We have established explicit relations between the alignment of the intrinsic octupole angular momentum and a new type of deformation which does not conserve the axial symmetry.

It seems that such an octupole deformation of the nuclear shape is indeed a general property of nuclear high spin states, whatever may become the axial octupole shape with increasing rotational frequency.

However the nature seems to be reticent to show such properties easily and fortunately there exist shell-effects which allow to study them at low spin but only in some exceptional cases.

One should push forwards both theory and experiments in this field. From the experimental point of view, one may expect much from the new multidetection techniques although one must overcome the experimental difficulties arising from the high rate of fission products in the (HI,xn) reactions in the actinide region.

From the theoretical point of view, the microscopic type calculations should reconcile themselves with the experiment in taking into account the new type of degree of freedom we have put in evidence. There also remain many interesting problems to be solved in the description of the electric dipole transitions. Especially important is the study of the nature of dipole electric polarization effects.

This work has been performed in the framework of the JINR (USSR) - IN2P3 (France) collaborative agreement.

REFERENCES

1) A. Bohr and B.R. Mottelson, "Nuclear Structure", Vol. 2. Ch. P W.A. Benjamin INC (1974)

2) Ch. Briançon and I.N. Mikhaïlov, Proc. of the Int. School on Nuclear Structure, Alushta (1985), p. 245

3) E.B. Balbutsev, I.N. Mikhaïlov and Z. Vaishvila, J.I.N.R.-Dubna, Publ. P4-85-876

4) R.S. Simon, R.P. DeVito, H. Emling, R. Kulessa, Ch. Briançon and A. Lefebvre, Phys. Lett. 108B (1982) 87
and A. Lefebvre, Thèse d'Etat (1984) Orsay

5) A. Celler, Ch. Briançon, J.S. Dionisio, A. Lefebvre, Ch. Vieu, J. Zylicz, R. Kulessa, Ch. Mittag, J. Fernandez-Niello, Ch. Lauterbach, H. Puchta and F. Riess, Nucl. Phys. A432 (1985) 421

6) V.V. Pashkevich, Proc. Int. Conf. on Heavy Ion Physics, Dubna (1983) p. 405

7) N. Nazarewicz, Proceedings Niels Bohr Centemial Conf. (1985), Ed. R. Broglia, G. Hagemann, B. Herskind, North-Holland (1985) p. 263

8) S. Cwiok, Ch. Briançon, J. Dudek and W. Nazarewicz, to be published.

9) I.N. Mikhailov, R.Ch. Safarov, Ph.N. Usmanov and Ch. Briançon, J.I.N.R. Dubna Publ., E4-82-489 and Sov. J. Nucl. Phys. 38 (1983) 297

10) S. Chandrasekhar, "Ellipsoïdal figures of Equilibrium", Yale Univ. Press (1969), Ch. 7

11) I.N. Mikhailov and Ch. Briançon, Dubna Preprint E4-81-402
and Izv. Akad. Nauk. SSSR, Ser. Fiz. 46 (1982) 849

12) Ch. Briançon an I.N. Mikhailov,
Sov. J. Part. Nucl. Phys. 13 (1982) 101

13) I.N. Mikhailov, E.B. Balbutsev and Ch. Briançon, Proc. Int. Summer School on Nuclear Collective Dynamics, Poiana-Brasov, Ed. World Scientific (1982) p.263

14) I.N. Mikhailov and Ch. Briançon,
Proc. 5th Nordic Meeting on Nuclear Physics, Jyväskylä (1984)

15) V. Strutinsky, Atominaya Energiya 4 (1956) 150
and J. Nucl. Energy 4 (1957) 523.

16) G.A. Leander, Proc. Niels Bohr Centennial Conf. (May 1984), Ed. R. Broglia, G. Hagemann, B. Herskind, North-Holland (1985) p.249

This work has been performed in collaboration with I.N. Mikhailov of the Laboratory of Theoretical Physics - JINR, in the framework of the JINR (USSR) - IN2P3 (France) collaborative agreement.

SECTION II

BOSE-FERMI SYMMETRIES
and
SUSYs

D. WARNER

Reviewer

R. CASTEN

Discussion Moderator

Session Moderator

O. Scholten

BOSE FERMI SYMMETRIES AND SUPERSYMMETRY

David D. Warner
Brookhaven National Laboratory
Upton, New York, 11973, USA

ABSTRACT

An overview of current problems and new results and directions in the study of bose-fermi symmetries and supersymmetries is given.

1. INTRODUCTION

The group theoretical approach to nuclear collective structure as represented by the Interacting Boson Approximation[1] (IBA) is centered around the recognition and use of dynamical symmetries as starting points for the development of a collective Hamiltonian. The recognition[2] that such symmetries can be constructed not only in a purely bosonic basis, but also in bose-fermi systems, has greatly broadened the scope of the framework by allowing it to encompass both even-even and odd-even nuclei. These extensions have also allowed the concept of dynamical supersymmetry[3] to be explored, in which the properties of the two types of system are assumed to stem from a single "supergroup" and to be described by a common group chain. Although there are no observables which give direct evidence for the existence of this higher supergroup, the assumption of its existence places rather severe constraints on the form and magnitude of the terms appearing in the Hamiltonians for both even-even and odd-even nuclei.

The empirical evidence for bose-fermi symmetries and the applicability of supersymmetry is most convincing in the neighborhood of the Pt nuclei with A = 194, 196, where the relevant group chain is O(6). This evidence, and the associated theoretical frameworks, are well documented in the literature[4-9] and it is not the purpose of this review to go over this ground again. Rather, the intention here is to highlight some of the underlying questions and problems which need to be considered in applying these methods, and also to mention some of the new directions currently being studied. Limitations implied by space will necessitate a rather brief description of the different aspects but, in many cases, further details can be gleaned from separate contributions to these proceedings.

2. BOSE-FERMI SYMMETRIES

2.1 Basic Ideas and Physical Requirements

The group chain representing a bose-fermi dynamical symmetry will commence from a structure of the form $U_B(m_B) \times U_F(m_F)$ and end with the group Spin(3) whose Casimir operator represents the total angular momentum of the system. Thus at some point in the chain, the boson and fermion systems have to be coupled. Here $m_B(m_F)$ represents the dimensionality of the boson (fermion) space and the current discussion will be limited to the case of $m_B = 6$, representing an s-d boson space, while $m_F = \Sigma_i (2j_i+1)$ will depend on the single particle space assumed for the odd fermion. Within these confines, there are a plethora of possible bose-fermi symmetry schemes already studied and undoubtedly many more possible, albeit of ever increasing complexity. However, initially at least, these represent only algebraic constructs. To consider where, and if, a particular symmetry scheme might be valid, it is most instructive to think in terms of the effective Hamiltonian which each represents. In group theoretical language this corresponds to a sum over the Casimir operators appropriate to the specific group chain, but it is more useful to regroup the various interactions implied by these operators into the more familiar form of a core-particle (boson-fermion) Hamiltonian, viz

$$H_{sym} = H_F + H_B + H_{BF} \tag{1}$$

For a specific symmetry, the form and relative strengths of the three components of H_{sym} will be linked, and thus give rise to very specific constraints on the core structure, single particle energies, and core-particle interaction required.

These ideas are best illustrated by a specific example. A single particle space incorporating $j = 1/2, 3/2,$ and $5/2$ orbits implies $m_F = 12$ and an SU(3) dynamical symmetry chain can be constructed as follows:

$$U_B(6) \times U_F(12) \supset U_B(6) \times U_F(6) \times SU_F(2)$$

$$\supset \genfrac{}{}{0pt}{}{U_{BF}(6) \times SU_F(2)}{SU_B(3) \times SU_F(3) \times SU_F(2)} \supset SU_{BF}(3) \times SU_F(2) \tag{2}$$

$$\supset O_{BF}(3) \times SU_F(2) \supset Spin(3)$$

Equation 2 shows the first two possible levels of coupling between the boson and fermion degrees of freedom, at U(6) or SU(3). Moving the point of B-F coupling down the chain corresponds to a progressively weaker core-particle coupling interaction in the Hamiltonian. For example, coupling at the level of $O_{BF}(3)$ would correspond to an interaction which generates no mixing between core states, since the presence of $O_B(3)$ in H_{sym} would demand that the angular momentum of the core states remain a good quantum number. Coupling at $SU_{BF}(3)$ maintains the SU(3) quantum numbers of the core, and hence implies no mixing between configurations built on the ground band, and on higher (e.g., β,γ) intrinsic excitations. This coupling scheme thus yields the core basis of the Nilsson model. Coupling at U(6) allows, <u>a priori</u>, mixing between all core states in the final odd-A wave function.

$U^{BF}(6)$	$SU^{BF}(3)$	$O^{BF}(3)$	Spin(3)
$[N_1 N_2]$	(λ,μ)	L	J
			——— 5/2
[1,0]	(2,0)	2	
			——— 3/2
[1,0]	(2,0)	0	——— 1/2

Fig. 1   Fermionic quantum numbers in the SU(3) chain.

The core Hamiltonian H_B is invariably that of one of the three limiting symmetries, and hence in this example, SU(3). The fermionic part H_F can be deduced by considering the various groups in the chain and the <u>fermionic</u> quantum numbers appropriate to each. The implicit spherical quasiparticle energy of each orbit is then obtained from the eigenvalue expression in the usual way.

The values of these quantum numbers for the chain (2) are illustrated in fig. 1, along with the resulting single particle energy scheme. For coupling at the level of U(6), the L = 0-2 splitting in fig. 1 is identical to the energy of the first 2^+ state implied by H_B. For coupling at SU(3) however, the additional presence of the operators $C_2[SU^B(3)]$ and $C_2[SU^F(3)]$ would relax this constraint. Nevertheless, it has been shown[10] that, in practice, the $C_2[U^{BF}(6)]$ operator is necessary to obtain the correct ground state structure in the odd W nuclei with A≈185 where the relevant shell model orbits are $p_{1/2}$, $p_{3/2}$, and $f_{5/2}$. Moreover, the same comments apply to the treatment of the O(6) odd Pt nuclei. There are two reasons for this. Firstly, this operator contributes exchange terms to the boson-fermion interaction which serve to adjust the position of the effective Fermi surface in the deformed field[11]. Secondly, the ordering of single particle states shown in fig. 1 is clearly inappropriate for odd W nuclei, and indeed remains so through to the Pt region. A BCS calculation in the spherical limit would demand that the quasiparticle energy of the $p_{1/2}$ orbit lie <u>above</u> that of the $p_{3/2}$, $f_{5/2}$ pair throughout this region. This correction is in fact supplied by an indirect contribution to H_F which arises from the $U^{BF}(6)$ Casimir operator. As pointed out in ref. 12, this operator gives rise to monopole terms of the form

$$(s^+s)^{(0)}(a^+_{1/2}\tilde{a}_{1/2})^{(0)}, \quad (d^+\tilde{d})^{(0)}(a^+_{3/2}\tilde{a}_{3/2})^{(0)}, \quad (d^+\tilde{d})^0(a^+_{5/2}\tilde{a}_{5/2})^{(0)}$$

However, the term involving the s boson can be removed by use of the boson number operator

$$N = (s^+s)^{(0)} + \sqrt{5}\,(d^+\tilde{d})^{(0)} \qquad (3)$$

and this procedure yields an additional contribution to the j = 1/2 single particle energy, proportional to N. Studies[13] of odd A nuclei from ^{185}W to ^{195}Pt reveal the need for a steadily decreasing value of $C_2[U^{BF}(6)]$, consistent with the expected decrease in the j = 1/2 quasiparticle energy.

2.2 Problems and New Directions

The remainder of this section will be devoted to some brief comments on an assortment of new results and outstanding problems related to the study of bose-fermi symmetries.

2.2.1 <u>The quadrupole operator in SU(5)</u>. It has long been recognized that the description of B(E2) data in vibrational nuclei requires the use of a quadrupole operator of the form $(s^+\tilde{d}+d^+s)^{(2)}$. Nevertheless, this choice is in direct contrast to the usual philosophy of the IBA, which dictates the use of a generator of the relevant symmetry structure. However, the generator of SU(5) would be $(d^+\tilde{d})^{(2)}$, and the use of this would predict non-zero quadrupole moments and zero B(E2) values for transitions connecting states differing by one d boson--the inverse of the known empirical situation. This anomaly has direct consequences for the formulation of U(5) bose-fermi symmetries, since the core-particle interaction implicit in this scheme will also involve matrix elements of the boson quadrupole operator which will, of course, be based on the SU(5) generator. Thus in the odd-even wave functions, there will be <u>no mixing</u> of states with different n_d, in contrast to the usual weak coupling treatment, where the dominant mixing effects in the <u>low-lying spectrum</u> are generated by a matrix element proportional to $\sqrt{B(E2:0^+_1 \to 2^+_1)}$. It would seem prudent, therefore, to examine the predictions of such symmetry schemes more closely, particularly with regard to single particle structure and B(E2) values, to ascertain whether the use of the SU(5) generator in the Hamiltonian is indeed appropriate.

2.2.2 <u>O(6); new results</u>. The nucleus ^{195}Pt is probably the best empirical example[14] of an O(6), U(6/12) symmetry chain. Recently a Coulomb excitation study[15] has been performed which allows the E2 selection rules inherent in the predicted scheme to be tested and the

Table I: Values of $B(E2; \text{g.s.} \to J^\pi_F)$ in ^{195}Pt

Level Energy	J^π_f	$B(E2)[e^2b^2]$	Selection Rule a)	Theory 1 b)	2 c)
98.9	$3/2^-$	0.076(12)	σ		0.07
129.8	$5/2^-$	0.198(13)	σ		0.10
199.5	$3/2^-$	0.051(3)	σ, τ		0
211.4	$3/2^-$	0.38(2)	A	0.35	0.35
239.3	$5/2^-$	0.51(2)	A	0.53	0.53
389.1	$5/2^-$	0.020(2)	σ, τ	0	0
419.7	$3/2^-$	0.030(2)	τ	0	0
455.3	$5/2^-$	$<1 \times 10^{-4}$	τ	0	0
524.9	$3/2^-$	0.033(2)	σ, τ	0	0
589.9	$1/2^-, 3/2^-$	$<7 \times 10^{-4}$	σ, τ	0	0

a) A = allowed; σ = σ-forbidden, τ = τ-forbidden.
b) With $e_B = e_F$ 0.11 e.b.
c) With $e_B = -e_F = 0.15$ e.b.

results are summarized in Table I and fig. 2. The selection rule for E2 transitions is $\Delta(\sigma_1\sigma_2\sigma_3) = (0,0,0)$ and $\Delta(\tau_1\tau_2) = (\pm 1,0)$ or $(0,\pm 1)$ meaning that only states with labels (7,0,0) (1,0) should be populated directly from the $J^\pi = 1/2^-$ ground state. These correspond to the levels observed at 211 and 239 keV, and it can be seen from Table 1 that, empirically, they are indeed the most strongly connected to the ground state. However, the population of other levels, in particular those at 99 and 130 keV, suggests a certain degree of symmetry breaking and in fact appears to single out the σ-selection rule as the one most strongly violated. This feature suggests a possible symmetry breaking mechanism which involves relaxing the constraint that the boson and fermion effective charges in the total E2 operator are equal. It is this feature which generates the σ-rule in the O(6) basis, since it leads to an exact <u>cancellation</u> of the boson and fermion E2 matrix elements. The τ-rule, on the other hand, arises because both matrix elements are identically zero. The results of the specific choice of $e_B = -e_F = 0.15$ eb are shown in the last column of Table 1 and also in fig. 2. The change in sign of e_F can be justified by considering the effect of the BCS occupation amplitudes on the single particle E2 operator in this region[7,14].

The results of fig. 2 for the excited states highlight one serious discrepancy in the case of the 525 keV level, where the allowed transitions to the 99 and 130 keV levels are both observed to be weak. This suggests that the assigned quantum numbers of this state are incompatible with its observed E2 decay. In fact, a new study[16] has suggested a re-arrangement of many of the previously adopted associations between theory and experiment in ^{195}Pt, which actually

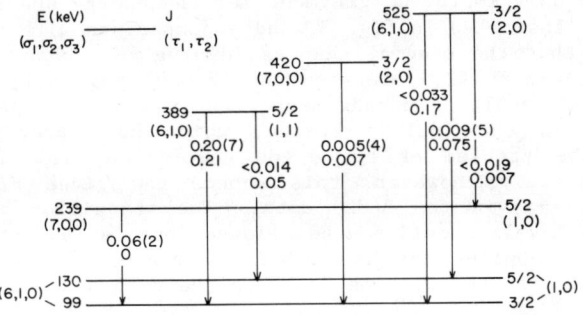

Fig. 2 Experimental (upper) and predicted (lower) B(E2) values between excited states in ^{195}Pt [ref. 14].

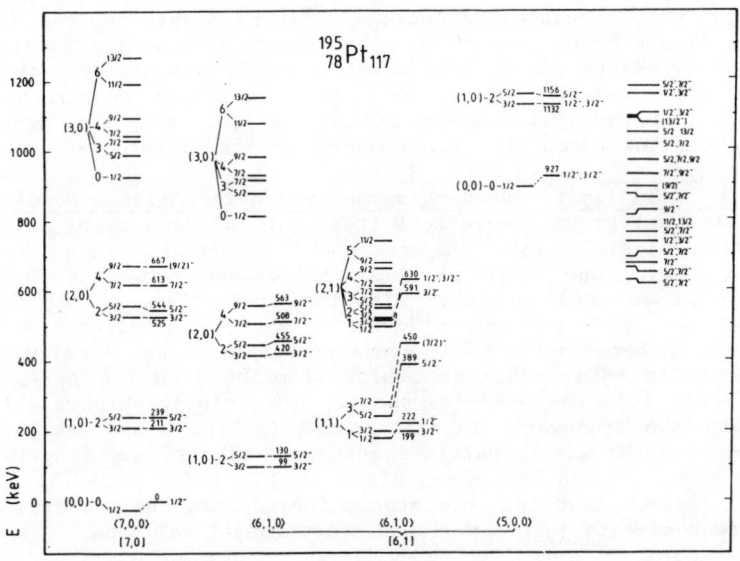

Fig. 3 The revised O(6) symmetry scheme versus ^{195}Pt using the parameterization of Mauthofer et al.[16]

gives a considerably improved agreement for the energy spectrum. This new scheme is illustrated in fig. 3, and a comparison with the earlier version[7] shows that the crucial changes involve an exchange of states with τ quantum numbers (2,0) between the (7,0,0) and (6,1,0) σ representations. The resulting arrangement greatly improves the predicted energy spacing in the (6,1,0) states, which have previously been plagued with the problem of being too compressed, relative to the data. Unfortunately, however, this scheme now gives rise to two discrepancies vis-a-vis the B(E2) data. The transition between the 525 and 239 keV levels should now be allowed, but is in fact weak, and the same comment applies to the 420→99 keV transition. This latter problem in particular throws some doubt on the proposed reassignments.

The other well studied region of O(6) symmetry is represented by the odd proton Ir nuclei. In this case, the symmetry chain stems from a U(6/4) group structure which incorporates a single j = 3/2 orbit in the single particle space. New results from extensive Coulomb excitation studies[17] of ^{191,193}Ir appear to confirm the validity of the E2 selection rules in these nuclei. Another new result of interest in this region is a study[18] by Cizewski and co-workers, who used the very high neutron flux available at the High Flux Reactor of the Institut Laue-Langevin to populate the nucleus ^{195}Ir by double neutron capture on ^{193}Ir. The significance of the ^{195}Ir study lies in the identification of an extensive set of levels which can be associated with the σ = σ_{max}-1 O(6) representation. This gives a direct confirmation of the validity of the O(6) symmetry group itself, in that the pattern of states in the lower representation recurs in the higher one.

2.2.3 <u>The SU(3) limit, pseudo-L symmetry and the Nilsson Model.</u> One problem with any group theoretical treatment is the lack of any obvious physical insight into the predicted structure. For deformed nuclei at least, one route to this type of understanding is via a comparison of an SU(3) symmetry with the predictions of the Nilsson Model and one such comparison[10] has recently demonstrated a clear correspondence between the SU(3) symmetry scheme of eq. 2 and the band structure built on the Nilsson orbits stemming from the $p_{1/2}$, $p_{3/2}$ and $f_{5/2}$ states in the N=82-126 shell. The relationship established between the two frameworks is illustrated in fig. 4 and was obtained by considering the single particle structure of the wave functions in the two bases. A choice of coupling at the level of SU(3) in the chain (2) ensures that the core states included in the two frameworks each corresponded to just the ground state rotational band.

An example of the comparison of single particle amplitudes is given in fig. 5 where the quantity $C_{j\ell}^{eff}$ is compared for the lowest three bands in the SU(3) scheme and the lowest three Nilsson orbits. The $C_{j\ell}^{eff}$ coefficients are defined in the Nilsson basis as

$$C_{j\ell}^{eff} = \sum_i a_i C_{j\ell}^i \qquad (4)$$

where $C_{j\ell}^i$ are the usual spherical amplitudes of the Nilsson basis for the state I = j and a_i are the Coriolis mixing amplitudes. An earlier study by Paar and co-workers[19] of the coupling of a particle to an SU(3) core within the general IBFM framework indicates that Coriolis coupling should be incorporated automatically in the SU(3) scheme. The quantities equivalent to $C_{j\ell}^{eff}$ in the symmetry wave functions correspond, with suitable normalization, simply to the amplitudes for the single particle orbit of spin j coupled to the ground state of the odd A core.

Nilsson $C_{j\ell}^{eff}$ coefficients for both the unperturbed and Coriolis mixed orbits are shown in fig. 5, the latter having been obtained from the calculations of ref. 20 for the odd W nuclei. The connection between the two frameworks is evident. The 1/2[521] and (λ,μ) = (2N+2,0) bands show an almost identical single particle structure throughout the W isotopes, although the absolute values in the SU(3) case are larger because of the neglected contributions from the $f_{7/2}$ and $h_{9/2}$ orbits. In the case of the K = 1/2 and 3/2 bands from the (2N,1) representation, the SU(3) structure agrees with that of the 1/2[510] and 3/2[512] bands, after inclusion of a <u>specific</u> Coriolis interaction which corresponds to that found empirically in ^{185}W. In fact, a comparison of the predicted B(E2) values in the two bases allows the explicit Coriolis mixing included in SU(3) to be deduced numerically, yielding, for the 3/2 members of each band

$$|I=3/2, K=3/2\rangle_{SU(3)} = 0.82|K'=3/2\rangle + 0.57|K'=1/2\rangle$$

$$|I=3/2, K=1/2\rangle_{SU(3)} = -0.57|K'=3/2\rangle + 0.82|K'=1/2\rangle$$

where the K' values on the right refer to the notional unperturbed bands.

A recent (n,γ) study[21] of ^{185,187}W considers the energy spectra and B(E2) values in more detail and reinforces the above conclusions, in particular, the fact that ^{185}W is the best empirical example of an SU(3) boson-fermion symmetry.

2.2.4 <u>Pseudo-L symmetry in deformed nuclei</u>. The discussion of the preceeding section shows that the SU(3) bose-fermi symmetry scheme includes a specific Coriolis interaction and points out that it is possible to numerically estimate its strength from the predicted B(E2) values. A similar approach has been used recently[22] within the framework of a general IBFM Hamiltonian, and seems to indicate a correlation of the effective Coriolis mixing strength with the magnitude of the exchange term. It would obviously be advantageous to obtain a more general and analytic understanding of these effects, and a useful step in this direction can be made[23] by considering a decomposition of an SU(3) bose-fermi Hamiltonian of the general form (2) into intrinsic and rotational degrees of freedom. Note that the step $U_F(12)$ $\supset U_F(6) \times SU_F(2)$ in the group chain involves treating the single

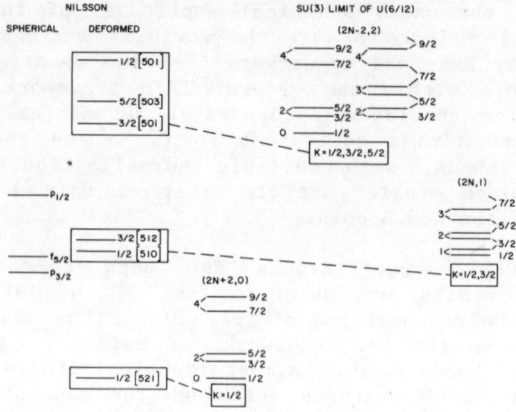

Fig. 4 Relationship between Nilsson orbits and the band structure in a U(6/12) SU(3) scheme. SU(3) states are labelled by L (left), J (right), and (λ,μ) (above).

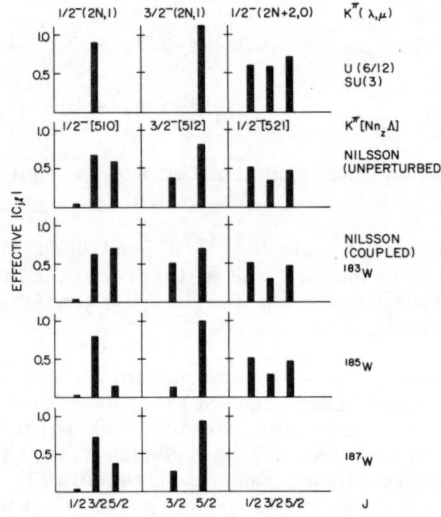

Fig. 5 Single particle amplitudes in the SU(3) and Nilsson schemes. Values for the latter were obtained from ref. 20.

particle space as arising from pseudo-orbital angular momenta $\ell=0,2$ coupled to pseudo-spin $s=1/2$. The use of a pseudo-ℓ decomposition of the single particle space in the treatment of deformed nuclei has, in fact, been studied[24] in some detail in the past. The SU(3) Hamiltonian can now be written as

$$H_{SU(3)} = H_0 + H_{ROT} \tag{6}$$

where H_0 represents the Casimir operators up to and including that for $SU_{BF}(3)$, and

$$H_{ROT} = A_L C_2[O^{BF}(3)] + A_J C_2[Spin(3)] \tag{7}$$

with

$$C_2[O^{BF}(3)] = \vec{L}\cdot\vec{L} \quad ; \quad C_2[Spin(3)] = \vec{J}\cdot\vec{J} \tag{8}$$

Here L represents the total pseudo orbital angular momentum operator and $J = L+s$. The low lying intrinsic configurations characterized by H_0 can be defined as $|(\lambda,\mu)K\sigma\rangle$ where K and σ give the projections of J and s on the symmetry axis. The mixing of intrinsic configurations can then be understood by a treatment entirely analogous to that used in the Nilsson scheme. Thus eq. 7 is rewritten as

$$H_{ROT} = A_L(\vec{J}-\vec{s})^2 + A_J\vec{J}^2$$
$$= (A_L+A_J)\vec{J}^2 + A_L\vec{s}^2 - 2A_L J_3 s_3 - A_L(J_+s_- + J_-s_+) \tag{9}$$

When acting between projected states with good angular momentum, the first three terms of (9) are diagonal, while the last represents a pseudo Coriolis interaction V_{pC} where the operators $j_\pm$ of the Nilsson version have been replaced by $s_\pm$ and the multiplicative constant is A_L rather than $(A_L + A_J)$. It is the latter which represents the inertial parameter of the bosonic core.

The pseudo-Coriolis term will not mix the different intrinsic representations of SU(3). Furthermore, band mixing will be limited to cases in which $\Delta K=1$, and $\Delta\sigma = 1$ and hence $\Delta K_L = 0$ where $K = K_L+\sigma$. It can also give rise to diagonal contributions (i.e., decoupling parameters) for K = 1/2 bands by virtue of the two components with $\sigma = \pm 1/2$ in the projected wave functions. The matrix elements of V_{pC} between projected wave functions take the form, in the large N limit,

$$\langle(\lambda,\upsilon)K+1,1/2;JM|V_{pC}|(\lambda,\mu)K,1/2;JM\rangle = -A_L\sqrt{(J-K)(J+K+1)}\langle\sigma'|s_+|\sigma\rangle \tag{10}$$

where the magnitude of the intrinsic spin matrix element is unity for $\Delta\sigma=1$. The K = 1/2 diagonal matrix elements are given by

$$A_L(-1)^{J+1/2}(J+1/2), \text{ for } K_L=0 \text{ (1/2[521] and 1/2[501] bands)}$$
$$0 \quad , \text{ for } K_L=1 \text{ (1/2[510] band)}.$$

leading in the former case to the energy expression

$$E_{ROT} = (A_L+A_J)J(J+1) + A_L[(-1)^{J+1/2}(J+1/2) + 1/4] \quad (11)$$

This expression is similar, but not identical, to the one which would arise from a Nilsson treatment of the same basis, viz

$$E_{ROT}(\text{Nilsson}) = (A_L+A_J)[J(J+1) - 1/4 + a(-1)^{J+1/2}(J+1/2)] \quad (12)$$

where a is the decoupling parameter, and hence the diagonal matrix element of j_- in the intrinsic basis. Thus a comparison of (11) and (12) shows that the effective decoupling parameter a_{pL} in the pseudo-L treatment is given by

$$a_{pL} = A_L/(A_L+A_J). \quad (13)$$

It now remains to treat the off-diagonal coupling between intrinsic configurations in the K_L = 2 and K_L = 1 bands of fig. 4. In each case, the unperturbed energies of the two bands with K = $K_L \pm 1/2$ can be obtained from the diagonal terms of eq. 9, resulting in an energy separation of $A_L(2K+1)$. A simple two band mixing analysis with the matrix element of eq. 10 then yields the original rotational energy spectrum implied by (7).

Again a similar, but not identical, result would have been obtained from a Nilsson treatment. In that case, the s_- matrix element would be replaced by j_-, and the full inertial parameter, A_L+A_J, would have been used. In the specific SU(3) basis where bose fermi coupling is done at the level of SU(3), the values of the intrinsic matrix elements of j_- and s_- are identical so that the strength of the Coriolis coupling in the pseudo-L scheme is related to that in the Nilsson scheme simply by the factor $A_L/(A_L+A_J)$. Depending on the sign of A_J relative to that of A_L, this will lead to an effective suppression or enhancement of the Coriolis coupling matrix elements. It is, of course, a well-known feature of the Nilsson model that the Coriolis coupling matrix elements required to phenomenologically describe the mixing of bands are usually weaker than those that arise theoretically[25]. The pseudo-L treatment can also incorporate coupling between intrinsic core states, since no adiabatic assumption is made. For example, in the SU(3) description proposed earlier for ^{185}W, it was shown[10] that the $U^{BF}(6)$ scheme, which generates a mixing between the first two SU(3) representations of the core, is required to describe the data. Nevertheless, the description of the rotational degrees of freedom remains unchanged since it depends only on good total pseudo-L quantum number, and not on the assumption of a good K value for the core. The separation of the rotational motion into pseudo orbital and spin contributions represented by the L·L and J·J terms in fact constitutes a treatment of certain additional non-adiabatic interactions which are not considered in the Nilsson framework, and which thus lead to a modification of the effective intrinsic coupling matrix elements.

2.2.5 <u>Transitional regions; the CQF in odd A nuclei</u>. One of the important roles which symmetries play in the general IBA approach is as benchmarks in developing a treatment for regions of transitional structure. An example of this is given by the extension[26] of the Consistent Q Formalism (CQF)[27] to provide a simple description of the changing structure of <u>odd A</u> nuclei between the SU(3) and O(6) limits of U(6/12). In practice, therefore, this offers a possibility to treat the region from ^{185}W to ^{195}Pt and, in particular, the transitional odd-neutron Os nuclei. The framework is constructed by writing the quadrupole operators for both bosons and fermions in terms of a single parameter χ. Thus for the bosons

$$Q_B = (s^+\tilde{d} + d^+s)^{(2)} + (\chi/\sqrt{5})(d^+\tilde{d})^{(2)} \quad (14)$$

while for the fermions, the pseudo-ℓ decomposition of the single particle space allows Q_F to be parameterized in an entirely analogous fashion in terms of fermion operators $G_F(\ell,\ell')$, viz

$$Q_F = G_F(0,2) + G_F(2,0) + (\chi/\sqrt{5}) G_F(2,2) \quad (15)$$

The $G_F(\ell,\ell')$ are defined elsewhere[5,7]. Then the quadrupole interaction contained in the Casimir operator of $SU^{BF}(3)$ is given by

$$Q \cdot Q = (Q_B + Q_F)(Q_B + Q_F) \quad (16)$$

with $\chi = -\sqrt{35}/2$. For $\chi=0$, however, the O(6) symmetry emerges since

$$C_2[SU^{BF}(3)] (\chi=0) \rightarrow C_2[O^{BF}(6)] - C_2[O^{BF}(5)] \quad (17)$$

Note the restriction that the O(6) and O(5) terms must now have the same magnitude. The earlier empirical fits[7] to ^{195}Pt did indeed yield almost equal constants for these two terms.

The simplicity of this approach results in the attractive feature that, as in the even-even version, the wave functions of the odd A Hamiltonian are uniquely determined by χ (and N) so that a number of properties such as relative B(E2) values and relative single particle structure factors depend only on this single parameter, and their behavior across the transition region can therefore be essentially

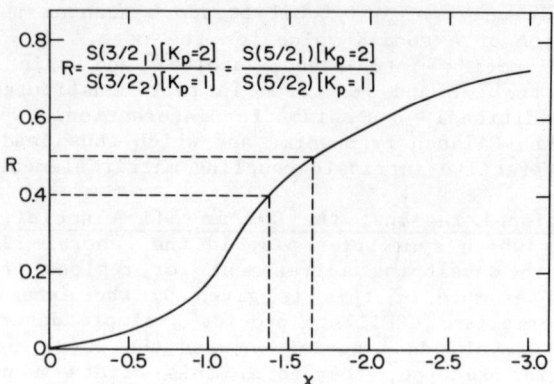

Fig. 6 Predicted ratio of (d,t) structure factors versus χ for N=9 (K_p corresponds to K_L of section 2.2.4). The dashed lines give the empirical range of values in ^{189}Os.

predicted. An example is shown in fig. 6. Also, the other quantum numbers of the group chain, and in particular the pseudo-L symmetry, are preserved throughout. It has been shown[26] that the dominant characteristic of the low lying structure of the odd Os nuclei are reproduced in this framework and a more detailed look at the case of ^{189}Os, following new experimental studies, has been discussed at this conference[28].

2.2.6 <u>The single particle transfer operator.</u> The outstanding unsolved problem in the treatment and testing of bose-fermi symmetries, and indeed in the overall IBFM approach, is the lack of a generally applicable form for the single particle transfer operator. As yet, the treatment of this operator[29] is only valid in near-spherical nuclei. This problem is intimately connected with the development of a microscopic understanding of the boson structure in strongly deformed nuclei, and it is therefore to be hoped that attempts to develop a mapping procedure in such regions, some of which are discussed in these proceedings[30], may also throw some light on the question of the transfer operator.

In practice, the existing uncertainties mean that the single particle structure of IBFM wave functions for deformed nuclei cannot be tested by direct comparison with single particle transfer results since, although the single particle amplitudes are obviously well defined, the effective occupation amplitudes for each state in the deformed field are not. Thus the only comparisons that can currently be made in a meaningful way are of the type shown in fig. 5, where a Nilsson analysis has been used to deduce the empirical single particle amplitudes. Clearly this is less than satisfactory, since the real challenge to the bose-fermi approach lies in predicting features outside the Nilsson basis, such as the coupling with and between intrinsic excitations of the core.

An alternative approach to the problem, which might also prove attractive, is via the core particle coupling method of Donau and Frauendorf[31]. In this framework the quadrupole interaction is treated explicitly within a particle-hole basis defined in the spherical limit, and the resulting deformed odd A wave functions carry both particle and hole amplitudes which give directly the stripping and pick up transfer strengths. A comparison of the results for the energy spectra resulting from this approach and that of a general IBFM Hamiltonian has recently been obtained[22]. An extension of this type of study to encompass the predicted single particle transfer strengths in each framework would therefore be of considerable interest.

3. SUPERSYMMETRY

3.1 Basic Ideas

The introduction of the concept of dynamical supersymmetry involves the postulate of a higher group structure of the form $U(6/12)$ from which a chain such as (2) is assumed to originate. Thus, the chain now begins

$$U(6/12) \supset U_B(6) \times U_F(12) \supset \ldots \ldots \quad (18)$$

The representation labels of $U_B(6)$ and $U_F(12)$ are N_B, the number of bosons and N_F, the number of fermions. Moreover, for $N_F = 0$, the chain (2) reduces to the usual $SU(3)$ chain for the purely bosonic system. The inclusion of the supergroup introduces the additional label N, where

$$N = N_B + N_F \quad (19)$$

Now the treatment of an even-even nucleus with N_B bosons, and an odd-even nucleus with N_B-1 bosons and 1 fermion, etc., can be viewed as the description of different representations of $U_B(6) \times U_F(12)$, contained within the same representation N of $U(6/12)$. The specific Casimir operators of the groups in (18) contribute only to the binding energy of the system and yield a quadratic dependence on N_B which is consistent with other models and, indeed, with the semi-empirical mass formula. However, the other consequence of this approach is that the remainder of the group chain is common to both the even-even and odd-even members of the $U(6/12)$ representation. Thus both must display the properties of the assumed dynamical symmetry, and the energy spectrum of each must be described by the same parameters in the eigenvalue expression. It is important to recognize that these <u>two</u> criteria must be satisfied <u>simultaneously</u> to be consistent with supersymmetry. It is certainly possible to envisage neighboring even-even and odd-even nuclei which both display the properties of a dynamical symmetry, and hence obey selection rules associated with electromagnetic transition and single particle transfer operators, but whose energy spectra cannot be described by a common set of parameters in the eigenvalue expressions.

It is worthwhile to consider the significance and utility of the idea of dynamical supersymmetry in nuclear structure. The bosons entering the problem here are, of course, not fundamental entities but rather represent a mathematical representation of the underlying fermion pairs. Thus the concept of supersymmetry is an algebraic rather than a physical one. It can also be argued that the fact that neighboring even-even and odd-even nuclei can be described by the same Hamiltonian is hardly surprising and only reflects the fact that the addition of a particle has little effect on properties of the core, such as the moment of inertia. However, it is also clear from the discussion of the preceding section that the constraints implied by supersymmetry link the structure of the core with the single particle Hamiltonian and with the strength of the core-particle interaction. The existence of this latter correlation in particular is not so easily dismissed, at least in systems which depart from spherical symmetry.

3.2 Recent Extensions

One problem in addressing questions such as those above by empirical tests is the very limited applicability of supersymmetry schemes which require the existence of exact dynamical symmetries. However, it has recently been pointed out[32] that such a constraint is not an inherent necessity. On the contrary, it is possible, in principle, to extend the concept of supersymmetry to any system where the Hamiltonian can be written as a sum of Casimir operators, even if these operators originate from different dynamical symmetry limits. That is, while the three symmetry chains may mix, they can still be assumed to originate from a single supergroup and hence to describe even-even and odd-even nuclei simultaneously. An example of this procedure is given in these proceedings[33]. The even-even Ru nuclei have been fitted with a Hamiltonian involving a combination of boson Casimir operators from the $U(5)$ and $O(6)$ group chains of $U(6/12)$. The identical Hamiltonian is then used to essentially <u>predict</u> the structure of the odd proton Rh isotopes. Although this specific example is not ideal, in that the Ru and Rh nuclei which are best studied experimentally are essentially vibrational and hence can be most easily described by such an approach, it nevertheless has two extremely important consequences. Firstly, it greatly expands the regions of nuclei which can be used to test the general validity and significance of the supersymmetry framework. Secondly, on a more practical note, it offers a method to generate at least a first approximation to the entire odd-even Hamiltonian simply from fitting neighboring even-even nuclei. Of course, the method still suffers from a limited single particle space, but the role of the $U^{BF}(6)$ Casimir operator in adjusting the $j = 1/2$ single particle energy should nevertheless allow a number of transitional regions to be studied. The most challenging of these would undoubtedly be the odd neutron W-Pt nuclei, and will hopefully be attempted in the near future.

Another example of the use of symmetry and supersymmetry bases has been developed recently by Sunko, Paar and co-workers[34] for deformed nuclei. They suggest that it is always possible to make an at least approximate reduction of the fermion structure $SU^F(2j+1)$ to $SU^F(3)$ and hence generate an SU(3) symmetry of the form of chain (2), with coupling at the level of SU(3). The condition for the existence of the $SU^{BF}(3)$ group then translates into a specific relationship between the $Q_B \cdot Q_B$ and $Q_B \cdot Q_F$ interaction strengths in the odd A Hamiltonian and different signs of the ratio of these strengths yield the strong coupling, and decoupled band structures of the rotational model. This feature emerges because the change in sign corresponds for a single j orbit to moving the effective Fermi surface in the deformed single particle potential from the vicinity of the high K to the low K Nilsson orbits, or vice-versa. The attraction of this approach is that it may expand the region of applicability of the general SU(3) framework by lifting the constraints on the single particle structure. The concept of supersymmetry associated with it seems less important since it only corresponds to an equality of moments of inertia in neighboring even-even and odd-even nuclei.

A further extension[33] of the supersymmetry approach has been discussed at this meeting[33], namely, the combining of odd-proton and odd-neutron group structures to provide a framework which, in principle, can describe a quartet on nuclei: even-even, odd-proton, odd-neutron, and odd-odd[35]. Moreover, the structure of the last is entirely determined by the fit to the first three. This must be recognized as a very ambitious step since it simultaneously links the core structure with the strength and form of the core particle interactions for odd-proton and odd-neutron nuclei, as well as defining the neutron-proton interaction in the odd-odd nucleus. The first empirical test of this idea[36] involved the quartet of ^{196}Pt, ^{197}Pt, ^{197}Au, and ^{198}Au and thus a crucial test of the predictions centers on the odd-odd ^{198}Au. Very recently, a (n,γ) study was undertaken using the technique of Average Resonance Capture (ARC) to study the low-lying negative-parity states in this nucleus[37]. The unique and most valuable feature of the ARC method lies in its ability to identify complete sets of low-lying states within a given range of spin and excitation energy. In this instance, this completeness extended to states with $J^\pi = 0^-$ to 3^- below 1 MeV.

The predicted energy spectrum is compared to the experimental one in fig. 7. It is clear that the agreement is poor, to the extent that the association of theoretical and empirical levels is largely speculative. The most encouraging feature that emerges is the ability to establish, albeit tentatively, a one-to-one correspondence between the two sets of states. This feature may indicate that the neglected s1/2 strength in the proton space does not appreciably affect the low-lying spectrum. The large discrepancies evident in fig. 7 are not really surprising, given the fact that the nuclei ^{197}Pt and ^{197}Au are not the best empirical examples of the O(6) symmetry for U(6/4) and U(6/12). The most promising quartet for this type of study would be that involving ^{194}Pt, ^{195}Pt, ^{195}Au, and ^{196}Au but, as yet, hardly any experimental information is available for the odd-odd member, ^{196}Au. A new experimental investigation is, however, in progress[38].

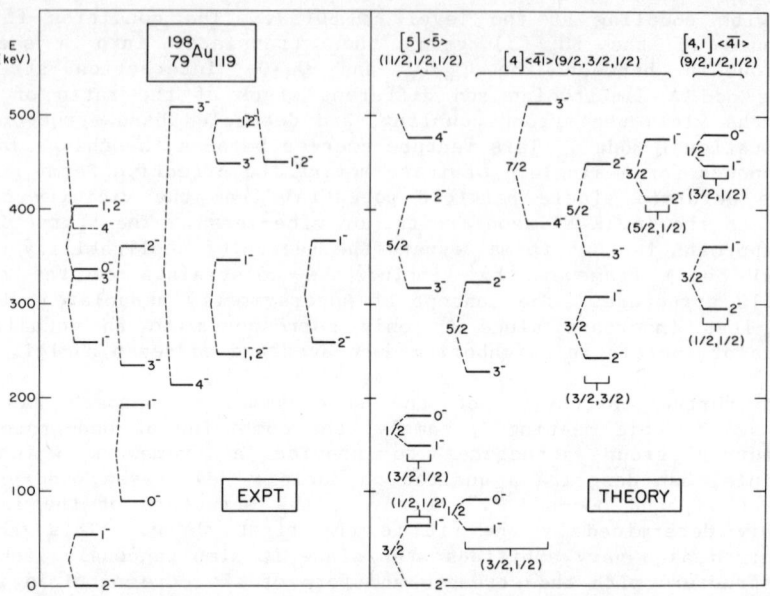

Fig. 7 Empirical and predicted energy spectrum for ^{198}Au. The group labels on the r.h.s. are explained in ref. 36. The experimental states are associated with the theoretical ones simply in order of increasing energy for each spin.

4. CONCLUDING REMARKS

The study of bose-fermi symmetries and supersymmetries has developed rapidly and extensively over the years since its inception. It is by now clear that the quality of the physical descriptions offered by many of these schemes is sufficient to indicate that their basis, though conceptually algebraic, must be founded on some degree of reality. Nevertheless, the purely mathematical origins of a symmetry chain necessitate a careful examination of the underlying physical basis implied by it, and an improved understanding of how to extract this type of information is now emerging. There must surely be a danger that, with the introduction of increasingly complex group structures and coupling schemes, an acceptable description of a nucleus or nuclei may be generated in a framework which is manifestly unphysical. Many of the existing schemes remain severely limited in their overall applicability by virtue of the joint constraints of a limited single particle space and a dynamical symmetry. It is therefore important to attempt to overcome these restrictions by the development of techniques to include entire major shells in the single particle Hamiltonian, and also to break the exact symmetries in a well-defined way.

Many of the studies cited in this review illustrate the potential power of the symmetry-based approach to the collective structure of odd A nuclei. Rather than choosing a specific core-particle interaction and varying its strength to fit the data, the form of the interaction, and in the case of supersymmetry its strength, is determined by the structure of the even-even nucleus alone. Moreover, this capability extends to regions where simplifying assumptions of spherical or axially symmetric potentials are no longer valid. Clearly, the eventual hope is to link both even-even and odd-even Hamiltonians to the underlying shell model basis. While one continuing approach to this problem centers on the development of a mapping procedure to connect the bosonic and fermionic systems, another has emerged recently, based on the recognition of symmetries in the fermionic basis itself. The SO(8) model of Ginocchio[39], and its subsequent extension[40], generates symmetries within a purely shell model basis which, in many cases, correspond to those of the IBA and, moreover, link their empirical observation to specific regions of the nuclear chart. In addition, symmetries appear which do not have counterparts in the IBA, and one example of this, SO(7), is discussed in these proceedings[41].

Finally, the treatment of odd A nuclei within a framework based on dynamical symmetries can be viewed in the broader context of the many other extensions of this approach to very different physical situations. These include molecular systems, scattering problems, dipole and clustering modes, and particle physics, and the common theme in many of these is the repeated occurrence of specific types of motion (vibrational, rotational, etc.) which can be treated in a unified mathematical framework.

ACKNOWLEDGEMENT

Many of the results and ideas described here stem from collaborations and discussions with colleagues, whom I would like to thank. They include R. Bijker, A. M. Bruce, R. F. Casten, A. E. L. Dieperink, D. H. Feng, A. Frank, W. Gelletly, W. R. Philipps, S. Pittel, O. Scholten, H.-Z. Sun, and P. van Isacker. Research supported under contract DE-AC02-76CH00016 with the United States Dept. of Energy.

REFERENCES

1. Arima, A. and Iachello, F., Ann. Phys. (N.Y.) **99**, 253 (1976); **111**, 201 (1978); **123**, 468 (1979).

2. Iachello, F., Phys. Rev. Lett. 44, 772 (1980).

3. Balantekin, A. B., Bars, I., and Iachello, F., Phys. Rev. Lett. 47, 19 (1981); Nucl. Phys. A370, 284 (1981).

4. Balantekin, A. B., Bars, I., Bijker, R., and Iachello, F., Phys. Rev. C27, 1761 (1983).

5. van Isacker, P., Frank, A., and Sun, H.-Z., Ann. Phys. (N.Y.) 157, 183) (1984) and Bijker, R. and Iachello, F., Ann. Phys. (N.Y.) 161, 360 (1985).

6. Hubsch, T., Paar, V., Vretenar, D., Phys. Lett. 151B, 320 (1985).

7. Bijker, R., Ph.D. Thesis, University of Groningen, 1984.

8. Iachello, F. and Kuyucak, S., Ann. Phys. (N.Y.) 136, 19 (1981).

9. Bijker, R. and Kota, V. K. B., Ann. Phys. (N.Y.) 156, 110 (1984).

10. Warner, D. D., Phys. Rev. Lett. 52, 259 (1984) and Warner, D. D. and Bruce, A. M., Phys. Rev. C30, 1066 (1984).

11. Scholten, O. and Warner, D. D., Phys. Lett. 142B, 315 (1984).

12. Bijker, R. and Scholten, O., Phys. Rev. C32, 591 (1985).

13. Bruce, A. M., Ph.D. Thesis, University of Manchester, 1986.

14. Warner, D. D., Casten, R. F., Stelts, M. L., Borner, H. G., and Barreau, G., Phys. Rev. C26, 1921 (1982).

15. Bruce, A. M., Gelletly, W., Lukasiak, J., Phillips, W. R., and Warner, D. D., Phys. Lett. 165B, 43 (1985).

16. Mauthofer, A., Stelzer, K., Gerl, J., Elze, Th. W., Happ, Th., Eckert, G., Faestermann, P., Frank, A., van Isacker, P., to be published.

17. S. J. Mundy, J. Lukasiak, and W. R. Phillips, Nucl. Phys. A426, 144 (1984).

18. Cizewski, J. A., Colvin, G., Borner, H. G., Hoyler, F., and Kerr, S. A., Yale preprint and contribution to this conference.

19. V. Paar, S. Brant, L. F. Canto, G. Leander, and M. Vouk, Nucl. Phys. A378, 41 (1982).

20. Casten, R. F., Kleinheinz, P., Daly, P. T., and Elbek, B., K. Dan Vidensk Selsk Mat. Fys. Medd. 38, No. 13 (1972).

21. Bruce, A. M., Hicks, D., Warner, D. D., Nucl. Phys., to be publ.
22. Semmes, P. B., Leander, G. A., Lewellen, D., and Donau, F., Phys. Rev. C33, 1476 (1986).
23. Frank, A., Pittel, S., Warner, D. D., and Engel, J., to be publ.
24. R. D. Ratna Raju, J. P. Draayer, and K. T. Hecht, Nucl. Phys. A202, 433 (1973).
25. Bohr, A. and Mottelson, B. R. in Nuclear Structure, Vol. II (W. A. Benjamin, New York, 1975).
26. Warner, D. D., van Isacker, P., Jolie, J., and Bruce, A. M., Phys. Rev. Lett. 54, 1365 (1985).
27. Warner, D. D. and Casten, R. F., Phys. Rev. C28, 1798 (1983).
28. Bruce, A. M., contribution to this conference.
29. Scholten, O., Ph.D. Thesis, University of Groningen, 1980.
30. See, for instance, S. Pittel, contribution to this conference.
31. Donau, F. and Frauendorf, S., Phys. Lett. 71B, 253 (1977).
32. Frank, A., van Isacker, P., and Warner, D. D., to be published.
33. van Isacker, P., contribution to this conference.
34. Sunko, D. K. and Paar, V., Phys. Lett. 146B, 279 (1984), and Sunko, D. K., Brant, S., Paar, V., Dadic, I., and Nielsen, H. B., preprint.
35. Bars, I., in Bosons in Nuclei, eds. D. H. Feng, S. Pittel and M. Vallieres (World Scientific, 1984), p. 155.
36. van Isacker, P., Jolie, J., Heyde, K., and Frank, A., Phys. Rev. Lett. 54, 653 (1985).
37. Warner, D. D., Casten, R. F., and Frank, A., to be published.
38. Aprahamian, A., private communication.
39. Ginocchio, J. N., Ann. Phys. (N.Y.) 126, 234 (1980).
40. Wu, G. L., Feng, D. H., Chen, X. G., Chen, J. Q., and Guidry, M. W., Phys. Lett. 168B, 313 (1986).
41. Casten, R. F., Wu, C. L., Feng, D. H., Ginocchio, J. N., and Han, X. L., contribution to this conference.

BOSE-FERMI SYMMETRIES AND SUSY IN NUCLEI

Discussion Session
Following Review Lecture
by
D. Warner

Moderator

RICHARD F. CASTEN
Brookhaven National Laboratory
Upton, Long Island, New York, USA

Moderator: Well, I think we had a very comprehensive talk which should give us a lot to discuss. Are there any questions that directly relate to Dave's talk? We'll postpone the general discussion until after the coffee break.

Talmi: Since Alejandro Frank isn't here, I want to represent him and say that the agreement obtained in his odd-odd nucleus is spectacular. Anyone who thinks it is not should try any theory for an odd-odd nucleus. I think it's remarkable. It's beautiful.

Warner: Alejandro Frank thinks it's bad. I wouldn't have said that if he didn't agree with it.

Donau: I want to comment on the remark concerning the core-particle coupling model. I think one should call it in a way also a boson-fermion coupling model. I would like to remind you that odd-A nuclei are a mean to probe the boson structure, in the sense that the system of equations allows us to treat an odd particle microscopically, and to probe in that way also the empirical description in terms of bosons or anything else, used as input for the core quasiparticle coupling model. Let's consider the profit you have from it. It is a matter of fact that two fundamental pieces of interaction exist, one piece is the pairing force as you can see in the odd-even staggering of the nuclear binding energy. The other piece is particle hole force which is responsible for the collective modes. We are studying in even-even nuclei the quadrupole mode and I think that what at the moment is really seen, it is more or less the coupling to the quadrupole force, not so much the coupling to the pairing modes which in a way is maybe much more complicated than we think now. I remind you there was a talk claiming that pairing is surface-peaked and if you have a vibration then you have to expect not only a vibration in the spatial part, your pairing field will also vibrate in a way. It is possible to implement this into the scheme shown before. The other profit is that you can directly obtain the transfer amplitudes from it. I want to mention also that this approach was now extended to describe two-quasiparticle modes. It includes, let's say the backbending on top of boson modes in transitional nuclei and it involves also the possibility to describe odd-odd nuclei. At the moment it's not applied the possibility to describe odd-odd nuclei. I think that I would be so glad to have so much support from many hands handling this approach then I could show you much more results than available at the moment. Thank you.

Moderator: Let's take a break for coffee, and when we come back, we'll continue the discussion.

Dave has beautifully summarized the status and direction of Bose-Fermi summetries. I would like to just to reiterate some of Dave's points and suggest a few topics for discussion. So far, in this field, most of the comparison with theory has compared energy levels and we have seen many beautiful examples of one-to-one level correspondences, sometimes supported with a few B(E2) values. However, what we really need to check, I think, is the structural correspondence, to make sure these levels really correspond to each other, and that the energy level agreement is not just accidental; for that we need to look at transfer reactions, and more B(E2)'s. This brings up the very important question that Dave mentioned and other people have pointed out: that of the transfer operator. I hope some people may have some comments on that. That it is important can be seen in recent cases where a few B(E2)'s or a few transfer strengths have substantially changed the correspondence between theoretical and experimental levels even though the overall energy level agreement is neither better or worse. So it's clearly sensitive to that question. Also we've seen cases now where several different supergroups have been applied to the same regions, U(6/4) and U(6/20) for example, to the mass 130 region, and so the question of the single-particle spaces and the single-particle energies is an important one. The question of microscopic understanding of the parameters and the interactions, these bose-fermi symmetries is important since it probes the underlying physical basis. And finally there have been some very interesting, what I call "exotic" extensions of bose-fermi symmetry ideas presented at this meeting and which Dave talked about. One is the extension to odd-odd nuclei, another is the generalized SUSY extension that can apply to transition regions, and this is the interesting beta decay calculations of Dobes that were reported yesterday, and probably some others that I've missed.
Who would like to start the discussion?

Sunko, Zagreb: Before the general discussion gets started, I would just like to make a comment on what the speaker has said about some of our work. It has been implied that this supersymmetric limit of IBFM that can accommodate any particle spin is in some way connected with an approximate reduction to an SU(3) group. This is really not true. What we have found out exists intrinsically in the IBFM model and actually anyone here who has an IBFM code can reproduce our results. It is only in order to understand what is going on that we have done some approximate reductions to the SU(3) but even if you do not like that we still have those features. And those features are in there in the model and if anyone can offer a more conservative interpretation than supersymmetry, I'll be quite happy to hear it. They are there and they need to be explained someway. It just shows that there is something in the model that we don't know how it works.

Szpikowski, Poland: I would like to mention that if we would judge the supersymmetry by the number of dimensions our contribution would be the best one because we use the supersymmetry group on boson side of 36 dimensions and on fermion-24 dimensions. But still it's rather simple because we take bosons, s and d, of six dimensions and then to put to the boson degrees of freedom like in the IBM-4 model, spin and isospin degrees of freedom. That gives you 36 dimensions. And on a fermion side we apply this scheme to the region of light nuclei in sd shell where protons and neutrons fill the same levels. Hence s-d shell six spatial dimensions are multiplied by spin and isospin degrees and 6 x 2 x 2 = 24. Although the dimension is rather big and there are many subgroups in the dynamical chain, the contribution to the relative energy of excited levels comes from only three of four Casimir operators because the irreducible representations of the other subgroups are chosen in a unique way. That work is underway and our first contribution was just given in this conference on yesterday session. We plan to extend and to deepen our model.

Stuart Pittel, Bartol Research Foundation: I was very interested in the contribution of Piet Van Isacker with regards to the extensions of supersymmetry away from dynamical supersymmetry limits. One of the points I didn't think was made in the talk was the fact that the analysis as I understood it was restricted to a group such as U(6/12). First of all I'd like to clarify that this is true and ask whether there's any possible way or any possible thoughts as to how one might extend these kinds of ideas to other kinds of supergroups?

Moderator: Any comments on that? Piet?

Van Isacker, Sussex: So indeed it's only confined to U(6/12)'s and about other applications of it for the moment I have no idea. But for instance one example also of deviations from exact symmetries has been explained by Jan Jolie in the sense that he perturbed these single-particle energies. This is another example of deviations from exact dynamical symmetries.

Warner: I think despite the fact that it's only restricted to U(6/12)'s there's still a possible problem that you could tackle and that's the tungsten-osmium-platinum region because U(6/12) is certainly suitable for that, and to look at the osmium nuclei in that framework.

Leviatan, Weizmann Institute: I have one question. You've shown the form of the intrinsic state, when you explained why the x-value of the bosons is equal to the x-value of the fermion. My question is, is there an indication how should the intrinsic states look like when you are away from the dynamical symmetries? In case of SU(3) symmetry one has a similar structure for the fermion and boson parts of the intrinsic state. I guess this will change when you are away from the dynamical symmetry limit.

Warner: The answer is that that was sort of just a suggestion. I don't really know but I thought it was true from the work of Ginocchio and Kirson and so forth on the intrinsic state in even-even nuclei that you can go over for instance from SU(3) to O(6) by changing the beta from group 2 down to 1 or whatever. I would have thought too naively that you could do the same thing with the boson fermion state. I haven't tried to actually work it out. I probably couldn't.

Morrison: I have two comments to make. The first one was that at the student talks one of the talks was about an attempt at calculating within a group theoretical framework the transfer operator in a deformed limit. This technique was also applicable in the light systems where bosons had spin, zero isospin one and also spin one isospin zero. So there has been at least a start in trying to algebraically work these operators in the deformed limit. The second comment I have is that in comparison to what I said in my talk, the microscopic justification seems now to be on the other side. The "beautiful" side of the IBA, in SUSY, is heavily dependent on microscopics. If you look at the underlying BCS justification and Dave Warner brought this up well, the structure of the supersymmetries depends very much where the fermi level is and on occupancies for each shell. Since U2 and V2's can differ drastically, it's such that the interaction between the odd quasiparticle in the core should in some sense change depending on the number of shells and can differ for each shell. Supersymmetries probably have a justification for working better where you have small numbers of shells rather than large numbers of shells because the chances of getting some compatible sets of these, generating direct and exchange contributions consistent with the supersymmetries should diminish with increasing shell number.

Moderator: Stuart, did you want to go back to the other topic?

Stuart Pittel, Bartol Research Foundation: I guess the question was asked about the intrinsic states when you go away from the SU(3) symmetry. The only thing that I did want to add was that to the extent that you maintain the pseudo-L symmetry the intrinsic states that Dave wrote down are precisely the same as the intrinsic states if you had an IBM2 system with N bosons of type 1 and one boson of the other type so you can basically write down the intrinsic states exactly as you would for IBM-2. Of course, as soon as you break the pseudo-L symmetry then you cannot make the separation that Dave did for the fermion. You have to just write down the entire deformed single particle wave function for the odd fermion.

Moderator: More comments, discussion? Is the question of the transfer operator reasonably settled? Is that what people really think? That we now have an idea of how to go about that, or not?

Morrison: My comment is that it's far from settled. But at least some start has been made at this conference.

Moderator: There was a plea by Michel Vergnes earlier today that in order for experimentalists to go on and do much more in this area they really need some theoretical work on how to approach the transfer strengths.

Peter von Brentano, Koln: I just don't understand why you say more work on the transfer operator is needed. Donau always claims that he solved that problem. Why do you say we need more work if the problem is solved?

Moderator: I'm not going to answer that. That may mean I didn't understand either your comment or his.

Vergnes: Apart from the joke, I feel that the question of solving the transfer operator problem and also the other problems which have been discussed this morning is not only necessary but is urgent. Because as Meyer has said at the beginning of the conference, all these boson models have started a sort of "Renaissance" in the field of nuclear spectroscopy. And this was true also, very strikingly for transfer reactions. And this stream has been slowing down now just because, I would say, of the difficulties encountered and therefore people are waiting for doing new transfer experiments for some sort of improvement of the theory, particularly of the transfer operator. And we should do that as fast as possible because as you all know probably: everywhere in the world people are a little foolishly I feel shutting down the machines which presently permit to do this kind of reactions. If we don't do something rapidly, when the problem will be solved, there will possibly be no machine left to do more tests.

Moderator: Any comment from funding agencies?

Draayer, Louisiana State: I think it's important to point out that the problem with the transfer operator arises because of the boson assumption. There are models around which treat nuclei as fermions in which case there is no trouble handling the transfer operator. It is true that these fermion methods don't work in the transition regions, but if you're interested in looking at strongly deformed nuclei they do work.

Iachello: I wanted again to come back to the question of transfer operators because this is indeed an important point. But the important point is not so much what Gerry just said because in deformed nuclei we can use the Nilsson model and that works very well. What we do not understand is how to compute the transfer intensities in transitional nuclei. And that is where we need correct forms of operators. Some work has been done but so far higher order terms have not been tested. Here is, for example, Vladimir Paar who has stressed several times the importance of higher order terms. What is needed now is that some calculations are done in order to test the higher order terms. That's one possible approach. The other possible approach is to use Fritz Donau's core and to see what happens by putting an IBA core and coupling to that the odd particle. Again, this would be very interesting. So we have two alternatives, one is the usual IBFM with higher order terms and the other is the Donau code with explicit 2 x 2 matrices shown here before. But the crucial point is transitional nuclei because we know to compute very well transfer in spherical nuclei and in strongly deformed nuclei. We don't need another model for that. We could do just with the Nilsson model and the spherical shell model. Where we need it is for transitional nuclei.

Moderator: Fritz Donau has a comment, or maybe even a viewgraph it looks like.

Donau: I wanted to present a transparency (no figure submitted for publication...Ed.) to show you the quality for a typical transitional nucleus. Forget about the left hand side. It's the usual core particle coupling model. But you see the experiment in the middle and the result of the 2 x 2 matrix problem on the right hand side. You have also the lifetime of the first excited state. You can explain the band structures quite nice. This is not the only example. We have three nuclei. In that case it's 123, 125, and 127 xenon. Unfortunately, I can't say something about transfer amplitudes because they are not measured here and not calculated. But there was a recent publication in Nuclear Physics in particular made to calculate transfer amplitudes. And it turned out that if you compare to the IBFA approach which can describe reasonably energies and maybe B(E2) too. But not the transfer amplitudes as good as this model can do it. I think we should really as Franco Iachello said span our forces together, i.e. to apply here the IBA method to describe the core properties and to couple a particle in a microscopical manner to the core.

Scholten: I would like to make a last comment -- no it's not the last probably -- one comment on the single-particle transfer operator. People are complaining about the single-particle transfer operator. For example, that the normalization is not correct. There exists a program which has no input parameters, only wave functions from the IBFA model. It calculates single-particle transfer amplitudes using the two terms that are originally proposed and normalization conditions are automatically taken care of. I feel most people simply stick to the very first term in the expression of the single-particle operator and don't bother about calculating the other terms. I don't really know why, but it's almost as easy. You will have a better chance to get agreement with experiment. As far as the comparison between let's say the Donau approach and the IBFA approach, I think none of the two really works perfectly. Each of them has problems. For example, in the iridium region Blasi et al. have tried to calculate the spectroscopic factors from the odd to the even. There are problems in the IBFA and also in the core quasiparticle model. It's good to reinvestigate what is giving the problem.

Moderator: Have you done any calculations with both terms?

Scholten: Yes, and, for example, in the case of the transfer from iridium to platinum I can quite easily get something that follows experiment, I mean an equal amplitude to both 2^+ states. There are some other things that I am not really happy with, but at least that feature is not a big problem.

Warner: I think the reason people have not used the full transfer operator is because most of us understood that it wasn't supposed to be valid except in regions near closed shells, near vibrational nuclei. Now if you're saying in fact that it should work in transitional nuclei I'm sure there are a lot of people who would start trying to use it.

Scholten: I don't say it should work, but it is certainly a much better approximation than taking only a first term.

Draayer: To follow up on my earlier comment, I want to emphasize that I did not mean to imply a Nilsson model treatment for the core. I meant to imply a theory where the core is treated in terms of a fermion picture. Within such a framework one can also treat two-particle transfer and we've heard about some nice experimental results on that subject at this conference.

Moderator: Franco, you have a comment?

Iachello: My comment is still on the one nucleon transfer but from a different point of view. One other important point I think is that the since the one nucleon transfer is either a proton or a neutron that using the corresponding of IBA-2 which is called IBFA-2 is probably better. However, some calculations have been done with the first order term in IBFA-2 and again they are not very good. So really you must go to higher order terms but certainly the proton neutron degrees of freedom in one nucleon transfer is very important. Any collective model with only one type of particles will always have difficulties in trying to describe the one nucleon transfer. And the second point is that the situation is similar for Ml transitions. There has been some discussion of Ml transitions. The Ml transitions very crucially depend on whether the odd particle is a proton or a neutron. It depends on details of the single-particle levels. So one cannot hope to get the Ml transitions right with just average properties. So in a way one should not overdo it one should not overestimate the ability of a certain model to do certain things. One has to understand the limitations of that model. The symmetry concepts in my opinion are very important because they provide a first-order understanding of the situation. But if one really wants to go and compare with the experimental data in detail one must go beyond the symmetry concept, especially for those properties which do not depend on global features but depend on details of the wave functions. And among these are certainly M1, one nuclear transfer and so on. So this is just a word of caution. I mean one can try but I think that the basic ingredients have to be put in before we can get some good agreement with the experiment.

Sunko, Zagreb: For someone who is not working right now in transfer reactions, the reason why this doesn't work in the transition region very well should be almost obvious because the transitional region is by definition the region in which mean field doesn't work very well. So the question is how do we go beyond the mean field and boson expansion is obviously one way. But for example the solid state people have other ways on their shelves, they have renormalization group, they have parquet expansions. So maybe we should just try to do it some other way. And in this context the connection between a zero coupled pair and an s boson just looks like a zeroth-order boson expansion. We should maybe go to higher orders.

Moderator: Any more comments?

Draayer, Louisiana State: I was glad to hear Franco comment on M1's. I think it's important to point out that at the UNISOR facility of the Holifield accelerator at Oak Ridge they're planning on doing some in-beam magnetic moment measurements. This should prove to be very important in determining the structure of these states because as Fraundorf showed almost ten years ago, the magnetic moments are extremely sensitive to the single-particle character of these odd-A nuclei.

Moderator: Other comments? Discussions? Questions? I think we should end this session and thank Dave Warner again.

Experimental evidences for Boson-Fermion symmetries and Supersymmetries in Nuclear Physics

Jean Vervier,

Institut de Physique Nucléaire,
Université Catholique de Louvain,
2, chemin du Cyclotron, B-1348 Louvain-la-Neuve
BELGIUM

ABSTRACT

The predictions of the Boson-Fermion symmetries and Supersymmetries developed so far are compared with the available experimental data. While these symmetries are generally successful for "collective-like" properties-level excitation energies, electric quadrupole properties, two-nucleon transfer reactions with "inactive" nucleons -, they face problems for "single-particle-like" properties-magnetic dipole properties, one-nucleon transfer reactions, two-nucleon transfer reactions with "active" nucleons -.

1 Introduction

This paper presents the main conclusions of a more extensive review article[1], which deals with a systematic comparison between the predictions of the Boson-Fermion (BF) symmetries and Supersymmetries developed so far, and the available experimental data on level excitation energies, electric-quadrupole and magnetic-dipole properties, one- and two-nucleon transfer reactions. It only gives a brief description of the theoretical developments of these symmetries, referring to the original papers or to other review articles for further details. It is restricted to pure BF symmetries and Supersymmetries, without distinction between proton and neutron bosons, and to even- even and odd-A nuclei only. It does not cover the intermediate situations between the considered symmetries, the IBM-2 and IBFM-2, and recent developments on odd-odd nuclei.

2 Theoretical developments of the Boson-Fermion symmetries and Supersymmetries

The BF symmetries are dynamical symmetries of the Interacting Boson Fermion Model (IBFM) which aims at a description of odd-A nuclei. They rest on the hypotheses that the even-even cores of the odd-A nuclei can be described by one of the 3 dynamical symmetries of the Interacting Boson Model (IBM), and that the odd nucleon can be restricted to 1, 2, 3 or 4 single-particle orbits with angular momenta j = 1/2, 3/2, 5/2, 7/2. Those cases which have been developed so far are listed in Table 1, together with those nuclei to which they have been applied. The most extensive comparisons between experiment and calculation have been carried out : in the Ru, Rh, Pd, Ag isotopes with $SU(5) \times 1/2 \supset Spin(3)$; in the Os, Ir, Pt, Au isotopes, with $O(6) \times 3/2 \supset Spin(6)$; in the even-even and odd-A Pt isotopes, with $O(6) \times 1/2, 3/2, 5/2$.

The Supersymmetries aim at a unified description of even-even and odd-A nuclei, which are associated together in supermultiplets of nuclei. The parameters needed to describe the excitation energies and the other level properties should be the same for all nuclei of the same supermultiplets. The number of bosons changes within the supermultiplets so as to conserve the total number N + M of bosons (N) and fermions (M) ; this implies that N is one unit smaller in the odd-A nuclei (M=1) than in the even-even nuclei (M=0). The supersymmetries are associated to supergroups $U\{6/[\sum_j (2j+1)]\}$.

The predicted level energies are functions of various parameters, which can be determined by least squares fitting theoretical states to experimental ones. The other observables also depend on parameters, but it is often possible to make parameter-free predictions on them : there are selection rules, which are associated with the quantum numbers which characterize the various symmetries ; the ratios between those transitions which are allowed by these selections rules often reduce to ratios between "geometrical" coefficients and are thus parameter-free.

3 Comparison with experiment

The main conclusions which can be drawn from an extensive comparison between the available experimental data and the predictions of the BF symmetries and supersymmetries are the following[1]. "Collective-like" properties, i.e. level excitation energies, electric quadrupole properties (quadrupole moments and B(E2)'s) and two-nucleon transfer reactions with "inactive" nucleons (i.e. those in even number for odd-A nuclei) are generally rather well reproduced by the considered symmetries. The latter however face major problems for "single-particle-like" properties, i.e. magnetic dipole properties (magnetic moments and B(M1)'s), one-nucleon transfer reactions and two-nucleon transfer reactions with "active" nucleons (i.e. those in odd number for odd-A nuclei).

These conclusions may be illustrated by the following examples. The levels of ^{191}Ir, as studied by many different spectroscopic tools, can be grouped into multiplets characterized by the quantum numbers σ_1 and τ_1 of the $O(6) \times 3/2 \supset Spin(6)$ symmetry[2]; the excitation energies of 20 of them, assigned to the $\sigma_1 = 17/2$, $\tau_1 = 1/2, 3/2, 5/2, 7/2, 9/2, 11/2$ multiplets, can be fitted with 2 parameters, with an average percentage deviation : $\phi = 13\%$. The levels of ^{195}Pt, also investigated by various methods, display many $J = L \pm 1/2$ doublets, with the values of L and of the other quantum numbers predicted by the $O(6) \times 1/2, 3/2, 5/2$ symmetry[3]; the energies of 18 levels, fitted with 4 parameters, yield : $\phi = 10\%$. The supersymmetric descriptions of the 2 pairs of nuclei $^{190}Os - ^{191}Ir$ and $^{194,195}Pt$, which are associated together in supermultiplets of $U(6/4)$ and $U(6/12)$, respectively, are also reasonably successful, in the sense that the descriptions of the even-even and odd-A nuclei, separately and together, yield reasonably similar values for the parameters involved and the average percentage deviation ϕ.

The experimental values of B E2)'s in the nuclei [$^{103}Rh, ^{107,109}Ag$] and [$^{191,193}Ir, ^{197}Au$] can be compared with the predictions of the corresponding symmetries, $U(6/2) \supset SU(5) \times 1/2 \supset Spin$ 3) [ref.[4]] and $U(6/4) \supset O(6) \times 3/2 \supset Spin(6)$ [ref.[2]], respectively ; this can be carried out, in a parameter-free way, by adopting the boson effective charge e_b deduced from $B(E2, 2_1^+ - 0_1^+)$ in their even-even supersymmetric partners, [$^{102}Ru, ^{106,108}Pd$] and [$^{190,192}Os, ^{196}Pt$], respectively, and by taking into account the decrease of N noticed before between the even-even and odd-A nuclei. The comparison is generally quite successful for [$^{103}Rh, ^{107,109}Ag$] and [^{197}Au], but not at all for [$^{191,193}Ir$]. In the former case, the B(E2)'s between the $n_d = 1$ and $n_d = 0$, $n_d = 2$ and $n_d = 1$ levels (for $^{103}Rh, ^{107,109}Ag$), and between the $\tau_1 = 3/2$ and $\tau_1 = 1/2$, $\tau_1 = 5/2$ and $\tau_1 = 3/2$ levels (for ^{197}Au), are nicely reproduced, sometimes within the small ($\leq 5\%$) experimental uncertainties[5]. The selection rules $\Delta n_d \leq 1$ (for $^{103}Rh, ^{107,109}Ag$), and $\Delta\sigma_1 = 0$, $\Delta\tau_1 \leq 1$ (for ^{197}Au), are also well obeyed in the experimental data, except for the $3/2_2^+ - 3/2_1^+$ transition in ^{197}Au.

The two-nucleon transfer reactions with "inactive" nucleons, i.e. (p,t) and (t,p) on even-even and odd-Z targets, with an orbital angular momentum transfer L=0, mainly populate the ground-states of the final nuclei, in good overall agreement with the predictions of the considered symmetries, but also of other nuclear models.

The experimental values of magnetic moments μ and B(M1)'s in the nuclei [$^{103}Rh, ^{107,109}Ag$] and [$^{191,193}Ir, ^{197}Au$] can be compared with the predictions of the above-mentioned symmetries, which are functions of 2 parameters, the boson and fermion gyromagnetic ratios, g_B and g_F respectively ; this can be carried out, in a parameter-free way, by adopting the values of g_B deduced from $\mu(2_1^+)$ in their even-even supersymmetric partners, and of g_F deduced from the ground-state magnetic moment of the odd-A nuclei. The comparison display large discrepancies in the 2 regions of the periodic table, especially for $\mu(5/2_1^-)$ and $B(M1, 5/2_1^- - 3/2_1^-)$ in

[$^{103}Rh, ^{109}Ag$], $\mu(1/2_1^+)$ and the B(M1)'s in [$^{191,193}Ir, ^{197}Au$].

The experimental spectroscopic factors for one-proton transfer reactions, stripping and pick-up, leading to or starting from ^{193}Ir and ^{197}Au, can be compared with the predictions of the $U(6/4) \supset O(6) \times 3/2 \supset Spin(6)$ symmetry, in a parameter-free way, by normalizing the spectroscopic factors to the ground-state transitions. The following conclusions can be drawn. While this symmetry is reasonably successful for ^{193}Ir, except for the $^{193}Ir(^3He, d)^{194}Pt(2_2^+)$ transition which is strong and should be "forbidden", there are large discrepancies for ^{197}Au, in particular the strong intensities of the "forbidden" transitions to 2_2^+ in $^{196}Pt, ^{198}Hg$. The experimental spectroscopic factors for the one-neutron transfer reactions, (d,p) and (p,d), leading to or starting from ^{195}Pt, can be compared with the predictions of the $U(6/12) \supset O(6) \times 1/2, 3/2, 5/2$ symmetry. There is a good agreement for the reactions $^{194}Pt(d,p)^{195}Pt$ and $^{196}Pt(p,d)^{195}Pt$, with however 5 and 3 fitted parameters, respectively ; there are large discrepancies for the reactions $^{195}Pt(p,d)^{194}Pt$, especially for 2_2^+, and $^{195}Pt(d,p)^{196}Pt$, for 2_2^+ and other levels[6].

The experimental cross sections for two-nucleon transfer reactions with "active" nucleons, i.e. $(^3He, n)$ on $^{106,108}Pd$ and $^{107,109}Ag$, (p,t) and (t,p) on $^{194,195}Pt$, can be compared with the predictions of the $U(6/2)$ and $U(6/12)$ supersymmetries, respectively ; this can be carried out, in a parameter-free way, by computing the ratios between the ground-state to ground-state L=0 transitions in the odd-A and even-even members of the supermultiplets, and by taking into account the decrease of N noticed before. The comparison shows that these supersymmetries do not reproduce the large decrease of these cross sections between even-even and odd-A nuclei, the discrepancies being about factors of 2 [ref.[7]].

4 Summary and conclusions

The successes and failures of the symmetries considered in the present paper suggest that the BF symmetries represent good starting points to study the spectra of odd-A nuclei, as the IBM dynamical symmetries are for even-even nuclei. The supersymmetries, on the other hand, allow to relate the spectra of neighbouring even-even and odd-A nuclei, at least for what concerns the excitation energies and the electric quadrupole properties. Possible ways of improving the description of these nuclei are also suggested by the problems encountered : distinction between proton and neutron bosons (IBM2 and IBFM2), departures from pure IBM and BF symmetries, use of other operators in the observables considered ... In any case, these symmetries have played - and will hopefully continue to do so - an important role in the spectroscopy of odd-A non-deformed nuclei, by allowing an approximate description of their spectra, a classification of their levels into various multiplets, possible reasons for the apparent selection rules in their feedings and decays, and, more importantly, by suggesting new experiments to test their predictions. This is the main role of any nuclear model.

References

1. Vervier, J., Revista del Nuovo Cimento (to be published).

2. Iachello, F., Phys. Rev. Lett. 44, 772 (1980) ; Iachello, F. and Kuyucak, S., Ann. Phys. (N.Y.) 136, 19 (1981) ; Balantekin, A.B., Bars, I. and Iachello, F., Phys. Rev. Lett. 47, 19 (1981), Nucl. Phys. A370, 384 (1981).

3. Van Isacker, P., Frank, A. and Sun H.Z., Ann. Phys. (N.Y.) 157, 183 (1984) ; Bijker, R. and Iachello, F., Ann. Phys. (N.Y.) 161, 360 (1985).

4. Bijker, R. and Kota, V.K.B., Ann. Phys. (N.Y.) 156, 110 (1984) ; Vervier, J. and Janssens, R.V.F., Phys. Lett. 108B, 1 (1982).

5. Vervier, J., Holzmann, R., Janssens, R.V.F., Loiselet, M. and Van Hove, M.A., Phys. Lett. 105B, 343 (1981) ; Loiselet, M., Holzmann, R., Van Hove, M.A. and Vervier, J., Phys. Lett. 146B, 187 (1984) ; Loiselet, M., Naviliat, O., Holzmann, R. and Vervier, J. (to be published).

6. Vergnes, M., Berrier-Roisin, G. and Bijker, R., Phys. Rev. C28, 360 (1983) ; Vergnes, M., Berrier-Roisin, G., Rotbard, G., Vernotte, J., Maison, J.M. and Bijker, R., Phys. Rev. C30, 517 (1984).

7. Vervier, J., and Mareschal, P., Zeits. f. Phys. A323, 179 (1986).

Symmetries	Nuclei
$U(6/2) \supset SU(5) \times 1/2 \supset Spin(3)$	Ru, Rh, Pd, Ag
$U(6/2) \supset SU(3) \times 1/2 \supset Spin(3)$	$^{171,172}Yb$
$U(6/2) \supset O(6) \times 1/2 \supset Spin(3)$	Hg, Tl
$U(6/4) \supset SU(5) \times 3/2 \supset Spin(5)$	Zn, Cu
$U(6/4) \supset O(6) \times 3/2 \supset Spin(6)$	Os, Ir, Pt, Au
$U(6/6) \supset SU(3) \times 1/2, 3/2$	$^{168}Er, ^{169}Tm$
$U(6/10) \supset SU(5) \times 3/2, 5/2$	$^{76}Se, ^{75}As$
$U(6/12) \supset SU(5) \times 1/2, 3/2, 5/2$	$^{76}Se, ^{75}As, ^{63}Zn, ^{195,197}Hg$
$U(6/12) \supset SU(3) \times 1/2, 3/2, 5/2$	$^{184,185}W$
$U(6/12) \supset O(6) \times 1/2, 3/2, 5/2$	$^{193-199}Pt$
$U(6/20) \supset O(6) \times 1/2, 3/2, 5/2, 7/2$	$^{195-203}Au$
	Te, Xe, Ba, Ce

Table 1: Boson-fermion symmetries and Supersymmetries developed so far, and nuclei to which they have been applied.

EXPERIMENTAL STUDY OF SYMMETRIES WITH TRANSFER REACTIONS

Jolie A. Cizewski
A.W. Wright Nuclear Structure Laboratory
Yale University, New Haven, CT 06511 U.S.A.

ABSTRACT

One- and two-particle transfer reactions can provide additional tests of the symmetries and supersymmetries of the IBA framework. These include the relationship between neighboring nuclei with regard to changes in shape and the tests of the supersymmetries, where single-particle transfer provides information on the fermion structure.

1. INTRODUCTION

The Interacting Boson Approximation (IBA) model was originally formulated to understand the quadrupole collective properties of medium and heavy mass nuclei. With basic building blocks of s and d bosons, the group structure is SU(6); three limiting symmetries occur: the SU(5) (corresponding to an anharmonic vibrator), the SU(3) (corresponding to a symmetric rotor) and the O(6) (corresponding to a γ-unstable rotor) limits. The identification of nuclei which are candidates for symmetry structures has been an ongoing search for many years. Most of the tests of the symmetry character of even-A nuclei have concentrated on the properties of a single nucleus, e.g., energy levels and electromagnetic transition rates. However, two-neutron transfer strengths can also provide a measure of the **change** in the structure of adjacent nuclei.

The new boson-fermion symmetries and supersymmetries in the Interacting Boson-Fermion Approximation (IBFA) model provide predictions of the structure of a single nucleus, such as electromagnetic transition rates, as well as predictions of energy levels and particle transfer for adjacent even- and odd-mass nuclei. The predicted fermion structure can only be probed via single-particle transfer reactions; predictions for single-particle transfer strengths exist for both even- and odd-A targets.

2. TWO-PARTICLE TRANSFER

For the limiting symmetries of the IBA, analytical expressions for two-particle transfer strengths exist[1] and have been applied frequently to the structure of ranges of isotopes of various elements. An example is given in Fig. 1 for the two-neutron transfer reactions on Pt and Os isotopes. The data[2] are compared with the predictions of the O(6) and SU(5) limits. First, this comparison shows that the Pt isotopes with A=192-200 can be described with the O(6) limit of the IBA, a conclusion which has been extensively supported[3,4] by the energy level and electromagnetic properties of excited levels in Pt nuclei and, most especially, in ^{196}Pt. Second, the smooth trend in the data, and that it can be reproduced by a single symmetry prediction, indicates that no rapid change in structure is occuring in these isotopes. Therefore, while some of the spectroscopic properties of ^{198}Pt have been suggested[5] to indicate a vibrational spectrum, the two-neutron transfer data do not support a drastic change in shape between ^{196}Pt(O(6)-like) and ^{198}Pt(SU(5)-like).

While the two-neutron transfer results for Pt nuclei show no evidence for a drastic shape change as a function of neutron number, there are several regions of the periodic table where such shape changes do occur. In the Sm isotopes a rapid change from SU(5) → SU(3) character is well-known and was studied in ref. 1. Another region of rapid shape change is in the Mo region; the corresponding two-neutron transfer data[6] are presented in Fig. 2. While the data are relatively flat as a function of neutron number, a vibrational description would predict a rapid increase in the two-neutron transfer strengths as a function of neutron number, because of the large change in valence particle number from the singly-magic ^{92}Mo to ^{100}Mo. The change in shape of the Mo isotopes has been modeled[7] in terms of the coexistence of spherical, with N_π=1, and deformed configurations, with N_π=3 or one-hole-two-particle boson configurations. The change in structure in the Mo isotopes then occurs because the deformed N_π=3 configuration, very high-lying in ^{94}Mo, becomes a dominant component of the ground state in ^{104}Mo.

For mixed configurations the two-neutron transfer strength can be approximated by[6]

$$I(N_\nu \to N_\nu+1) \approx \alpha_\nu^2 \{C_1(N_\nu)C_1(N_\nu+1)[(N_\nu+1)(\Omega_\nu - N_\nu)]^{1/2}$$
$$+ C_3(N_\nu)C_3(N_\nu+1)[(N_\nu+1)\frac{(2N+3)}{3(2N+1)}]^{1/2}[\Omega_\nu - N_\nu - \frac{4(N-1)}{3(2N-1)}N_\nu]^{1/2}\}^2$$

where Ω_ν is the degeneracy of the neutron shell, $C_1(N_\nu)$ is the amplitude of the N_π=1 component in the ground state of the nucleus with N_ν, and $C_3(N_\nu)$ is the amplitude of the N_π=3 component. The mixing amplitudes were obtained by Sambataro and Molnar (ref. 7) from fits to the energy levels and electromagnetic transition rates in Mo nuclei. As shown in Fig. 2 these calculations properly account for the two-neutron transfer results.

Another region in which "intruder" configurations play an important role in low-lying excitations is near the Z=50 closed shell, in Sn and Cd nuclei. The ground-state relative transfer strengths have recently been measured for Cd nuclei[8] on all stable even-A=106-116 targets. The mixing between normal and intruder configurations has been treated by Aprahamian, et al. and by Heyde, et al. (ref. 9,10) within the IBA-2 framework, and a mixing matrix element between $N\pi$=1 and intruder N_π=3 configurations of ≈ 100 keV was obtained. An alternative, more microscopic, approach which explictly examines particle-hole fermion configurations, calculation A of ref. 10, has a mixing matrix element of ≈ 400 keV. A comparison between the experimental results[8], the predictions of the SU(5) limit of the IBA (solid line), and predictions with a large mixing matrix element and unperturbed 0^+ energies of ref. 9 (dotted line) is shown in Fig. 3. While the effect on the ground state from the intruder configuration should be largest in ^{114}Cd, in fact, it is for this state that the largest transfer strength is observed and, therefore, a large mixing matrix element is ruled out by these data. The measurements are not sufficiently

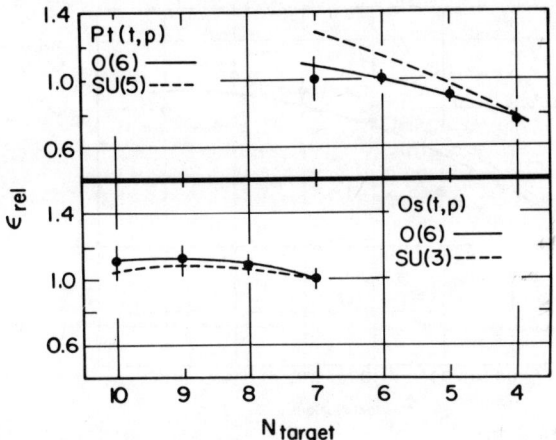

Fig. 1   Comparison between IBA predictions and experimental Pt,Os (t,p) reaction strengths.

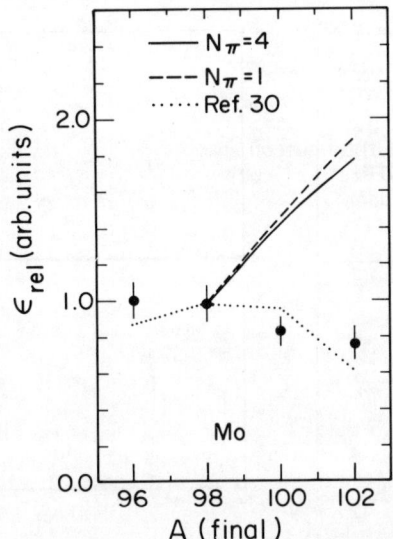

Fig. 2   Comparison between calculated and experimental Mo(t,p) reaction strengths.

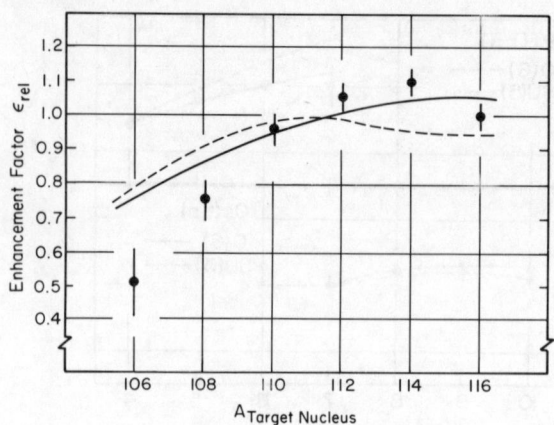

Fig. 3 Comparison between calculated and experimental Cd(p,t) reaction strengths.

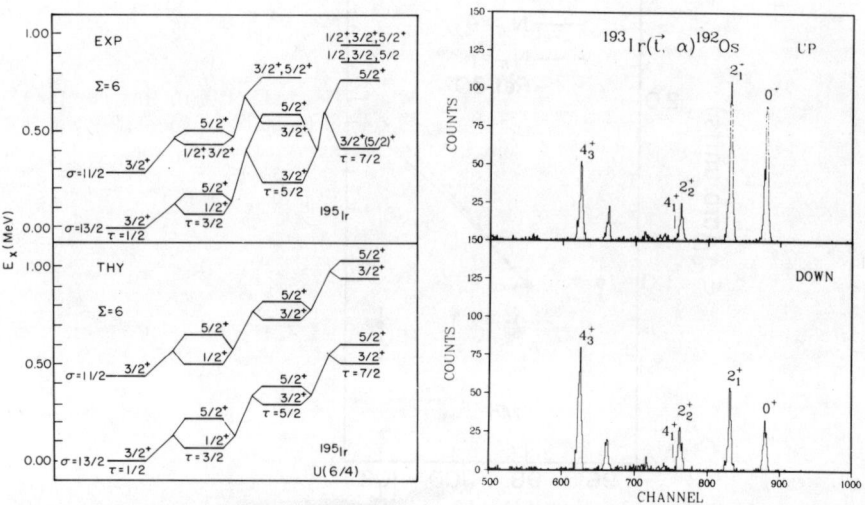

Fig. 4 Comparison between U(6/4) predicted and ^{195}Ir level spectra

Fig. 5 Spectrum of ^{193}Ir(t,α)^{192}Os reaction

sensitive to rule out mixing with a rather small matrix element (≈ 100 keV) between intruder and normal ground-state configurations.

3. SINGLE-PARTICLE TRANSFER

The new boson-fermion symmetries and supersymmetries in the IBFA require a symmetry for the boson core, particular single-particle orbitals that the valence particle can occupy, and a particular form of the boson-fermion interactions. Most tests of these boson-fermion symmetries have concentrated on nuclei near the O(6) limit of the IBA, in the Os-Ir-Pt-Au nuclei. The first dynamical supersymmetry to be proposed was the U(6/4) Spin(6) symmetry chain[11], with an O(6) boson core and a j=3/2 fermion. Several isotopes of Ir and Au nuclei have been proposed as candidates for this boson-fermion symmetry structure. For example, for ^{193}Ir the observed excitation energies, electromagnetic deexcitation and single-particle transfer (both stripping and pick-up reactions) are well-reproduced by the symmetry predictions[12].

A more detailed knowledge of levels in ^{195}Ir has only recently become available. From earlier (t,α) reaction studies[13], most of the $d_{3/2}$ strength was observed in the ground state with additional strength at ≈ 250 keV in excitation, and a relatively small fraction of the $s_{1/2}$ strength in the $1/2^+_1$ state, as expected for this boson-fermion structure. Following two-neutron capture measurements[14] on ^{193}Ir, several new low-spin positive-parity levels were located. A comparison between the positive-parity levels of ^{195}Ir and the predictions of the U(6/4) supersymmetry, with energy parameters taken from an earlier study[4] of ^{194}Os, is shown in Fig. 4. The states are labeled by the (σ, τ) quantum numbers of this symmetry. What is remarkable in this comparison is that an entire sequence of $\sigma < \sigma_{max}$ levels have been identified; it is these levels which are the key signature of an O(6) symmetry structure.

4. SINGLE-PARTICLE TRANSFER AND ODD-A TARGETS

While single-particle transfer reactions with even-A targets have frequently provided good agreement between the predictions of the new boson-fermion symmetries and experiment, no measurement of single-particle transfer starting with an **odd**-mass target can be well reproduced by the symmetry predictions. As examples are the $(d,^3He)$ reactions on Ir targets[15], in which the 2^+_2 states are populated with a sizeable fraction of the strength to the 2^+_1 states, while the symmetry would predict the former to be forbidden, and the latter to be allowed. In the symmetry, only $d_{3/2}$ transfer is allowed and it is assumed that the transfer reactions proceed by single-step processes.

To further test the mechanism of the pickup reactions on odd-mass targets in the A=190 region, we have studied the Ir(t,α) reactions with **polarized** tritons[16]; a spectrum is shown in Fig. 5 for the ^{193}Ir target. For single-step $d_{3/2}$ transfer the analyzing powers are large and negative (≈ -0.80 at $\geq 30^\circ$), while for $d_{5/2}$ transfer the analyzing powers are positive ($\approx +.40$). While the analyzing powers to the ^{192}Os ground and 2^+_1 states are negative, the analyzing power to the 2^+_2 state is consistent with isotropy at one angle and positive at another. These results indicate that the 2^+_2 state in ^{192}Os is **not** populated by single-step $d_{3/2}$ transfer and, therefore, the transfer results for this state can not be used as a definitive test of the supersymmetry predictions. A more detailed understanding of the exact nature of this transition will have to await more extensive transfer reaction studies with polarized projectiles.

5. CONCLUSION

The present paper has presented some specific examples of how light-ion induced transfer reaction studies can provide spectroscopic information that complements that obtained in reaction studies that investigate the electromagnetic properties of medium- and heavy-mass nuclei. In particular, two-particle transfer studies can provide information on the change in structure of adjacent **ground-states,** while single-particle transfer studies provide excellent probes of the specific **single-particle** character of states. For transfer reactions, however, it is important to properly account for the reaction mechanism in extracting the spectroscopic factors.

The measurements reported in this paper were done in collaboration with scientists at Los Alamos National Laboratory, McMaster University, the University of Pennsylvania, the Institut Laue-Langevin, and Livermore National Laboratory; I thank all of you for your invaluable assistance. This work was supported by the U.S. Department of Energy under contract DE-AC02-76ER03074.

6. REFERENCES

1. A. Arima and F. Iachello, Phys. Rev. **C16**, 2085 (1977).
2. J.A. Cizewski, E.R. Flynn, R.E. Brown, D.L. Hanson, S.D. Orbesen and J.W. Sunier, Phys. Rev. **C23**, 1453 (1981).
3. J.A. Cizewski, **et al.**, Nucl. Phys. **A323**, 349 (1979).
4. R.F. Casten and J.A. Cizewski, Nucl. Phys. **A309**, 477 (1978).
5. S.W. Yates, **et al.**, Phys. Rev. **C23**, 1993 (1981).
6. E.R. Flynn, **et al.**, Phys. Rev. **C24**, 2475 (1981); Phys. Rev. **C25**, 2850(E) (1982).
7. G. Sambataro and G. Molnar, Nucl. Phys. **A376**, 201 (1982).
8. R.W. Bauer, **et al.**, submitted to Phys. Rev. C.
9. A. Aprahamian, **et al.**, Phys. Lett. **140B**, 12 (1984).
10. K. Heyde, **et al.**, Phys. Rev. **C25**, 3160 (1982).
11. F. Iachello, Phys. Rev. Lett. **44**, 772 (1980); A.B. Balantekin, **et al.**, Nucl. Phys. **A370**, 284 (1981).
12. F. Iachello and S. Kuyucak, Ann. Phys. (NY) **136**, 19 (1981) and S. Kuyucak, Ph.D. Thesis, Yale University (1982) unpublished.
13. J.A. Cizewski, D.G. Burke, E.R. Flynn, R.E. Brown and J.W. Sunier, Phys. Rev. **C27**, 1040 (1983).
14. J.A. Cizewski, G.G. Colvin, H.G. Borner, F. Hoyler, S.A. Kerr and K. Schreckenbach, submitted to Phys. Lett.; G.G. Colvin, **et al.**, to be published.
15. N. Blasi, **et al.**, Nucl. Phys. **A388**, 77 (1982).
16. J.A. Cizewski, D.G. Burke, R.E. Brown and J.W. Sunier, to be published.

EXPERIMENTAL TESTS OF SUSYS USING TRANSFER REACTIONS

Michel Vergnes

Institut de Physique Nucléaire
91406 Orsay
FRANCE

ABSTRACT

The main types of supersymmetries proposed for nuclei in the last six years are briefly presented and compared when possible to transfer reactions data. Zones of realization are found, both for U(6/4) and for U(6/12). Difficulties subsist, however, in applying these models to particle transfers.

1. INTRODUCTION

The search for supersymmetry in nuclei was initiated[1] by F. Iachello in 1980. Since then, many developments have been suggested and worked out, many experimental tests have been performed and the extension of the supersymmetry models is still continuing. The present paper considers mainly the applications of these models to particle transfer reactions. Due to pages limitations, it is only possible here to summarize rather briefly the main results of a large review performed on this subject. For the same reason, and I apologize in advance for it, it will be possible to give only very few references. It is suggested to the reader interested by more details and references, to consult ref. 2.

2. THE SPIN (6) B-F AND U(6/4) SUPERSYMMETRY MODELS

The tremendous success of the IBM-1 model for even-even nuclei, with its 3 limits : U(5), SU(3) and O(6) has led to extend it to the odd-A nuclei.

In this case, like in the case of even-even nuclei, it is particularly interesting to look for dynamical symmetries, or limits, called Bose-Fermi (B-F) symmetries.

F. Iachello has shown[1] that such a symmetry exists in a very special case where the odd-A nucleus can be described as a j = 3/2 fermion coupled to an O(6) boson core. In this case, the group structure of the problem is $U^B(6) \otimes U^F(4)$. The groups SU(4) and O(6) being isomorphic, it is possible to combine the boson and fermion groups into a unique chain :

$$U^B(6) \otimes U^F(4) \supset O^B(6) \otimes SU^F(4) \supset Spin(6) \supset Spin(5) \supset Spin(3) \supset Spin(2) \quad (1)$$

A further extension consists to attempt to consider both the even-even and the odd-A nuclei as a multiplet of a higher symmetry described by a larger group. This extension was worked out by Balantekin, Bars and Iachello[3] who found that the B-F chain (1) is indeed embedded into the graded Lie group, or super-group, U(6/4) :

$$U(6/4) \supset U^B(6) \otimes U^F(4) \ldots \quad (2)$$

The nuclei can no longer - as in the B-F Spin (6) symmetry - be fitted separately, but they form supermultiplets[2] and must be described by the same parameters. It appears[3] that the model works quite satisfactorily for energies and B(E2) in the Os, Ir, Pt, Au region, where the even-even nuclei have been rather well described by the O(6) limit of IBM-1.

In the case of transfer reactions, the problem is that we test both the SUSY scheme itself and the transfer operators used.

A first test concerns the two nucleon transfer reaction. In the region of interest here, the Os, Ir, Pt, Au, Hg region, the transfer strengths are predicted equal, for the even-even and the odd-A partners, within a few percent. Tests of this kind were performed[2], first by J. Cizewski et al. and later by us, and found in agreement with U(6/4).

For single nucleon transfer reactions the model predicts[2] transfer intensities I_{if} and selection rules. The excellent agreement between experiment and theory is shown in Fig. 1 for the proton reactions leading to the $J^\pi = 3/2^+$ levels of ^{193}Ir.

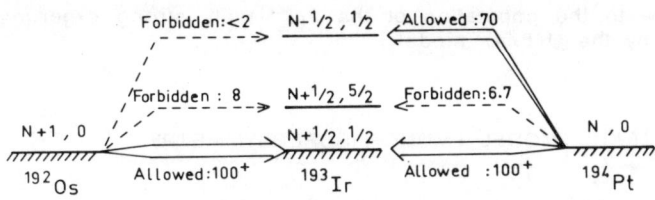

Fig. 1 : Experimental confirmation of the validity of Bose-Fermi symmetry and supersymmetry selection rules in reactions leading to $J^\pi = 3/2^+$ levels of ^{193}Ir.

However, difficulties[2] have been shown for other reactions leading to final odd-A nuclei and for most of the reactions leading to final even-even nuclei. The disagreement between theory and experiment can be characterized, for a p.u reaction, by the ratio

$$R_S = \sum_f \left| S_f^{exp.} - S_f^{th.} \right| \Big/ \sum_f S_f^{exp.} \qquad (3)$$

(and by a similar ratio for a stripping reaction, with S replaced by G).

Table 1 R_s values[2] for two "triads" of <u>isotones</u> in the Pt region (comparison with U(6/4))

Final	Same supermultiplet			Same supermultiplet		
	^{192}Os	^{193}Ir	^{194}Pt	^{196}Pt	^{197}Au	^{198}Hg
Odd-A	<26.7%(9.1%)	21%(7.4%)		64%(44%)	58%(39%)	
Even-even	30.5%*		44%*	56%*		58%*

within parenthesis : only the three lowest $3/2^+$ levels are considered
without parenthesis : as above, plus the $5/2_1^+$ and $5/2_2^+$ levels
* 4 levels = 0_1^+, 2_1^+, 2_2^+, 0_2^+.

The values of R_s are presented in Table 1 (only the l = 2 transitions are considered) for all the reactions performed up to now.

To summarize, the reactions in the first "triad" leading to $^{193}_{77}$Ir are rather well described by the U(6/4) SUSY. Always in this first "triad", an appreciable breaking is observed for the reactions on the ^{193}Ir target. This breaking, generally observed for odd-A targets as we shall see later, is mainly due to the population of the 2_2^+ level, strong experimentally but forbidden by the U(6/4) model.

3. THE U(6/12) MODEL (MULTI-j SUPERSYMMETRY)

It is possible to extend the SUSY idea, in a rather general way. If j, j', j'' are the angular momenta of the orbitals considered, it is always possible to split them into a pseudo-spin s = 1/2 and a pseudo-orbital momentum k. If k happens to sit in an irreducible representation of the boson group $G^B \equiv U(6)$ it is possible to combine G^B and G^F in a common theoretical framework. This has been done[4] for the orbitals j = 1/2, 3/2, 5/2 with k = 0,2. It is now possible to use the 3 cores of the IBM-1 model and the super-group is U(6/12). The model has first been applied with an O(6) core. The chain is then :

$$U(6/12) \supset U^B(6) \otimes U^F(12) \supset U^B(6) \otimes U^F(6) \otimes SU^F(2) \supset U^{B+F}(6) \otimes SU^F(2)$$

$$\supset O^{B+F}(6) \otimes SU^F(2) \supset O^{B+F}(5) \otimes SU^F(2) \supset O^{B+F}(3) \otimes SU^F(2) \qquad (4)$$

$$\supset Spin(3) \supset Spin(2)$$

This model has been applied to the Pt isotopes and the R_s values for single neutron transfers are shown in Table 2.

Table 2 : R_s values[2] for two "triads" of <u>isotopes</u> in the Pt region (comparison with U(6/12)).

	Same supermultiplet			Same supermultiplet		
Final	^{194}Pt	^{195}Pt	^{196}Pt	^{196}Pt	^{197}Pt	^{198}Pt
Odd-A	⟶ 14% (11 levels)		⟵ 20.3% (11 levels)	⟶ 29% (7 levels)		⟵ 39% (7 levels)
Even-even	⟵ 33%*		33%* 47%(7 levels) ⟶	+		+

+ no target   * = 4 levels = 0_1^+, 2_1^+, 2_2^+, 0_2^+.

To summarize, the reactions in the first "triad" leading to $^{195}Pt_{117}$ are rather well described (Rs = 16% for 22 transitions) by the U(6/12) model. The situation is appreciably worse in the second "triad" and, in the first one, for the reactions on the ^{195}Pt target (anomalous population of the 2^+_2 levels).

The U(6/12) model has also been applied[5], with a U(5) core, to the Hg isotopes. An analysis of our (p,d) results shows a reasonably good agreement[2] with the model for the reaction to $^{197}Hg_{117}$. It is striking that the agreement is observed at 117 neutrons both for Pt and for Hg, although the cores are different. Possible explanations are suggested in ref. 2, where the application[6] of the same model to ^{76}Se and ^{75}As is also discussed.

The U(6/12) model, with a SU(3) core, has been applied[7] rather successfully ($R_s \approx 25\%$ for 8 levels) to the p.u reaction to ^{185}W, and to the level scheme of this nucleus. It is not evident[7], however, that the level scheme of the even-even partner, ^{184}W, is described as well.

4. EXTENSION OF SUSY TO ODD-ODD NUCLEI

If the even-even core is no longer described by IBM-1, but instead by IBM-2, differentiating neutron and proton bosons ($U^B(6)$ replaced by $U^B_\nu(6) \otimes U^B_\pi(6)$), the SUSY concept can be naturally extended[8] to simultaneously describe both the isotone and isotope odd-A neighbours and the odd-odd nucleus. I suggest[2], from the results of sections 2 and 3, that the best such "quartet" of nuclei could be : ^{192}Os, $^{193}Os_{117}$, $^{193}Ir_{77}$, $^{194}Ir_{117}$. Transfer tests are planned.

5. DIFFICULTIES

It has been recalled above that transfer data for final even-even nuclei are not well described by the SUSY models.

Recent IBFM-2 calculations in the Pt region[9] do not exhibit a better agreement with transfer data[2] than the one obtained with U(6/4) and U(6/12).

A sum rule analysis of transfer data in Pt shows a big discrepancy[10] for U(6/12) with an O(6) core. The discrepancy appears much smaller[2] for Hg, with a U(5) core.

It has been suggested[11] that there is a "missing term" in the IBFM transfer operator.

It therefore appears that, although zones of realization for both U(6/4) and U(6/12) SUSYS have been shown, difficulties subsist when applying these models, as well as the IBFM model, to transfer reactions. These difficulties could be related to the transfer operator but also perhaps to the boson-fermion exchange term of the IBFM model.

6. REFERENCES

1) Iachello, F., Phys. Rev. Letters $\underline{44}$, 772 (1980)

2) For more details and references, see :
Vergnes, M., "Tests of the Supersymmetry Models by Transfer Reactions", Proc. Workshop on Interacting Boson-Boson and Boson-Fermion Systems, Scholten, O. ed., World Scientific, Singapore (1984), p 91 and references therein.
Vergnes, M., "Review of Bose-Fermi and Supersymmetry Models : Problems in Particle Transfer Tests", Institut de Physique Nucléaire Internal Report, Orsay, IPNO-DRE 86-08 (unpublished) and references therein.

3) Balantekin, A.B., Bars, I. and Iachello, F., Nucl. Phys. $\underline{A370}$, 284 (1981) and references therein.

4) Balantekin, A.B., Bars, I., Bijker, R. and Iachello, F., Phys. Rev. $\underline{C27}$, 1761 (1983)

5) Sun, H.Z., Frank, A. and Van Isacker, P., Phys. Letters, $\underline{124B}$, 275 (1983)

6) Vervier, J., Van Isacker, P., Jolie, J., Kota, V.K.B and Bijker, R., Phys. Rev. $\underline{C32}$, 1406 (1985)

7) Warner, D.D. and Bruce, A.M., Phys. Rev. $\underline{C30}$, 1066 (1984)

8) Van Isacker, P., Jolie, J., Heyde, K. and Frank, A., Phys. Rev. Letters $\underline{54}$, 653 (1985)

9) Arias, J.M., Alonso, C.E. and Lozano, M., Phys. Rev. $\underline{C33}$, 1482 (1986) and private communication

10) Pinkston, W.T. and Feng, D.H., Phys. Rev. $\underline{C30}$, 1431 (1984)

11) Paar, V. and Brant, S., Phys. Letters $\underline{143B}$, 1 (1984)

OVERVIEW OF BOSE-FERMI SYMMETRIES IN ODD-EVEN NUCLEI

Roelof Bijker
Department of Physics, University of Pennsylvania
Philadelphia, Pennsylvania 19104-6396, USA

ABSTRACT

The Bose-Fermi symmetries connected with each of the dynamical (Bose) symmetries of the interacting boson model, the U(5), SU(3) and SO(6) limits, will be discussed. By comparing a description of odd mass nuclei in terms of Bose-Fermi symmetries with a description in terms of the semi-microscopic IBFA model and the Nilsson model, it has been possible to obtain a physical interpretation of these symmetries.

1. INTRODUCTION

The Interacting Boson-Fermion Model (IBFM) was introduced in 1979 by Iachello and Scholten[1,2] as a possible unified framework to discuss properties of a wide variety of odd-even nuclei, irrespective of their nature (spherical, axially deformed, gamma unstable or triaxial). In addition the algebraic structure of the IBFM Hamiltonitn opened the possibility to study special cases in which the energy eigenvalues of the Hamiltonian can be obtained in closed form. The limiting situations correspond to dynamical Bose-Fermi (BF) symmetries. The first example of such a symmetry was found[3] in the energy spectrum of the positive parity states in ^{191,193}Ir, which was described by coupling an odd proton in the $2d_{3/2}$ orbit to the adjacent even-even ^{192,194}Pt core nuclei with SO(6) symmetry. Since then many different examples of BF symmetries have been discussed and compared with experimental data. In this talk I will present an overview of BF symmetries associated with each of the dynamical symmetries of the Interacting Boson Model (IBM), the U(5), SU(3) and SO(6) limits, respectively, and suggest possible experimental examples. To obtain a physical interpretation of BF symmetries, the relation with a description of odd mass nuclei in terms of the semi-microscopic IBFA model and the Nilsson model is studied.

2. DYNAMICAL BOSE-FERMI SYMMETRIES

In the IBFM odd-even nuclei are described by coupling the single particle (fermion) degrees of freedom of the odd particle to the collective (boson) degrees of freedom of the even-even core nucleus. The group structure G of the IBFM Hamiltonian is the direct product of the boson and fermion symmetry groups, $G_B = U(6)$ and $G_F = U(m)$, respectively, where m specifies the dimension of the single particle space. When the single particle and collective part have the same symmetry, G_B and G_F have a common chain of subgroups which can then be combined into a boson-fermion subgroup chain. Whenever the Hamiltonian is expressed in terms of Casimir invariants of this boson-fermion subgroup chain only, a dynamical Bose-Fermi symmetry arises and the energy spectrum as well as electromagnetic transition rates and spectroscopic factors can be derived in closed analytic form using standard group theoretical techniques. These analytic expressions provide a simple framework to analyze and interpret experimental data.

In principle many different BF symmetry coupling schemes can be constructed. However, there is no a priori reason why they should occur in actual nuclei. The single particle structure, i.e. the angular momenta and the relative energies of valence shell model orbits, combined with a specific core structure (U(5), SU(3) or SO(6) symmetry) determines in which region of the nuclear mass table a certain BF symmetry might be expected to occur. The case in which the odd nucleon can occupy single particle orbits with angular momentum $j = 1/2, 3/2, 5/2$ is of particular interest, since this situation seems to be encountered in several regions of the nuclear mass table, e.g. in the Pt region in which the 3 $p_{1/2}$, 3 $p_{3/2}$ and 2 $f_{5/2}$ neutron orbits are well separated from the other negative parity levels in the 82-126 shell and in the 28-50 neutron and proton major shells where the $2p_{1/2}$, 2 $p_{3/2}$ and 1 $f_{5/2}$ orbits are the only levels with negative parity.

The first step in the construction of a BF symmetry is to decompose the fermion angular momenta $j = 1/2, 3/2, 5/2$ into a pseudo-orbital part with k = 0,2 and a pseudo-spin part with s = 1/2.[4]) The corresponding fermion group reduction $U(12) \supset U(6) \otimes U(2)$ can be decomposed further into

$$U(12) \supset U(6) \otimes U(2) \supset \begin{cases} U(5) \otimes U(2) & (1a) \\ SU(3) \otimes U(2) & (1b) \\ SO(6) \otimes U(2) & (1c) \end{cases}$$

Each of the above fermion group chains can be combined with its bosonic counterpart, the U(5), SU(3) and SO(6) limits of the IBM[4-8]), respectively, into a common boson-fermion coupling scheme. The corresponding BF symmetries will be referred to as the $U(5) \otimes U(2)$, the $SU(3) \otimes U(2)$ and the $SO(6) \otimes U(2)$ limits, respectively.

2.1 The U(5) Limit (Anharmonic Vibrator)

As a first example I will discuss BF symmetries in spherical odd-even nuclei. The energy eigenvalues of the Hamiltonian in the $U(5) \otimes U(2)$ limit are given by

$$\begin{aligned} E = {} & \alpha\{(N+1-i)(N+6-i) + i(i+3)\} \\ & + \varepsilon_1 (n_1+n_2) + \varepsilon_2 \{n_1(n_1+4) + n_2(n_2+2)\} \\ & + \beta\{\tau_1(\tau_1+3) + \tau_2(\tau_2+1)\} + \gamma_1 L(L+1) + \gamma_2 J(J+1) \end{aligned} \quad (2)$$

In fig. 1 the energy spectrum of the negative parity states in ^{63}Zn is compared with the theoretical spectrum in the $U(5) \otimes U(2)$ limit [9]). There is a one to one correspondence between calculated and experimental states below 1.5 MeV. The overall structure of the energy spectrum is determined by the terms proportional to α, ε_1 and ε_2. The term with strength ε_1 alone gives a spectrum which is very similar to that of a weak-coupling scheme. The levels are then grouped into a series of multiplets labeled by (n_1,n_2) and separated by constant energy spacings. The term with strength ε_2 introduces anharmonic effects in the spacings between the different multiplets, whereas the second term proportional to α shifts multiplets with different values of i with respect to each other without disturbing the internal structure. The splitting of the energy levels within each multiplet is given by β, γ_1 and γ_2. The $U(5) \otimes U(2)$ limit has also been applied to the spectra of the Hg[5]) and As[10]) isotopes.

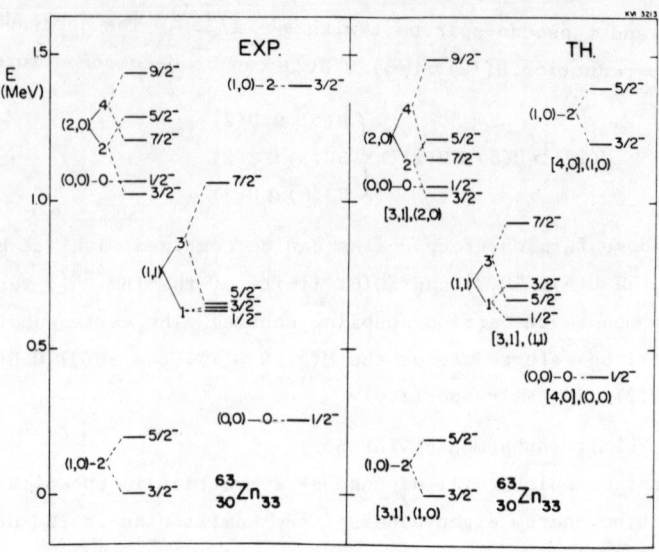

Fig. 1. Comparison between the experimental energy spectrum of ^{63}Zn and the theoretical spectrum in the U(5) ⊗ U(2) limit[9] calculated using eq. (2) with α = 150 keV, ε_1 = 325 keV, ε_2 = 100 keV, β = 0, γ_1 = -12 keV and γ_2 = 19 keV. The number of bosons is N = 3.

In addition to the coupling scheme discussed above, several other examples of BF symmetries in spherical odd-even nuclei have been studied. The simplest case is that of a particle in a single orbit with j = 1/2, which has been applied to the Rh[6,9] and Ag[11] isotopes. Furthermore the spectrum of the negative parity states in ^{63}Cu has been interpreted[6,9] in terms of the Spin (5) limit which arises from the coupling of the 2 $p_{3/2}$ orbit to a spherical core nucleus.

2.2 The SU(3) Limit (Axially Symmetric Rotor)

Another interesting set of BF symmetries arises when the odd nucleon is coupled to a core nucleus with SU(3) symmetry which in the geometrical model corresponds to an axially symmetric rotor. Whenever the single particle space consists of all levels in an oscillator shell j = j_1, j_1 - 1,..., 1/2, it is possible to construct a BF symmetry. The simplest non-trivial case is that of j = 3/2, 1/2[9]. A more realistic case is that of j = 5/2, 3/2, 1/2.[7]) It was shown by Warner and Bruce [12,13] that this coupling scheme corresponds to a very specific coupling of bands in the Nilsson model, a situation which seems to be realized in ^{185}W.

The energy eigenvalues of the Hamiltonian in the SU(3) ⊗ U(2) limit are given by

$$E = \alpha\{(N+1-i)(N+6-i) + i(i+3)\}$$
$$+ \kappa\{\lambda^2 + \mu^2 + \lambda\mu + 3(\lambda+\mu)\} + \gamma_1 L(L+1) + \gamma_2 J(J+1) \quad (3)$$

The spectrum consists of a series of rotational bands labeled by $(\lambda,\mu)K$, whose relative excitation energies are determined by the first two terms in (3). The splitting within each rotation band only depends on the last two terms. In fig. 2 we show a comparison with the spectrum of the negative parity states in ^{185}W.[13]) The first term in (3) is essential to reproduce the correct ordering of rotational bands which suggests that its strength α is related to the position of the Fermi surface in the deformed potential. A distinctive feature of the energy spectrum in the SU(3) ⊗ U(2) limit is the repeated occurrence of doublets of states with $J = L \pm 1/2$, which can only be split by the last term in (3). This gives rise to two nearly degenerate rotational bands with K = 1/2 and 3/2 which both belong to the SU(3) ground state band with $(\lambda,\mu) = (2N,1)$, and a series of doublets of states in the first excited SU(3) band with $(\lambda,\mu) = (2N+2,0)$ and K = 1/2. Both features are indeed observed in ^{185}W.

The experimental (d,t) cross sections shown in fig. 2 show a very distinct pattern. The j = 3/2 and 5/2 strengths to the ground state configuration are each concentrated in a single state, while the J=1/2$^-$ state is not excited at all. In the IBFM the lowest order operator for the ^{186}W (d,t) ^{185}W reaction is $\zeta_j a_j^\dagger$, which has selection rules $\Delta L = 0$, for j = 1/2 and $\Delta L = 2$ for j = 3/2, 5/2, where L is the label that characterizes the doublet structure. The theoretical cross sections calculated with $\zeta_{1/2}^2 = 0.64$ and $\zeta_{3/2}^2 = \zeta_{5/2}^2 = 0.36$ are in good agreement with the observed data.

The same characteristic features for both the energy spectrum and

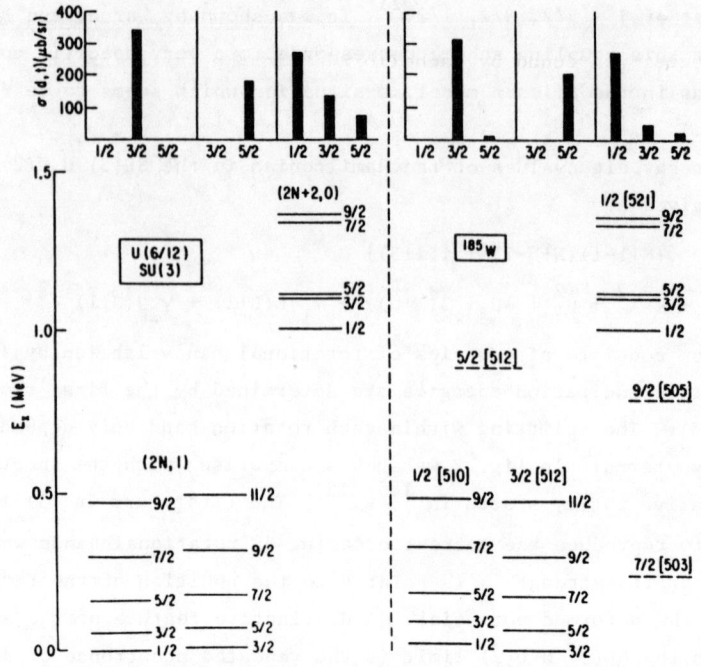

Fig. 2. Comparison between the experimental energy spectrum of ^{185}W and the theoretical spectrum in the SU(3) ⊗ U(2) limit[13] calculated using eq. (3) with α = 103.3 keV, κ = -20.0 keV, γ_1 = 15.0 keV and γ_2 = 2.5 keV. The number of bosons is N=11. On the top the experimental and theoretical (d,t) cross sections are shown.

the (d,t) cross sections arise in the Nilsson model from a very specific coupling of rotational bands. However, the single particle space in the U(6) ⊗ U(12) scheme was restricted to the 3 $p_{1/2}$, 3 $p_{3/2}$ and 2 $f_{5/2}$ orbits. Therefore the contributions from other single particle levels like the 2 $f_{7/2}$ and 1 $h_{9/2}$ orbits to the energy spectrum (dashed levels in fig. 2) and the single particle structure of the Nilsson orbits cannot be accounted for in this simple scheme.

2.3 The SO(6) Limit (Gamma Unstable Rotor)

Finally the BF symmetries that have attracted most attention are the ones associated with the SO(6) (gamma unstable) limit of the IBM.

The first experimental evidence for the existence of BF symmetries in odd mass nuclei was found by Iachello[3,14] in the energy spectrum of the positive parity states in the nuclei ^{191,193}Ir. This symmetry called the Spin(6) limit arises from the coupling of an odd proton in the 2 $d_{3/2}$ orbit to the ^{192,194}Pt core nuclei, which have SO(6) symmetry. The breaking of the Spin (6) limit due to the presence of other single particle levels, most notably the 3 $s_{1/2}$ orbit has been studied in perturbation theory[14], numerically[15] and from a more microscopic point of view[16].

Another example of a BF symmetry associated with the SO(6) limit was found in the spectrum of the negative parity states of ^{195}Pt.[4,8,9,17] These states arise from the coupling of an odd neutron occupying mainly the 3 $p_{1/2}$, 3 $p_{3/2}$ and 2 $f_{5/2}$ orbits to the ^{196}Pt core nucleus. The energy spectrum in the SO(6) ⊗ U(2) limit

$$E = \alpha\{(N+1-i)(N+6-i) + i(i+3)\}$$
$$+ \eta\{\sigma_1(\sigma_1+4) + \sigma_2(\sigma_2+2) + \sigma_3^2\}$$
$$+ \beta\{\tau_1(\tau_1+3) + \tau_2(\tau_2+1)\}$$
$$+ \gamma_1 L(L+1) + \gamma_2 J(J+1) \qquad (4)$$

is characterized by a series of bands labeled by $(\sigma_1,\sigma_2,\sigma_3)$. The energy splitting between the different bands depends on α and η while the splitting within each of the bands is determined by the last three terms in (4). In fig. 3 the spectrum of the negative parity states in ^{195}Pt is compared with that in the SO(6) ⊗ U(2) limit. Although the general structure is reproduced quite well there are several deviations. The main discrepancy is the energy splitting in the first excited band. The term proportional to α gives an additional energy splitting between bands labeled by different values of i and is essential in reproducing the observed ordering of bands. A more complete analysis of the properties of the negative parity states in ^{195}Pt in terms of the SO(6) ⊗ U(2) limit of the IBFM which includes electromagnetic transition rates and one nucleon transfer reactions can be found in refs. 8,9.

In addition to the Pt-Os region there are two more regions of the

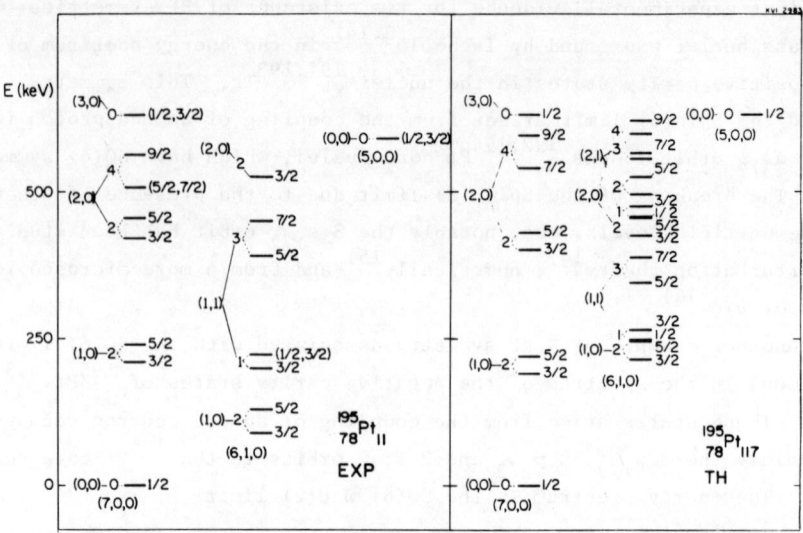

Fig. 3. Comparison between experimental energy spectrum of $^{195}_{78}Pt$[8,9] and the theoretical spectrum in the SO(6) ⊗ U(2) limit calculated using eq. (4) with = 31.6 keV, η = -16.75 keV, β = 17.5 keV, γ_1 = 2.5 keV and γ_2 = 3 keV. The number of bosons is N = 6.

nuclear mass table where the SO(6) limit has been successful, namely the Xe-Ba and the Kr-Sr nuclei. Possible evidence for the occurrence of BF symmetries in these regions has been discussed by Gelberg[18] for ^{79}Kr.

3. TRANSITION REGIONS

In the previous sections we have studied the U(5), the SU(3) and the SO(6) limits of the U(6) ⊗ U(12) coupling scheme of the IBFM in which the single particle space is restricted to orbits with j = 1/2, 3/2, 5/2. They provide a simple framework to analyze and interpret properties of odd-even nuclei. Dynamical BF symmetries correspond to a very specific coupling scheme with restrictions on both the single particle structure, the core Hamiltonian and the core-particle interaction. Nevertheless several nuclei have been suggested as experimental examples. A much larger class of nuclei can be described in terms of a transition between any of these limiting situations.

3.1 The U(5) → SO(6) Transition

The even-even nuclei in the Ru-Pd mass region have been interpreted[19] in terms of a transition between the U(5) and the SO(6) limits of the IBM. Similarly the spectra of the negative parity states in the Rh isotopes which are built on the $2 p_{1/2}$, $2 p_{3/2}$ and $1 f_{5/2}$ proton orbits have been studied[20] in terms of a U(5) → SO(6) transition in the U(6) ⊗ U(12) coupling scheme of the IBFM. Since the Hamiltonian for the odd mass Rh isotopes was derived from the one used in the description of the even-even Ru and Pd isotopes, it describes simultaneously a transition region of even-even and odd-even nuclei in the framework of a U(6/12) ⊃ U(6) ⊗ U(12) supersymmetric coupling scheme. In ref. 20 it was shown that the experimental data in the $^{101-109}$Rh isotopes for energy spectra, E2 transitions and one proton transfer reactions can be interpreted in terms of a U(5) → SO(6) transition.

3.2 The SO(6) → SU(3) Transition

Another well known transition region is that of the W-Os-Pt nuclei[21], which shows a transition from SO(6) like (gamma unstable) nuclei e.g. ^{196}Pt towards SU(3) like (axially deformed) nuclei e.g. ^{186}Os and ^{186}W. Since the low-lying negative parity states in the odd neutron nuclei in this mass region are built mainly on the $3 p_{1/2}$, $3 p_{3/2}$ and $2 f_{5/2}$ orbits, it has been suggested by Warner et. al[22], that these states can be described as a transition between the SO(6) (^{195}Pt) and the SU(3) (^{185}W) limits of the U(6/12) ⊃ U(6) ⊗ U(12) supersymmetric coupling scheme. The transitional region was studied in the consistent-Q framework of the IBFM, in which the quadrupole operator used in the Hamiltonian and in the E2 operator are identical. An additional constraint used for odd mass nuclei is that the structure of the collective and single-particle part of the quadrupole operator is the same. It was shown[22] that the observed increase in the single particle transfer strengths to the first excited doublet of states with (L=2 and) $J = 3/2^-$, $5/2^-$, going from ^{185}W to ^{189,191}Os and ^{195}Pt arises naturally in the transition from the SU(3) to the SO(6) limit of the IBFM. In the SU(3) limit these states arise solely from the coupling to the ground state rotational band in the core nucleus. In the

SO(6) limit, however, they arise from the coupling to all the states with $(\sigma_1, \sigma_2, \sigma_3) = (N,0,0)$ which also includes sidebands.

4. INTERPRETATION OF BF SYMMETRIES

The structure of the Hamiltonian in BF symmetries is completely determined by group theoretical arguments and does not provide a clear interpretation of the physical origin of the various terms. A common feature of the U(5), the SU(3) and the SO(6) limits of the U(6) ⊗ U(12) scheme is the presence of a U(6) Casimir invariant which gives rise to the first term in the energy formulas of eqs. (2) - (4). This interaction shifts bands of states belonging to different U(6) representations [N+1=i,i] with respect to each other without changing the internal structure of the bands. In all three applications discussed in section 2 it was found necessary to include this term to reproduce the observed ordering of bands in the experimental spectra of ^{63}Zn, ^{185}W and ^{195}Pt. A comparison between the SU(3) limit and the Nilsson model shows[12,13] that the interplay between the U(6) ($\sim\alpha$) and the SU(3) ($\sim\kappa$) term in (3) corresponds to changing the position of the Fermi surface in the Nilsson coupling scheme. In the semi-microscopic IBFA model the same effect can be obtained by adjusting the strengths of the quadrupole and exchange terms[1,2]. The SU(3) term corresponds to the quadrupole force, while the U(6) term is closely related to the exchange interaction in the IBFA.[23]

In a similar analysis the relation between the SO(6) limit and the semi-microscopic IBFA model has been investigated.[9,24] Restricting the values of the IBFA parameters to a physically allowed region it is possible to obtain a set of conditions for the occupation probabilities v_j^2 of the single particle orbits and the parameter χ that determines the structure of the collective quadrupole operator in the core-particle interaction. As a result the SO(6) ⊗ U(2) symmetry can be expected to occur for (i) $\chi \sim -1.0$, $v_{1/2}^2 \sim 0.1$ and $v_{3/2}^2 = v_{5/2}^2 \sim 0.6$ or (ii) $\chi \sim +1.0$, $v_{1/2}^2 \sim 0.9$ and $v_{3/2}^2 = v_{5/2}^2 \sim 0.4$. For the case of ^{195}Pt we have $\chi = -0.80$ and BCS calculations yield $v_{1/2}^2 = 0.16$, $v_{3/2}^2 = 0.67$ and $v_{5/2}^2 = 0.46$. The large breaking of the symmetry in the neighboring nuclei ^{193}Pt and ^{197,199}Pt[25,26] can then be understood in terms of a change in the occupancies of the single particle levels.

5. SUMMARY AND CONCLUSIONS

In this talk I presented an overview of dynamical Bose-Fermi symmetries in odd-even nuclei. Apart from the elegant mathematical structure an attractive feature of these symmetries is that they provide a simple analytic framework to classify and analyze the properties of odd mass nuclei. The symmetries that arise when the odd nucleon can occupy single particle orbits with $j = 1/2, 3/2, 5/2$ were discussed in some detail and compared with experimental data. Furthermore the transitional regions between any of the three limits (U(5), SU(3) or SO(6)) of the U(6) ⊗ U(12) scheme can be studied numerically by combining the Hamiltonians of the three limits.

A studying of the relation between BF symmetries and the Nilsson model and/or the semi-microscopic IBFA model made it possible to obtain a better understanding of the physical interpretation of the various terms in the symmetry Hamiltonian.

In conclusion, the concept of dynamical symmetries has proven to be a useful tool in studying the properties in both even-even and odd-odd nuclei. More recently it has been extended even further to odd-odd nuclei.[8,27,28]

REFERENCES

1. Iachello, F. and Scholten, O., Phys. Rev. Lett. <u>43</u>, 679 (1979).
2. See e.g. "Interacting Bose-Fermi Systems in Nuclei", edited by F. Iachello, Plenum Press, N.Y. (1981).
3. Iachello, F., Phys. Rev. Lett. <u>44</u>, 772 (1980).
4. Balantekin, A.B., Bars, I., Bijker, R. and Iachello, F., Phys. Rev. <u>C27</u>, 1761 (1983).
5. Sun, H.Z., Frank, A., and Van Isacker, P., Phys. Lett. <u>124B</u>, 275 (1983).
6. Bijker, R. and Kota, V.K.B., Ann. Phys. (NY) <u>156</u>, 110 (1984).
7. Van Isacker, P., Frank, A. and Sun, H.Z., Ann. Phys. (NY) <u>157</u>, 183 (1984).
8. Bijker, R. and Iachello, F., Ann. Phys. (NY) <u>161</u>, 360 (1985).
9. Bijker, R., Ph.D. Thesis, University of Groningen.

10. Vervier, J., Van Isacker, P., Jolie, J., Kota, V.K.B. and Bijker, R., Phys. Rev. $\underline{C32}$, 1406 (1985).
11. Vervier, J. and Janssens, R.V.F., Phys. Lett. $\underline{108B}$, 1 (1982).
12. Warner, D.D., Phys. Rev. Lett. $\underline{52}$, 259 (1984).
13. Warner, D.D., and Bruce, A.M., Phys. Rev. $\underline{C30}$, 1066 (1984).
14. Iachello, F. and Kuyucak, S., Ann. Phys. (NY) $\underline{136}$, 19 (1981).
15. Cizewski, J.A., Burke, D.G., Flynn, E.R., Brown, R.F., and Sunier, J.W., Phys. Rev. $\underline{C27}$, 1040 (1983).
16. Scholten, O., Brant, S., and Paar, V., Phys. Lett. $\underline{171B}$, 335 (1986).
17. Sun, H.Z., Frank, A., and Van Isacker, P., Phys. Rev. $\underline{C27}$, 2430 (1983).
18. Gelberg, A., Z. Phys. $\underline{A315}$, 119 (1984).
19. Stachel, J., Van Isacker, P. and Heyde, K., Phys. Rev. $\underline{C25}$, 650 (1982).
20. Van Isacker, P., Jolie, J., Heyde, K., Waroquier, M., and Moreau, J., Phys. Lett. $\underline{149B}$, 26 (1984).
21. Casten, R.F., and Cizewski, J.A., Nucl. Phys. $\underline{A309}$, 477 (1978).
22. Warner, D.D., Van Isacker, P., Jolie, J., and Bruce, A.M., Phys. Rev. Lett. $\underline{54}$, 1365 (1985).
23. Scholten, O., and Warner, D.D., Phys. Lett. $\underline{142B}$, 315 (1984).
24. Bijker, R. and Scholten, O., Phys. Rev. $\underline{C32}$, 591 (1985).
25. Casten, R.F., Warner, D.D., Gowdy, G.M., Rofail, N., and Lieb, K.P., Phys. Rev. $\underline{C27}$, 1310 (1983).
26. Vergnes, M., Berrier-Ronsin, G., and Bijker, R., Phys. Rev. $\underline{C28}$, 360 (1983).
27. Hubsch, T., Paar, V., and Vretenar, D., Phys. Lett. $\underline{151B}$, 320 (1985).
28. Van Isacker, P., Jolie, J., Heyde, K., and Frank, A., Phys. Rev.Lett. $\underline{54}$, 653 (1985).

A NEW U(6/20) SUPERSYMMETRY AND ITS APPLICATION TO THE
A = 130 MASS REGION

J.Jolie
Institute for Nuclear Physics,Proeftuinstraat 86
B-9000 Gent (Belgium)

ABSTRACT

We construct a new dynamical supersymmetry appropriate for the single-particle orbits j = 1/2, 3/2, 5/2 and 7/2. This limit is applied to the A=130 mass region. We discuss possible evidence for the existence of a U(12/20) supersymmetry in this region.

1. INTRODUCTION

In the past few years, the Interacting Boson Model[1,2,3] (IBM), in which dynamical symmetries play an important role, has shown to be a simple, yet remarkably versatile model for the description of low-lying energy states in medium-heavy and heavy nuclei. In order to describe the collective excitations in more detail and to treat more complex systems, this model has been extended in several ways. An important extension of the IBM is the Interacting Boson-Fermion Model (IBFM-1)[4] where a fermion is coupled to a boson core,allowing a description of odd-mass nuclei. Also here the concept of dynamical symmetries has been introduced yielding Bose-Fermi symmetries, where the odd-mass nucleus is described in a space spanned by the irrep [N] x [1] of $U^B(6) \times U^F(m)$, where B(F) stands for boson (fermion) and m = Σ_j (2j+1) is the dimension of the single-particle space of the unpaired nucleon.[5] Since one has many possible combinations of single particle orbitals, a large variety of dynamical symmetries can be constructed. By embedding $U^B(6) \times U^F(m)$ into a graded Lie algebra U(6/m), one obtains a supersymmetry which contains both the dynamical symmetry for the even-even nucleus and the Bose-Fermi symmetry for the odd-mass nucleus, within the same supersymmetric irrep [N} of U(6/m).

In this contribution we will describe odd-mass nuclei with a single-particle space consisting of the j = 1/2, 3/2, 5/2 and 7/2 single-particle orbitals and their even-even partners, so we need a U(6/20) supersymmetry. In the work done by Ling et al.[6] a U(6/20) scheme was constructed by reducing $U^B(6) \times U^F(20)$ to the, already in U(6/4) used, $U^B(6) \times U^F(4)$ chain but now considering the [2,1] irrep of $U^F(4)$ instead of the [1] irrep. Here we will use a more complex group chain in which the fermion groups are coupled at two different levels to the boson groups. We indicate that this chain will be appropriate for the A=130 mass region while the one of ref.6 is used for the description of the Pt and Au isotopes.

2. THE $\overline{SPIN(5)}$ LIMIT OF U(6/20)

The basic group structure of a odd-A nucleus whose valence shell contains the single-particle orbits j = 1/2, 3/2, 5/2 and 7/2 in the IBFM is $U^B(6) \times U^F(\Sigma_j(2j+1)) = U^B(6) \times U^F(20)$. If we now consider the single-particle orbits j = 1/2, 3/2, 5/2 and 7/2, as resulting from the coupling of a pseudo orbital ℓ = 2 part and a pseudo spin

$s = 3/2$, and couple first the pseudo spin part to zero we get the generators of an $U^F(5)$ Lie algebra. Coupling the pseudo orbital part to zero we obtain the generators of $SU^F(4)$ if we omit the $k = 0$ generator. So we obtain the group reduction:

$$U^F(20) \supset SU^F(4) \times U^F(5). \tag{1}$$

The group chain for the bosons in the $O^B(6)$ limit reads:

$$U^B(6) \supset O^B(6) \supset O^B(5) \supset O^B(3) \tag{2}$$

Since energy levels of nuclei have a good angular momentum we have to construct a chain starting from $U^B(6) \times U^F(20)$ and ending with $SU^{B+F}(2)$, the group which has the angular momentum J as quantum number. Since $U^F(4)$ is isomorphic with $O^B(6)$, $SU^F(4) \simeq O^B(6) \simeq Spin(6)$, we can combine the generators of $O^B(6)$ with the generators of $SU^F(4)$ and obtain the generators of Spin(6). A subset of the generators of $U^F(5)$ closes under commutation forming the group $O^F(5) = Sp^F(4)$ which is isomorphic to Spin(5), the first subgroup of Spin(6). By adding these generators we obtain a new Spin(5)group, which we shall denote $\overline{Spin(5)}$ where the bar indicates that its generators are different of those of Spin(5). Finally we go from $\overline{Spin}(5)$ to $Spin(3) = SU^{B+F}(2)$. By embedding now $U^B(6) \times U^F(20)$ into the supergroup $U(6/20)$, we obtain the following group chain:

$$U(6/20) \supset U^B(6) \times U^F(20) \supset U^B(6) \times U^F(4) \times U^F(5) \supset O^B(6) \times U^F(4) \times U^F(5) \supset$$
$$[N\} \quad [N] \quad [1] \quad [N] \quad [1] \quad [1] \quad \langle\Sigma\rangle \quad [1] \quad [1]$$

$$Spin(6) \times U^F(5) \supset Spin(5) \times O^F(5) \supset \overline{Spin}(5) \supset Spin(3), \tag{3}$$
$$\langle\sigma_1,\sigma_2,\sigma_3\rangle \quad [1] \qquad (\tau_1,\tau_2) \quad (1) \qquad (\overline{\tau}_1,\overline{\tau}_2) \qquad J$$

where we have indicated also the associated quantum numbers, which characterize the irreducible representations (irreps) of the groups.

Using the Casimir operators of the group chain(3)we obtain the following simple Hamiltonian,relevant for the low-lying spectrum:

$$H = A\, C_2(Spin(6)) + B\, C_2(Spin(5))$$
$$+ B'\, C_2(\overline{Spin}(5)) + C\, C_2(Spin(3)), \tag{4}$$

which has the eigenvalues:

$$E = A(\sigma_1(\sigma_1+4) + \sigma_2(\sigma_2+2) + \sigma_3^2) + B(\tau_1(\tau_1+3) + \tau_2(\tau_2+1))$$
$$+ B'(\overline{\tau}_1(\overline{\tau}_1+3) + \overline{\tau}_2(\overline{\tau}_2+1)) + C\, J(J+1) \tag{5}$$

From the reduction and multiplication rules, which give us the quantum numbers to be used in (5), we obtain the energy spectrum of this U(6/20) chain.

The Hamiltonian (4) is of the form:

$$H = H^B + H^F + V^{BF}. \tag{6}$$

where H^B is the boson part of the Hamiltonian which is given by:

$$H = A \ C_2(O^B(6)) + (B + B') \ C_2(O^B(5)) + C \ C_2(O^B(3)). \tag{7}$$

H^F describes the fermion part of the hamiltonian and V^{BF} the boson-fermion interaction. For even-even nuclei, only the boson part of H^B is of importance of which the spectra is described in ref 3. In figure 1 we show the low lying spectrum belonging to the $<N+1/2,1/2,1/2>$ irrep of Spin(6). Due to the particular coupling of the fermion groups

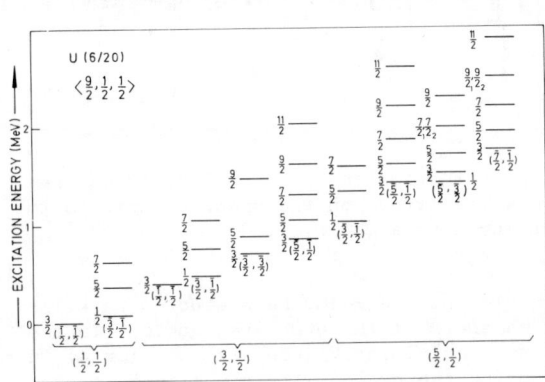

Figure 1: The energy spectrum of the $<9/2,1/2,1/2>$ irrep of Spin(6) with the quantumnumbers of chain(3).

to the boson groups in the cain, the contribution of the different Casimir operators to the fermion Hamiltonian allows for a range of different energy spectra associated with one even-even nucleus. This is a result of the possibility to move the 3/2 quasiparticle energy relatively to the 1/2, 5/2 and 7/2 energies which results in different spectra. Independent of the energy spectrum for the even-even nucleus, we can choose the strenght of the Spin(5) and Spin(5)Casimir operators, with the only constraint that their sum is constant.

3. APPLICATION OF U(6/20) TO THE A = 130 MASS REGION

Recently, evidence for an extensive region of O(6) nuclei near A=130, was presented by Casten and von Brentano[7]. In the A=130 region the odd neutron hole fills the $3s_{1/2}, 2d_{3/2}, 2d_{5/2}$ and $1g_{7/2}$ orbits below the N=82 closed neutron shell, if we only consider positive parity states. The two lowest orbits are $3s_{1/2}$ or $2d_{3/2}$ and then the $2d_{5/2}$ and $1g_{7/2}$ appear higher in energy. So we expect that the present U(6/20) shceme could give a reasonable description of the positive-parity odd-neutron spectra in the A=130 region. However as we shall see the energies of the $2d_{5/2}$ and the $1g_{7/2}$ orbits do not follow the J(J+1) rule. The spectra for these odd neutron nuclei are very complex, since as a result of the deformation of the even-even core in this region, all bandheads of these orbits occur below or around 500 keV. As supersymmetry directly relates the boson-boson interaction, describing the deformed even-even nucleus, with the boson-fermion interaction, this region is a good test for supersymmetry as a new

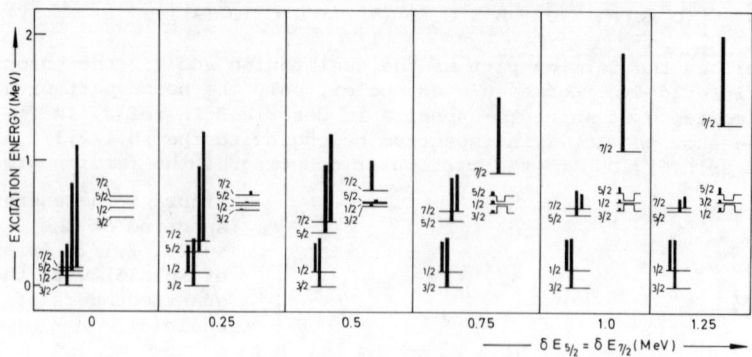

Figure 2: The effect of the perturbation δE_j for $J = 5/2, 7/2$ on the energy spectrum. The bars show the effect on the spectroscopic factor C^2S_j for pick-up reaction strength to each level.

symmetry concept.

Since the even-even nuclei have already been studied, we will concentrate on the odd-mass partners of the even-even nuclei. In the light of the existence of F-spin multiplets[8] (see also section 4) in the A=130 region we will describe the nuclei with the same N in one calculation.

3.1 Odd-even nuclei with $N=5$

The two nuclei which are well known are ^{127}Te and ^{131}Xe. Also for ^{135}Ba some data are known, which will be discussed in chapter 4. The nuclei with $N = 5$ are near to the closed N = 82 shell and thus the 1/2 and 3/2 orbitals will dominate the low-lying levels with the 5/2 and 7/2 singel-particle orbits occuring at higher energy. Recently a pick-up experiment [9] using the ^{128}Te(d,t)^{127}Te and the ^{128}Te(3He,α)^{127}Te reactions showed that in ^{127}Te the second $7^+/2$ level ar 926 keV is strongly populated and that the $5^+/2$ levels at 783 keV and 1140keV take most of the $2d_{5/2}$ spectroscopic strength. Since these levels occur arount 1 MeV in the spectrum of ^{127}Te and the U(6/20) predicts them much lower, we will need a perturbation to describe the nucleus. The most transparant way to perturb the Hamiltonian (4) without disturbing the supersymmetry condition too much is as follows. Since the even-even nucleus together with the $3^+/2_1$, $1^+/2_1$ spacing fully determine the parameters in (4), they also determine the quasiparticle energies E_j. Numerically we can introduce a perturbation δE_j for the $j = 5/2$ and $7/2$ orbital which corrects the too low quasiparticle energy such that $E_j + \delta E_j = E_j^{BCS}$, the quasiparticle energies obtained after solving the BCS equation. Since this correction occurs only in H^F we only correct the fermion-fermion interaction and preserve a simultaneous description of the boson-fermion interaction and the boson-boson interaction as obtained using the supersymmetry prescription. The effect of this δE_j on the calculated energy spectrum and the spectrocopic factors for pick-up reactions is shown in Figure 3. We observe a redistribution of the strength of the lowest levels towards the higher-lying levels.

In the present calculations we determine the parameters of (4) from the even-even nuclei ^{126}Te, ^{130}Xe and ^{134}Ba and the 1/2, 3/2 spacing. The corrections δE_j were obtained from the BCS calculation done by Cunningham for ^{131}Xe.10) This yields $\delta E_{5/2}$ = 960 keV and $\delta E_{7/2}$ = 814 keV. With the parameters now completely determined, we calculate the energy levels in ^{127}Te and the spectroscopic strength using:

$$T_{lj}(N, m = 0 \to N, m = 1) = \Sigma_j \; r_j a_j^+ , \qquad (8)$$

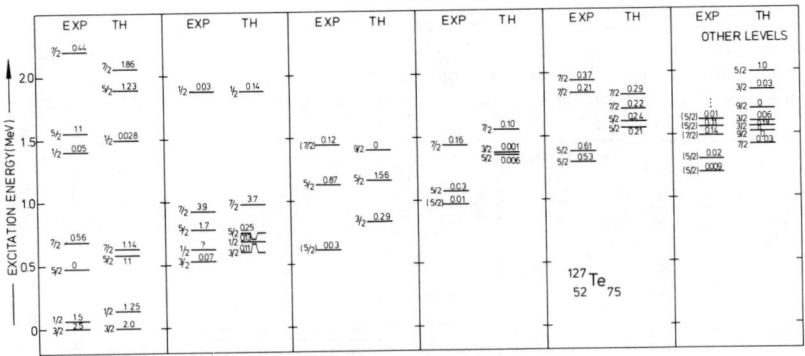

Fig.3: The experimental[9] and theoretical positive-parity levels for ^{127}Te. The parameters used are A=-96 keV, B=111 keV, B'=19 keV and C = 5 keV.

with $r_j = v_j$ taken from the same BCS calculation[10]. The comparison with experimental data[9] is given in Figure 3. There is a large discrepancy for the lowest two 5$^+$/2 levels, where the theoretical distribution is opposite to the experimental. The other levels show a one-to-one relation between the theoretical and experimental levels up to high energy, except for the levels shown on the righthand side of the figure. These levels are only seen in the study of ref 9 and have not yet a unique spin assignment. The authors have for example not yet an experimental basis to choose between 3$^+$/2 and 5$^+$/2 and assume the 5$^+$/2 assignment. Following our calculation we assume the 3$^+$/2 assignment for these levels. With the same parameters we also performed a calculation for ^{131}Xe as given in Figure 4. We also show the calculated and experimental B(E2)[10,11,12] values. The E2 transition operator used was:

$$T^{(E2)} = e_b(s^+\tilde{d} + d^+s)^{(2)} + e_f S^{(2)} + e_f, A^{(2)}(2,2), \qquad (9)$$

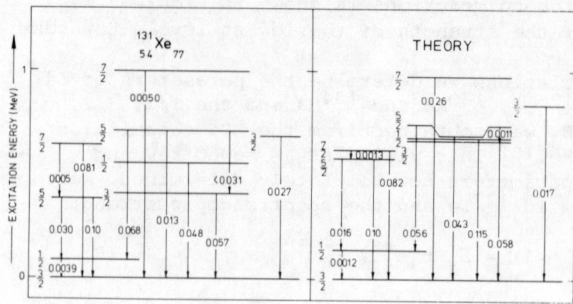

Figure 4: The experimental and theoretical positive-parity levels of ^{131}Xe and the known B(E2) values. The parameters are the same as in Fig.3.

with $S^{(2)}$ and $A^{(2)}$ (2,2) the k = 2 generators of $U^F(4)$ and $U^F(5)$. In the calculated energy spectrum the second band is too compressed, but the electromagnetic E2 transitions are well reproduced. For the electromagnetic transition operator we assumed $e_b = -e_f = -e_{f'} = 0.148$ e.b.

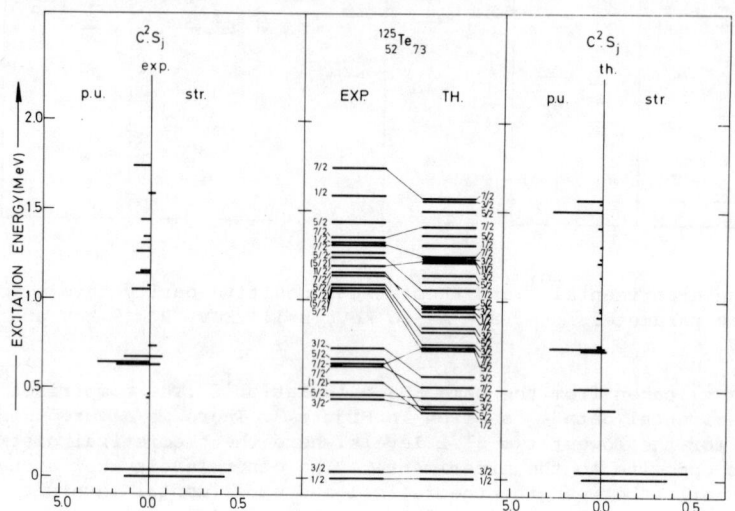

Fig.5: The experimental and theoretical positive-parity level scheme for ^{125}Te. In the center both energy spectra are given. The left-hand side of the figure gives the known C^2S_j for both stripping (str.) and pick-up (p.u.) reactions. The right-hand side gives the theoretical values for C^2S_j. The parameters are A = -64 keV, B = 76 keV, B' = -10 keV and C = 18 keV. The experimental spin assignment is taken from 13.

3.2   Odd-even nuclei with N = 6

Applying the same procedure as described before we now obtain $\delta E_{5/2}$ = 802 keV and $\delta E_{7/2}$ = 488 keV. In figure 5 we give a comparison between the experimental energy spectrum with the pick-up[13] and stripping[14] spectroscopic factors and the theoretical prediction. Analogous to the calculation of the pick-up transfer we take as parameters in the transfer operator $r'_j = u_j$, with the u_j's taken from the BCS calculation[10]. For the stripping reaction if we neglect more complex exchange processes, a transfer operator which consists of the summation of the operators $s_{1j,j}$ $(a_j^+, b_1^+)^{(j)}$ results. Simple analytic expressions are obtained if we choose $s_{1j,j}$ such that the transfer operator has a definite tensorial character under Spin(6) and its subgroups. We will take the lowest Spin(6) representation which has the following tensorial character under the groups of chain (3) :

$$T_{1j} (N, m=0 \to N-1, m=1) =$$
$$_T|1||1|;<1><1/2,1/2,1/2>|1|;<1/2,1/2,1/2>;(v,1/2)(1);(\tau1,\tau2);jm. \quad (10)$$

and in (10) we replace all a_j^+ by $u_j a_j^+$. In the application to ^{125}Te the absolute value of the spectroscopic factors needs to be normalized to the experimental value. The spectroscopic factors are normalized to the experimental factors by putting $\Sigma_j C^2 S_j^{exp} = \Sigma_j C^2 S_j^{th}$ for each j. In table 1 we compare the experimental B(E2) values[10] for ^{129}Xe with the theoretical values obtained with $e_b = -e_f = -e_{f'} = 0.134$ e.b.

$J_i^+ \to J_f^+$	B(E2) ($e^2 b^2$)	
	Experiment	Theory
$\frac{3}{2}_1 \to \frac{1}{2}_1$		0.007
$\frac{3}{2}_2 \to \frac{3}{2}_1$	< 0.0005	0.013
$\frac{3}{2}_2 \to \frac{1}{2}_1$	0.12(1)	0.12
$\frac{5}{2}_1 \to \frac{3}{2}_1$	0.22(3)	0.10
$\frac{5}{2}_1 \to \frac{1}{2}_1$	0.077(7)	0.039
$\frac{1}{2}_2 \to \frac{3}{2}_1$	0.044(11)	0.12
$\frac{5}{2}_2 \to \frac{1}{2}_1$	0.057(4)	0.071
$\frac{3}{2}_3 \to \frac{1}{2}_1$	0.0032(2)	0.0056
$\frac{3}{2}_4 \to \frac{1}{2}_1$	0.0030(2)	0.0004

Table 1: The experimental and theoretical B(E2) values for the positive-parity values in ^{129}Xe. The indices correspond to the appearance of the levels in the experimental spectrum.

3.3 Odd-even nuclei with N=7

The situation of the nuclei with N = 7 is similar to the one described above, yielding $\delta E_{5/2}$ = 451 keV and $\delta E_{7/2}$ = 137 keV. In Figure 6 we show the results for the pick-up and stripping reaction to ^{123}Te. For the stripping reactions, we used the same renormalization as in 3.2.

4. F-SPIN MULTIPLETS AND SUPERSYMMETRY IN THE A = 130 REGION

In a recent article Harter et al.[8] pointed out that the even-even spectra with the same total number of bosons in the A=130 region have remarkably similar spectra and that this can be understood in the context of F-spin multiplets. Following their study we have examined the combination of F-spin symmetry and supersymmetry in U(6/20)

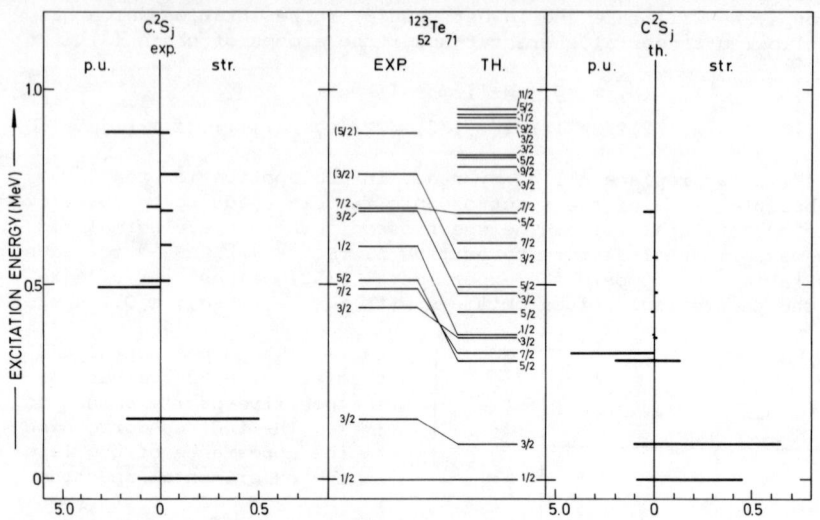

Figure 6: Same caption as Figure 6 for the nucleus ^{123}Te. The parameters are A= -60 keV, B = 80 keV, B' = -20 keV and C = 18 keV. The spin assignment is taken from 15.

and shown that this can approximately be applied to the odd neutron nuclei in the A = 130 region[16]. This U(12/20) supersymmetry predicts the same spectra for all odd-A nuclei with the same $N = N_\nu + N_\pi + 1$, where N is the quantum number of the U(12/20) irrep and $N_\nu(N_\pi)$ the number of neutron (proton) bosons of the core. So a particular multiplet contains the odd-mass nuclei ^{139}Ce, ^{135}Ba, ^{131}Xe and ^{127}Te. Since the Hamiltonian (4) is used to describe the symmetric representation of $U^B_{\nu+\pi}(6)$, the results of the present calculations remain true in U(12/20) and can be used to test if the occurence of F-spin multiplets in even-even nuclei in the A = 130 region can be extended to the odd-A nuclei. In the calculations carried out here we have broken the supersymmetry at the level of Spin(6) by adding the correction δE_j to the quasi-particle energies in the fermion Hamiltonian. This modification does not break the prediction of the same energyspectra for the odd-A nuclei. In this section we will examine the results obtained in the detailed calculation of section 3 in the context of F-spin multiplets. Also here we restrict ourselfs to those nuclei of which more is known than just the energy spectra. For these spectra classified according to N, it is clear that they are much more similar compared with the spectra classified by N_ν or N_π. As we have already implicitly taken the same parameters for nuclei with the same boson number, we have shown in section 3 that at least in the Te and Xe odd-mass

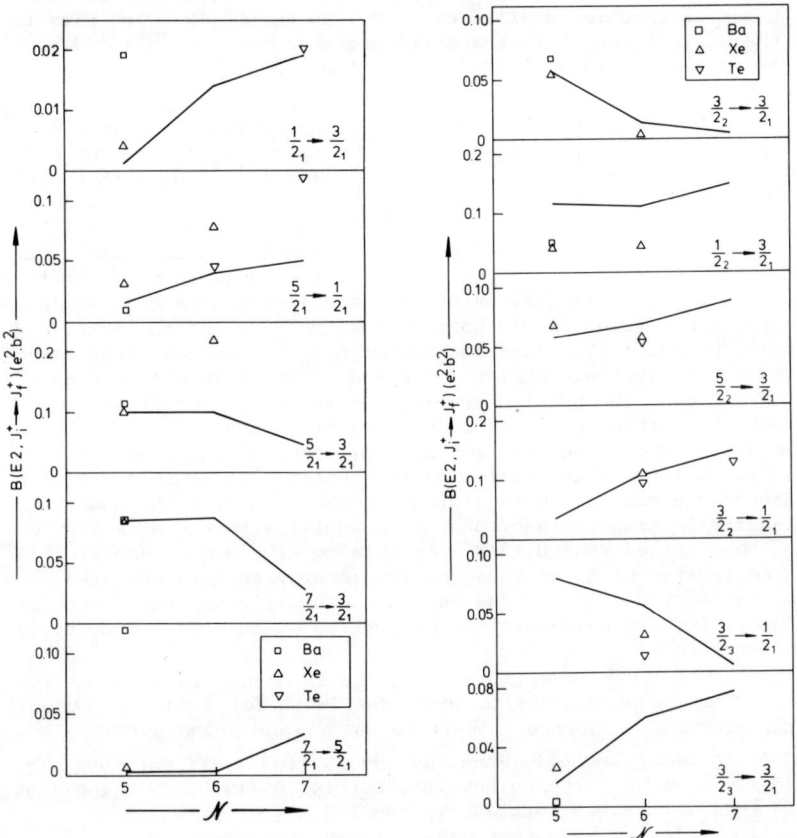

Fig.7 : The known experimental B(E2, $J_i^+ \to J_f^+$) values vs. N. The solid lines give the theoretical predictions as obtained in the broken U(12/20) scheme.

isotopes the U(12/20) scheme gives a good description of all known experimental data. However we have not compared the experimental data of the different nuclei belonging to a multiplet. Experimentally B(E2) values are known for several odd-mass nuclei in this region. In figure 7 we compare these experimental values with the theoretical values as obtained in U(12/20). First it seems that, with the exception of the $1/2_1 \to 3/2_1$ and $7/2_1 \to 5/2_1$ transitions in ^{135}Ba, there is a striking similtarity between the experimental B(E2) values of the different nuclei with the same N, as is imposed by F-spin invariance. Secondly the theoretical B(E2) values give a good description of the behaviour of these values as a function of N, taking into account the different scales in Figure 7. In section 3 have shown that U(6/20) can give a good description of the neutron single-particle transfer reaction amplitudes in Te. To test U(12/20) we have to go to the Ba isotopes where the stripping reactions Ba^{130,132}(d,p) Ba^{131,133} have been investigated. In table 2 we compare the C^2S_{1j}

J_i^π	^{131}Ba	^{123}Te	^{127}Te	^{133}Ba	^{125}Te	^{129}Te
$1/2_1$	0.265	0.39	0.25	0.225	0.42	0.22
$3/2_1$	0.257	0.505	0.38	0.31	0.46	0.34

Table 2: Comparison of the experimental stripping spectroscopic factor C^2S_j for ^{131,133}Ba with these in ^{123,125}Te following U(12/20) and with the spectroscopic factors in the isotonic nuclei ^{127,129}Te.

values with the experimental values in Te. Following the U(12/20) hypothesis the C^2S_{1j} value should be the same in ^{131}Ba and in ^{123}Te, and also in ^{133}Ba and ^{125}Te. This is however not the case and they resemble more to the isotonic nuclei ^{127}Te and ^{129}Te. This can be understood as follows. The particular experiments from a sensitive probe to measure the single particle components of the wavefunction which makes that the spectroscopic factors C^2S_{1j} depend highly on the occupation probability of the last unpaired neutron. So it is expected that isotonic nuclei are more similar to each other, with respect to single-particle transferreactions, than nuclei with the same N. The picture then arises that U(12/20) is able to give a good description of the collective parts of these nuclei, energy spectra and B(E2) values, but not, due to the changes of the underlying single-particle structure, of the single-particle properties of the odd-A members of a F-spin multiplet.

The U(6/20) scheme will be discussed in detail in a forthcoming paper[17]. The author whishes to thank the IWONL for financial support and this work was supported in part by NATO grant nr.RG.85/0699.

1-3) A.Arima and F.Iachello, Ann.Phys.99,253(1976);111,201(1978);123, 468(1979)
4) F.Iachello and O.Scholten,Phys.Rev.Lett.44,67(1979).
5) F.Iachello, Phys.Rev.Lett.44,772(1980).
6) Y.S.Ling, et al., Phys.Lett.148B,13(1984).
7) R.F.Casten and P.von Brentano,Phys.Lett.152B,22(1985).
8) H.Harter, et al.,Phys. Rev. C32,631(1985).
9) T.Rodland, et al., Phys. Scripta,32,201(1985).
10) M.A.Cunningham,Nucl.Phys.A385,221(1982).
11) D.C.Palmer, et al.,J.Phys.G4,1143(1978).
12) A.D.Irving, et al.,J.Phys.G5,1595(1979).
13) T. Rodland, et al.,Phys.Scripta,29,629(1984).
14) A.Graue, et al.,Nucl.Phys.A136,513(1969).
15) M.A.G. Fernandes and M.N.Rao,J.Phys.G3,1397(1977).
16) J.Jolie, et. al.,Phys.Rev.Lett.55,1457(1985).
17) J.Jolie,K.Heyde,P.Van Isacker and A.Frank, subm. to Nucl.Phys.A

THE SO(7) FERMION DYNAMICAL SYMMETRY AND THE Ru,Pd TRANSITIONAL REGION

R. F. Casten[*], C. L. Wu[†,**], D. H. Feng[†],
J. N. Ginocchio[††], and X. L. Han[**]
[*]Brookhaven National Laboratory, Upton, New York, 11973, USA
[**]Jilin University, Changchun, People's Republic of China
[†]Drexel University, Philadelphia, Pennsylvania, 19104, USA
[††]Los Alamos National Laboratory, Los Alamos, New Mexico, 87545, USA

ABSTRACT

The possibility of Fermion Dynamical Symmetries is discussed and empirical evidence for an SO(7) symmetry in the Ru, Pd region is presented. This symmetry has the property that it describes (analytically), not a static structure, but rather a phase transitional region from vibrator towards γ-soft rotor.

1. INTRODUCTION

In the last decade there has been intense interest in the application of symmetry ideas and algebraic techniques to nuclear structure. Most of this work has centered on the Interacting Boson Approximation Model[1] and its extensions. Empirical examples of all three boson dynamical symmetries of the U(6) group have been identified. This is important even though few nuclei actually satisfy the strict criteria of these idealized limits since the symmetries give a structural framework for the systematic understanding of heavy nuclei. Despite this, there is the unsettling aspect that these boson symmetries are essentially phenomenological: they correspond to specific choices of the parameters of the IBA-1 Hamiltonian. Ultimately, those parameter values are consequences of fermion shell structure and residual interactions and it is not yet known whether the particular values giving rise to dynamical symmetries occur more or less accidentally from variations in the underlying microscopy or whether they reflect deeper origins. The repeated appearance, throughout heavy nuclei, of examples or near examples of boson symmetries, however, does suggest a fundamental basis and hints at the possibility of fermion symmetries.

There has recently been considerable theoretical interest in such fermion symmetries. This has primarily centered on a pseudo-SU(3) approach[2-3] and on one[4-5] based on the SO(8) and Sp(6) groups. The former emphasizes the symmetries as appropriate bases in terms of which reasonable model Hamiltonians can be diagonalized while the latter stresses the nuclear structure of the symmetries themselves. In it the SO(8) and Sp(6) groups give rise to chain decompositions leading to dynamical symmetries denoted SU(3), SO(6), SO(5) ⊗ SU(2), SO(3) ⊗ SU(2) and SO(7). The first two correspond to the IBA symmetries SU(3) and O(6) while the others do not have exact counterparts although their energy spectra resemble U(5). While empirical examples of near O(6) or SU(3) nuclei can serve as indirect evidence for the corresponding fermion symmetries, a more convincing result would be to identify empirically one of the fermion symmetries exhibiting properties <u>different</u> from any of the boson symmetries. Aside from the inherent interest attached to any symmetry, this would suggest the relevance of this approach[4-5] as a fermionic model for the IBA.

The purpose of the present paper is to present just such evidence, for the SO(7) symmetry, in the Pd and Ru region. This symmetry is, in fact, particularly interesting since, unlike others, it does not correspond to a static structure (e.g., γ-unstable rotor or anharmonic vibrator) but rather to one which inherently varies with mass: the SO(7) fermion dynamical symmetry therefore does not correspond to an IBA symmetry but rather describes a <u>transitional sequence</u> which evolves between structures closely related to two of those symmetries, U(5) and O(6).

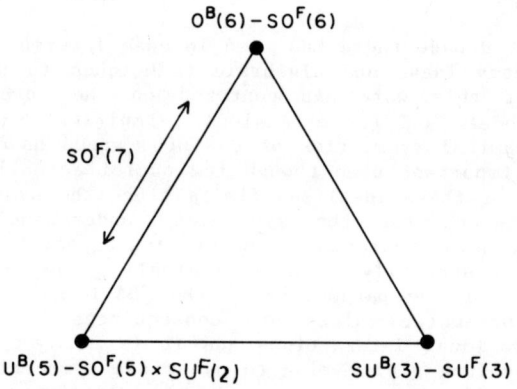

Fig. 1 The symmetry triangle modified to show the approximate correspondence of boson and fermion symmetries and to indicate the structural sequence embodied by the SO(7) limit. Strictly speaking, the $SO^F(5)$ symmetry differs from $U^B(5)$ because of Pauli factors. Note that, here and below, the SO(5) vibrational symmetry is a shorthand notation for SO(5) x SU(2).

The details of the Ginocchio model[4] are available in the literature and will not be repeated here. The essential step involves rewriting the fermion angular momenta, j, in a pseudo orbital and spin decomposition as $\bar{j} = \bar{k}+\bar{i}$. Specific values of k and i lead to sequences of j values ranging from j = 1/2 to (k+i) in unit increments. A major difficulty arises in the actual application to heavy nuclei, however, since each physical shell involves a unique parity orbit and, therefore, a gap of two units in j, at the upper end. This problem has been recently circumvented by the development of the FDS (Fermion Dynamical Symmetry) model, in which the unique parity orbit is decoupled, forming its own, seniority zero, subspace whose only effect in the low energy, low spin regime is to consume a certain number of valence nucleons. The FDS model also avoids the so-called "fatal flaw" of the original Sp(6) symmetry when applied to the SU(3) chain. Moreover, it establishes a unique correspondence between the shell model and the k-i basis so that the normal parity orbits in each major shell in heavy nuclei then have either SO(8) symmetry (if i = 3/2) or Sp(6) symmetry (if k=1). Thus a direct relation between symmetry and shell structure is obtained: this link is analogous to the complementary one between the IBA and collective excitation modes of geometric models.

As just noted, the SO(8) ⊃ SO(7) symmetry thus requires i = 3/2 and, if k=1, describes a shell with j = 1/2, 3/2, 5/2, while, if k=2, then j = 1/2, 3/2, 5/2, 7/2. The former would apply to the 28-50 shell or the Pt region (where these orbits are effectively isolated). The latter corresponds to the 50-82 shell. In Pd, Ru, the protons are in the former shell, the neutrons in the latter. Since both can therefore exhibit SO(7) this region offers an ideal candidate, and all the more so since recent[6] numerical IBA calculations have identified these nuclei as intermediate between U(5) and O(6).

The structure of the SO(7) symmetry can be easily understood by considering the particular Hamiltonian underlying it which contains residual interactions written in terms of fermion operators corresponding to pairing and multipole interaction terms. When the fermion residual interaction is dominated by monopole pairing only, it leads to the SO(5) symmetry, when the quadrupole interaction dominates, it leads to SO(6) and when monopole and quadrupole terms are of <u>equal</u> strength, an SO(7) symmetry is generated. The corresponding IBA Hamiltonian has boson energy and quadrupole terms and describes a structure <u>intermediate between U(5) and O(6)</u> as indicated schematically by the symmetry triangle of fig. 1.

This structural evolution is clearly evident in fig. 2 which shows the effective intrinsic state deformation β_{int} (in approximate units of A/2N $\beta_{nucleus}$). β_{int} values of 0 and 1 describe the U(5) (SO(5)) and O(6) (SO(6)) limits. The evolution towards the latter is clearly evident. The lower part of fig. 2 shows the ratio of the overlaps of the SO(7) ground state wave function with those of SO(6) and SO(5): again, for large N, the similarity to SO(6) grows rapidly.

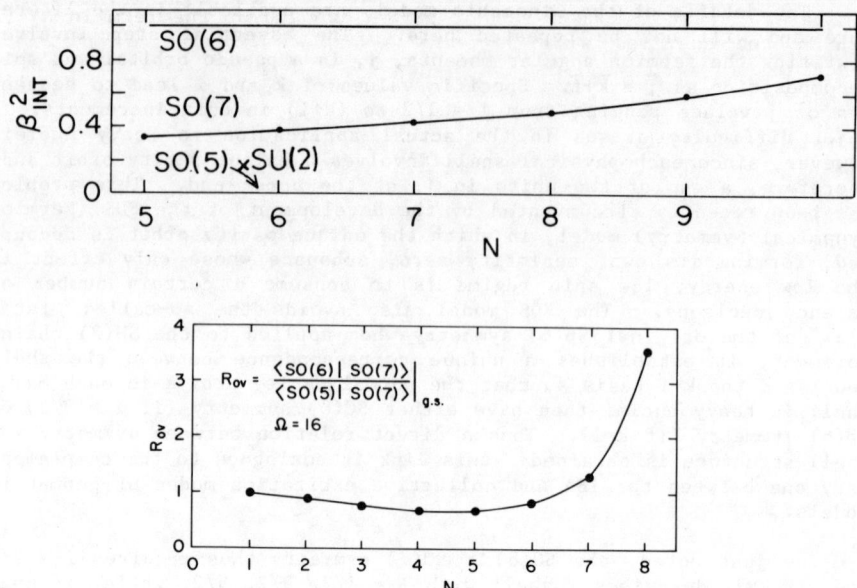

Fig. 2 (top): β values for the intrinsic state; (bottom): ratio of overlaps of the SO(7) ground state with those of SO(6) and SO(5).

The SO(7) chain decomposition is $SO(8) \supset SO(7) \supset SO(5) \supset SO(3)$ and the excitation energies are therefore given by[4]

$$E(N_1,\bar{\kappa},\tau,J) = -G_0\bar{\kappa}(\Omega_1 - 2N_1 + \bar{\kappa} + 5) + b_3\tau(\tau+3) + 1/5(b_1-b_3)J(J+1) \qquad (1)$$

In eq. 1, Ω_1 is the pair degeneracy of the normal parity proton and neutron orbits (for Pd, Ru Ω=16) and N_1 is the effective fermion valence pair number defined by $N_1 = \alpha N$ where α represents the fraction of the 2N valence nucleons that occupy the normal parity levels, $\bar{\kappa}$ is a phonon-like quantum number and τ corresponds to the O(5) IBA quantum number. In the present approach the proton and neutron shells are combined (as in IBA-1): a more microscopic study would separately consider N_{1_π} and N_{1_ν}.

The energy level spectrum of eq. 1 consists of multiplets, with successively larger $\bar{\kappa}$ values, whose spacing is controlled by G_0. Within a multiplet the b_3 and (b_1-b_3) terms, which originate from odd multipole terms in the Hamiltonian, introduce degeneracy splitting. Extensive experience with shell model calculations suggests that the first term, which stems from monopole and quadrupole terms, should dominate. A crucial aspect is that, on account of the term in N_1 in eq. 1, the energies will <u>decrease</u> linearly with increasing mass, reaching a minimum at mid shell. This feature resembles a transitional region and, again, is a reflection of the evolving structure of the wave functions from vibrator toward γ-soft rotor. It is an appealing one since transitional regions are especially difficult to treat, and yet here the properties are obtained analytically. However, it also presents a difficulty in attempting to identify SO(7) empirically: it is essential to distinguish possible SO(7) regions from phase transitions towards a deformed rotor. Thus, for example, the Ba-Gd nuclei near A=150 are merely a prelude to the deformed rare earth region. However, in such a region the $E_{4^+_1}/E_{2^+_1}$ ratio will attain values >3.0 and E2 branching ratios such as $B(E2:2^+_2 \to 0^+_1)/B(E2:2^+_2 \to 2^+_1)$ approach the Alaga rule value (0.7) while they vanish in SO(7). In the $^{104-110}$Pd and $^{98-104}$Ru candidates for SO(7), this ratio is ≤ 0.05. Similarly, the ratio $B(E2:3^+_1 \to 2^+_1)/B(E2:3^+_1 \to 4^+_1)$ approaches 2.5 in a transition to the rotor, but vanishes in SO(7): it is <0.1 (0.2) in these Pd (Ru) isotopes.

The application of the SO(7) symmetry to Pd and Ru is shown in figs. 3-5. The parameter values adopted for Ru (Pd) were: $-G_0$ = 45 (47.5) keV, α = 0.96 (0.91), b_3 = 5.3 keV and $1/5(b_1-b_3)$ = 7.2 keV (the latter two for both Ru and Pd). Note that, as expected, b_3, $1/5(b_1-b_3) \ll G_0$. The agreement with experiment in both figs. 3 and 4 is quite good. The $E_{4^+_1}/E_{2^+_1}$ ratio is closely reproduced as are the level patterns and systematics for each element. It is important to emphasize that the parameters, which are nearly identical in Ru and Pd, were held <u>constant</u> for each set of isotopes. The most characteristic aspect of SO(7), the smooth decrease in energies with mass, nicely reproduces the empirical trend. The slope changes predicted (and partially observed) for ^{106}Ru, ^{112}Pd arise because the normal parity orbits are effectively filled past midshell (i.e., $N_1 > \Omega/2$). Of course, the midshell point is somewhat less ascertainable in the present approach since it depends on the number of nucleons assigned to the unique parity orbit and is parameterized by α. The empirical upturn tends to occur slightly later than predicted.

Figure 5 shows the $B(E2:2^+_1 \to 0^+_1)$ values and the ratio R_0 = $B(E2:0^+_2 \to 2^+_1)/B(E2:2^+_1 \to 0^+_1)$. The SO(7) predictions here involve no new parameters (except for a normalization of the former quantity at N=6). The predicted $B(E2:2^+_1 \to 0^+_1)$ values are in excellent agreement with the data for N≤8. This B(E2) value does not actually distinguish SO(7) from SO(6) but the ratio R_0 shown below does. R_0 is in fact a crucial indicator since the decay of the 0^+_2 level is one of the cleanest discriminators between vibrator and O(6)-like limits. The SO(7) predictions reproduce the data remarkably well and point to the transition from SO(5)-like towards SO(6)-like. Other observables such as R_4 = $B(E2:4^+_1 \to 2^+_1)/B(E2:2^+_1 \to 0^+_1)$ are in good agreement with the SO(7) values as are numerous branching ratios such as $B(E2:3^+_1 \to 2^+_1)/B(E2:3^+_1 \to 4^+_1)$ and $B(E2:2^+_2 \to 0^+_1)/B(E2:2^+_2 \to 2^+_1)$. As noted above, these latter distinguish SO(7), not from SO(6) or SO(5), but from the

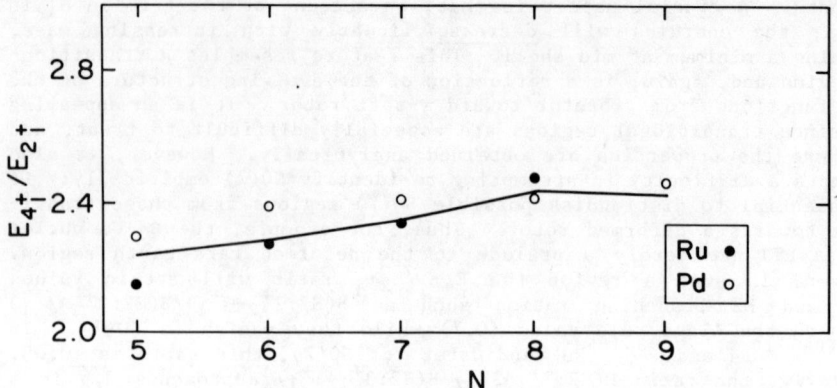

Fig. 3 Empirical and predicted $E_{4^+_1}/E_{2^+_1}$ ratios in Pd and Ru.

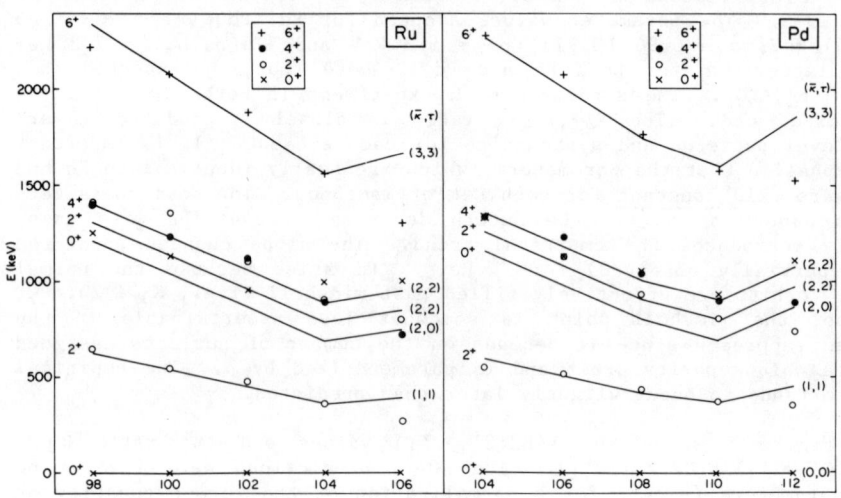

Fig. 4 Empirical[6-8] and predicted energy levels for Ru and Pd. The parameters, given in the text, are <u>constant</u> for each element.

symmetric rotor SU(3) limit.

Finally, it is worth commenting that the large α values imply that nearly all the valence neutrons and proton holes are in the normal parity orbits and, thus, in particular, that the $g_{9/2\pi}$ orbit

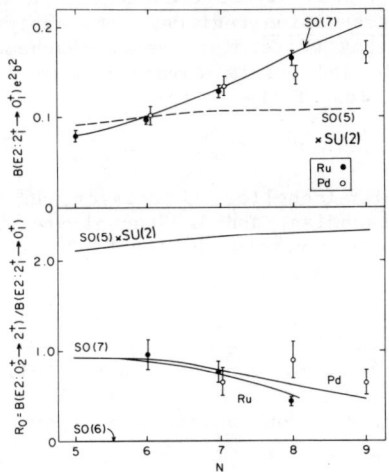

Fig. 5 Empirical[9-14] and predicted B(E2) values. Separate curves are shown for Ru and Pd where if they differ significantly.

is nearly full in Ru and Pd. Though this is contrary to naive expectations of a subshell gap near Z=40, single particle transfer data is not in fact in serious disagreement and would suggest $\alpha \approx 0.8$. Such an α value actually produces fits nearly as good as those just discussed in figs. 3-5.

Of course, there are some deviations from this idealized picture. The triplet energy levels, especially 0^+_2, are predicted at slightly lower energies than observed, and the yrast levels in the heaviest isotopes shown drop well below the predictions. This latter tendency and the rise in $E_{0^+_2}$ signal a deviation from SO(7) towards a deformed character. In Ru, the same evolution is indicated by the ratio $B(E2:2^+_2 \to 0^+_1)/B(E2:2^+_2 \to 2^+_1)$ which reaches 0.1 in ^{108}Ru. Moreover, the $E_{4^+_1}/E_{2^+_1}$ ratio in ^{106,108}Ru is 2.65 and 2.75 whereas for $^{104-110}$Pd and $^{98-104}$Ru it lies in the narrow interval from 2.14-2.48. Also, the empirical values of $R_2 = B(E2:2^+_2 \to 2^+_1)/B(E2:2^+_1 \to 0^+_1)$ are about half the SO(7) predictions and only slightly closer to SO(7) than to SU(3). All in all, however, SO(7) produces an excellent representation of the data.

To summarize, empirical evidence has been presented which shows that $^{98-104}$Ru and $^{104-110}$Pd isotopes are good empirical realizations of the SO(7) symmetry. This is the first example of a <u>fermion</u> dynamical symmetry in heavy nuclei that is not simply analogous to an established boson symmetry. Moreover, it is the first observed example of a symmetry that achieves an analytic treatment of a variable structure applicable to nuclear transition regions. Its empirical verification encourages continued study of fermion symmetry schemes as microscopic foundations for the IBA and points toward a more fundamental origin for the observed symmetries of the latter.

ACKNOWLEDGEMENTS

Discussions with F. Iachello, J. Draayer, D. D. Warner, J.-Q. Chen, M. Guidry, A. Aprahamian, and J. Stachel are gratefully acknowledged. Research has been supported by the USDOE, the Chinese Science Foundation and the NSF, and by the BMFT.

REFERENCES

1. Arima, A. and Iachello, F., Ann. Rev. Nucl. Sci. 31, 75 (1981).

2. Arima, A., Harvey, M., and Shimizu, K., Phys. Lett. 30B, 517 (1969).

3. Hecht, K. T. and Adler, A., Nucl. Phys. A137, 129 (1969) and Raju, R. D., Draayer, J. B., and Hecht, K. T., Nucl. Phys. A202, 433 (1973); Draayer, J. B. and Weeks, K. J., Ann. of Phys. 156, 41 (1984), Phys. Rev. Lett. 51, 1422 (1983).

4. Ginocchio, J. N., Ann. Phys. (N.Y.) 126, 234 (1980).

5. Wu, C.-L. et al., Phys. Lett. 168B, 313 (1986), Phys. Rev. C, submitted), Guidry, M., Wu, C. L., Feng, D. H., Ginocchio, J. N., Chen, Xuan-Gen and Chen, Jin-Quan, Phys. Lett, submitted.

6. Stachel, J., Van Isacker, P., and Heyde, K., Phys. Rev. C25, 650 (1982); Stachel, J. et al., Z. Phys. 316A, 105 (1984) and private communication.

7. Sakai, M., At. Data and Nucl. Data Tables 31, 399 (1984).

8. Bucurescu, P. et al., preprint, Cent. Inst. of Phys. Bucharest, and Luontama, M. et al., Jyvaskyla preprint JYFL 12/85.

9. Endt, P. M., At. Data and Nucl. Data Tables 26, 47 (1981).

10. McGowan, F. K. et al., Nucl. Phys. A113, 529 (1968).

11. Landsberger, S. et al., Phys. Rev. C21, 588 (1980).

12. Robinson, R. et al., Nucl. Phys. A124, 553 (1969).

13. Hasselgren, L. et al., UUIP-957 (1977).

14. Christy, A. et al., Nucl. Phys. A142, 591 (1970).

DYNAMICAL SYMMETRIES FOR ODD-ODD NUCLEI*

A. B. Balantekin[†]
Physics Division, Oak Ridge National Laboratory
Oak Ridge, Tennessee 37831 U.S.A.

ABSTRACT

Recent work for developing dynamical symmetries and supersymmetries is reviewed.

1. INTRODUCTION

The Interacting Boson-Fermion Model (IBFM) has greatly facilitated both experimental and theoretical studies of odd-mass nuclei. This model provides a simple phenomenological description to analyze and classify the experimental data[1]. In addition, the IBFM has considerable theoretical importance in two ways. First, the coupling of single particle degrees of freedom to bosons, invoked in this model, provides an excellent theoretical laboratory to investigate boson-fermion mapping. Second, when the even-even core is described by one of the symmetry chains of the Interacting Boson Model and the odd fermion is in a configuration with particular j values, the solutions of the IBFM Hamiltonian exhibit dynamical supersymmetries[2], thus providing the first experimentally observed example of a supersymmetry in nature. Consequently, the IBFM has been widely used to analyze data for odd-even nuclei.

---

*Research sponsored by the Division of Nuclear Physics, U.S. Department of Energy under contract DE-AC05-84OR21400 with Martin Marietta Energy Systems, Inc.

[†]Eugene P. Wigner Fellow.

The situation for odd-odd nuclei is completely different. The excitation spectra are much more compressed than those of odd-even nuclei, and the information on the electromagnetic transitions is very scarce. Nevertheless, there are some recent attempts to search for dynamical symmetries describing odd-odd nuclei[3,4,5]. One major motivation is the simultaneous success of the U(6/4) dynamical supersymmetry for ^{194}Pt and ^{195}Au, and the U(6/12) dynamical supersymmetry for ^{194}Pt and ^{195}Pt. The existence of these two schemes together raises the question whether or not the odd-odd nucleus ^{196}Au can be described by a suitable combination of them. It turns out that the simplest such scheme[4,5], incorporating the direct product supergroup U(6/4) × U(6/12), cannot account for the correct ground state spin.

Before we study algebraic approaches to the odd-odd nuclei, let me emphasize that such nuclei are systems with two fermions. Furthermore, since very little data are available for the excited states in odd-even nuclei with two unpaired nucleons, odd-odd nuclei are the only candidates for a mixed system of many bosons and two fermions that we can investigate experimentally and theoretically. In order to be able to describe odd-even nuclei properly, it is sufficient to incorporate the correct form of the boson-fermion exchange interaction in the Interacting Boson-Fermion Hamiltonian. However, an accurate description of odd-odd nuclei requires the correct form of the fermion-fermion force (the residual interaction) as well. Correspondingly, the algebraic modeling of the spectra of odd-odd nuclei is intrinsically more difficult.

One major step towards a proper account of the residual neutron-proton residual force is to algebraically distinguish between the fermion configurations which are particle-like and those which are hole-like. To illustrate how this is done, let me assume for simplicity that the unpaired proton (π) and the unpaired neutron (ν) occupy single j orbitals j_π and j_ν. The dimension of unpaired proton space is $m_\pi = 2j_\pi+1$, and that of the unpaired neutron space is $m_\nu = 2j_\nu+1$. The odd-proton is placed in an m_π-dimensional representation of the group $U_F^\pi(m_\pi)$ and the odd-neutron in an m_ν-dimensional representation of

$U_F^\nu(m_\nu)$. The group U(N) has two N-dimensional representations: the fundamental (particle) representation (denoted by □ in the Young tableau notation), and its conjugate (antiparticle) representation (denoted by ⊡). The unpaired fermions which are mostly particle-like ($u_j > v_j$) are placed in the fundamental representation □ of the appropriate fermionic group, and the unpaired fermions which are hole-like in the conjugate representation ⊡ of the associated fermionic group[6]. I will denote the group realized in the conjugate representation by $\overline{U}(m)$. Consequently, if the odd-proton is hole-like and the odd-neutron is particle-like, the group structure of the Hamiltonian is $U_B(6) \times \overline{U}_F^\pi(m_\pi) \times U_F^\nu(m_\nu)$. In a similar way, when they both are hole-like, the group structure is $U_B(6) \times \overline{U}_F^\pi(m_\pi) \times \overline{U}_F^\nu(m_\nu)$.

2. REALIZATION OF THE PARABOLIC RULE

Some time ago it was shown that[7] the energies of the lowest-lying states of the proton-neutron multiplet in odd-odd nuclei are quadratic functions of L(L+1), where L is the angular momentum of such states. Furthermore, if neutron and proton are both particle-like or both hole-like, the parabola is open down, and if one of them is hole-like, but the other one is particle-like, the parabola is open up[7]. Let me now demonstrate that an approximate parabolic dependence readily follows from the scheme described in the previous paragraph. To do so, I will assume that $j_\pi = j_\nu = 3/2$. For the two cases, when both unpaired nucleons are particle-like (or hole-like), or when one of them is particle-like and the other one is hole-like, the energy spectrum is described by the same energy formula. However, since the groups are realized in different representations, the quantum numbers, hence the level schemes are different for these two cases. Typical spectra for these two possibilities are shown in Fig. 1, where the same parameters are used to calculate both spectra. The energies of four low-lying states as a function of L(L+1) are plotted in Fig. 2. We observe that these states approximately lie on a parabola, which is inverted appropriately when particle or hole character of the configuration changes[6].

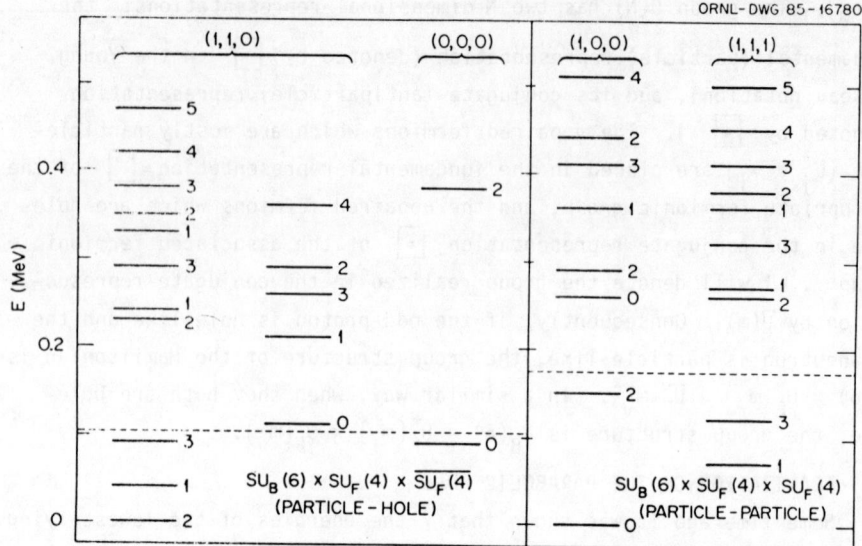

Fig. 1. Typical spectra for odd-odd nuclei when $j_\pi = j_\nu = 3/2$. The $\mathrm{Spin}_F^{\pi,\nu}(6)$ representations are shown at the top of the figure.

3. EMBEDDING IN A SUPERGROUP

So far I have only talked about the Bose-Fermi symmetries for odd-odd nuclei. Now let me briefly explain how to embed a group structure like $U_B(6) \times U_F^\pi(m_\pi) \times \overline{U}_F^\nu(m_\nu)$ into a supergroup. First if $m_\pi = m_\nu \equiv m$, we observe that the embedding[8]

$$Sp_F(2m) \supset SU_F(m) \tag{1}$$

places the odd-odd nucleus, and the two quasi-particle states of the even-even nuclei in the same representation of $Sp_F(2m)$ since the $\langle 1^2 \rangle$ representation of $Sp(2m)$ decomposes into $SU(m)$ representations as

$$\langle 1^2 \rangle = \boxed{}\!\!\boxed{} \oplus \boxed{\bullet}\!\!\boxed{\bullet} \oplus \boxed{\bullet} \tag{2}$$

The decomposition $U_F^\pi(m) \times U_F^\nu(m) \supset U_F^{\pi,\nu}(m)$ yields the adjoint ($\boxed{\bullet}$) and the singlet representations of $U_F(m)$. The adjoint representation is already included in Eq. (IV.2). The singlet comes from the

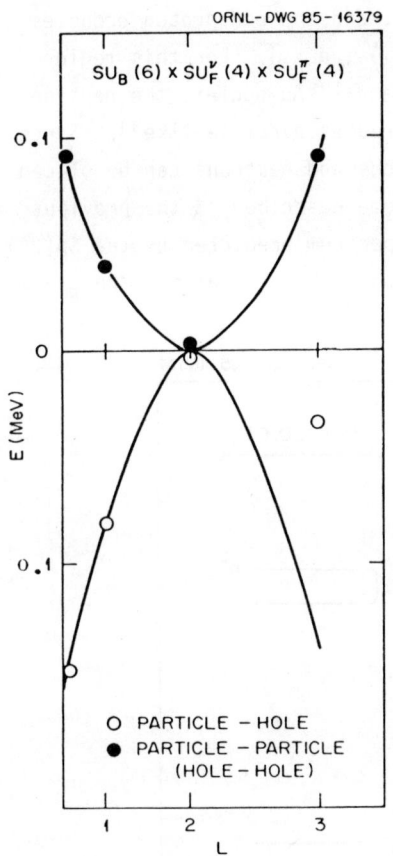

Fig. 2. Realization of the parabolic rule in the Bose-Fermi symmetry for odd-odd nuclei.

decomposition $U_F(2m) \supset Sp_F(2m)$, since the two-fermion representation $\{1^2\}$ of $U_F(2m)$[2] contains $\langle 1^2 \rangle$ and $\langle 0 \rangle$ of $Sp_F(2m)$. Hence the dynamical supersymmetry starts with the chain

$$U(6/2m) \supset U_B(6) \times U_F(2m) \supset U_B(6) \times Sp_F(2m) \supset U_B(6) \times U_F(m) \qquad (3)$$

and can be continued in the standard fashion[2].

4. APPLICATION TO ODD-ODD GOLD ISOTOPES

In the Pt-Au region, the odd-neutron occupies mostly the levels

$j^\nu = 1/2^-, 3/2^-, 5/2^-$ ($p_{1/2}, p_{3/2}, f_{5/2}$), and the odd-proton occupies the levels $j = 1/2^+, 3/2^+, 5/2^+$ ($s_{1/2}, d_{3/2}, d_{5/2}$). For this region $m_\pi = m_\nu = 12$. We assume that for ^{196}Au and ^{198}Au nuclei, the neutron orbitals enumerated above can be considered as particle-like[9]. Since the protons are hole-like, unpaired protons and neutrons can be placed in conjugate representations and the scheme described in the previous sections may be applicable. A typical spectrum predicted by the Sp(24) scheme is shown in Fig. 3, where the boson number is taken to be n = 5

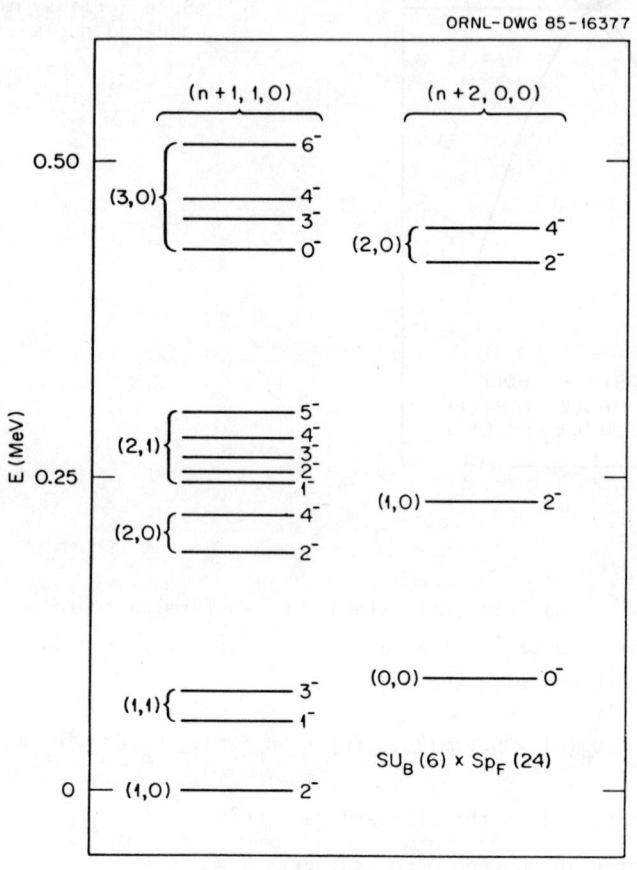

Fig. 3. A typical spectrum with $U_B(6) \times Sp_F(24)$ symmetry.

(same as the ^{196}Au core). Unfortunately, there is very little data available on the level scheme of ^{196}Au. If one assumes that the structure of the low-lying levels would not considerably change from ^{198}Au, one might expect to get a rough idea about the applicability of this scheme by examining the level scheme of ^{198}Au. A comparison of Fig. 3 with the experimental level scheme for ^{198}Au is encouraging. In this figure a third band is not shown. The two lowest levels of this band has L = 1 and 3, and the bandhead (L=1) state can be placed at ~200 keV by choosing the parameters appropriately. Except for a low-lying 3$^-$ state, there is a reasonable correspondence between the experimental spectra and the levels predicted for E < 300 keV. In particular, the ground state spin is correct, the experimentally observed three 1$^-$ states and the 4$^-$ state are accounted for. On the contrary, the scheme presented in Refs. 4 and 5 predicts the wrong ground state spin, and an additional low-lying 0$^-$ state which is not experimentally seen. Furthermore, it cannot account for the experimentally observed 4$^-$ state. These states cannot originate from the coupling of the positive-parity $i_{13/2}$ neutron orbit and the negative-parity $h_{11/2}$ proton orbit either as was also pointed out in Ref. 5. Further experimental exploration of a low-lying 3$^-$ state in ^{198}Au and a study of the low-lying negative parity states in ^{196}Au is requisite to establish the validity of our scheme in this region. Obviously, a study of the energy spectrum alone is not sufficient in assessing the significance of a new symmetry, especially since the energy differences are very small (~75 keV). It is also essential to study experimentally the electromagnetic transition rates, since such a study provides the best test of the wavefunctions of the model. I should indicate, however, a potential difficulty if one wants to extend the present scheme to a dynamical supersymmetry describing the neighboring even-even, odd-even, and even-odd nuclei in this region. Namely, the odd-proton isotopes ^{195}Au and ^{197}Au would not be satisfactorily described in such an extended scheme. In particular, the appropriate Spin(6) limit, which was shown to be successful in describing these isotopes, cannot be obtained when the odd proton occupies three orbitals with j = 1/2, 3/2, 5/2. Hence, no attempt has been made to extend this dynamical Bose-Fermi symmetry for odd-odd nuclei to a supersymmetry.

The Sp(24) scheme outlined above motivated further work on the dynamical symmetries for odd-odd nuclei[11]). These authors found that although there is considerable difference between particle-hole and hole-hole representations of the dynamical symmetry $U_B(6) \times U_F^\pi(12) \times U^\nu(12)$, there is negligible difference between the hole-hole and particle-hole representations of the dynamical symmetry $U_B(6) \times U_F^\nu(12) \times U_F^\pi(4)$. Much further work is needed before we achieve a successful algebraic description of odd-odd nuclei. However, I believe that the algebraic distinction between the fermions with particle and hole character, and the following realization of the parabolic rule are the right steps in this direction.

I would like to thank V. Paar, in collaboration with whom most of the work described here is done.

5. REFERENCES

1. For a recent review, see Scholten, O., Prog. in Part. Nucl. Phys. $\underline{14}$, 189 (1985).

2. Balantekin, A. B., Bars, I., and Iachello, F., Nucl. Phys. $\underline{A370}$, 284 (1981).

3. Hubsch, T. and Paar, V., Z. Phys. $\underline{A318}$, 355 (1984); Hubsch, T., Paar, V., and Vretenar, D., Phys. Lett. $\underline{151B}$, 320 (1985).

4. Bars, I., in "Bosons in Nuclei", Feng, D. H., Pittel, S., and Vallieres, M., Eds. (World Scientific, Singapore, 1984).

5. Van Isacker, P., Jolie, J., Heyde, K., and Frank, A., Phys. Rev. Lett. $\underline{54}$, 653 (1985).

6. Balantekin, A. B. and Paar, V., Phys. Lett. $\underline{169B}$, 9 (1986).

7. Paar, V., Nucl. Phys. $\underline{A331}$, 16 (1979).

8. Balantekin, A. B., Hubsch, T., and Paar, V., in preparation.

9. Sorensen, R. A., Nucl. Phys. $\underline{A420}$, 221 (1984).

10. Balantekin, A. B. and Paar, V., in press.

11. Feng, D. H., Sun, H. Z., and Vallieres, M., Phys. Rev. $\underline{C33}$, 1471 (1986).

EXTENSION OF SUPERSYMMETRY TO TRANSITIONAL AND ODD-ODD NUCLEI

P. Van Isacker

School of Mathematical and Physical Sciences
University of Sussex, Brighton BN1 9QH
ENGLAND

ABSTRACT

Two possible extensions of supersymmetry in nuclei are discussed: (i) an extension to regions of transitional nuclei, and (ii) an extension to odd-odd nuclei. Examples of the two approaches are discussed. A combination of the two approaches might possibly lead to a unified treatment of odd-odd nuclei in a broad range of the nuclear table.

1. INTRODUCTION

The simultaneous description of even-even and odd-A nuclei via supersymmetric considerations, has been one of the many fruitful ideas which have been proposed in the context of the interacting boson model (IBM) of Arima and Iachello.[1,2] The idea was presented by Iachello in Ref. 3, and sprang from his observation that hamiltonians corresponding to adjacent nuclei (differing by one nucleon) and described by dynamical symmetries in the IBM and the IBFM (interacting boson-fermion model, Ref. 4), have very similar parameters. In its original version, supersymmetry was applicable only in mass regions where the odd nucleon is restricted to one single-particle orbit, and this represented a severe limitation to the idea. Subsequently, it was extended to situations with several orbits via a decomposition of the angular momentum in a pseudo-spin and a pseudo-orbital part.[5] This improved its range of applications, of which an overview is given in Ref. 6. In spite of these successes in providing a common description for sets of neighbouring nuclei, the fact that supersymmetry always has been linked to the idea of a dynamical symmetry, has restricted its applicability to a few regions, where these symmetries are able to describe the experimental observations. Is, however, well known that the vast majority of medium-mass and heavy even-even nuclei corresponds to transitional hamiltonians, which are in between the three dynamical symmetries of the IBM.

As a first extension of supersymmetry, it is suggested[7] in this contribution that supersymmetry may in fact have a wider applicability than previously realized, and that its (approximate) validity must be considered independent of the particular IBM hamiltonian needed for the description of the even-even nucleus. It is thus argued that the existence of a supersymmetry is not dependent on the realization of a specific dynamical symmetry. The arguments justifying this claim will be presented in Sect. 2, and an example of its application will be given in the Ru-Rh region.

A second extension of supersymmetry concerns odd-odd nuclei. As was demonstrated in Refs. 8-11, a description of odd-odd nuclei can be achieved via the explicit treatment of the neutron and proton degrees of freedom in the IBFM. Application of the idea of supersymmetry in this case links the properties of <u>quartets</u> of nuclei (even-even, even-odd, odd-even and odd-odd), and in Ref. 9, the nuclei ^{196}Pt, ^{197}Pt, ^{197}Au and ^{198}Au were tentatively proposed as an example of such a quartet. In Sect. 3, some further results concerning the $U_\nu^B(6) \times U_\pi^B(6) \times U_\nu^F(12) \times U_\pi^F(4)$ scheme of Ref. 9 are given, and alternative schemes, applicable in other mass regions, are proposed. The application to odd-odd nuclei of the approach presented in Sect. 2, will be qualitatively assessed. Finally, in Sect. 4 the conclusions of this work are discussed.

1. SUPERSYMMETRY EXTENDED TO TRANSITIONAL NUCLEI

The procedure to generalize supersymmetry to transitional nuclei can best be explained by writing all hamiltonians in terms of Casimir operators. It is well known that the most general IBM hamiltonian for even-even nuclei can be cast into the following form:[12]

$$H = k_1 C_1(U^B(6)) + k_2 C_2(U^B(6)) + k_3 C_1(U^B(5)) + k_4 C_2(U^B(5))$$
$$+ k_5 C_2(O^B(6)) + k_6 C_2(SU^B(3)) + k_7 C_2(O^B(5)) + k_8 C_2(O^B(3)), \quad (2.1)$$

where the notation $C_n(G)$ is used for the linear ($n=1$) or quadratic ($n=2$) Casimir operator of the group G. The k_i are coefficients and the superscript B indicates that G has a realization in terms of boson operators. The boson hamiltonian (2.1) is diagonalized in the symmetric representation $[N]$ of $U^B(6)$ and since the energy eigenvalues are functions of the coefficients k_i, these can be obtained by a least-squares fit to the energy levels of a specific even-even nucleus or set of even-even nuclei.

To make the connection with a boson-fermion hamiltonian, the single-particle space needs to be specified. In the following it will be assumed to consist of orbits with angular momenta $j = 1/2$, $3/2$ and $5/2$, which fixes the model space for the odd-A nucleus to the representation $[N'] \times [1]$ of $U^B(6) \times U^F(12)$. Other combinations of single-particle angular momenta may well be treated in a similar way, but have not been investigated to date. A transitional boson-fermion hamiltonian of $U^B(6) \times U^F(12)$ reads

$$H = k_1' C_1(U^{B+F}(6)) + k_2' C_2(U^{B+F}(6)) + k_3' C_1(U^{B+F}(5))$$
$$+ k_4' C_2(U^{B+F}(5)) + k_5' C_2(O^{B+F}(6)) + k_6' C_2(SU^{B+F}(3))$$
$$+ k_7' C_2(O^{B+F}(5)) + k_8' C_2(O^{B+F}(3)) + k_9' C_2(Spin(3)). \quad (2.2)$$

The explicit form of the Casimir operators appearing in Eq. (2.2) can be found in Ref. 13 or, alternatively, in Ref. 14. Supersymmetry now imposes the following relations between the coefficients k_i in an even-

even nucleus with N bosons and k_i' in an odd-A nucleus with N-1 bosons and 1 fermion: $k_i = k_i'$ (i = 1,...,7) and $k_8 = k_8' + k_9'$. Supersymmetry indeed requires the use of the same hamiltonian for the odd-A system $[N-1] \times [1]$ as for the even-even system $[N]$, and in the latter space the hamiltonian (2.2) effectively reduces to (2.1), provided the above relations hold. The cases of supersymmetry discussed in Refs. 3, 5, 6, 13 and 14 can be viewed as particular examples of this general phenomenon, in which comparisons among the bosonic and fermionic systems become especially simple, because of the diagonal energy eigenvalues and the analytic expressions available for other spectroscopic properties.

These ideas have been applied to the Ru and Rh isotopes.[7] The coefficients k_i are obtained from a fit to the energy spectra of $^{102-108}$Ru, and are subsequently used in the hamiltonian (2.2) to <u>predict</u> the properties of the $^{103-109}$Rh isotopes. Details of this calculation are reported in Ref. 7. Similar calculations are presently being performed in other regions of the nuclear table.

3. SUPERSYMMETRY EXTENDED TO ODD-ODD NUCLEI

The boson-fermion symmetry considered in Ref. 9 is appropriate for situations in which the neutrons occupy orbits with angular momenta j = 1/2, 3/2 and 5/2, and the protons have j = 3/2. In the following, the notation of Ref. 9 will be adopted, but for the sake of clarity the group chain and its quantum numbers are repeated here:

$$U_\nu^B(6) \times U_\pi^B(6) \times U_\nu^F(12) \times U_\pi^F(4) \supset U_{\nu+\pi}^B(6) \times U_\nu^F(6) \times SU_\nu^F(2) \times SU_\pi^F(4) \supset$$
$$[N_\nu] \quad [N_\pi] \quad [1^{M_\nu}] \quad [1^{M_\pi}] \quad [N_1 N_2] \quad [1^{M_\nu}] \quad S \quad [1^{M_\pi}]$$

$$U_{\nu+\pi}^{B+F}(6) \times SU_\nu^F(2) \times SU_\pi^F(4) \supset \overline{O}_{\nu+\pi}^{B+F}(6) \times SU_\nu^F(2) \times SU_\pi^F(4) \supset$$
$$[N_1 N_2 N_3] \quad S \quad [1^{M_\pi}] \quad \langle \sigma_1 \sigma_2 \sigma_3 \rangle \quad S \quad [1^{M_\pi}]$$

$$O_{\nu+\pi}^{B+F}(6) \times SU_\nu^F(2) \supset O_{\nu+\pi}^{B+F}(5) \times SU_\nu^F(2) \supset$$
$$\langle \sigma_1 \sigma_2 \sigma_3 \rangle \quad S \quad (\tau_1 \tau_2) \quad S$$

$$O_{\nu+\pi}^{B+F}(3) \times SU^F(2) \supset O(3) \supset O(2). \qquad (3.1)$$
$$J \quad S \quad L \quad M$$

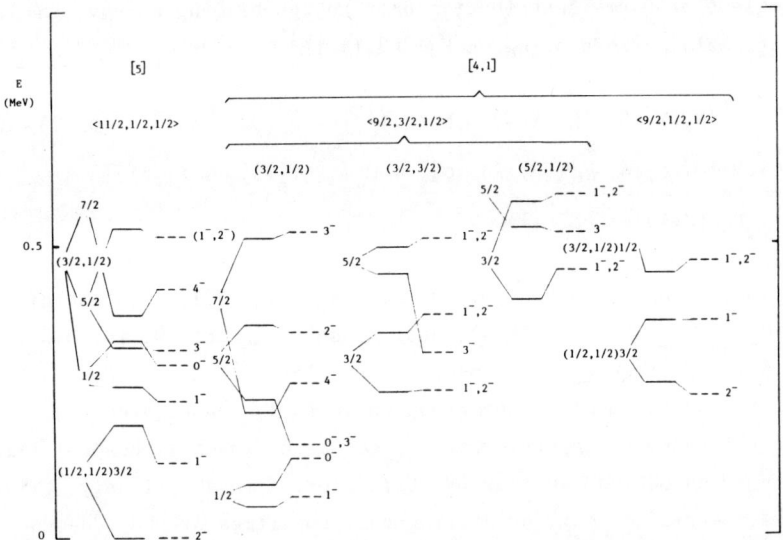

Fig. 1. Comparison of experimental (solid) and calculated (dashed) energies in ^{198}Au. States are labelled with $[N_1 N_2]$, $<\sigma_1 \sigma_2 \tfrac{1}{2}>$, $(\tau_1 \tau_2)$, J and L, and all have $\bar{\sigma}_1 = N_1$, $\bar{\sigma}_2 = N_2$.

splitting, however, is large and negative, and as such different from the value needed in ^{197}Pt. The large, negative value of E in ^{198}Au is needed, not only to obtain the correct spin for the ground state, but also to reproduce the two low-lying 4^- states at 215 and 382 keV, respectively.

The group-theoretical characterization of the levels of ^{198}Au proposed in Fig. 1, must be tested on the basis of properties other than energies. Specifically, the particle structure of the states is of importance, and can be probed in particle-transfer reactions. In Ref. 16 it is shown that the wavefunctions of states in the odd-odd nucleus classified according to Eq. (3.1), can be written as a sum over odd-neutron states coupled to a $j = 3/2$ proton:

With neglect of terms contributing only to the binding energy, the energy formula corresponding ot Eq. (3.1) is

$$E = A\left(N_1(N_1+5) + N_2(N_2+3) + N_3(N_3+1)\right) + \bar{B}\left(\bar{\sigma}_1(\bar{\sigma}_1+4) + \bar{\sigma}_2(\bar{\sigma}_2+2) + \bar{\sigma}_3^2\right)$$
$$+ B\left(\sigma_1(\sigma_1+4) + \sigma_2(\sigma_2+2) + \sigma_3^2\right) + C\left(\tau_1(\tau_1+3) + \tau_2(\tau_2+1)\right)$$
$$+ D\ J(J+1) + E\ L(L+1). \tag{3.2}$$

Supersymmetry comes about by embedding the chain (3.1) into the direct-product group $U_\nu(6/12) \times U_\pi(6/4)$, and, as shown in Ref. 9, in this way the properties of <u>quartets</u> of nuclei are related to each other. With respect to odd-odd nuclei, the group chain (3.1) can be used (i) <u>without</u> reference to supersymmetry, i.e., as a dynamical boson-fermion symmetry for an odd-odd nucleus by itself, or, more ambitiously, (ii) in a supersymmetric context, predicting the properties of the odd-odd nucleus from those of the three other nuclei in the quartet. In ref. 9 the second approach was followed, which gave acceptable agreement for ^{196}Pt, ^{197}Pt and ^{197}Au, but which was inconclusive for the odd-odd member of the quartet (^{198}Au) because of ambiguities in the spin-parity assignments. Since then the energy spectrum of ^{198}Au has been measured again in an (n,γ) experiment,[15] making the test of the boson-fermion symmetry for the odd-odd nucleus possible. The results of the experiment and the calculation are shown in Fig. 1. The theoretical levels are obtained by a least-squares fit to the experimental excitation energies using the formula (3.2). The resulting parameters are (in keV): $A = 70$, $\bar{B} = 46$, $B = -100$, $C = 51$, $D = 52$, and $E = -32$. (The states shown in Fig. 1 depend on $A + B$ and not on A and B separately since they all have $\bar{\sigma}_1 = N_1$ and $\bar{\sigma}_2 = N_2$; the above values for A and B fix the $[41]<4><\tfrac{3}{2}\tfrac{1}{2}\tfrac{1}{2}>(\tfrac{1}{2}\tfrac{1}{2})3/2\ 2$ state at $E_x = 520$ keV). Up to an energy of 500 keV a one-to-one correspondence between the observed and calculated spectrum is established, the largest deviation being the 3^- state at 450 keV which is predicted at 315 keV. The parameter values A, $\bar{B}$, B, C, and $D + E$ are fairly close to those obtained in the fit of Ref. 9 to the even-even, even-odd and odd-even nuclei, and hence satisfy the supersymmetry requirement. The parameter E which determines the doublet

$$|[N_1N_2]<\bar{\sigma}_1\bar{\sigma}_2><\sigma_1\sigma_2\tfrac{1}{2}>(\tau_1\tau_2)LJM>$$

$$= \sum_{\{-\}} (-1)^{J-\bar{J}} ((2J+1)(2\bar{J}+1))^{\tfrac{1}{2}} \left\{ \begin{array}{ccc} \bar{L} & 3/2 & J \\ L & 1/2 & \bar{J} \end{array} \right\}$$

$$\times \left\langle \begin{array}{c} <\bar{\sigma}_1\bar{\sigma}_2><\tfrac{1}{2}\tfrac{1}{2}\tfrac{1}{2}> \\ (\bar{\tau}_1\bar{\tau}_2)\,(\tfrac{1}{2}\tfrac{1}{2}) \end{array} \bigg| \begin{array}{c} <\sigma_1\sigma_2\tfrac{1}{2}> \\ (\tau_1\tau_2) \end{array} \right\rangle \left\langle \begin{array}{c} (\bar{\tau}_1\bar{\tau}_2)(\tfrac{1}{2}\tfrac{1}{2}) \\ \bar{L} \quad 3/2 \end{array} \bigg| \begin{array}{c} (\tau_1\tau_2) \\ J \end{array} \right\rangle$$

$$\times \, (A_\nu^\dagger([N_1N_2]<\bar{\sigma}_1\bar{\sigma}_2>(\bar{\tau}_1\bar{\tau}_2)\overline{LJ}) \times a_\pi^\dagger(3/2))_M^{(L)}|0\rangle, \tag{3.3}$$

where $\{-\} = \{\bar{\sigma}_1,\bar{\sigma}_2,\bar{\tau}_1,\bar{\tau}_2,\bar{L},\bar{J}\}$ and $A_\nu^\dagger$ creates a state of an odd-neutron nucleus with quantum numbers defined according to the O(6) limit of $U^B(6) \times U^F(12)$.[13] Expressions for the generalized coupling coefficients $\langle \vdots | \vdots \rangle$ are derived in Ref. 16. Using Eq. (3.3) simple predictions can be found for single-proton transfer amplitudes. For instance, in proton pick-up reactions which have ^{198}Au as a final nucleus and which conserve the number of bosons, two 1^- - 2^- doublets are predicted to be excited, and the spectroscopic strengths are given by

$$<[N+1]<N+1><N+\tfrac{3}{2}\tfrac{1}{2}\tfrac{1}{2}>(\tfrac{1}{2}\tfrac{1}{2})3/2L\,\|a_\pi^\dagger(3/2)\|\,GS(\nu);1/2>^2$$
$$= \frac{(2L+1)(N+5)}{2(N+3)}, \tag{3.4a}$$

$$<[N+1]<N+1><N+\tfrac{1}{2}\tfrac{1}{2}\tfrac{1}{2}>(\tfrac{1}{2}\tfrac{1}{2})3/2L\,\|a_\pi^\dagger(3/2)\|\,GS(\nu);1/2>^2$$
$$= \frac{(2L+1)(N+1)}{2(N+3)}, \tag{3.4b}$$

where $GS(\nu)$ represents the ground state of the odd-neutron nucleus. In ^{198}Au the 2^- and 1^- states belonging to the doublet (3.4a) are – according to the fit in Fig. 1 – the ground state and the excited state at E_x = 193 keV, respectively; the doublet (3.4b) is not shown in Fig. 1 and occurs at an energy E_x = 1400 keV.

In Ref. 16 the structure of the wavefunctions of an odd-odd nucleus in the $U_\nu^B(6) \times U_\pi^B(6) \times U_\nu^F(12) \times U_\pi^F(4)$ scheme is derived and further results concerning particle-transfer reactions are given.

Among other symmetry schemes for odd-odd nuclei the possibility

$U_\nu^B(6) \times U_\pi^B(6) \times U_\nu^F(12) \times U_\pi^F(12)$, appropriate for mass regions where both neutrons and protons occupy the single-particle orbits $j = 1/2$, $3/2$ and $5/2$, stands out. A preliminary analysis indicates that ^{76}As, as the odd-odd partner of the quartet ^{76}Se, ^{77}Se, ^{75}As and ^{76}As, conceivably could be interpreted in such a scheme. The even-even member of the quartet, ^{76}Se, has vibrational characteristics.[17] The odd-proton member, ^{75}As, has been analyzed recently[18] in the context of the U(5) limit of $U^B(6) \times U^F(12)$. The odd-neutron member, ^{77}Se, is fairly well described in the same limit, but with an exchange interaction, different from the one in ^{75}As. The choice of a subgroup chain of $U_\nu^B(6) \times U_\pi^B(6) \times U_\nu^F(12) \times U_\pi^F(12)$, which allows for this difference in exchange interaction between the two odd-A nuclei and which might possibly lead to a proper description of ^{76}As, will be discussed in Ref. 19.

4. CONCLUDING REMARKS

It should be realized that the boson-fermion symmetry for odd-odd nuclei proposed in Refs. 8-11 and discussed in Sect. 3, implies some definite relations between the boson-boson, the boson-fermion and the fermion-fermion interaction. It remains to be investigated whether realistic shell-model interactions are consistent with these relations. By combining the extension proposed in Sect. 2 with the treatment of odd-odd nuclei discussed in Sect. 3, the constraints which are imposed on the different interactions, are relaxed and hence in this way a more generally applicable model for odd-odd nuclei could be constructed.

I am indebted to A. Frank, J. Jolie, and D. D. Warner for useful discussions. This research was supported by NATO grant n° RG.85/0699.

REFERENCES

1. Arima, A. and Iachello, F., Adv. Nucl. Phys. __13__, 139 (1984).
2. Elliott, J. P., Rep. Progr. Phys. __48__, 171 (1985).
3. Iachello, F., Phys. Rev. Lett. __44__, 772 (1980).
4. Iachello, F. and Scholten, O., Phys. Rev. Lett. __43__, 679 (1979).
5. Balantekin, A. B., Bars, I., Bijker, R.,and Iachello, F.,

Phys. Rev. C27, 176 (1983).
6. Vervier, J., to be published.
7. Frank, A., Van Isacker, P., and Warner, D. D., to be published.
8. Bars, I., in Bosons in Nuclei, edited by Feng, D. H., Pittel, S., and Vallieres, M. (World Scientific, Singapore, 1984), p. 155.
9. Van Isacker, P., Jolie, J., Heyde, K., and Frank, A., Phys. Rev. Lett. 54, 653 (1985).
10. Balantekin, A. B. and Paar, V., Phys. Lett. B, to be published.
11. Balantekin, A. B. and Paar, V., to be published.
12. Castaños, O., Chacón, E., Frank, A., and Moshinsky, M., J. Math. Phys. 20, 35 (1979).
13. Van Isacker, P., Frank, A., and Sun, H.-Z., Ann. Phys. (N.Y.) 157, 183 (1984).
14. Bijker, R. and Kota, V. K. B., Ann Phys. (N.Y.) 156, 110 (1984); Bijker, R. and Iachello, F., Ann. Phys. (N.Y.) 161, 360 (1985); Iachello, F. and Bijker, R., Ann. Phys. (N.Y.), to be published.
15. Warner, D. D., Casten, R. F., and Frank, A., to be published.
16. Van Isacker, P., to be published.
17. Subber, A. R. H., Robinson, S. J., Hungerford, P., Hamilton, W. D., Van Isacker, P., Kumar, K., Park, P., Schreckenbach, K., and Colvin, G., to be published.
18. Vervier, J., Van Isacker, P., Jolie, J., Kota, V. K. B., and Bijker, R., Phys. Rev. C32, 1406 (1985).
19. Hoyler, F., et al., to be published.

APPROXIMATE SUPERSYMMETRY IN ODD-EVEN AND ODD-ODD NUCLEI

D.K. Sunko, S. Brant, D. Vretenar and V. Paar

Department of Theoretical Physics
University of Zagreb, 41000 Zagreb,
YUGOSLAVIA

ABSTRACT

It is shown that an analytic solution for the wave functions of PTQM/IBFM in the SU(3) limit has supersymmetric properties. It is extended to the odd-odd case and an algebraic approximation scheme is discussed.

1. INTRODUCTION

Since the classical article[1] on nuclear supersymmetries, many others have been published. All of these have one basic idea in common: they presume the PTQM/IBFM Hamiltonian to be imbedded in a higher structure that is explicitly supersymmetric. This structure then limits the form of the spectrum.

The essential distinction of the approach of this work is that it imposes no a priori conditions on the Hamiltonian. Rather, it is shown that a certain special limit of the standard version of the model possesses a much higher degree of symmetry than would be expected. It is defined by conditions on the parameters, like the various limits of TQM/IBM[2]. One feature of its high symmetry is the equality of moments of inertia of the even and odd systems, which was explicitly predicted[1] as a consequence of supersymmetry.

Details of these developments will be published elsewhere[3].

2. THE HAMILTONIAN AND ITS SOLUTION

The standard PTQM/IBFM Hamiltonian consists of three parts:

$$H = H_{IBM} + H_{DYN} + H_{EXC} \qquad (1)$$

Although the calculations have been performed in the PTQM formulation[4], we use here the more widely known equivalent IBM notation. The limit we consider is obtained by taking the SU(3) limit of the core term and dropping the exchange term. This corresponds to the particle-rotor case in the geometric model. Since only the states built on the ground-state band of the core will be considered, and since the quadrupole interaction does not mix the bands of the core, the Casimir operator gives only an additive constant, and the relevant Hamiltonian is

$$\delta H' = \delta I^B \cdot I^B + \Gamma (G_2^B G_2^F)_0 \qquad (2)$$

where I^B is the core (IBM) angular momentum, and the dynamic interaction couples the quadrupole operators of the core (B) and of the odd fermion (F). To define our limit completely, we need to fix Γ/δ. This is done by noting that there exists an analytic solution[4] of (2):

$$|KJM\rangle = \mathcal{N} P_{MK}^J \{|jK\rangle |c\rangle\} \qquad (3)$$

where P is the Peierls-Joccoz operator and $|c\rangle$ a coherent state of the core[4]. For K=j (3) is a solution of (2) only for a particular value of Γ/δ, to be called $(\Gamma/\delta)_{SUSY}$, and this fixes our limit.

3. PROPERTIES OF THE SOLUTION

For any value of Γ/δ, (3) may be used as a basis to diagonalize (2). It turns out that in this basis both terms of the matrix (2) are tridiagonal, i.e. the only nonzero matrix elements are those for which $\Delta K = |K - K'|$ is zero or one. This is not unexpected for the $I^B \cdot I^B$ term, which contains the Coriolis force[5], but it is certainly surprising for the interaction term. A priori one would expect the explicit presence of the fermion operator to break any selection rule valid in the boson space.

The value $(\Gamma/\delta)_{SUSY}$ now appears as the one which cancels the only offdiagonal term connecting the $K=j$ and $K=j-1$ columns in the matrix (2). It is remarkable that $(\Gamma/\delta)_{SUSY}$ does not depend either on J or on N, the total number of s and d bosons. It depends only on j, the spin of the odd particle. Numerical values are given in Table 1.

The J-independence means that for $(\Gamma/\delta)_{SUSY}$ the whole $K=j$ band will be uncoupled from the others. The spectrum of this band may be written

$$E_J(K=j) = \delta J(J+1) \qquad (4)$$

with the same value of δ as in (2). The ground-state band of the even system and the $K=j$ band of the odd system have the same moment of inertia.

4. EXTENSION TO THE ODD-ODD CASE

The Hamiltonian considered for the odd-odd case is a simple extension of (2):

$$\delta H = \delta I^B \cdot I^B + \Gamma_p (G_2^B G_2^F)_0 + \Gamma_n (G_2^B G_2^F)_0 \qquad (5)$$

Table 1
Exact and approximate values of $(\Gamma/\delta)_{SUSY}$

j	3/2	5/2	7/2	9/2	11/2	13/2	15/2	17/2	19/2	21/2
exact	12.6	19.3	27.3	36.5	46.3	57.2	68.4	80.5	93.7	106.9
eq.(9)	12.1	20.0	29.0	38.4	48.1	58.0	67.8	77.6	87.4	97.1

Following (5), we construct the projection

$$|(\Omega_p \Omega_n) KJM\rangle = \mathcal{N} P^J_{MK} \{|j_p \Omega_p\rangle |j_n \Omega_n\rangle |C\rangle\} \quad (6)$$

For $\Omega_p = j_p$, $\Omega_n = j_n$ and $K = j_p + j_n$, the wave function (6) is an exact solution of (5) when $(\Gamma_p/\delta) = (\Gamma_p/\delta)_{SUSY}$ and $(\Gamma_n/\delta) = (\Gamma_n/\delta)_{SUSY}$ with the SUSY values being taken from the odd-even model (Table 1) for the corresponding proton (neutron) angular momenta. The band $K = j_p + j_n$ is then also perfectly rotational, with the energies given by (4) with the same moment of inertia as for the even and odd-even system.

5. THE ROTATION-ALIGNED CASE

In the geometrical model, the strong coupling spectrum is obtained when a particle is coupled to an oblate core, or a hole to a prolate core[6]. Stephens has shown[7] that the opposite, decoupled, case may be described by rotating the strong-coupling wave function by $\pi/2$. It can be shown[8] that the analogous wave function (see (3))

$$|\alpha JM\rangle = \sum_K \langle 2N\ 0\ jK|2N+\alpha\ K\rangle |KJM\rangle \quad (7)$$

with $\alpha = j$ describes the yrast band of (2) for the case $(\Gamma/\delta) = -(\Gamma/\delta)_{SUSY}$ and that the Stephens energy rule is satisfied to a good approximation. Note that the Clebsch-Gordan coefficient in (7) goes over into $d^j_{K\alpha}(\pi/2)$ as $N \to \infty$.

6. CALCULATION OF THE SUSY COUPLING CONSTANT

There is an approximate method to calculate $(\Gamma/\delta)_{SUSY}$ proceeding directly from the algebra. The idea is to simulate supersymmetry by truncating the $SU^F(2j+1)$ fermion algebra to an $SU(3)$ form; an effective renormalization factor is introduced in the structure constant of the resulting $SU^F(3)$ algebra.

The connection between the coupling constant Γ and

the structure constants is made by requiring that the odd-even Hamiltonian be written in the same form as the even one. The renormalization factor is fixed by requiring that

$$\langle j \| [Q^F, Q^F] \cdot [Q^F, Q^F] \| j \rangle \quad (8)$$

should not be changed by the truncation. The resulting algebra is not semisimple for $j \to \infty$. To correct this, an additional factor is introduced which does not change the structure of the algebra for finite j but amounts to algebra expansion[9] in the asymptotic limit. The final result is

$$(\Gamma/\delta)_{SUSY} \approx \frac{4\sqrt{5}}{3} \sqrt{j(2j+1)(2j+2)} \left[1 + 4\frac{(2j-2)(2j+4)}{(2j-1)(2j+3)}\right]^{-1/4} [1+k(2j-3)]^{-1} \quad (9)$$

where k is a free parameter, reflecting the ambiguity inherent in algebra expansions. Setting k=1/36, the results in Table 1 are obtained.

7. DISCUSSION

The limit of PTQM/IBFM considered in this work has two remarkable properties: the moments of inertia of the even and odd systems are equal and the same selection rule is observed in the boson and fermion sector. The first of these has explicitly been predicted[1] as a consequence of supersymmetry. We interpret both as supersymmetric properties, obtained with no a priori assumption of supersymmetry constraints on the Hamiltonian.

The distinguishing features of this approach are that it may equally well be applied for any spin of the odd particle, and that it is naturally extended to odd-odd nuclei.

A typical value of δ in nuclei is 0.015 MeV, corresponding to $\Gamma \sim 1$ MeV according to Table 1. This is a realistic value for fitting data, but far from strong. The supersymmetric limit corresponds to an intermediate value of the particle-core coupling. For the same value, but opposite sign, a decoupled spectrum results that was shown[8] to be the truncated analogue of the Stephens rotation-aligned solution.

Finally, it should be emphasized that the results of sections 2-5 are intrinsic features of the standard version of the model, obtainable by anyone with an IBFM computer code. They are independent of the supersymmetric interpretation that we have given.

8. REFERENCES

1) Balantekin, A.B., Bars, I. and Iachello, F. Nucl. Phys. A370, 284 (1981)
2) Arima, A. and Iachello, F. Ann. Phys. (N.Y.) 99, 253 (1976); 111, 201(1978); 123, 468(1979)
3) Sunko, D.K., Brant, S., Paar, V., Dadić, I., Nielsen, H.B. and Balantekin, A.B. to be published
4) Paar, V. Brant, S., Canto, L.F., Leander, G. and Vouk, M. Nucl. Phys. A378, 41 (1982)
5) Ring, P. and Schuck, P. "The Nuclear Many-Body Problem", Springer, 1980., p.108
6) Alaga, G. and Paar, V., Phys. Lett. 61B, 129 (1976)
7) Stephens, F.S. Rev. Mod. Phys. 47, 43 (1975)
8) Sunko, D.K. and Paar, V. Phys. Lett. 146B, 279(1984)
9) Gilmore, R. "Lie Groups, Lie Algebras, and Some of Their Applications", John Wiley, 1974

EXTENSION OF SUPERSYMMETRY FOR THE DESCRIPTION OF ODD-ODD NUCLEI IN THE VIBRATIONAL LIMIT

Zoltán Árvay

Institute of Nuclear Research of the Hungarian Academy of Sciences, Debrecen, P.O.Box 51, H-4001, HUNGARY

Dynamical symmetries of an odd-odd nucleus with one extra proton and neutron in $j=1/2$ states can be characterised by the group $U^{(B)}(6) \otimes U^{(F\pi)}(2) \otimes U^{(F\nu)}(2)$. Boson-fermion excitation spectra in a quartet of ^{102,103}Ru-^{103,104}Rh nuclei have been described according to the supersymmetry principle.

The interacting boson model (IBM) was introduced by Arima and Iachello[1-5] for the description of collective states in even-even nuclei. In the IBM, nucleons outside the closed-shell core form a system of s- and d-bosons characterised by the U(6) dynamical symmetry. This U(6) symmetry can be decomposed into three limiting cases that are described by the groups U(5), SU(3) and O(6) which refer to the vibrational, rotational and γ-unstable situations. The model has been developed further and extended for the description of mixed systems of nuclear bosons and fermions[6-8]. In the framework of the corresponding model referred to as the interacting boson-fermion model (IBFM), the group structure of the Hamilton operator of an odd-A nucleus containing N bosons and one fermion (M=1) is described by the direct product $U^{(B)}(6) \otimes U^{(F)}(\Omega)$, where $\Omega = 2j+1$ and j labels the fermion angular momentum. Iachello, Bars and Balantekin[9,10] have extended the $U^{(B)}(6) \otimes U^{(F)}(\Omega)$ symmetry to supersymmetry by imbedding it into a larger group, which is the graded Lie group U(6/Ω). In this model both boson and fermion degrees of freedom can be treated within the same framework. Recently Hübsch et al.[11] and Van Isacker et al.[12] have extended the supersymmetry for the description of odd-odd nuclei. In this extended supersymmetry scheme authors have made a distinction between neutrons and protons.

When in an even-even nucleus occur vibrational excitations the next group chain describes the boson symmetry

$$U^{(B)}(6) \supset U^{(B)}(5) \supset SO^{(B)}(5) \supset SO^{(B)}(3) \supset SO^{(B)}(2). \quad (1)$$

The symmetries of a j=1/2 fermion are characterised by the following decomposition

$$U^{(F)}(2) \supset SU^{(F)}(2) \supset SO^{(F)}(2). \quad (2)$$

The symmetry group of a system consisting of an even-even core plus one proton and one neutron both in j=1/2 states, which occur in an odd-odd nucleus, is the direct product $U^{(B)}(6) \otimes U^{(F_\pi)}(2) \otimes U^{(F_\nu)}(2)$. Combining the group chains (1) and (2) we can decompose the symmetry group of an odd-odd nucleus

$$U^{(B)}(6) \otimes U^{(F_\pi)}(2) \otimes U^{(F_\nu)}(2) \supset U^{(B)}(5) \otimes SU^{(F_\pi)}(2) \otimes$$
$$\otimes SU^{(F_\nu)}(2) \supset SO^{(B)}(5) \otimes SU^{(F_{\pi\nu})}(2) \supset SO^{(B)}(3) \otimes \quad (3)$$
$$\otimes SU^{(F_{\pi\nu})}(2) \supset Spin^{(BF_{\pi\nu})}(3) \supset Spin^{(BF_{\pi\nu})}(2)$$

where $SU(2) \simeq SO(3)$, $Spin(3) \simeq SU(2)$ and $Spin(2) \simeq SO(2)$.
The basis states are classified by the set of labels

$$\left| \begin{array}{l} U^{(B)}(6); \; U^{(F_\pi)}(2); \; U^{(F_\nu)}(2); \; SU^{(F)}(2); \; U^{(B)}(5); \\ \text{[N]} \quad ; \; \{M_\pi\} \quad ; \; \{M_\nu\} \quad ; \quad \xi \quad ; \quad n_d \quad ; \quad (4) \\ SO^{(B)}(5); \; SO^{(B)}(3); \; Spin(3); \; Spin(2) \\ v, n_\Delta \quad ; \quad L \quad ; \quad J \quad ; \quad M_J \end{array} \right.$$

The Hamiltonian can be expressed in terms of Casimir invariants of the groups appearing in chain (3) and the corresponding energy eigenvalues can be written as

$$E(N, M, \xi, n_d, v, L, J) = E_o(N,M) + A \cdot n_d + B \cdot n_d \cdot M + C \cdot n_d \cdot (n_d+4) +$$
$$+ D \cdot v(v+3) + E \cdot L(L+1) + F \cdot \xi(\xi+1) + G \cdot J(J+1). \quad (5)$$

Here N is the total number of bosons, M is the total number of fermions ($M = M_\pi + M_\nu$, where M_π (M_ν) is the number of the odd proton(neutron)), n_d is the number of d-bosons, v labels the boson seniority, ξ labels the irreps of $SU^{(F_{\pi\nu})}(2)$ arising from the decomposition of $SU^{(F_\pi)}(2) \otimes SU^{(F_\nu)}(2)$ and L,J stands for the orbital and total angular momenta, respectively. By imbedding the dynamical symmetry $U^{(B)}(6) \otimes$ $\otimes U^{(F_\pi)}(2) \otimes U^{(F_\nu)}(2)$ into a larger supersymmetry characterised by the graded Lie group U(6/2+2) one can treat the boson and fermion degrees of freedom in the same framework. In U(6/2+2) the $U^{(F_\pi)}(2) \otimes U^{(F_\nu)}(2)$ group characterises the fermion sector which is devided into proton and neutron subsectors. The number of odd protons(neutrons) $M_\pi=1(M_\nu=1)$ labels the totally antisymmetric irreps of $U^{(F_\pi)}(2)(U^{(F_\nu)}(2))$. One of the consequences of the supersymmetry principle is, that all nuclei belonging to the same supermultiplet labelled by [$\mathcal{N}$] (where $\mathcal{N}$ =N+M) must be described by the same A,B,C,D,E,F,G coefficients. Discrepancies are signals of supersymmetry breaking.

In the Ru, Rh and Mo, Tc region there are good vibrational-like even-even nuclear cores which can therefore be described by the U(5) limit. Furthermore, in the same region proton $p_{1/2}$ and neutron $s_{1/2}$ fermion states there are present in the corresponding odd-A nuclei, that are candidates for the Spin(3) symmetry.

The energy expression given by Eq.(5) describes the levels in the even-even member of the super-quartet, when M=0. Taking M=1, Eq.(5) gives the levels in the neighbouring odd-A isotone ($M_\pi=1$, $M_\nu=0$) or isotope ($M_\pi=0$, $M_\nu=1$). By fitting the parameters in Eq.(5) to the experimental spectra of the even-even core and the odd-A isotone and isotope we have predicted, applying the supersymmetry principle, states in the odd-odd member of the quartet corresponding to the irrep [N-2] $\otimes$ {1} $\otimes$ {1}. Because of the lack of relevant experimental data in the odd-odd nuclei F=0 has been choosen for the coefficient of the energy term depending on the ξ quantum number in Eq.(5). The value

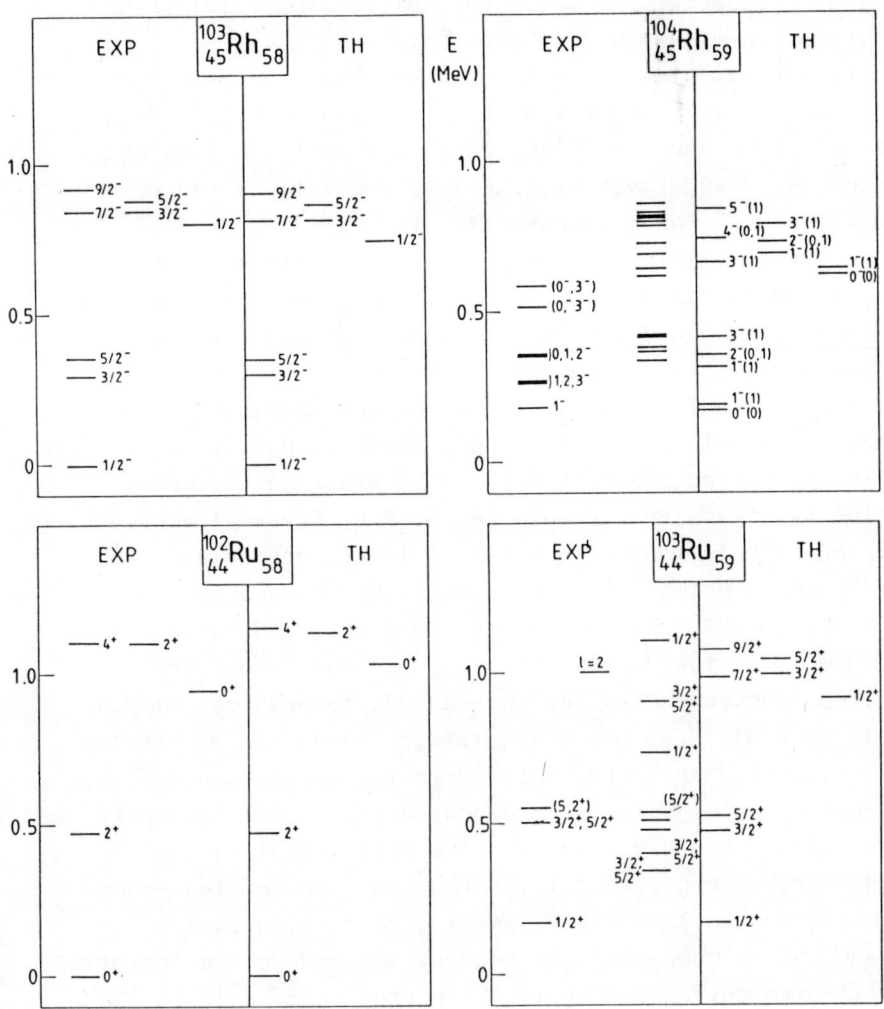

Fig.1. Selected excited states in ^{102,103}Ru and ^{103,104}Rh nuclei. Experimental data are taken from Refs.15-25).

of the F parameter can serve for setting the relative position of levels with $\xi=0$ and $\xi=1$.

Experimental and theoretical level spectra connected with the U(6/2+2) supersymmetry are shown in Fig.1. for the ^{102,103}Ru and ^{103,104}Rh quartet of nuclei belonging to the $[N=7]$ supermultiplet. Bijker and Kota[13]) and Vervier and Janssens[14]) have investigated the validity of the U(6/2) supersymmetry scheme for the description of the corresponding states in ^{102}Ru and ^{103}Rh. They have succesfully reproduced the experimental data and even predicted a 3/2$^-$ state at about 820 keV energy. G. Kajrys et al.[22]) have found in a high-spin experiment in ^{103}Rh a 3/2$^-$ state at 847.6 keV energy, which we can identify as the predicted state. For the 7/2$^-$ state we have accepted 847.5 keV energy obtained by R.O. Sayer et al.[21]). For the calculation of theoretical levels presented in Fig.1. we used the next parameter values A=-8.5 keV, B= =-145 keV, C=87.5 keV, D=10 keV, E=-9 keV, F=0, G=10 keV.

In Fig. 1. the values of the ξ quantum number are written in paranthesis for the odd-odd theoretical spectra.

In case of ^{103}Ru where the fermion state was an excited one we applied an energy shift with the excitation energy. Degeneracies in ξ in the odd-odd spectrum will disappear when using a nonzero value for the coefficient F, but it needs further experimental information. The applied model also reproduces the energy lowering of fermion states having phonon component. In case of the odd-odd nucleus the Hamiltonian corresponding to the energy formula in Eq.(5) incorporates a residual proton-neutron interaction, which is responsible for the splitting of proton-neutron multiplet states.

To summarize we have proposed an extension of U(6/2) supersymmetry to describe a quartet of nuclei belonging to the same supermultiplet. On the base of the extended

model one can predict low-lying levels, which refer to U(6/2+2), in the odd-odd member of the supermultiplet by fitting those of the even-even, odd-even, and even-odd members of the quartet. As can be seen from Fig.1., numerous experimental data support the presence of U(6/2+2) supersymmetry in the considered even-even and odd-A nuclei but further experiments are required to determine the corresponding U(6/2+2) states in the odd-odd member.
The expected supersymmetry can be broken by an admixture of the $d_{3/2}$ and $d_{5/2}$ neutron orbits to the $s_{1/2}$ one.

REFERENCES
1. Arima, A. and Iachello, F., Phys. Rev. Lett. $\underline{35}$, 1069 (1975)
2. Arima, A. and Iachello, F., Ann. Phys. $\underline{99}$, 253 (1976)
3. Arima, A. and Iachello, F., Ann. Phys. $\underline{111}$, 201 (1978)
4. Scholten, O. et al., Ann. Phys. $\underline{115}$, 325 (1978)
5. Arima, A., and Iachello, F., Ann. Phys. $\underline{123}$, 468 (1979)
6. Iachello, F. and Scholten, O., Phys. Rev. Lett. $\underline{43}$, 679 (1979)
7. Scholten, O., Ph.D. Thesis (University of Groningen, The Netherlands, 1980).
8. Iachello, F., Nucl. Phys. $\underline{A347}$, 51 (1980)
9. Iachello, F., Phys. Rev. Lett. $\underline{44}$, 772 (1980)
10. Balantekin A.,B., et al., Nucl. Phys. $\underline{A370}$, 284 (1981)
11. Hübsch T., et al., Phys. Lett. $\underline{151B}$ 320 (1985)
12. Van Isacker, P. et al., Phys. Rev. Lett. $\underline{54}$ 653 (1985)
13. Bijker, R. and Kota, V.K.B., Ann. Phys. $\underline{156}$, 110 (1984)
14. Vervier, T. and Janssens, R.V.F., Phys. Lett. $\underline{108B}$, 1. (1982)
15. Harmatz, B., Nuclear Data Sheets $\underline{28}$, 403 (1979)
16. Auble R.L.,et al., Nuclear Data Sheets $\underline{19}$, 1 (1976)
17. Blachot T.,et al., Nuclear Data Sheets $\underline{41}$, 325 (1984)
18. Bieber, E., Z. Phys. $\underline{189}$, 217 (1966)
19. Hagn E.,et al., Z. Phys. A$\underline{299}$, 353 (1981)

20. Gelder de P., et al., Nuclear Data Sheets <u>35</u>, 443 (1982)
21. Sayer R.O., et al., Nucl. Phys. <u>A179</u>, 122 (1972)
22. Kajrys G., et al., Phys. Rev. <u>C28</u>, 2335 (1983)
23. Fortune H.T.,et al., Phys. Rev. <u>C3</u>, 337 (1971)
24. Berg G.P.A.,et al., Nucl. Phys. <u>A379</u>, 93 (1982)
25. Haste, T.J. and Thomas, B.W., J. Phys. <u>G1</u>, 981 (1975)

THEORY OF UNIFIED NUCLEI ?

Tristan Hübsch
Department of Physics and Astronomy
University of Maryland, College Park, MD 20742
U.S.A.

ABSTRACT

The notion of dynamical symmetries is extended to describe the energy spactrum of nuclei with more than one off-core fermion. Generalyzing the Boson-Fermion symmetries of even-odd nuclei, this method is applied to the case of odd-odd nuclei, hypernuclei with one or two different off-core nucleons and double hypernuclei. It is argued that in all these cases the off-core fermions span a maximal $U(\Pi_n \Omega_n)$ Fermion dynamical symmetry, with Ω_n being the multiplicity of the state of the nth off-core fermion. The relevant group-theoretical methods of embedding the angular momentum Spin(3) in $U(\Pi_n\Omega_n)$, with the background of the Boson SO(6), U(5) or SU(3) dynamical symmetries are discussed. The generality of this approach suggests relations between the nuclei built of successive numbers of fermions (nucleons and hyperons) - extended supersymmetry. This makes it possible to envision a theoretical framework for describing the energy spectra of a wide class of nuclei.

1. INTRODUCTION

Possible dynamical symmetries of the valence nucleons with all but one of them paired into quasi-particles (bosons, hereafter) have lead[1] to a very fruitful approach to even-odd nuclei. This approach was generalized for the case of odd-odd nuclei[2] and hypernuclei[3]. The principal idea was the straight-forward generalization of the treatment of the single odd nucleon of the even-odd nuclei, where the maximal fermion dynamical symmetry (F-symmetry) was found to be $U(\Sigma_i\Omega_i)$, with $\Omega_i=2j_i+1$ and j_i being the total angular momentum of the ith configuration of the odd nucleon. In other words, the space of all possible states of the odd nucleon was identified with the fundamental representation of the unitary group of F-symmetry of the entire one-fermion j-subshell and is the largest one that preserves the number of particles[4].

Consider now the subsystem of two odd nucleons in an odd-odd nucleus, assuming that the individual interactions of these nucleons with those forming the bosons is negligible with respect to the one-nucleon average energy and with respect to the two-nucleon interaction of the two odd nucleons. Clearly, the space of all possible states of the two odd nucleons is labelled by $\oplus_\ell J_\ell^{\pi\nu} = (\oplus_i j_i^\pi) \otimes (\oplus_k j_k^\nu)$, where j_i^π (j_k^ν) is the total angular momentum of the ith (j^{th}) configuration of the odd proton (neutron). Thus the maximal group of F-symmetries, that does not change the number of particles, becomes $G \equiv U(\Sigma_\alpha \Omega_\alpha^{\pi\nu}) = U(\Sigma_{i,k} \Omega_i^\pi \cdot \Omega_k^\nu)$. This symmetry is easily seen to be realized in the following way. Let $\pi_{jm}^\dagger$ ($\nu_{jm}^\dagger$) be the secon-quantized creation operator of the odd proton (neutron) with angular momentum j and projection m. For a single odd nucleon, $U(\Omega)$ would have the generators[1] :

$$A_\mu^{(\lambda)}(j,j') = [\pi_j^\dagger \tilde{\pi}_{j'}]_\mu^{(\lambda)}, \qquad (1)$$

with $\tilde{\pi}_{jm} := (-)^{j+m} \pi_{j,-m}$ [4] and j (j') is in general reducible (which we suppress hereafter). For a two-fermion system this manifestly generalizes :

$$A_\mu^{(\lambda)}(J,J') = [[\pi^\dagger \nu^\dagger]_J [\widetilde{\pi\nu}]_{J'}]_\mu^{(\lambda)}, \qquad (2)$$

which is simply the statement that, for any $U(n)$, $\text{Adj}\{U(n)\} \in \mathbf{n}^* \otimes \mathbf{n}$.

For a system of N different fermions this is readily generalized by substituting the $\pi\nu$ product operators by the $\Pi_\alpha c_\alpha$ product operators, where α is labeling the species. This has a straight-forward application to hypernuclei, where the Λ hyperon (and others) represent the third (etc.) odd valence fermion.

2. SOME RELEVANT SUBGROUPS

Note that the set of generators in Eq.(2) containing the factor $[\pi_j^\dagger \tilde{\pi}_{j'}]_0^{(0)}$ generates the $U(\Omega^\nu)$ subgroup, and similarly those containing the factor $[\nu_k^\dagger \tilde{\nu}_{k'}]_0^{(0)}$ generate the $U(\Omega^\pi)$ subgroup. Thus, the collection of these operators generate the tensor product of these two, the :

$$G \supset U(\Omega^\pi) \otimes U(\Omega^\nu) \qquad (3)$$

subgroup.

Furthermore, note that the "transfer" operators in Eq.(2) (those which contain factors $[\pi_j^\dagger \tilde{\nu}_{k'}]_0^{(0)}$ or $[\nu_k^\dagger \tilde{\pi}_{j'}]_0^{(0)}$), together with the previous subset, generate the $U(\Omega^\pi + \Omega^\nu)$ subgroup of F-symmetries. This subgroup can exist only if $j^\pi = j^\nu$, since otherwise the "transfer" operators cannot exist. In this case the rather important subgroup chain is established :

$$G \equiv U(\Omega^\pi \cdot \Omega^\nu) \supset U(\Omega^\pi + \Omega^\nu) \supset U(\Omega^\pi) \otimes U(\Omega^\nu). \qquad (4)$$

Another suitable subgroup of G can be found by noting that $J^{\pi\nu} = \oplus_\ell J_\ell^{\pi\nu}$ spans the fundamental representation of the product $\otimes_\ell U(\Omega_\ell^{\pi\nu})$, which is clearly a subgroup of G. It is generated by operators of the form $A_{\mu,\ell}^{(\lambda)} = A_\mu^{(\lambda)}(J_\ell, J_\ell' = J_\ell)$.

Further interesting subgroups are found in the case when $j^\pi = j^\nu$, since then $J^{\pi\nu} \supseteq 0$. This state of the $(\pi-\nu)$ system does not perturb the core angular momentum properties and one may consider the rest of the $(\pi-\nu)$ system separately.

With or without the singlet state ($J^{\pi\nu}=0$) the set of available angular momenta $J^{\pi\nu}$ may be decomposed into the quasi-orbital and quasi-spin part $J^{\pi\nu} = K^{\pi\nu} \otimes Q^{\pi\nu}$ spanning $U(\Omega_K^{\pi\nu}) \otimes U(\Omega_Q^{\pi\nu})$ (or even with more factors : $J^{\pi\nu} = K^{\pi\nu} \otimes \ldots \otimes Q^{\pi\nu}$).

Guided by the work of Dynkin[5] on the classification of all Lie algebras, we further look for a unitary subgroup for which $\Omega^{\pi\nu}$ (with or without the singlet) represents the symmetrical (q=1) or the antisymmetrical (q=-1) representation and the fundamental representation of which is spanned by the symmetric (p=1) or the antisymmetric (p=-1) product $j^\pi \otimes j^\nu$ ($j^\pi = j^\nu$ here, defining $\Omega = \Omega^\pi = \Omega^\nu$) and we have, for $\Omega^{\pi\nu}$ including the singlet :

$$(\Omega + p)(\Omega^2 + p\Omega + 2q) - 8\Omega = 0 \qquad (5)$$

and

$$\Omega(\Omega + p)(\Omega^2 + p\Omega + 2p) - 8\Omega^2 + 8 = 0 \qquad (6)$$

for $\Omega^{\pi\nu}$ not including the singlet. Eq.(5) has only the trivial ($\Omega=1$) positive integer solution (for $p=q=1$), while Eq.(6) has two non-trivial solutions : $\Omega = 2$ ($p=-q=1$) and $\Omega = 4$ ($p=q=-1$). The first solution corresponds to :

$$(j^{\pi} \otimes j^{\nu})_S = [3]_{SU(3)}, \qquad (3 \otimes 3)_A = [3]_{SU(3)}. \qquad (7)$$

The second solution is really non-trivial, corresponding to $j^{\pi} = j^{\nu} = 3/2$ and then :

$$(j^{\pi} \otimes j^{\nu})_A = [6]_{U(6)}, \qquad (6 \otimes 6)_A = [15]_{U(15)} \qquad (8)$$

where the $[6]_{U(6)}$ is decomposing exactly as the **6** of $U(6)^B$! This embedding was discovered in Ref. [2] and it is the only one (as seen from Eq.(5-6) with one intermediate subgroup of this type. More-fold iterations of this kind are of course possible, but they lead to Diophantine-like equations of degree 5 or more, and are therefore unsolvable in general.

This outlines the methods utilized[2] to reveal possible F-symmetries of the 2-fermion system. Next, one looks for isomorphisms (homomorphisms) between the various subgroups in the chains of F-symmetries and in the corresponding Boson subgroup-chains allowing thus for the coupling of the fermionic system to the bosonic "background". (Actually, when looking for possible F-symmetry subgroups, one already aims at an isomorphism with a certain bosonic subgroup.) These methods are straight-forwardly applicable to 3- and more-fermion systems and have already led to results for hypernuclei[3].

As of the bosonic "background", it is consistent to choose the SO(6), U(5) or SU(3)-limit nuclei. This is easily seen by observing that upon β-decay, the odd neutron (ν) turns into a proton (π') pairing with the odd-proton into a π-boson ($j^{\pi} = j^{\nu}$). Thus in the above described scheme the $(\pi-\nu)$ wave function, and all operators correspondingly, should reduce according to :

$$(j^\pi \otimes j^\nu) \to (j^\pi \otimes j^{\pi'})_A . \tag{9}$$

It is an experimental test of the approach to observe transitions from states in $(j^\pi \otimes j^\nu)_S \to (j^\pi \otimes j^{\pi'})_A$, since the symmetric states (upon the β-decay) are Pauli-forbidden. Similar experimental tests can be designed for hypernuclei as well.

3. CONCLUSIONS

All these symmetries conserve the number of fermions and bosons and thus the total number of nucleons as well. To link nuclei of different subshells, one must invoke particle creation (annihilation) symmetry operators, the simplest of which are one-fermion operators, i.e. supersymmetry operators. Introduced to link even-even and even-odd nuclei[6], they have to be generalized[2] into extended supersymmetries to link even-even, π-even-odd, ν-even-odd and odd-odd nuclei (here one defines π- and ν-supersymmetries, i.e. N=2 extended supersymmetry). Naively, one may think that linking hypernuclei through supersymmetry requires Λ-bosons. Whereas that would make the supersymmetry manifest, it is possible to circumvent this obstacle (even though, with the advent of double hypernuclei Λ-bosons may become an experimental fact). Consider :

$$(\Lambda,\pi) \to (\pi,\pi) \to (\pi.\pi), \tag{10}$$

where the first transition is through emitting a π^--meson, the second one is the π-supersymmetry transition and $(\pi.\pi)$ denote the formation of a π-boson. This sequence establishes an effective $\Lambda - \pi$-boson supersymmetry, which is obviously going to be broken worse than the π-supersymmetry, inducing thus approprialtelly larger discrepancy in the parameters of the energy formulae.

Note finally that the creation of the S and D bosons is a consequence of the consecutive application of two supersymmetry operators, demonstrating that the dynamical symmetries[7] involving creation and annihilation operators are subgroups of supersymmetry-augmented Boson-Fermion symmetries of the approach presented here. Encouraged by the generality of the methods of this approach, it is foreseeable that many nuclei (if not all) may be described by it, leading to a unified phenomenological framework for nuclei.

REFERENCES

[1] Arima, A., Iachello, F.: Ann.Phys.(N.Y.)123(1979)468; see also Bijker, R.: Thesis, University of Groningen, The Netherlands (1984), that contains an extensive list of relevant references.

[2] Hübsch, T., Paar, V.: Z.Phys.A319(1984)111;
Hübsch, T., Paar, V., Vretenar, D.: Phys.Lett.151B(1985)320;
Balantekin, A.B., Hübsch, T, Paar, V.: to be published.

[3] Hübsch, T., Paar, V.: Phys.Lett.151B(1985)1, Z.Phys.A320(1985)351, Fizika17(1985)211.

[4] Wybourne, B.G.: "Classical Groups for Physicists" (John Wiley & Sons, New York, 1974).

[5] Dynkin, E.B.: Transl.Amer.Math.Soc.(1)9(1962)328, ibid.(2)6(1957)111,245.

[6] Balantekin, A.B., Bars,I., Iachello, F.: Nucl.Phys.A379(1981)284.

[7] Ginnochio, J.N.: Ann.Phys.126(1980)234.

SYMMETRIES IN ODD-EVEN NUCLEI: $U^B(6) \otimes U^F(20)$ AND $U^B(15) \otimes U^F(30)$ BOSON-FERMION SYMMETRY SCHEMES

V. K. B. KOTA
Physical Research Laboratory, Ahmedabad 380009, India

ABSTRACT

Two new boson-fermion symmetry schemes $U^B(6) \otimes U^F(20)$ and $U^B(15) \otimes U^F(30)$ are discussed, the later including g-bosons. For the various groups in the symmetry limits of these schemes we obtained the generators, irreducible representations and casimir operators. Analytic results for energies, B(E2) values etc. are derived. Analysis of data indicates that ($_{48}$Cd, $_{54}$Xe, $_{69}$Tm and $_{79}$Au) and $_{74}$W isotopes provide examples for some of the symmetry limits of $U^B(6) \otimes U^F(20)$ and $U^B(15) \otimes U^F(30)$ schemes respectively.

1. INTRODUCTION

The description of band structures and related properties of the lowlying states in odd-A nuclei, in a systematic fashion, from the standpoint of symmetry principles, has become possible with the development of the interacting boson-fermion model (IBFM)[1,2] which has the group structure $U^B(6) \otimes U^F(m)$. Here, with the odd-particle occupying singleparticle j-orbits, $m = \Sigma_j (2j+1)$. In this model the dynamical symmetries in odd-even nuclei (or the boson-fermion(BF) symmetry schemes) correspond to the subgroup chains G in $U^B(6) \otimes U^F(m) \supset G \supset \text{Spin}(3)$. Over the last few years BF symmetries related to the coupling of a j =1/2[3,4]; j=3/2[4,5]; j=1/2, 3/2[6]; j=3/2,5/2[7,8] and j=1/2,3/2,5/2[6,8,9-12] particle to the core nucleus described by IBM or one of its limiting symmetries (U(5), SU(3) or O(6)), are developed in detail and the isotopes of ($_{45}$Rh, $_{47}$Ag), ($_{77}$Ir, $_{79}$Au, $_{29}$Cu), $_{69}$Tm, $_{33}$As and ($_{30}$Zn, $_{36}$Kr, $_{76}$Pt, $_{80}$Hg, $_{74}$W) respectively are shown to provide emperical examples for them. Going further the coupling of a j=1/2,3/2,5/2,7/2 and j=1/2,3/2,5/2,7/2,9/2 particle to IBM and gIBM core nucleigives raise to the BF symmetry schemes $U^B(6) \otimes U^F(20)$ and $U^B(15) \otimes U^F(30)$ respectively. Note that gIBM is the interacting boson model including g bosons and similarly the gIBFM. A study of these two new BF symmetry schemes form the subject matter of this article.

2. THE $U^B(6) \otimes U^F(20)$ BF SYMMETRY SCHEME

The BF symmetry schemes can be broadly classified into three categories[4,13], (1) spinor symmetries or BF-1 class (2) pseudospin symmetries or BF -2 class and (3) mixure of pseudospin and spinor symmetries or BF-12 class. The $U^B(6) \otimes U^F(20)$ scheme has examples for all the three classes. With the fermion group spin^F(6) irreducible representation(IRR), $[3/2\ 1/2\ 1/2]$ containing $j=1/2$-$7/2$, leads to the spin^{BF}(6) symmetry scheme which belong to BF-1 class. The $j=1/2$-$7/2$ can be decomposed into pseudo orbital angular momentum $\tilde{l}=1,3$ and pseudo spin $\tilde{s}=1/2$ and this correspond to the subgroup $U^B(6) \otimes U^F(20) \supset U^B(6) \otimes U^F(10) \otimes U^F(2)$. Now the boson and fermion subgroups can be coupled to give a wide variety of BF-2 class symmetry chains shown in fig.1. Similarly the decomposition of $j=1/2$-$7/2$ into $\tilde{l}=2$, $\tilde{s}=3/2$ correspond to the subgroup $U^B(6) \otimes U^F(5) \otimes U^F(4)$ and it generates BF-12 class symmetries shown in fig.2. A brief discussion of the properties of some of the chains shown in figs.1,2 is given below together with the emperical examples available for them.

2.1. Symmetries Related To The U(5) Limit

The $U^{BF}(5) \otimes U^F(2)$ (chain ① in fig.1), $U^{BF}(5) \otimes \text{spin}^F(5)$ and spin$^{BF}(5) \otimes U^F(5)$ (chains II and III respectively in fig.2) symmetries are related to the U(5) limit of IBM. For these three limits the basis states and energy (E) formula are

$$U^{BF}(5) \otimes U^F(2) : \left| \begin{array}{ccccccc} U^B(6) & U^B(5) & U^F(5) & U^{BF}(5) & O^{BF}(5) & O^{BF}(3) & \text{Spin}(3) \\ N & n_d & \{1^f\} & \{f_1 f_2 f_3\} & \alpha[v_1 v_2] & \beta \tilde{L} & J \end{array} \right\rangle$$

$$E = E_0 + a_1 n_d + a_2 n_d (n_d+4) + a_3 \sum f_i(f_i+6-2i) + a_4[v_1(v_1+3) + v_2(v_2+1)] + a_5 \tilde{L}(\tilde{L}+1) + a_6 J(J+1) \tag{1}$$

$$U^{BF}(5) \otimes \text{Spin}^F(5): \left| \begin{array}{cccccc} U^B(6) & U^B(5) & U^{BF}(5) & O^{BF}(5) & \text{Spin}^{BFF}(5) & \text{Spin}(3) \\ N & n_d & \{n_1 n_2\} & [v_1 v_2] & [\tilde{r}_1 \tilde{r}_2] & \alpha J \end{array} \right\rangle$$

$$E = E_0 + b_1 n_d + b_2 n_d(n_d+4) + b_3[n_1(n_1+4) + n_2(n_2+2)] + b_4[v_1(v_1+3) + v_2(v_2+1)] + b_5[\tilde{r}_1(\tilde{r}_1+3) + \tilde{r}_2(\tilde{r}_2+1)] + b_6 J(J+1) \tag{2}$$

$$\text{Spin}^{BF} \otimes U^F(5) : \left| \begin{array}{cccccc} U^B(6) & U^B(5) & O^B(5) & \text{Spin}^{BF}(5) & \text{Spin}^{BFF}(5) & \text{Spin}(3) \\ N & n_d & v & [r_1 r_2] & [\tilde{r}_1 \tilde{r}_2] & \alpha J \end{array} \right\rangle$$

$$E = E_0 + c_1 n_d + c_2 n_d (n_d+4) + c_3[v(v+3)] + c_4[r_1(r_1+3) + r_2(r_2+1)] + c_5[\tilde{r}_1(\tilde{r}_1+3) + \tilde{r}_2(\tilde{r}_2+1)] + c_6 J(J+1) \tag{3}$$

In (1)-(3) α, β are multiplicity labels. The U(5), O(5) and spin(5) IRR appearing in (1)-(3) can be obtained using the following rules,

$$\{n_d\} \times \{1\} = \{n_d+1\} \oplus \{n_d,1\}, \quad \{n_d\} \times \{1^f\} = \{n_d+1,1\} + \{n_d,1,1\}$$
$$[r] \times [1] = [r+1] \oplus [r-1] \oplus [r,1], \quad [r] \times [1^f] = [r+1,1] \oplus [r,1] \oplus [r-1,1] \oplus [r]$$
$$[r\tfrac{1}{2}] \times [1] = [r\pm\tfrac{1}{2},\tfrac{1}{2}] \oplus [r+\tfrac{1}{2},\tfrac{3}{2}] \oplus [r+\tfrac{3}{2},\tfrac{1}{2}], \quad [r] \times [\tfrac{1}{2}\tfrac{1}{2}] = [r\pm\tfrac{1}{2},\tfrac{1}{2}]$$
$$\text{and} \quad [r,1] \times [\tfrac{1}{2}\tfrac{1}{2}] = [r\pm\tfrac{1}{2},\tfrac{1}{2}] \oplus [r\pm\tfrac{1}{2},\tfrac{3}{2}] \tag{4}$$

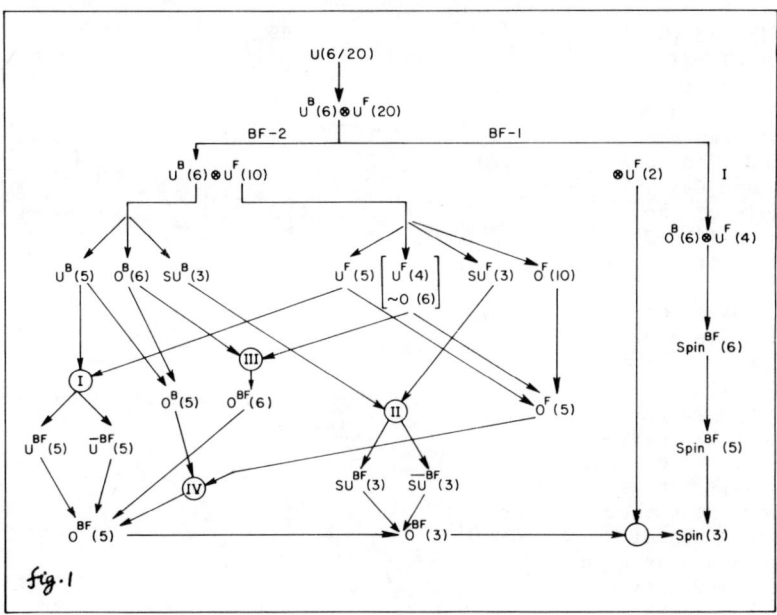

fig. 1

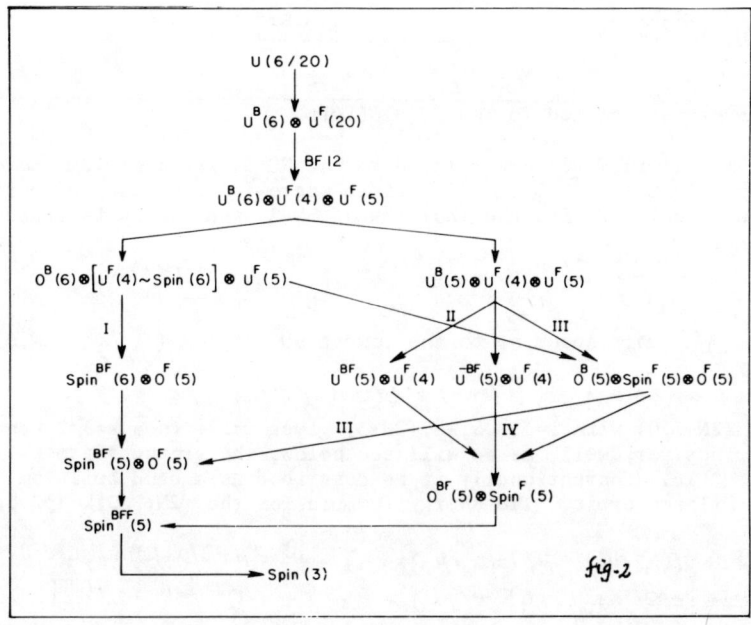

fig. 2

Using (4) and the results of ref.5, 9) the O(3) and spin(3) IRR in the above chains are obtained. The generators and Casimir operators of the groups in (1)-(3) are explicitly constructed and the basis states for $n_d=0,1,2$ are expanded in terms of the weak coupled basis states $|(n_d \nu L); j; J\rangle$ to obtain expressions for B(E2) values etc. The details can be had from the author. A first analysis of data shows that $_{48}$Cd isotopes may provide examples for the symmetries related to the U(5) limit.

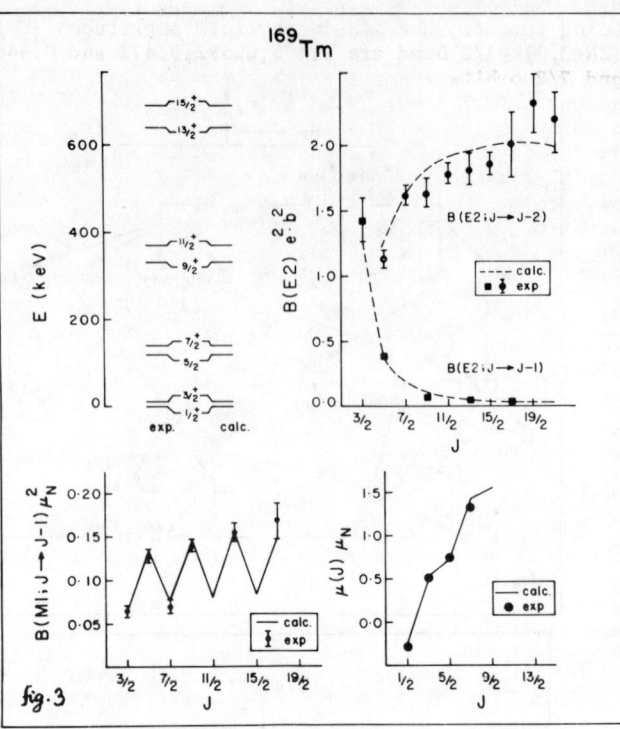

fig.3

2.2. Symmetries Related To The SU(3) Limit

Chains (II) in fig.1 are related to the SU(3) limit of IBM. Here we consider only the $SU^{BF}(3) \otimes U^F(2)$ limit where $l=1,3$ of the odd particle are associated with the (30) IRR of SU(3) and the basis states in this limit are

$$\left| \begin{array}{ccccc} U^B(6) & SU^B(3) & SU^{BF}(3) & O^{BF}(3) & Spin(3) \\ N & (\lambda_B \mu_B) & (\lambda \mu) & k\tilde{L} & J \end{array} \right\rangle$$

The IRR $(\lambda \mu)$ corresponding to the lowest SU(3) IRR $(\lambda_B \mu_B)=(2N,0)$ are given by

$$(\lambda \mu) \rightarrow (2N,0) \times (30) = (2N+3,0) \oplus (2N+1,1) \oplus (2N-1,1) \oplus (2N-3,3)$$

The IRR (2N+3,0) with $\tilde{L}=1,3,5,--,(2N+3)$ gives raise to a k=1/2 band and it describes very well, as we will see below, the ground state k=1/2 band of ^{169}Tm. Conventionally it is described as a band built on $[41\bar{1}]\frac{1}{2}$ Nilsson orbit. The energy formula for the (2N+3,0)k=1/2 band is

$$E = E_0 + e_1(\lambda_B^2 + \mu_B^2 + \lambda_B \mu_B + 3(\lambda_B + \mu_B)) + e_2(\lambda^2 + \mu^2 + \lambda \mu + 3(\lambda + \mu)) + e_3 \tilde{L}(\tilde{L}+1) + e_4 J(J+1)$$

$$\xrightarrow{k=\frac{1}{2} \text{ band}} E_0' + A[J(J+1) + a_0\{1 + (-1)^{J+\frac{1}{2}}(J+\frac{1}{2})\}] \quad (5)$$

where $A=e_3 + e_4$ and the decoupling parameter $a_0 = -e_3/(e_3 + e_4)$. Eq(5) is similar to the one in the geometric models[14]. With $e_3=9.42$ keV and

e_4=3.01 keV, the spectra[5] of ^{169}Tm fits very well with (5); see fig.3. Going further, the single particle amplitudes $|C j(\ell)|$ for the (2N+3,0)k=1/2 band are 0.475, 0.672, 0.472 and 0.546 for j=1/2, 3/2, 5/2 and 7/2 orbits respectively. They compare well with the Nilsson values[15] which are 0.274, 0.713, 0.473 and 0.413 respectively. This study clearly points out the inadequacy of Bijker's analysis[6] where only j=1/2, 3/2 orbits are considered to be active. However the expressions for B(E2) and $Q_2(J)$ values derived by Bijker[6] remain unaltered, we have to put λ =2N+3 while λ =2N+1 for Bijker. These results are fitted with ^{169}Tm data in fig.3. The expressions for magnetic properties are derived taking the M1-operator to be,

$$T^{M_1} = g_B \vec{L}_B + g_\ell \vec{\ell} + g_s \vec{s} + g_Q (Q^B \times J^{BF})^1_\nu \qquad (6)$$

After considerable SU(3) algebra we obtained the expressions for B(M1) and μ(J) values and in the limit $\lambda \gg 1$, $\lambda \to \infty$ they are same as in the geometric models[14],

$$B(M1; J^I \to J^f = J^I - 1) = \frac{3}{4\pi} \frac{(A_\lambda - 2B_\lambda)^2}{4} |\langle J\tfrac{I}{2} 10 | J\tfrac{f}{2}\rangle|^2 \{1 + b_0 (-1)^{J^I + \tfrac{1}{2}}\}$$

and $\mu(J) = g_{BB} J + (A_\lambda - 2B_\lambda)\{1 + b_0 (2J+1)(-1)^{J+\tfrac{1}{2}}\}/4(J+1)$ (7)

where $A_\lambda = (g_s - g_B) - 3(g_\ell - g_B)/\lambda$, $B_\lambda = 3g_Q (2\lambda+3)/8\sqrt{5}$, $g_{BB} = g_B + 3(g_\ell - g_B)/\lambda + 2B_\lambda/3$ and the 'magnetic decoupling parameter' $b_0 = A_\lambda/(A_\lambda - 2B_\lambda)$. The above results together with that of ref.6) show the complete analogy between the SU^{BF}(3) $\otimes$ U^F(2) limit and the collective models[14]. The experimental data[16] on B(M1) and μ(J) values are fitted with theory and the results are shown in fig.3. We used $(A_\lambda - 2B_\lambda) = -2.056 \mu_N$, $g_{BB} = 0.4 \mu_N$ and $b_0 = -0.157$. As $b_0 = 1$ with $g_Q = 0$, the fitted value clearly shows the importance of the two body term in (6).

2.3. Symmetry Limits Related To The O(6) Limit

The spin^{BF}(6) (fig.1, BF-1 class), O^{BF}(6) $\otimes$ U^F(2) (fig.1 chain III) and spin^{BF}(6) $\otimes$ U^F(5) (fig.2 chain I) belong to the O(6) limit of IBM. Ling et.al[17] analyzed $_{79}$Au isotopes as example of spin^{BF}(6) limit. The basis states and energy formula here are

$$\left| \begin{array}{ccccc} U^B(6) & O^B(6) & Spin^{BF}(6) & Spin^{BF}(5) & Spin(3) \\ N & \sigma & [\sigma_1 \sigma_2 \sigma_3] & [\tau_1 \tau_2] & \alpha J \end{array} \right\rangle \qquad (8)$$

$$E = f(N) + \alpha \sigma(\sigma+4) + A[\sigma_1(\sigma_1+4) + \sigma_2(\sigma_2+2) + \sigma_3^2] + B[\tau_1(\tau_1+3) + \tau_2(\tau_2+1)] + C J(J+1)$$

The ^{197}Au (N=5, σ=5) data is fitted with (8) using A=64keV, B=44keV and C=20keV and the results are shown in fig.4; the r.m.s deviation for the fit is 136keV. Similarly the fit to B(E2) values is also shown in the insect figure. Here one sees that the $\Delta \tau_1 = 1$ transitions are reproduced well but not the $\Delta \tau_1 \neq 1$ transitions. See ref.17,18) for further details. Recently Jolie et.al[19] made a preliminary analysis of spectra and B(E2) values for Xe and Ba isotopes as examples of spin^{BF}(6) $\otimes$ U^F(5) limit. Analytical expressions derived for B(E2) values etc. could not be given due to lack of space. The algebra of the O^{BF}(6) $\otimes$ U^F(2) limit was given in some detail in ref.13) and experimental examples for this chain are yet to be found.

3. (BF)-SYMMETRIES WITH g-BOSONS

In gIBFM one can construct several BF symmetry schemes; for example by coupling a j=5/2 particle to the U(6) limit [20] of gIBM leads to the spinor symmetry $U^{BF}(6) \supset Sp^{BF}(6)$. With the odd particle occupying j=1/2-9/2 orbits, one has the $U^B(15) \otimes U^F(30)$ BF symmetry scheme. The analysis of the single particle structure of the rotational bands in ^{185}W nucleus show the relevance of the SU(3) limit of $U^B(15) \otimes U^F(30)$ scheme; see ref.21) for details. These schemes may yield valuable insight into the role of hexadecapole deformation in odd-A nuclei.

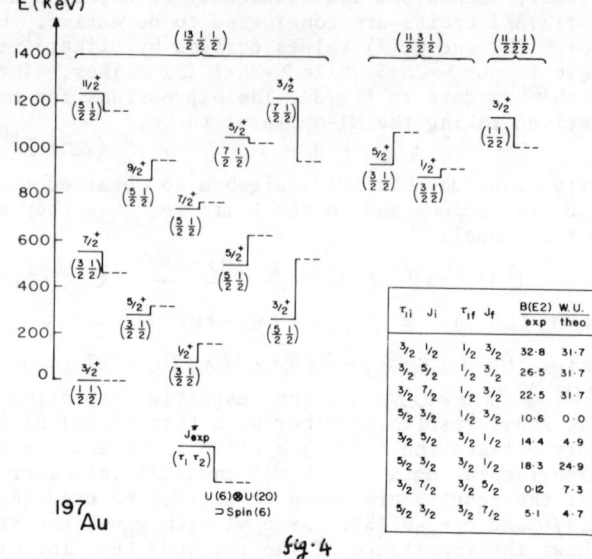

fig. 4

4. CONCLUSIONS

It will be exciting to find examples for all the chains in figs.1,2) and also the transitional nuclei within and between the various limits described in sec. 2.1 and 2.3. The subject of BF symmetry schemes with g-bosons is opened up and their potential usefulness remains to be seen in detail.

5. REFERENCES

1) Iachello,F.and Scholten,O.,Phys.Rev.Lett.43,679(1979). 2) Iachello,F, Phys.Rev.Lett.,44,772(1980). 3) Vervier,J. and Janssens,R.V.J., Phys. Lett.,108B,1(1982). 4) Bijker,R. and Kota,V.K.B., Ann.Phys.(N.Y)156,110 (1984). 5) Iachello,F. and Kuyucak,S., Ann.Phys.(N.Y)136,19(1981).
6) Bijker,R., Ph.D Thesis,University of Groningen,1984. 7) Kota,V.K.B, in "Group theoretical methods in physics" ed. W.W.Zachary(World Scientific,1984)p.404. 8) Vervier,J. et.al., Phys.Rev..C32,1406(1985).
9) Van Isacker,P.et.al., Ann.Phys.(N.Y)157,183(1984). 10) Bijker,R.and Iachello,F., Ann.Phys.(N.Y)161,360(1985). 11) Vergnes,M. et.al., Phys. Rev.,C31,2071(1985). 12) Warner,D.D. and Bruce,A.M., Phys.Rev.,C30,1066 (1984). 13) Kota, V.K.B., in the proceedings of the Workshop on "Recent trends in theoretical nuclear physics" held at MATSCIENCE,Madras(India), May 1985, to appear. 14) Bohr, A. and Mottelson,B., "Nuclear Structure" Vol.II. 15) Davidson,J.P.,"Collective models of the nucleus". 16) Taras, P. et.al.,Nucl.Phys.,A289,165(1977). 17) Ling,Y.et.al.,Phys.Lett.,148B, 13(1984). 18) "Interacting Boson-Boson and Boson-Fermion Systems", ed. O.Scholten (World Scientific,1984). 19) Jolie,J. et.al., Phys.Rev.Lett., 55,1457(1985). 20) Kota, V.K.B., in ref.18), De Meyer,H. et.al.(To appear in J.Phys.A). 21) Kota,V.K.B. (To appear in Phys.Rev.C).

FERMION-BOSON MODEL WITH ISOSPIN FORMALISM*

STANISŁAW SZPIKOWSKI, PIOTR KŁOSOWSKI and LESZEK PRÓCHNIAK
Institute of Physics, M. Curie-Skłodowska University,
20-031 Lublin, Poland

Abstract The Interacting Boson Model with isospin degrees of freedom has been applied to $s-d$ shell for even-even and odd-even pairs of nuclei. The fermion-boson supermultiplets seems to be in accord with experimental data.

INTRODUCTION

The success of an extension of the Interacting Boson Model toward fermion inclusion in the region of heavy nuclei[1,2,3] on one side and on the other - an application to light nuclei of the IBM-4 version with spin and isospin degrees of freedom[4,5] have been the ground of a Boson-Fermion Model for light nuclei on the shell $s-d$.

GROUP-THEORY STRUCTURE OF BOSON-FERMION SYMMETRIES

Bosons

In the standard IBM the bosons with angular momenta $l=0$ (s-bosons) and $l=2$ (d-bosons) have been taken. In an extension[4] the spin and isospin degrees of freedom $\sigma = 0,1$ and $\tau = 0,1$ respectively are also considered. Hence, the one boson basis is spanned by vectors

$$b^+_{lm_l \sigma m_\sigma \tau m_\tau} |o\rangle \tag{1}$$

where, according to the interpretation of bosons as correlated pairs of nucleon, there are the following possibilities

$$l = 0\,2 \;;\; \sigma = 1 \quad \tau = 0 \quad \text{or} \quad \sigma = 0 \quad \tau = 1 \tag{2}$$

---
*Partially supported under the contract CPBP 01,09.

There are altogether 36 vector states (1) and hence, the vector space is of 36-dimensions. The generators of unitary transformations $U(36)$ are formed by the second order creation and destruction boson-operators

$$b^+_{lm_l \, 6m_6 \, Tm_T} \, b_{l'm'_l \, 6'm'_6 \, T'm'_T} \tag{3}$$

Instead, we can take the coupled generators of the transformations $U(36)$ in the form

$$(b^+ b)^{L \Sigma T} \tag{4}$$

where $L = 0^2, 1, 2^3, 3, 4$ \hfill (5)

and $(\Sigma T) = (0,0)^2; (0,1); (0,2); (1,0); (1,1)^2; (2,0)$

Degeneracies in quantum numbers L, Σ, T follow from the different coupling ways

for $L = 0 : (l_1 \times l_2) = (0 \times 0)$ or (2×2)

for $L = 2 : (l_1 \times l_2) = (0 \times 2)$ or (2×0) or (2×2)

for $(\Sigma, T) = (0,0) : (1,0) \times (1,0)$ or $(0,1) \times (0,1)$

for $(\Sigma, T) = (1,1) : (1,0) \times (0,1)$ or $(0,1) \times (1,0)$

<u>Fermions</u>

For nucleons on the levels $s - d$, the single-particle quantum numbers take on the values

$$l = 0, 2 \; ; \; s = \tfrac{1}{2} \; ; \; j = \tfrac{1}{2}, \tfrac{3}{2}, \tfrac{5}{2}$$

and now the isospin is equal to $t = \tfrac{1}{2}$.

Twenty-four dimension space is spanned by vectors

$$a^+_{lm_l \, sm_s \, tm_t} |0\rangle \tag{6}$$

and the generators of the unitary transformations $U(24)$ read

$$a^+_{lm_l \, sm_s \, tm_t} a_{l'm'_l \, sm'_s \, tm'_t} \tag{7}$$

or in a coupled form

$$(a^+ a)^{LST} \tag{8}$$

where $L = 0^2, 1, 2^3, 3, 4$ and $S = 0, 1 ; T = 0, 1$ (9)

The Boson-Fermion Supergroup

If we take the boson generators (4), with fermion generators (8) adding the boson-fermion ones

$$(a^+ b)^{LS'T'} \quad \text{and} \quad (b^+ a)^{LS'T'} \tag{10}$$

where L is the same as before and

$$(S'T') = (\tfrac{1}{2}, \tfrac{1}{2}), (\tfrac{1}{2}, \tfrac{3}{2}), (\tfrac{3}{2}, \tfrac{1}{2}) \tag{11}$$

then we get altogether $(36+24)^2$ generators for the unitary supergroup

$$U(36/24) \supset U^B(36) \otimes U^F(24) \tag{12}$$

with further decoupling

$$U^B(36) \supset SU^B_L(6) \otimes SU^B_{ST}(6) \supset SU^B_L(6) \otimes SU^B_{ST}(4) \tag{13}$$

and $U^F(24) \supset SU^F_L(6) \otimes SU^F_{ST}(4)$ (14)

It is seen from (13-14) that we have got the formally identical groups in boson and fermion spaces and then the boson-fermion groups can be formed in the following further chain

$$SU^{BF}_L(6) \otimes SU^{BF}_{ST}(4) \supset X^{BF} \otimes SU^{BF}_S(2) \otimes SU^{BF}_T(2) \supset$$
$$\supset SO^{BF}_L(3) \otimes SU^{BF}_S(2) \otimes SU^{BF}_T(2) \supset$$
$$\supset SU^{BF}_J(2) \otimes SU^{BF}_T(2) \tag{15}$$

where there are three different choices for the subgroup X^{BF}, namely

$$X^{BF} = \left\{ \begin{array}{c} SO^{BF}(6) \\ SU^{BF}(5) \\ SU^{BF}(3) \end{array} \right\} \tag{16}$$

as in the standard IBM. In the application to the nuclei on the levels $s - d$ we have found that the first choice, $SO^{BF}(6)$, is the most suitable and in this case there is another subgroup which is to be taken into account

$$SO^{BF}(6) \supset SO^{BF}(5) \qquad (17)$$

From the generators (4, 8, 10) one can construct in a standard way the generators of the subgroups in the chains (13, 14, 15 and 16), and then the second order Casimir operators can be also constructed.

BOSON-FERMION HAMILTONIAN AND ENERGY

Generally speaking, the interacting boson-fermion Hamiltonian should be constructed, according to considered dynamical symmetry, as a second order operator of supergroup generators. However, as a first step - the simplest approximation is usually taken as limited combination of supergroup generators in the form of a sum of second order Casimir operators of symmetry groups under considerations. The states of lower energy levels usually belong to the unique representations of the first symmetry groups in the chain (12-16). Hence, eigenvalues of Casimir operators of those groups do not contribute to the relative energies of the excited states of nuclei. The relevant, for low excited states, Casimir operators are those for the subgroups $SO^{BF}(5)$; $SO^{BF}_L(3)$ and $SU^{BF}_J(2)$ and under this restriction the fermion boson Hamiltonian reads

$$\hat{H} = C + K\hat{C}_1 + D\hat{C}_2 + F\hat{C}_3 \qquad (18)$$

where C is a constant, $\hat{C}_i$ are the Casimir operators for the above groups respectively and K, D, F are adjusted model parameters. The eigenvalues of Casimir operators are expressed by means of the quantum numbers of irreducible representations respective groups and the eigenenergy of $\hat{H}$ reads

$$E = C + K\tau(\tau+3) + DL(L+1) + FJ(J+1) \qquad (19)$$

The crucial point, in the energy-level interpretation by the formulae (19), is fixing, by the group-theory consideration, the allowed quantum numbers τ and L while J will follow the coupling of L and S. One of the simplifying assumption in this respect is to take only a single-nucleon representation and to couple it to the completely symmetric representation [N-1] of N-1 bosons where N is the fixed number of particles in the supergroup irreducible representation. With some further minor assumptions to choose properly the irreducible representations of the subgroups, we get the following allowed values:

(i) Irreducible representations of the group $SO^{BF}(6)$

are $\langle \sigma,0,0 \rangle$ where $\sigma = N, N-2, \ldots, 1$ or 0 \hfill (20)

(ii) Irreducible representations of the group $SO^{BF}(5)$

are $(\tau, 0)$ where $\tau = \sigma, \sigma-1, \ldots, 0$ \hfill (21)

(iii) The allowed values of the angular momentum subgroup $SO^{BF}(3)$ are

$$L = 2\Delta\,;\, 2\Delta-2\,;\, 2\Delta-3, \ldots, \Delta+1, \Delta \hfill (22)$$

where

$\Delta = \tau, \tau-3, \tau-6, \ldots$

APPLICATIONS

One of the arguments of the presence of supersymmetry in nuclei[6] is the consideration of pairs of nuclei: even-even with N bosons and even-odd with N-1 bosons and one fermion. Such nuclei, according to assumed supersymmetry must belong to the same supermultiplet and hence, the strength parameters of the Hamiltonian (18) ought to be the same for both nuclei. The rules (20-22) give, of course, different sets of allowed quantum numbers for nuclei in the pair and hopefully the calculated energy levels for such two sets of quantum numbers are expected to follow the experimental data. There were chosen three pairs of nuclei: ^{29}Si and ^{30}Si; ^{31}P and ^{32}S; ^{33}S and ^{34}S. The comparison of calculated and experimental[7] energy levels are given on Figure 1. We have listed all of the even parity experimental levels below theoretically interpreted energies and as in seen some of them are beyond of the supersymmetry scheme.

The first comparison with theory seems to be rather encouraging if we take into account that we have considered the simplest Hamiltonian of our dynamic symmetry. There is also possible to consider hole-bosons and a hole-fermion or particle-bosons and a hole fermion or vice versa. Although the quantum numbers will not be changed very much, the pair

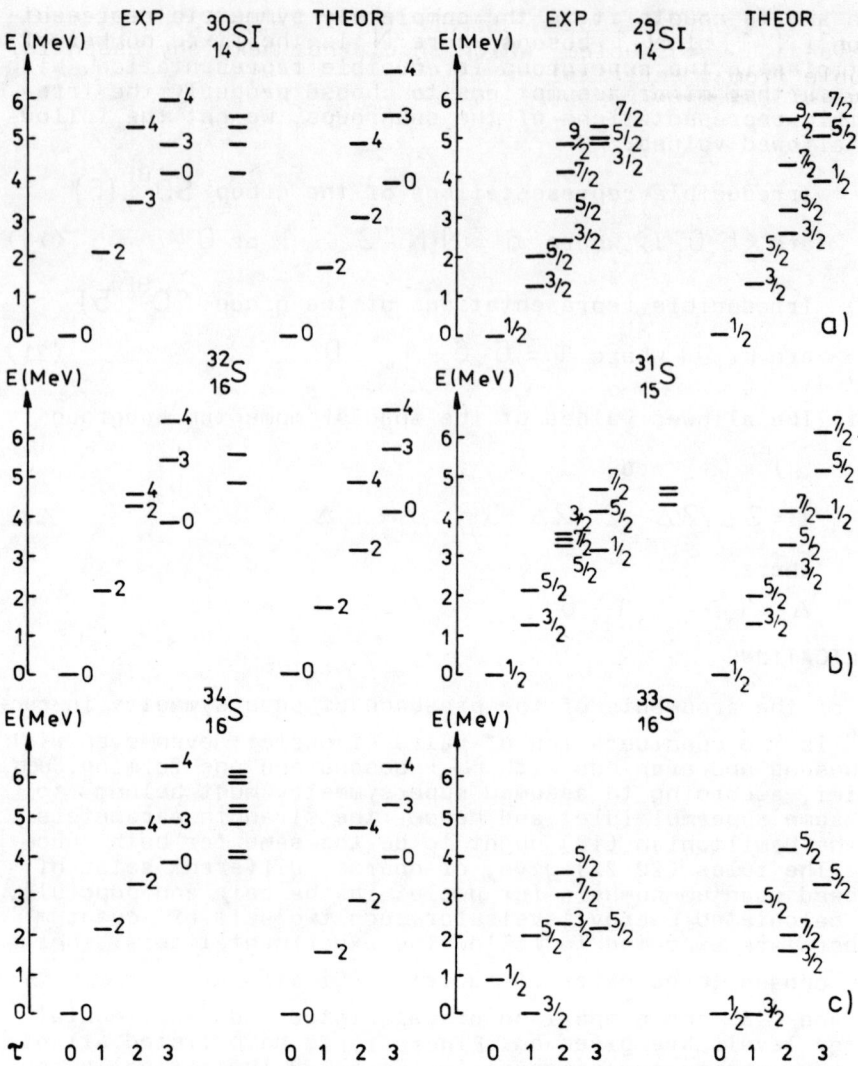

FIGURE 1. Comparison of theoretical and experimental levels for nucleus pairs of three supermultiplets:
a) K = 0.217; D = 0.010;   F = 0.127 MeV
b) K = 0,228; D =-0.012;   F = 0.143 MeV
c) K = 0.221; D =-0.420;   F = 0.541 MeV

of nuclei might be chosen differently. Another problem is to consider odd-odd nuclei, which should not make any essential trouble from theoretical point of view, because our scheme is an isospin one, however, as we know, odd-odd nuclei behave in experiment quite differently.

The present report is a very preliminary one but we plan to extend and deepen our calculation.

One of us (S.S.) is very grateful to Franco Iachello for encouraging comments in our effort.

REFERENCES

1. F. Iachello and S. Kuyucak, Ann. Phys. (N.Y.) 136, 19 (1981).
2. M. Vallieres et al., Phys. Lett. 135B, 339 (1984).
3. R. Bijker and F. Iachello, Ann. Phys. (N.Y.) 161, 360 (1985).
4. J. P. Elliott and J. A. Evans, Phys. Lett. 101B, 216 (1981).
5. P. Halse et al., Nucl. Phys. A417, 301 (1984).
6. A. B. Balantekin, I. Bars and F. Iachello, Nucl. Phys. A370, 284 (1981).
7. P. M. Endt and C. van der Leun, Nucl. Phys. A310, 1 (1981).

FERMION DYNAMICAL SYMMETRY AND HIGH SPIN PHYSICS

Mike W. Guidry

Department of Physics
University of Tennessee
Knoxville, TN 37996
U.S.A.

Da Hsuan Feng and Cheng-Li Wu[+]

Department of Physics
Drexel University
Philadelphia, PA 19104
U.S.A.

Joseph N. Ginocchio

Los Alamos National Laboratory
Los Alamos, NM 87545
U.S.A.

ABSTRACT

A Fermion dynamical symmetry model is described which appears to have a broad range of applicability in nuclear physics. One symmetry limit corresponds to a particle-rotor model. The relevance of this model to high spin physics is discussed.

1. INTRODUCTION

The use of dynamical symmetry in nuclear physics has generally been restricted by one or both of two difficulties. Either the symmetry is too limited to describe a broad range of phenomena, or the microscopic foundation of the symmetry is obscure. An example of the former difficulty occurs for the Eliott SU_3 model.[1] An example of the latter problem is the Interacting Boson Model (IBM), where the dynamical symmetries are implemented in terms of boson generators.[2] Recently, a fermion dynamical symmetry model has been introduced which has a number of dynamical symmetries, and which has a complete microscopic foundation. This model contains all of the dynamical symmetries of the IBM, but because it is a <u>fermion</u> model it contains symmetries which go far beyond those of the IBM.[3]

2. THE GINOCCHIO MODEL

The starting point for the fermion dynamical symmetry model (FDSM) is a shell model Hamiltonian for a single major shell. Then the single-particle angular momentum **j** is decomposed into an integer pseudo-orbital angular momentum **k** and a half-integer pseudospin **i**

$$\mathbf{j} = \mathbf{k} + \mathbf{i} \qquad (1)$$

This is not an approximation, since the two representations are related by a unitary transformation. This representation was originally introduced by Ginocchio, who noted that two choices of k and i ensure that there is a subspace of eigenstates composed only of S (J=0) and D (J=2) fermion pairs.[4] The first is to choose k=1, and i=anything consistent with eq. (1). This we refer to as k-k coupling, or the k-active coupling scheme. The second is to choose i=3/2, and k=anything consistent with eq. (1). This we term i-i coupling, or the i-active coupling scheme. With either of these choices, the fermion S-D subspace is decoupled from the remainder of the shell-model space, and one finds closed algebras for the S-pair, D-pair, and low-order multipole operators. If the multipole operators are restricted to order r=0,1,2,3, the i-active coupling scheme leads to an SO_8 dynamical symmetry, while the k-active coupling scheme yields an SP_6 algebra if the multipole operators are restricted to order r=0,1,2.[4] These symmetries have multichain decompositions which correspond to a variety of rotational and vibrational collective modes. Thus, the Ginocchio model has the desirable properties that it 1) realizes a variety of dynamical symmetries within the same model, and 2) is a fermion model which yields these symmetries without the explicit introduction of bosons.

However, the Ginocchio model is not applicable to real nuclei in the form just outlined for several reasons. 1) There is no clear connection to the shell model. 2) The k/i decomposition is not unique, since eq. (1) allows multiple combinations of k and i. 3) The axially symmetric rotor limit of the k-active scheme ($Sp_6 \supset SU_3$) has Pauli-principle restrictions on the allowed representations, which caused this symmetry to be abandoned in the original model.[4]

3. A UNIQUE k/i DECOMPOSITION

In fact, these problems with the Ginocchio model appear to be a single problem. The solution to all three is to properly relate the model to the underlying shell structure. The first point is the realization that the k/i decomposition can be made <u>unique</u> if we require that the number of states in the k/i space exactly equal the number in the j space. As discussed in ref. 3, this requires that within each major shell the normal parity orbitals have k=1 or i=3/2, and that the unique-parity intruder orbital have k=0, and i=j. The resulting reclassification of the shell model is shown in fig. 1, taken from ref. 3.

No	1	2	3		4	5		6		7			8		
n	0	1	2		3	3	4	4	5	5	5	6	6	6	7
k	0	1	1		0	1	0	2	0	1	1	0	1	1	0
i	1/2	1/2	3/2		7/2	3/2	9/2	3/2	11/2	1/2	7/2	13/2	3/2	9/2	15/2
SYM CONFIGURATION	$s_{1/2}$	$p_{1/2}$ $p_{3/2}$	$s_{1/2}$ $d_{3/2}$ $d_{5/2}$		$f_{7/2}$	$p_{1/2}$ $p_{3/2}$	$g_{9/2}$ $f_{5/2}$	$s_{1/2}$ $d_{3/2}$ $d_{5/2}$ $g_{7/2}$	$h_{11/2}$	$p_{1/2}$ $p_{3/2}$	$f_{5/2}$ $f_{7/2}$	$i_{13/2}$ $h_{9/2}$	$s_{1/2}$ $d_{3/2}$ $d_{5/2}$	$g_{7/2}$ $g_{9/2}$ $i_{11/2}$	$j_{15/2}$
			G_6 G_8 G_3			G_6 G_8 G_3		G_8		G_6			G_6		
Ω_0	0	0	0		0	5		6		7			8		
Ω_1	1	3	6		4	6		10		15			21		
Ω	1	3	6		4	11		16		22			29		
n	2	8	20		28	50		82		126			184		

$G_6 = (Sp_6^k \times SO_3^j) \times (SU_2 \times SO_3)$

$G_3 = (SU_3^k \times SO_6^i) \times (SU_2 \times SO_3)$

$G_8 = (SO_8^j \times SO_3^k) \times (SU_2 \times SO_3)$

Fig. 1 Reclassification of shell model in terms of pseudo-orbital (k) and pseudospin (i) quantum numbers

Notice that this reclassification associates particular values of k and i with particular shells, and that this in turn implies that the shells being filled by valence particles will determine which dynamical symmetries and which collective modes can occur.

4. THE FERMION DYNAMICAL SYMMETRY MODEL (FDSM)

The FDSM is implemented in terms of three assumptions, once the transformation to the k/i basis is done. 1) Only J=0,2 terms are kept in the pairing part of the effective interaction. 2) The pairing and multipole matrix elements are assumed to be proportional to the degeneracy of the shell. 3) The multipole terms of the effective interaction and the single-particle energy terms are approximated in such a way that the resulting Hamiltonian is invariant under a tractable dynamical symmetry. The corresponding symmetries are discussed in ref. 3, and are displayed in fig. 2.

The Sp_6 and SO_8 Ginocchio symmetry chains reappear in the FDSM, and there is a new $SU_3 \times SO_6$ symmetry. However, even in the chains corresponding to the original Ginocchio symmetries there is a very important difference. The inclusion of the unique-parity intruder orbital adds direct-product $S\mathcal{U}_2$ factors (unique-parity orbital symmetries are generally denoted by script symbols). This follows from the previous assumptions, which require that in the lowest-energy states the unique parity orbitals contain only J=0 pairs (S-condensate), and corresponds to a quasispin symmetry for those orbitals. The participation of the intruder orbital in the dynamical

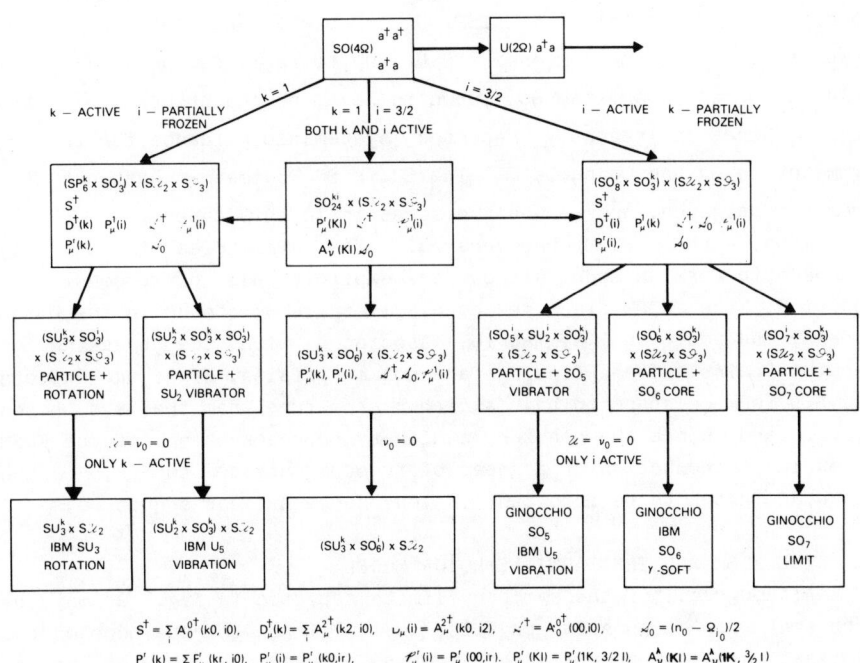

Fig. 2   Dynamical symmetries of the FDSM[3)]

symmetry has two important consequences. The first is that the Sp_6 symmetry is revived, because the Pauli principle difficulties become less severe when some of the valence particles fill the intruder orbital, since those particles don't participate in the Sp_6 symmetry. Secondly, the quasispin symmetry means that there are seniority quantum numbers associated with the particles in the unique-parity orbitals, and the model contains the possibility of breaking pairs in the high-j unique-parity orbitals (and in the normal-parity orbitals). Thus the FDSM resurrects the symmetry $Sp_6 \supset SU_3$ corresponding to an axially-symmetric rotor, and introduces the possibility of coupling broken pairs in both normal and unique-parity orbitals to a variety of rotational and vibrational cores.

If there are no broken pairs (seniority quantum numbers = zero) the SU_3, and SO_6 limits of fig. 2 have exactly the same spectrum, and transition probabilities similar to, the corresponding limits in the IBM. The zero-seniority representations of the SU_2 and SO_5 limits correspond essentially to IBM U_5 spectra, and the transition probabilities are the IBM ones for U_5, up to Pauli factors. The $SO_{24} \supset SU_3 \times SO_6$ chain and the SO_7 subchain have no counterpart in the IBM. There is recently reported systematic evidence for the SO_7 symmetry in Pd and Cd nuclei.[5] The $SU_3 \times SO_6$ symmetry has not yet been investigated, and we won't discuss it in this paper.

Thus, the FDSM recovers all IBM symmetries in a fully microscopic fermion model without the explicit use of bosons. In addition, the FDSM finds new core symmetries not found in the boson models, and additional symmetries associated with the breaking of fermion pairs. These findings are not surprising, since one expects the S,D fermion space to have a richer structure than the s,d boson space, and since the non-zero seniority representations of the FDSM allow the introduction of degrees of freedom outside the S,D space which are known to be important in low-energy nuclear structure.

5. THE FDSM AND THE PARTICLE-ROTOR MODEL

Although all of the symmetry limits depicted in fig. 2 may be important for high-spin physics, we will concentrate here on the simplest case: the coupling of broken pairs of particles to an axially symmetric SU_3 core. The spectrum in that case takes the simple form[6]

$$\Delta E = \Delta + \alpha R(R+1) + \sum_i \gamma_i I_i(I_i+1) - \sum_i \delta_i \mathbf{R} \cdot \mathbf{I}_i - \sum_{i<j} \delta_{ij} \mathbf{I}_i \cdot \mathbf{I}_j \qquad (2)$$

In this expression, the parameters Δ, δ, and γ are microscopically defined, R is the total pseudo-orbital angular momentum, and I_i is the pseudospin of the ith broken pair of particles. Physically, R is the angular momentum carried by the SU_3 core, and I_i represents the angular momentum of a broken pair, and Eq. (2) may be recognized as a microscopic particle-rotor Hamiltonian.

This particle-rotor limit gives rise to various rotational bands which may be characterized by the quantum numbers (λ,μ,u,ν_0), where λ,μ are SU_3 quantum numbers, u is the number of broken pairs in the normal-parity orbitals, and ν_0 is the number of broken pairs in the unique-parity orbitals. The representation $(2N_1,0,0,0)$, where N_1 is the number of pairs in normal-parity orbitals, corresponds to the ground band of a rotor. The representation $(2N_1,0,2,0)$ corresponds to a rotational band built on a broken pair in the normal-parity orbitals. This pair has a simple structure in the k/i representation, but corresponds to a linear combination of broken pairs in the original shell-model space. Therefore, this band may be identified with the Coriolis antipairing (CAP) effect in high-spin physics, with a structure corresponding to small amplitudes for breaking many pairs in the normal-parity orbitals.[6] The representation $(2N_1,0,0,2)$ corresponds to a band in which a broken pair of particles in the high-j intruder orbital contributes an angular momentum which adds to the core angular momentum generated by the S and D pairs. Thus, this type of band can be identified with the aligned 2-quasiparticle bands responsible for the first backbends observed in many rotational nuclei.

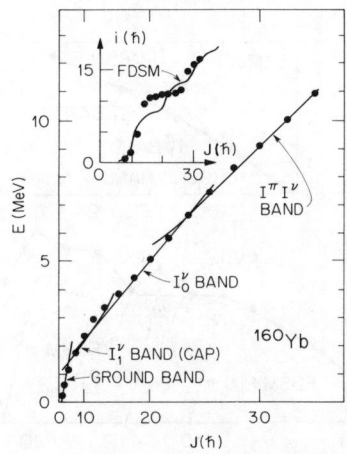

Fig. 3 Energy levels and alignment for ^{160}Yb. The calculation assumes SU_3 symmetry[6]

In figs. 3 and 4 we show some examples of the use of the SU_3 limit of the FDSM. Once an effective interaction for the truncated shell-model space is chosen, the FDSM parameters are determined microscopically. Such calculations are in progress. In these examples we have just set the parameters to physically reasonable values to illustrate the method.

In fig. 3 the energy of yrast states in ^{160}Yb is plotted as a function of angular momentum, and compared to the experimental values. The four crossing bands are, in order of increasing energy, a ground-state band; a band denoted by I_1^ν corresponding to the CAP effect; a band labled I_0^ν corresponding to the rotation alignment of two $i_{13/2}$ neutrons (first backbend); and a band labled $I^\pi I^\nu$ corresponding to the rotation alignment of a pair of unique-parity neutrons and a pair of unique-parity protons (second backbend).

Shown as an inset is the rotational alignment i of the unique-parity particles calculated in the FDSM and compared to the values deduced experimentally. In the symmetry limits the bands are orthogonal, and all crossings are sharp. A small amount of symmetry breaking, expected in any realistic case, will cause smoothing of the crossings, particularly at the crossing between the ground and CAP bands. Thus, symmetry-breaking is expected to generate a smooth variable moment of inertia (VMI) behavior in the low-spin region.

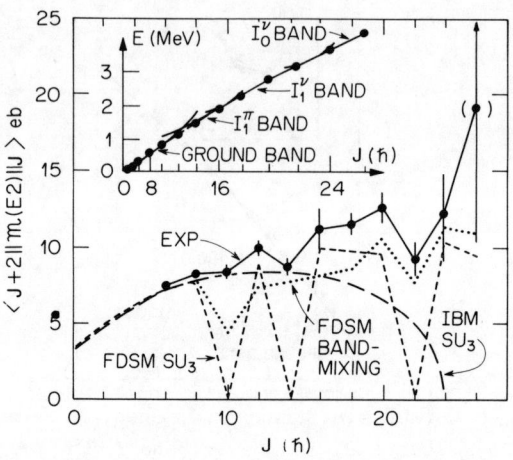

Fig. 4 Energy levels and E2 matrix elements in ^{232}Th. The FDSM and IBM calculations assume SU_3 symmetry.[6]

In fig. 4, a calculation of the yrast states for ^{232}Th is compared to the experimental data (inset). Also shown is a comparison of the yrast E2 matrix elements to experiment. As noted above, in the symmetry limit the bands are orthogonal. A small amount of symmetry-breaking will mix the bands and yield finite matrix elements at the band crossings. The curve marked FDSM band-mixing simulates the symmetry-breaking by mixing the 3 lowest-lying bands at a given spin with a constant interaction. One sees the important point that the FDSM matrix elements do not exhibit the disastrous early fall-off of the M(E2) matrix elements characteristic of the IBM. This is because in the FDSM the S,D core carries only part of the angular momentum, with the rest carried by broken pairs. Thus the truncation of the rotational strength characteristic of any shell model is pushed to higher spin in the FDSM, and the results are in much better agreement with the high-spin data than for the boson models.

In summary, we have described the application of a fermion dynamical symmetry model to the high-spin states of axially symmetric rotors. A particle-rotor model with a microscopic foundation results, and the basic phenomena of high-spin physics such as Coriolis forces, rotation alignment, band crossings, and multiple backbends emerge as a natural consequence of fermion dynamical symmetry. This has been achieved without the explicit introduction of macroscopic geometrical concepts such as deformation, and without explicit bosons. If the microscopic calculations currently underway deal kindly with the FDSM, we may expect it to become an important theoretical technique for the study of high-spin physics.

[+] Permanent address: Department of Physics, Jilin University, Changchun, People's Republic of China
1. J.P. Elliott, Proc. Roy. Soc. A245,128(1958); A245,562(1958).
2. A. Arima and F. Iachello, Ann. Rev. Nucl. Part. Sci.31,75(1981)
3. C.L. Wu, D.H. Feng, X.G. Chen, J.Q. Chen, and M.W. Guidry, Phys. Lett. 168B,313(1986); and submitted to Phys. Rev. C.
4. J.N. Ginocchio, Ann. Phys. 126,234(1980).
5. R.F. Casten, C.L. Wu, D.H. Feng, J.N. Ginocchio, and X.L. Han, Phys. Rev. Lett., in press; and R.F. Casten, this Proceedings.
6. M.W. Guidry, C.L. Wu, D.H. Feng, X.G. Chen, J.Q. Chen, and J.N. Ginocchio, Phys. Lett. B, in press. Data for figs. 3 and 4 are taken from H. Ower, et al., Nucl. Phys. A388, 421(1982), and from L.L. Riedinger, et al., Phys. Rev. Lett. 44, 568(1980).

U(12) MULTIPLETS

H. G. Solari*, R. Gilmore, and M. Vallieres
Department of Physics and Atmospheric Science
Drexel University, Philadelphia, PA 19104

Supersymmetry models have had some success in simultaneously describing the properties of pairs of nuclei using a single hamiltonian and a single set of transition operators[IAC80, BAL81]. These successes can be traced, ultimately, to the slow variation of parameter values over nearby nuclei in the periodic table.

The superalgebras themselves cannot be considered as fundamental symmetries of nuclear systems since they possess operators which violate baryon number conservation. However, the superalgebra establishes a Hilbert space, N, which decomposes, under the reduction $U(6|m) \supset U_B(6) \times U_F(m)$ ($m = \Sigma\, 2j+1$), into a direct sum of direct product Hilbert spaces $(n) \times (1^k)$, $N = n + k$, one for each nucleus in the supermultiplet. For reasons of energetics, usually only the lowest two Hilbert spaces [(N,0) and (N-1,1)] are retained.

An analogous theory can be established for even-even nuclei [HAR85, FRA85]. The appropriate dynamical group for the IBA-2 is $U_\pi(6) \times U_\nu(6)$ (analogous to $U_B(6) \times U_F(m)$ of the IBFA). This can be embedded in a larger group, U(12). This large group cannot be considered as a fundamental nuclear symmetry as the algebra possesses operators which violate charge conservation (analogous to U(6|m)). However, the large group establishes a Hilbert space which decomposes, under the reduction $U(12) \supset U_\pi(6) \times U_\nu(6)$, into a direct sum of direct product Hilbert spaces $(N_\pi) \times (N_\nu)$, $N_\pi + N_\nu = N$, one for each nucleus in a U(12) multiplet.

When both the proton bosons and the neutron bosons are either particle-like or hole-like, all members of a U(12) multiplet have the same mass and constitute part of an isospin multiplet. This multiplet is transverse to the valley of stability, and contains typically only two stable members, rendering tests of the model difficult. When the proton bosons and the neutron bosons are of opposite types, adjacent members of a

U(12) multiplet differ by "an α-particle." The U(12) multiplet of "generalized isobars" lies parallel to the valley of stability, and may contain four or more stable members.

We have tested this U(12) model on the multiplet containing ^{122}Te, ^{126}Xe, ^{130}Ba, ^{134}Ce, ^{138}Nd, and ^{142}Sm. The 59 identified excited levels in these six nuclei [SAK84] have been fitted with the nine parameter IBA-2 Hamiltonian

$$H = \varepsilon_\pi n_{d\pi} + \varepsilon_\nu n_{d\nu} + \kappa Q_\pi Q_\nu + (1/2) \Sigma C_\pi(L) \{[d_\pi{}^+ d_\pi{}^+]^{(L)} [d_\pi d_\pi]^{(L)}\}^{(0)}$$

$$(1/2) \Sigma C_\nu(L) \{[d_\nu{}^+ d_\nu{}^+]^{(L)} [d_\nu d_\nu]^{(L)}\}^{(0)} + (1/2) Un0 \{[d_\nu{}^+ d_\nu{}^+]^{(0)} s_\nu s_\nu + h.c.\}$$

The fitting was done using the NPBOS code followed by both first and second order least squares search routines. The following parameter values produced the best fit, with an RMS of 142 kev: $\varepsilon_\pi = 0.8954$, $\varepsilon_\nu = 0.5419$, $\kappa = -0.0884$, $C_\pi(0) = 0.2776$, $C_\pi(2) = -0.3605$, $C_\pi(4) = 0.0953$, $C_\nu(0) = 0.2839$, $C_\nu(2) = -0.2495$, $Un0 = 0.3380$. The comparison between the experimental levels and those predicted by this model is shown in Fig. 1 and preseented in Table 1.

The relative E2 values for transitions were computed using a consistent Q assumption. The effective charges were taken equal for both protons and neutrons. The comparison between the experimentally observed ratios [LED78] and those predicted from this model is given in Table 2. The transition matrix elements indicate clearly that the two excited 0^+ states have very different electromagnetic properties. Further, these levels "cross" between ^{122}Te and ^{126}Xe. The experimental data indicate clearly that the $0^+{}_2$ state which is observed corresponds to the $0^+{}_2$ state in ^{122}Te but to the $0^+{}_3$ state in ^{126}Xe.

Cizewski and Gülmez [CIZ86] have recently proposed that a shell gap occurs at Z=64 for N > 78. This suggests that ^{142}Sm is a member of a multiplet with N=2 rather than the multiplet containing ^{122}Te with N=7. If ^{142}Sm is removed from this multiplet, there is a further improvement in the fit.

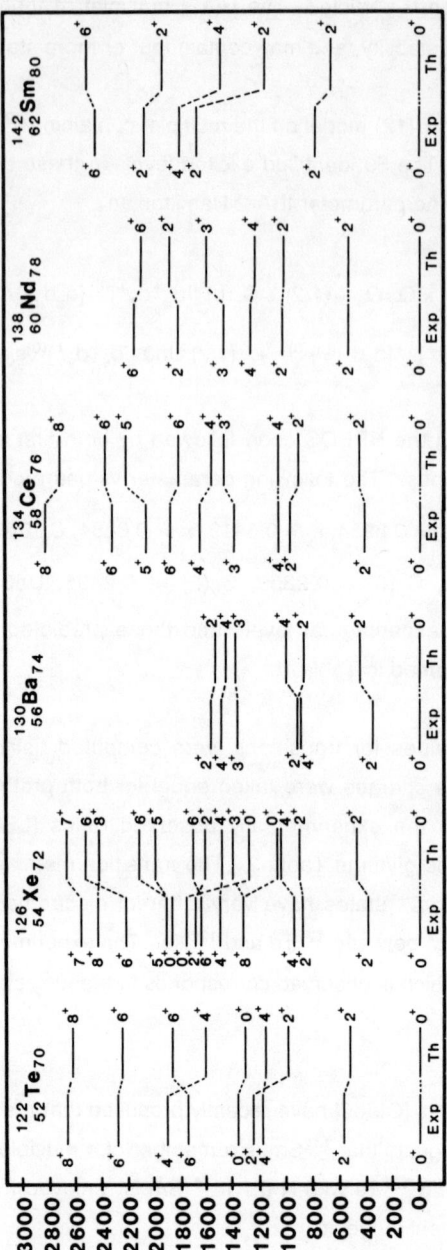

Figure 1. Comparison between the experimentally observed energies of the nuclei in the U(12) multiplet containing ^{122}Te with the energy level spectrum predicted by the Hamiltonian H.

Table 1. Comparison between the experimentally observed energy levels and those predicted using the Hamiltonian (1).

J^π	^{122}Te Exp	^{122}Te The	^{126}Xe Exp	^{126}Xe The	^{130}Ba Exp	^{130}Ba The	^{134}Ce Exp	^{134}Ce The	^{138}Nd Exp	^{138}Nd The	^{142}Sm Exp	^{142}Sm The
0(1)	0	0	0	0	0	0	0	0	0	0	0	0
0(2)	1357	1304	1314	1094		985		1038		1215	1450	1436
0(3)		1372	1759	1289		1297		1383		1476		1556
2(1)	564	521	389	484	357	462	409	475	521	547	768	682
2(2)	1257	975	880	916	908	888	966	932	1013	1047	1658	1213
2(3)		1752	1678	1624	1558	1527		1659	1800	1877	2056	1971
3(1)		1475	1318	1397	1361	1380	1383	1471	1452	1620		1785
4(1)	1181	1149	942	1056	902	1014	1049	1066	1249	1234	1791	1504
4(2)	1910	1580	1489	1482	1477	1461	1643	1565	1843	1748		1972
4(3)	1951	1949		1880		1921		2075		2224		2334
5(1)		2077	1904	1983	2012	2011	2050	2177	2262	2367		2552
6(1)	1751	1860	1635	1706	1592	1658	1863	1781	2134	2059	2420	2454
6(2)	2284	2238	1867	2113	2101	2134	2304	2320		2570		2841
6(3)		2533	2214	2486		2634		2858		3030		3123
7(1)		2715	2562	2627		2759		3003		3231		3429
8(1)	2670	2627	2436	2422	2395	2403	2811	2627	3105	3019		3524
8(2)		2931		2804	2799	2922	3017	3199		3510		3821

Table 2. Relative BE2 values for the six nuclei in the N=7 U(12) multiplet containing ^{122}Te. For each initial level one transition is assigned a value of 100. Data are taken from Refs. 10 and 11. The BE2 values indicate that the 0_2 and 0_3 levels exchange electromagnetic properties above ^{122}Te, and that the computed 0_3 in ^{126}Xe should be identified with the observed 0_2.

Transition		^{122}Te		^{126}Xe		^{130}Ba		^{134}Ce		^{138}Nd		^{142}Sm	
i	f	Exp	The	Exp	The	Exp	The	Exp	The	Exp	The	Exp	The
2(2)	2(1)	100	100	100	100	100	100	-	100	100	100	100	100
	0(1)	1	1.5	1.4	0.9	5.7	0.8	-	0.6	1	0.3	4.2	0.1
3(1)	2(2)	-	100	100	100	100	100	-	100	100	100	-	100
	4(1)	-	38	47	37	30	37	-	37	-	37	-	38
	2(1)	-	1.4	1.1	1.2	1.5	1.2	-	1	1.5	0.6	-	0.1
4(2)	2(2)	-	100	100	100	100	100	-	100	-	100	-	100
	3(1)	-	5.7	-	8	-	10	-	10	-	9	-	6
	4(1)	-	88	42	87	89	84	-	83	-	84	-	86
	2(1)	-	1.6	1.0	1.5	3.9	1.2	-	0.6	-	0.4	-	0.2
5(1)	4(2)	-	46	127	45	<57	44	-	45	-	45	-	45
	3(1)	-	100	100	100	100	100	-	100	-	100	-	100
	6(1)	-	42	-	40	381	40	-	40	-	40	-	41
	4(1)	-	1.6	4.9	1.4	6.7	1.2	-	0.9	-	0.6	-	0.2
0(2)	2(2)	-	100	100	100	-	100	-	100	-	100	-	100
	2(1)	-	16	9	13*	-	6.4*	-	4*	-	8*	-	67

* The computed 0_3 has been identified with the observed 0_2.

We thank R. F. Casten, B. H. Wildenthal, and J. Cizewski for useful discussions. This work is supported in part by NSF Grant PHY-844-1891.

*   Fellow of the Consejo Nacional de Investigaciones Cientificas y Tecnicas de Argentina.

Balantekin, A. B., I. Bars and F. Iachello (1981) Nucl. Phys. **A370**, 284

Cizewski, J. A. and E. Gülmez (1986) Yale preprint 3074-876

Frank, A. and P. van Isacker (1985) Phys. Rev. **C32**, 1770

Harter, H., P. von Bretano, A. Gelberg, and R. F. Casten (1985) Phys. Rev. **C32**, 631

Iachello, F. (1980) Phys. Rev. Lett. **44**, 772

Lederer, C. M. and V. S. Shirley (1978) *Table of Isotopes* (NY: Wiley)

Sakai, M. (1984) Atomic Data and Nuclear Data Tables **31**, 391

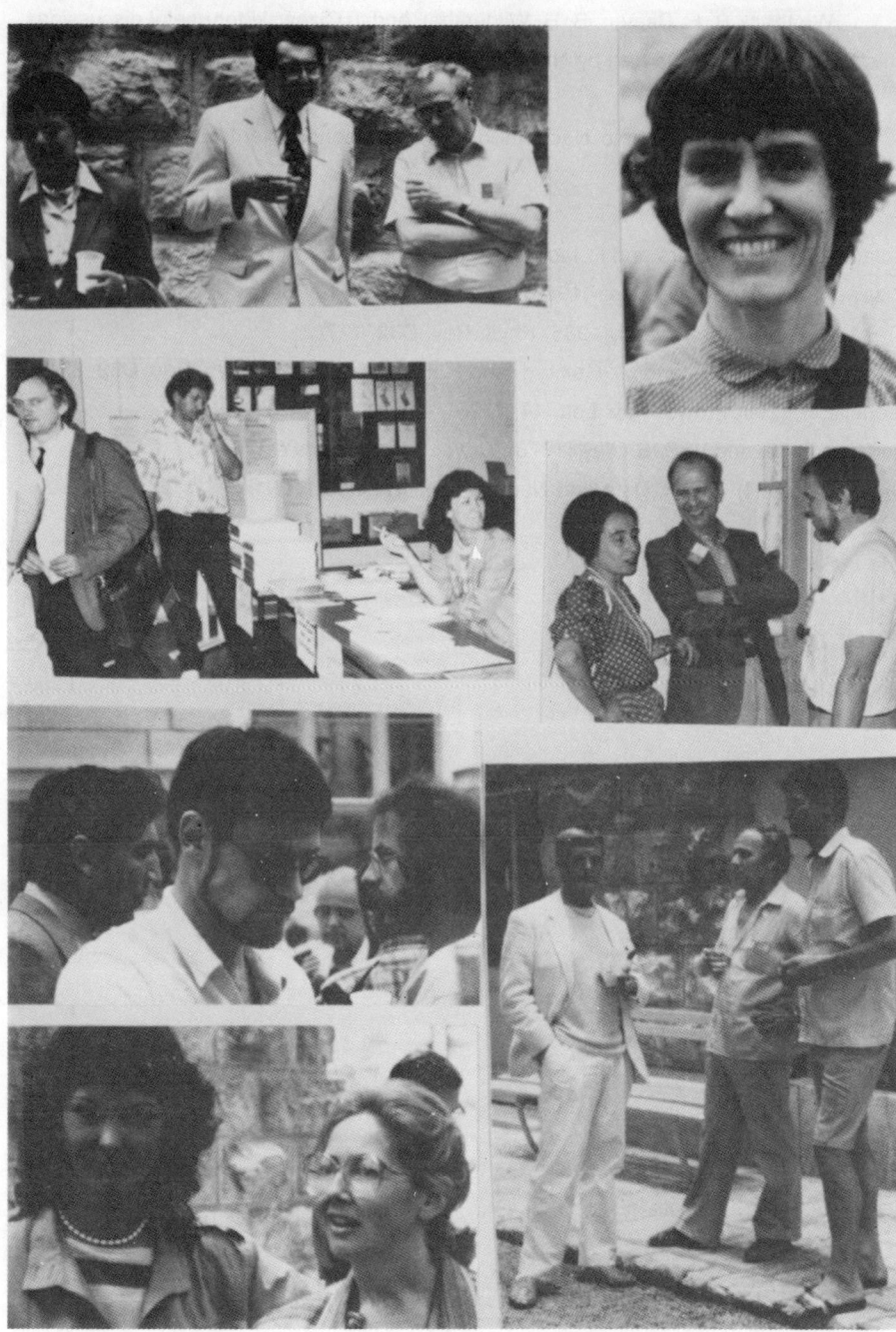

SECTION III

MIXED SYMMETRY STATES IN NUCLEI

K. HEYDE

Reviewer

I. TALMI

Discussion Moderator

Session Moderator

J. Ginocchio

MIXED-SYMMETRY STATES IN NUCLEI

K. Heyde
Institute for Nuclear Physics, Proeftuinstraat 86
B-9000 Gent (Belgium)

ABSTRACT

Properties of a new class of collective states with mixed-symmetry character in the proton-neutron variables are discussed and reviewed. A detailed discussion in terms of F-spin symmetry within the proton-neutron interacting boson model is given. Other approaches such as the two-fluid collective model, the shell-model and microscopic Random-Phase Approximation (RPA) calculations in a deformed potential are also presented. Some concluding remarks on future prospects are made.

1. INTRODUCTION

When reviewing mixed-symmetry states, one has to remind that the subject has actually been initiated very much by the experimental work of D. Bohle et al.[86], in which 1^+ levels were observed to be strongly excited in inelastic electron scattering on ^{156}Gd. Even if, up to now, a small number of experimental papers have been published[81-86], there has been a steady output of theoretical papers relating to different aspects of these mixed-symmetry states within the proton-neutron degree of freedom.

In the present contribution, we shall concentrate on some features that do not depend very much on the particular approach one is using (interacting boson model, two-fluid model, shell-model, RPA,...) and point out also possible experiments that can elucidate certain aspects of these mixed-symmetry states.

In many two-component systems, be it a proton-neutron two-rotor model (TRM), a simple proton-neutron coupled shell-model wave function (SM) or the proton-neutron interacting boson model (IBM-2), states that have a mixed-symmetry character under the permutation of the coordinates relating to the elementary modes of excitation for the uncoupled system will result. In the present discussion, we shall mainly concentrate on some general results within the framework of the IBM-2 and compare with existing data. We shall also discuss other approaches.

2. MIXED-SYMMETRY STATES IN IBM-2

2.1 General Aspects

In the IBM-2[1-14], although the total boson wave function has to be symmetric under the interchange of all variables, a mixed-symmetry character can arise for both the spatial (sd-part) and the proton-neutron charge part of it. The neutron-proton degree of freedom in IBM-2 is mathematically equivalent to that of a particle with "spin" 1/2 and thus one may classify the total boson wave function by its total "spin", which was called F-spin with $F_z = +1/2(-1/2)$ for proton(neutron) bosons [43-50]. The (anti)symmetric two-boson np states then have

$F_z=0$ and $F=(0)1$, respectively. For general N_π, N_ν where N_ρ denotes the boson number for charge ρ, the basis states can be classified according to the irrep. of either the group $U_\pi(6) \otimes U_\nu(6)$ or by $U(6) \otimes SU^{(F)}(2)$ where $SU^{(F)}(2)$ is the F spin group (because of the overal boson symmetry, the $U(6)$ and $SU^{(F)}(2)$ representations are characterized by the same Young diagrams with row lenght $N/2 + F$ and $N/2 - F$ ($N=N_\pi+N_\nu$)).

Many numerical studies of proton-neutron systems have been carried out for a general IBM-2 Hamiltonian using the NPBOS-code, as written by T.Otsuka[15-22]. Here, we concentrate on the analytic aspects inherent in the IBM-2 model and the symmetries, related to F-spin

i) F-spin is a good quantum number for Hamiltonians that are invariant under the interchange of neutron and proton variables. This implies that a general Hamiltonian

$$H = H_\pi + H_\nu + V_{\pi\nu} \quad , \tag{2.1}$$

or

$$H = \sum_{\alpha\beta\rho} \varepsilon^{(\rho)}_{\alpha\beta} G^{(\rho)}_{\alpha\beta} + \frac{1}{2} \sum_{\substack{\alpha\beta\gamma\delta \\ \rho,\rho'}} U^{(\rho,\rho')}_{\alpha\beta\gamma\delta} G^{(\rho)}_{\alpha\beta} G^{(\rho')}_{\gamma\delta} \quad , \tag{2.2}$$

with generators $G^{(\rho)}_{\alpha\beta} \equiv b^+_{\alpha\rho} b_{\beta\rho}$ has the following constraints

$$\varepsilon^{(\pi)}_{\alpha\beta} = \varepsilon^{(\nu)}_{\alpha\beta} \quad ,$$

$$U^{(\pi,\pi)}_{\alpha\beta\gamma\delta} = U^{(\nu,\nu)}_{\alpha\beta\gamma\delta} = U^{(\pi,\nu)}_{\alpha\beta\gamma\delta} = U^{(\nu,\pi)}_{\alpha\beta\gamma\delta} \quad . \tag{2.3}$$

Thereby, H of eq.(2.2) can be rewritten as

$$H = \sum_i a_i C^{(\pi+\nu)}_i \quad , \tag{2.4}$$

where $C^{(\pi+\nu)}_i$ are the linear and quadratic Casimir invariant operators associated with a particular subgroup chain of the product group $U_\pi(6) \times U_\nu(6)$ i.e.

$$U_\pi(6) \otimes U_\nu(6) \supset U_{\pi+\nu}(6) \supset \ldots \tag{2.5}$$

The general study of possible symmetries in (2.5) are studied elsewhere.

ii) States belonging to the symmetric irrep. $[N]$ of $U_{\pi+\nu}(6)$ have counterparts in the space of IBM-1 or, for a given F-spin symmetric Hamiltonian in IBM-2, a corresponding Hamiltonian of the IBM-1 can be found such that the energies of all states in the IBM-1 space coincide with those of the corresponding states in the irrep. $[N]$ of $U_{\pi+\nu}(6)$.

iii) It is possible to consider a Majorana operator in the Hamiltonians which are diagonal in the chain (2.5) and whose eigenvalues depend on the F-spin of the states. This is so, since the Majorana operator, conventionally written as

$$M = \xi_2 (d_\pi^+ s_\nu^+ - s_\pi^+ d_\nu^+)^{(2)} \cdot (\tilde{d}_\pi s_\nu - s_\pi \tilde{d}_\nu)^{(2)}$$

$$-2 \sum_{k=1,3} \xi_k (d_\pi^+ d_\nu^+)^{(k)} \cdot (\tilde{d}_\pi \tilde{d}_\nu)^{(k)} , \qquad (2.6)$$

is related (for $\xi_1 = \xi_2 = \xi_3 = a$) to the quadratic Casimir operator of $U_{\pi+\nu}(6)$ as

$$M = \frac{1}{2} \left[N(N+5) - C_2(U_{\pi+\nu}(6)) \right] , \qquad (2.7)$$

with eigenvalue

$$\langle M \rangle = (N/2-F)(N/2+F+1) , \qquad (2.8)$$

for states with $N = N_\pi + N_\nu$ and given F-spin. Since F varies from $F_{max} = N/2$ (symmetric states) to $F_{min} = 1/2 |N_\pi - N_\nu|$ in a given nucleus, one obtains families of levels with decreasing F-spin $F = F_{max}$, $F_{max}-1,...$ The F-spin scalar Hamiltonian of eq.(2.4) now splits the different multiplets but does not mix F-spin(the energy difference between corresponding $F = F_{max}$ and $F = F_{max}-1$ levels amounts to a N). In Fig.1,we

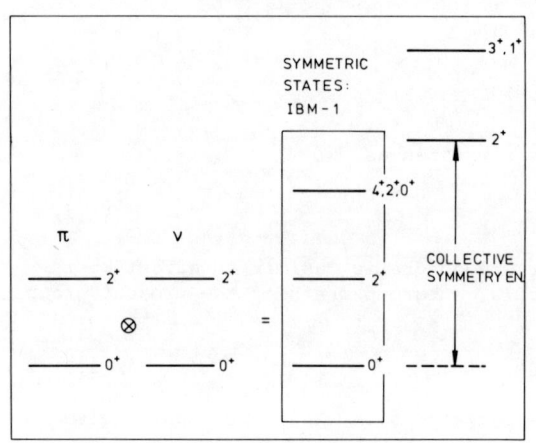

Fig. 1
Schematic representation of a coupled proton-neutron system with $N_\pi = 1$, $N_\nu = 1$. The symmetric states (F=1) and antisymmetric (F=0) states are drawn at the right-hand side

schematically give the levels for $N_\pi = 1$, $N_\nu = 1$.

iv) For the electromagnetic operators, general results can be formulated on the basis of F-spin symmetry[31,33]. In IBM-2, a transition operator T has in general two parts

$$T = t_\pi T_\pi + t_\nu T_\nu , \qquad (2.9)$$

with $t_\pi(t_\nu)$ the proton(neutron)effective charge. Before specifying T, but assuming that T_π and T_ν have the same structure (e.g $\chi_\pi = \chi_\nu$ for the quadrupole operator), T becomes a generator of $U_{\pi+\nu}(6)$ whenever

$t_\pi = t_\nu$. In many cases (e.g. the M1 operator where $g_\pi \neq g_\nu$; the E2 operator where in general $e_\pi \neq e_\nu$) both the T_π and T_ν matrix elements need to be calculated. One can still derive some general results ie.

$$\frac{<[N_\pi]\times[N_\nu];[N]\alpha|T_\nu|[N_\pi]\times[N_\nu];[N]\alpha'>}{<[N_\pi]\times[N_\nu];[N]\alpha|T_\pi|[N_\pi]\times[N_\nu];[N]\alpha'>} = \frac{N_\nu}{N_\pi} \qquad (2.10)$$

$$\frac{<[N_\pi]\times[N_\nu];[N-f,f]\alpha|T_\nu|[N_\pi]\times[N_\nu];[N-f',f']\alpha'>}{<[N_\pi]\times[N_\nu];[N-f,f]\alpha|T_\pi|[N_\pi]\times[N_\nu];[N-f',f']\alpha'>} = -1, \qquad (2.11)$$

(for $f \neq f'$ with $f = N/2-F$). The general trend of eqs (2.10)(2.11) are combined in fig.2 and are compared to more detailed calculations for

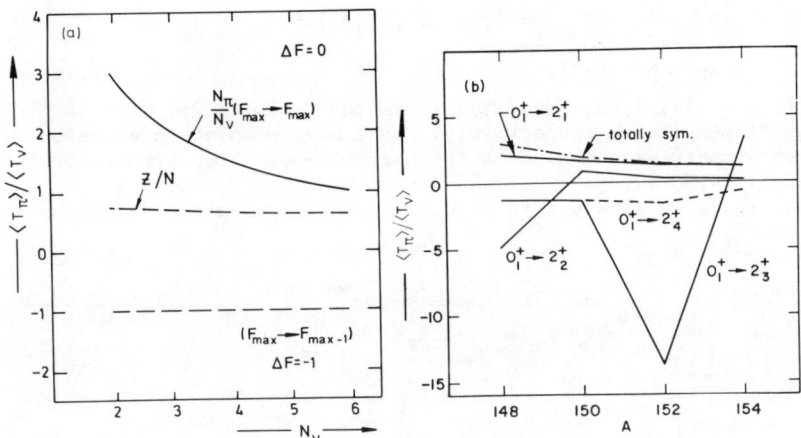

Fig.2.: The ratio of proton to neutron matrix elements $<T_\pi>/<T_\nu>$ as given for an F-spin scalar Hamiltonian (eqs.2.10,2.11)(a) and for the Sm nuclei with $148 \leq A \leq 154$ as obtained in ref. 38 (b)

the Sm nuclei as carried out by Otsuka and Ginocchio[38] pointing out that these rather general F-spin arguments are fulfilled to a good extend in actual nuclei.

From now on, we shall concentrate mainly on the mixed-symmetry levels with spin $J^\pi = 1^+$, 2^+, 3^+ on which also experimental data are available in different nuclear mass regions[79-89] (rare-earth nuclei, light nuclei, actinides).

2.2 Energy Spectra

Starting from an F-spin scalar Hamiltonian (which is not the most general one but contains the U(5), SU(3), O(6) limits), analytic expressions for the eigenvalues and wave functions can be obtained[31,33]. Thus, using the Hamiltonian

$$H = \varepsilon_d (n_{d_\pi} + n_{d_\nu}) + \kappa (Q_\pi + Q_\nu) \cdot (Q_\pi + Q_\nu) + aM, \qquad (2.12)$$

with

$$Q_\rho = (d^+ s + s^+ \tilde{d})_\rho^{(2)} + \chi_\rho (d^+ \tilde{d})_\rho^{(2)} \quad (\rho \equiv \pi, \nu), \qquad (2.13)$$

one obtains the energy eigenvalues

$$U(5): \varepsilon_d (n_1 + n_2) + a(N/2-F)(N/2 + F + 1), \qquad (2.14)$$

$$SU(3): 1/2\kappa(\lambda^2 + \mu^2 + \lambda\mu + 3\lambda + 3\mu) - \frac{3}{8}\kappa L(L+1) + a(N/2-F) \times$$
$$(N/2 + F + 1), \qquad (2.15)$$

$$O(6): \kappa \left[\sigma_1(\sigma_1 + 4) + \sigma_2(\sigma_2+2)\right] - \kappa\left[\tau_1(\tau_1+3) + \tau_2(\tau_2+1)\right]$$
$$+ a(N/2-F)(N/2+F+1), \qquad (2.16)$$

where $\{n_1,n_2\}, \{\lambda,\mu\}, <\sigma_1,\sigma_2>, (\tau_1,\tau_2)$ are the $U(5)$, $SU(3)$, $O(6)$ and $O(5)$ quantum numbers, respectively. For a more complete discussion of these symmetries and also on the particle-hole like symmetries see refs.1-14,30-42.

In fig.3, we show a typical $U(5)$ and $SU(3)$ spectrum (adding a

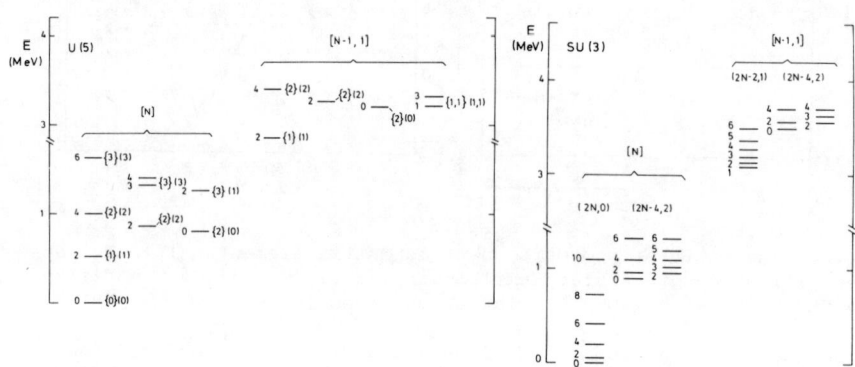

Fig.3.: Typical spectra in the $U(5)$ and $SU(3)$ limits of the IBM-2. The parameters are respectively
$U(5): N_\nu = 5, N_\pi = 1; \varepsilon_d = 400$ keV, $a = 400$ keV, $b = 10$ keV
$SU(3): N_\nu = 7, N_\pi = 6; \kappa = -12$ keV, $a = 200$ keV, $b = 5.5$ keV

small term $bL(L+1)$) for typical situations.
For the lowest 1^+ level with mixed-symmetry character($F=N/2 - 1$), simple values for $E_x(1_M^+)$ result

$$E_x(1^+_M, U(5)) = 2\varepsilon_d + a.N \qquad (2.17)$$

$$E_x(1^+_M, SU(3)) = (-3\kappa + a).N - \frac{3}{8}\kappa L(L+1), \qquad (2.18)$$

$$E_x(1^+_M, O(6)) = (-2\kappa + a)N - 6\kappa. \qquad (2.19)$$

In most realistic situations, starting from a general Hamiltonian, F-spin scalar, vector and tensor terms will be present: F-spin will not be a good quantum number but from numerical applications, at least for the lowest 1^+ level, a rather pure $F=N/2 - 1$ character remains in both the U(5), SU(3) and O(6) limits[30-42].

Here, we also shortly discuss the effect of modifying the Majorana operator ie. not taking ξ_1, ξ_2 and ξ_3 equal. Although the particular choice $\xi_1 = \xi_2 = \xi_3 = a$ makes the general Majorana operator of eq.(2.6) coincide with the quadratic $U_{\pi+\nu}(6)$ Casimir operator, there is no à priori microscopic reason for this [33,34]. Relaxing the condition towards $\xi_2 = 0$, $\xi_1 = \xi_3 = b$ (the other extreme), some interesting features occur in the SU(3) limit, as shown in fig.4. Here,

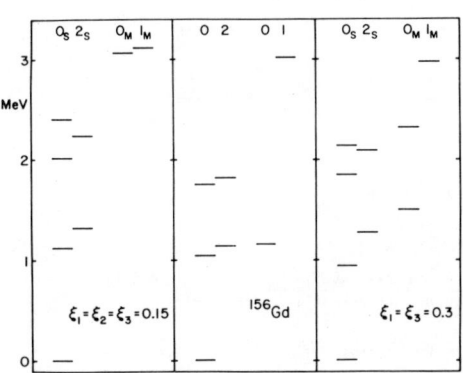

Fig.4: Calculated band head energies compared with the data in ^{156}Gd. The left hand side (I) uses a =0.15 the right hand side (II) b = 0.3 (ξ_2 = 0)

the bandheads and some experimental known band heads in ^{156}Gd are shown. Even though in both calculations, a and b have been chosen such that the 1^+ band lies near $E_x \simeq 3$ MeV, the spectrum of mixed-symmetry states is totally different. In case (I), the 1^+ band is the lowest mixed-symmetry band while in (II) the lowest mixed-symmetry band head (0^+) occurs near the position of the β and γ -band heads. This is near the experimental position of the well known 0^+_3 band. The present calculation indicates that an interpretation as a mixed-symmetry state might be possible(earlier interpretations were based on a 2 qp configuration or through the introduction of s',d' bosons). This would open the possibility of studying in some detail the properties of mixed-symmetry states, since the levels lie in a region of low level density. Much further investigation is needed here.

2.3 Electromagnetic Properties

Starting now from the specific structure of the M1, E2(and M3) operators, using an F-spin invariant Hamiltonian, reduced transition probabilities and static moments can be calculated using the general methods as outlined in ref.31,33. These apply as well to symmetric as to the mixed-symmetry states. In the rest of this contribution we concentrate mainly on 1^+ and 2^+ mixed-symmetry states.

2.3.1. <u>M 1-mode</u> (refs.51-59)

The magnetic dipole moment (in all limits) for the 1_M^+ level becomes

$$g(1_M^+) = 1/2(g_\pi + g_\nu) \quad , \tag{2.20}$$

which is a remarkable result but probably for some time to come outside the realm of experiments.

Because of the experimental information obtained on the excitation (via inelastic e^- scattering) and the γ decay of the 1_M^+ level in many nuclei, we have calculated reduced transition probabilities for the U(5), SU(3) and O(6) limits (see table 1 for the SU(3) limit and

TABLE I. Electromagnetic Decay of the 1_M^+ State in the SU(3) Limit

J_f^π	$T\lambda$	$B(T\lambda;1_M^+ \to J_f^\pi)$
0_1^+	M1	$\frac{3}{4\pi}(g_\nu - g_\pi)^2 \frac{8}{3(2N-1)} N_\nu N_\pi$
2_1^+	M1	$\frac{3}{4\pi}(g_\nu - g_\pi)^2 \frac{2(2N+3)}{3N(2N-1)} N_\nu N_\pi$
0_β^+	M1	$\frac{3}{4\pi}(g_\nu - g_\pi)^2 \frac{4(2N+1)}{3N(2N-3)(2N-1)} N_\nu N_\pi$
2_β^+	M1	$\frac{3}{4\pi}(g_\nu - g_\pi)^2 \frac{(4N^2-8N+7)^2}{3(N-1)N(2N-3)(2N-1)(4N^2-8N+1)} N_\nu N_\pi$
2_γ^+	M1	$\frac{3}{4\pi}(g_\nu - g_\pi)^2 \frac{8(N-2)(2N+1)}{N(2N-3)(4N^2-8N+1)} N_\nu N_\pi$
2_1^+	E2	$(e_\nu - e_\pi)^2 \frac{3(2N+3)}{4N(2N-1)} N_\nu N_\pi$
2_β^+	E2	$(e_\nu - e_\pi)^2 \frac{3(4N^2-8N-1)^2}{8(N-1)N(2N-3)(2N-1)(4N^2-8N+1)} N_\nu N_\pi$
2_γ^+	E2	$(e_\nu - e_\pi)^2 \frac{(N-2)(2N+1)}{N(2N-3)(4N^2-8N+1)} N_\nu N_\pi$
3_γ^+	E2	$(e_\nu - e_\pi)^2 \frac{N-2}{(N-1)N(2N-3)} N_\nu N_\pi$

and fig 5 where the classical limit ($N \to \infty$) for the expressions of table 1 was used). We recollect the U(5), SU(3) and O(6) $0_1^+ \to 1_M^+$ values

$$B(M1; 0_1^+ \to 1_M^+) = 0 \quad , \tag{2.21}$$

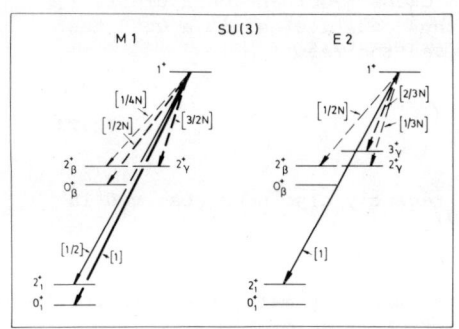

Fig.5.: The M1 and E2 decay of the 1^+_M state in the SU(3) limit. B(M1) and B(E2) values are normalized to the strongest one, respectively, for the classical limit ($N \to \infty$)

$$B(M1; O^+_1 \to 1^+_M) = \frac{3}{4\pi} \frac{8N_\pi N_\nu}{2N-1} (g_\nu - g_\pi)^2, \qquad (2.22)$$

$$B(M1; O^+_1 \to 1^+_M) = \frac{3}{4\pi} \frac{3N_\pi N_\nu}{N+1} (g_\nu - g_\pi)^2, \qquad (2.23)$$

respectively. In table 2, we compare $B(M1; O^+_1 \to 1^+_M)$ values, calculated

TABLE II: M1 strength in vare earth nuclei (exp.data and various models)

Nuclei	Ex (MeV)	B(M1; ↑)(μ^2_N)						
		Exp.	IBM-2		TRM		RPA	
			(I)	(II)	(I)	(II)	(I)	(II)
^{154}Sm	3.200	0.8±0.2	2.7	1.2	17.52	7.66	5.84	9.52
^{156}Gd	3.075	1.3±0.2	2.9	1.2	17.12	7.07	5.69	9.27
^{158}Gd	3.200	1.4±0.3	3.2	1.4	17.93	7.58	5.95	9.70
^{164}Dy	3.110	1.5±0.3	3.9	1.6	18.97	8.05	6.25	10.19
^{168}Er	3.390	0.9±.2	3.9	1.6	20.50	8.24	6.73	10.97
^{174}Yb	3.555	0.8±0.2	3.8	1.6	20.73	8.17	6.77	11.03

IBM.2:(I):ref 31 (II)ref.31 TRM(1):ref.121 (II)ref.94 RPA:(I)ref.119 (II)ref.126

using microscopically derived g_π, g_ν values[59] and g_π, g_ν values obtained from a fit to experimental $g(2^+_1)$ - factors with experimental values[33] This is still a simplified calculation since fragmentation has not been considered here.

Recently, the decay of the 1^+_M state in ^{156}Gd was measured by Berg et al.[84]. These results are consistent with the ratio $B(M1; 1^+_M \to O^+_1)/B(M1; 1^+_M \to 2^+_1) = 2$, shown also in fig.5. This ratio also results from the Alaga rules as the ratio of two Clebsch-Gordan coefficients i.e. $|<11,1-1|00>/<11,1-1|20>|^2=2$, and thus constitutes not a good test for the IBM-2, which, for finite N gives the result

$$\frac{B(M1;1_M^+ \to 0_1^+)}{B(M1;1_M^+ \to 2_1^+)} = \frac{4N}{2N+3} < 2. \tag{2.24}$$

Mixed-symmetry 1^+ states have recently also been observed in both light nuclei[81] and actinides[87].

2.3.2. E2-Mode

In rotational nuclei, the lowest mixed-symmetry state is the 1^+ state, now observed in a number of strongly deformed rare-earth nuclei[81-86]. In vibrational nuclei (see also fig.3), however, a $J^\pi=2^+$ state is expected to be the lowest member of the mixed-symmetry $F=F_{max}-1$ levels. This is the case in U(5) nuclei and probably also for O(6)-like nuclei. Recent experiments have indicated possible candidates (the 2_3^+ level in some N=84 nuclei[41]; the 2_2^+ and 2_3^+ levels in ^{56}Fe [32]) for such a 2^+ state.

In the three limits of the IBM-2, the E2 excitation from the ground state to the lowest 2_M^+ state is allowed and becomes for the U(5), SU(3) and O(6) limits [31,33]

$$B(E2;0_1^+ \to 2_M^+) = \frac{5N_\nu N_\pi}{N} (e_\nu-e_\pi)^2, \tag{2.25}$$

$$B(E2;0_1^+ \to 2_M^+) = \frac{3(N-1)N_\nu N_\pi}{N(2N-1)} (e_\nu-e_\pi)^2, \tag{2.26}$$

$$B(E2;0_1^+ \to 2_M^+) = \frac{2(N+2)N_\nu N_\pi}{N(N+1)} (e_\nu-e_\pi)^2, \tag{2.27}$$

respectively. These values depend in a critical way on the difference of effective charges $e_\nu-e_\pi$ for which a precise determination over a large region of nuclei is an important problem,[65-78] especially since in many IBM-2 numerical studies, it was customary to use $e_\pi = e_\nu$.

In fig.6, we show the ratio $B(E2;0_1^+ \to 2_M^+)/B(E2;0_1^+ \to 2_1^+)$ for the U(5), SU(3) and O(6) limits, as a function of e_ν. Even if this figure shows the rather profitable situation for U(5)-like nuclei (some suggestions along this line were made some time ago by F.Iachello[42] for a possible 2_M^+ level near $E_x \approx 2$MeV in ^{110}Pd), no firm experimental results are obtained as yet in the Pd, Ru region. In the light of the well-known 1_M^+ level in ^{156}Gd and since the $B(E2;0_1^+ \to 2_1^+)$ in strongly deformed nuclei are much larger than in vibrational like nuclei, experiments were carried out in Darmstadt[83] studying the E2 strength in ^{156}Gd(fig.7). Expecting the mixed-symmetry state (or states since splitting can well occur) above the 1_M^+ level, three candidates arise at E_x=3.096, 3.150 and 3.400 MeV (see ref.83 for a detailed discussion). Using the SU(3) expression of eq.(2.26), one deduces the following limits on the ratio e_π/e_ν in ^{156}Gd ie. $1.1 \leq e_\pi/e_\nu \leq 1.3$. In order to also reproduce the measured $B(E2;0_1^+ \to 2_1^+)$ value, one finally gets

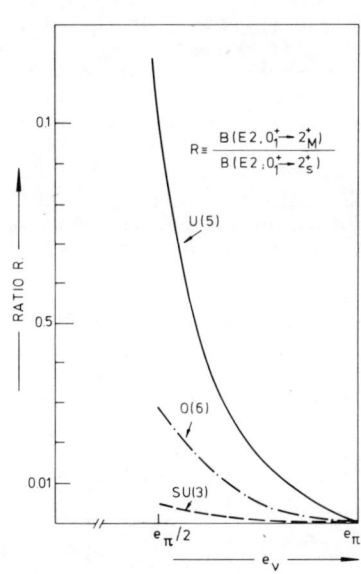

Fig.6.: The ratio $R=B(E2;0_1^+ \to 2_M^+)/B(E2;0_1^+ \to 2_1^+)$, for the U(5), SU(3) and O(6) limit as a function of e_ν (given in units of e_π)

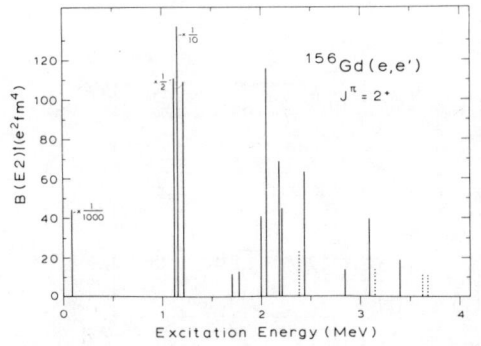

Fig.7.: The E2 strength distribution determined in ^{156}Gd(e,e') (ref. 83). The dotted lines indicate candidates for which a 2^+ assignment is not unambiguous.

$$10.8 \pm 0.3 \text{ efm}^2 \geq e_\nu \geq 10.0 \pm 0.3 \text{ efm}^2$$
$$12.4 \pm 1.4 \text{ efm}^2 \leq e_\pi \leq 13.0 \pm 1.4 \text{ efm}^2$$

or near equality of e_π and e_ν. Using the expression for $B(E2;0_1^+ \to 2_1^+)$ in the SU(3) limit and using experimental $B(E2;0_1^+ \to 2_1^+)$ values in ^{154}Gd, ^{154}Sm, ^{158}Gd, ^{160}Dy values of e_π and e_ν have been obtained by plotting the quantity $[N.B(E2)/(2N+3)]^{1/2}$ versus the neutron boson number, with as a result $e_\nu = 9.2 \pm 0.9$ efm^2, $e_\pi = 13.2 \pm 1.1$ efm^2, in good agreement with the above results. Starting from the latter charges, one then calculates $B(E2;0_1^+ \to 2_M^+) = 67$ e^2 fm^4 (pure SU(3) limit), a value in rather good agreement with the value of 40 ± 6 e^{2}fm^4

for the 2^+ state at E_x = 3.096 MeV. A possible splitting of the mixed-symmetry state can probably account for the differences.

At present, studies on the mixed-symmetry $J^\pi = 2^+_M$ state to other well-deformed nuclei and in vibrational-like nuclei is highly desirable. Also, study of the mass dependence of effective charges is of the utmost importance in near future.

2.4 Spreading of the Collective Strength

The mixed-symmetry states are situated in an energy region where the level density for states of a given J^π value becomes large. So, it is natural that spreading of collective strength into the background occurs, although only a few simple calculations within this respect have been carried out. Experimentally, splitting of 1^+ strength has been observed in nuclei such as ^{164}Dy, ^{174}Yb but with the increasing quality of data, also spreading of 1^+ strength in ^{156}Gd is observed.[79-89] For the E2 strength (see fig.7), from 2 MeV on, E2 strength is spread out rather strongly.[83]

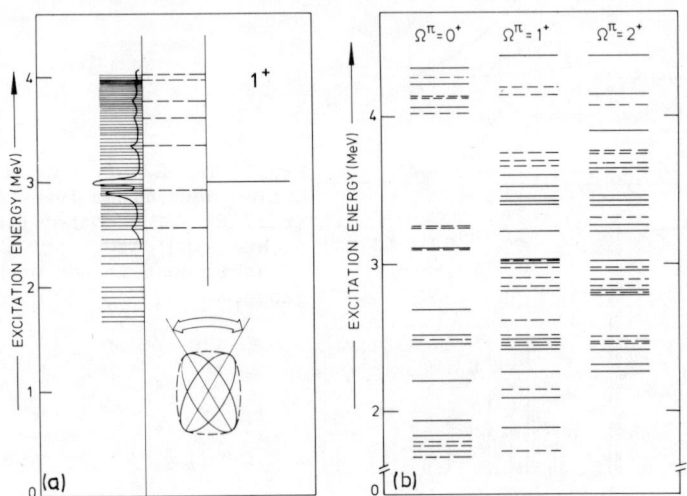

Fig.8.: Schematic representation (a) of a possible splitting and spreading of a mixed-symmetry 1^+ state (right handside) through doorway states containing hexadecapole configurations (middle part) into the quasi-particle background (left hand side). In part (b), the detailed 2 qp density of Ω^π = 0^+, 1^+ and 2^+ Nilsson states in ^{156}Gd is given. Full (dashed) lines are proton (neutron) 2qp states.

For the 1^+ state, hexadecapole states could act as natural doorway states since the mixed-symmetry state is expected to contain an important hexadecapole dynamic structure. These doorway states can then couple to the more complex 2qp states. This mechanism is shown in a schematic way in fig.8a, where besides (fig.8.b), also the detailed 2qp density of $\Omega=0^+$, 1^+, 2^+ levels for ^{156}Gd is shown.

For 2^+ levels, the different symmetric vibrations ($\beta,\gamma,\beta\beta,\beta\gamma,\gamma\gamma,..$) can probably act as doorway states to spread the E2 mixed-symmetry states over the dense quasi-particle background.

In the light of the above discussion, a serious problem remains with respect to the total summed M1 and E2 strength residing in the background "grass" starting around $E_x \simeq 2$ MeV.(see also 3.3.3).

3. OTHER APPROACHES TO MIXED-SYMMETRY STATES

Now, we present, although shortly, some of the other approaches in order to describe mixed-symmetry states.

3.1 Two-Fluid Models

Since an electric dipole mode, due to translational oscillations of proton against neutron densities exist in all nuclei(the electric giant dipole resonance),it was very natural to extend that approach to other modes within a two-fluid model[90-106].

Surface vibrations of protons and neutrons in phase and antiphase,coupled by the symmetry energy term were treated already in 1965 by Faessler[106] and Greiner[105]. Later on, rotational oscillations in a two-rotor model(TRM)[101,102] or in a vibrational potential model(VPM),[103] leading to $K^\pi = 1^+$ excitations in deformed nuclei were considered. The difficulty in these approaches lies,in particular,in a reliable calculation of the restoring force defining the frequency and thus the energy for this particular mode. In a recent letter by Faessler et al.[94], expanding the symmetry energy in powers of proton and neutron density deformations, fairly good agreement with the experimental 1^+ energies for deformed nuclei has been obtained. Thereby, from the too large 1^+ excitation energy in the original papers in 1966, a reduction to about the correct energy has been observed. The $B(M1;\uparrow)$ strength in this particular 1^+ mode however, still remains between a factor 5 to 10 too large compared with the data(see table 2) because it is the convection current associated with the precessional motion for the proton fluid of the total nucleus that determines this strength.

In studying the classical limit of the IBM-2 [60-64],a similar geometrical interpretation of the 1^+ mode in deformed nuclei has been obtained.

3.2 Shell-Model For Light Nuclei

Within the nuclear shell-model,especially for light nuclei,protons and neutrons can be handled in detail(exact within a reduced model space and for a given residual interaction). In particular, detailed $1f_{7/2}$ shell-model calculations were carried out in 1964 by Mc Cullen, Bayman and Zamick[114] (MBZ) and $1^+,2^+,3^+,..$ levels with a mixed-symmetry character resulted. Recently,Zamick has reinvestigated the excitation of 1^+ levels in Ti nuclei with $B(M1;\uparrow)$ values,typical of the order of $\simeq 2\mu_N^2$ in the energy region $3<E_x<5$ MeV and has thereby shown that mixed-symmetry states(in particular 1^+ levels)are a far more widespread phenomenon[111-113] than expected from the experimental results as obtained by Bohle et al.[86]. Recent experiments by Djalali et al.[81], have indeed observed important M1 strength to a 1^+ level in ^{46}Ti at E_x=4.32 MeV($B(M1;\uparrow)$=1.0 ± 0.2 μ_N^2). It is thus quite natural to suggest the existence of such mixed-symmetry states near $E_x \simeq 4$ MeV in even-even Ti, Cr and Fe nuclei for the $A \simeq 50$ mass region. Since in this particular mass region, the $1f_{7/2}$ orbital determines M1 excitation, both the spin and orbital parts will contribute in an important way to the $B(M1;\uparrow)$ value. This is quite natural since one can show(fig.9), from inspecting the diagonal single-particle M(L) matrix

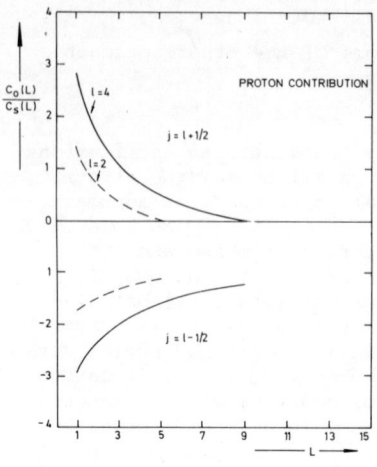

Fig.9.: Ratio of orbital to spin diagonal magnetic single-particle matrix elements as a function of the multipolarity L of the operator for protons. Values for $j=l\pm 1/2$ for $l=4$ and $l=2$ are shown.

elements[108] $<nlj\| M(L) \| nlj>$, that the spin part is non-negligible, especially for low l and higher L values.

Another interesting outcome of the MBZ shell-model calculations is the largely increasing fragmentation of $B(ML;\uparrow)$ strengths with increasing multipolarity L[112].

Shell-model calculations are clearly feasible, also in the neighbourhood of other closed shells (Z=50, N=50; Z=50, N=82;...) for vibrational-like nuclei and can shed light in obtaining good estimates for mixed-symmetry states in regions where the U(5) or O(6) limit of the IBM-2 can be applicable.

3.3 RPA in deformed nuclei

One can also expect that a microscopic approach, starting from a deformed single-particle basis and a residual proton-neutron interaction, using the RPA or TDA approximations, can describe the particular mixed-symmetry states, discussed before[115-127]. A number of such calculations have been performed varying from schematic ones using a single major oscillator proton and neutron shell[119] (neglecting spin-orbit effects and using residual quadrupole interactions) up to more realistic ones[118] (using a detailed Nilsson-model basis but still using schematic interactions). In these calculations, a major fragmentation of $B(M1;\uparrow)$ strength results with some concentration around $E_x \simeq 3$ MeV and 5 MeV. Moreover, a summed $B(M1;\uparrow)$ of almost 10 times the experimental value for the strongest 1^+ level is observed in the region of deformed nuclei(see table 2). Even if in a number of nuclei, experimental evidence for splitting of the M1 strenght in more than one peak exists (^{164}Dy, ^{174}Yb), the problem of fragmentation and how to reconcile the RPA results with the IBM-2 and other collective approaches, remains a major problem(see also 2.4).

4. CONCLUSIONS

Besides the study of mixed-symmetry state within the IBM-2 and other approaches, there remain many aspects of these states to be studied such as the problem of splitting and spreading of $1^+, 2^+$, strength, effects of triaxiallity in nuclei, relation to microscopic theories and the nuclear shell-model for a better determination of parameters in the collective approaches.

We mention a few more specific items that could be tackled with the present experimental facilities in near future:

-exploration of the 2^+ mixed-symmetry state in light nuclei in particular (Ti,Cr,Fe) since the level density is still rather low,

-determination of orbital versus spin contribution in the M1 excitation mode from light nuclei towards heavy nuclei($1f_{7/2}$ shell, Ru,Pd region, strongly deformed nuclei, Pt nuclei, Actinides)

-study of mixed-symmetry states in odd-A nuclei. In coupling an odd-particle to the mixed-symmetry 1^+ state, M1 strength will now become split. Therefore, for a first search, a look at deformed nuclei with well-known 1^+ mixed-symmetry levels in the core is at hand. The nuclei ^{165}Ho(^{164}Dy) with ground state spin $7/2^-$ and ^{169}Tm(^{168}Er) with ground state spin $1/2^+$ present such possibilities.

The work presented here grew out of a collaborative work over the last years in Gent, with P.Van Isacker, J.Jolie and A.Sevrin, on mixed-symmetry states in the IBM-2.

He is most grateful to U.E.P.Berg, D.Bohle, P.von Brentano, R.F.Casten, A.E.L.Dieperink, J.P.Elliott, A.Frank, D.Hamilton, A.Gelberg, J.N.Ginocchio, F.Iachello, T.Otsuka, A.Richter, O.Scholten, J.Vervier and D.D.Warner for useful discussions and communication of results prior to publication over the years. He is indebted to the NFWO, and IIKW for financial support

IBM-2 : General (review and conferences)

1. A.E.L.Dieperink and G.Wenes,Ann.Rev.Nucl.Sci.$\underline{35}$,77(1985)
2. J.P.Elliott, Rep.Progr.Phys.48,171(1985)
3. P.O.Lipas,Int.Rev.Nucl.Phys.$\underline{2}$,33(1984)
4. A.Arima and F.Iachello,Adv.in Nucl.Phys.$\underline{13}$,139(1984)
5. A.Arima and F.Iachello,Ann.Rev.Nucl.Sci.$\underline{31}$,75(1981)
6. A.E.L.Dieperink,Comm.Nucl.Part.Phys.$\underline{14}$,25(1985)
7. A.E.L.Dieperink,Nuclear Shell Models(World Scientific Publ. Co,1985),271
8. F.Iachello,Nuclear Shell Models,(World Scientific Publ. Co,1985)279
9. A.E.L.Dieperink,Theory of Nucl.Structure and Reactions, (World Scientific Publ.Co,1985),205
10. A.E.L.Dieperink,Nucl.Phys.$\underline{A421}$,189C(1984)
11. F.Iachello,Progr.Part.Nucl.Phys.$\underline{9}$,5(1983)
12. A.E.L.Dieperink,Prog.Part.Nucl.Phys.$\underline{9}$,121(1983)
13. A.E.L.Dieperink,Il Nuovo Cim.$\underline{76A}$, 377(1983)
14. A.E.L.Dieperink, Nuclear Collective Dynamics(World Scientific Publ.Co,1982),149

Books and Theses

15. Interacting Boson-Boson and Boson-Fermion Systems(ed.O.Scholten), World Scientific Publ.Co,1985
16. Interacting Bose-Fermi Systems in Nuclei(ed.F.Iachello) Plenum,N.Y.1981
17. Interacting Bosons in Nuclear Physics(ed.F.Iachello) Plenum, N.Y,1979
18. D.Bohle, Ph.D.Thesis,Technische Hochschule Darmstadt,1985
19. A.Sevrin, Lic.Thesis, University of Gent, 1985
20. P.Van Isacker, Thesis Hoger Aggr.University of Gent,1984
21. O.Scholten, Ph.D.Thesis,University of Groningen,1980
22. T.Otsuka, Ph.D.Thesis, University of Tokyo,1978

Specific problems

23. C.H.Bruce,S.Pittel,B.R.Barrett,P.D.Duval,Phys.Lett.$\underline{157B}$,115(1985)
24. A.Novoselsky and I.Talmi,Phys.Lett.$\underline{160B}$,13(1985)
25. A.Novoselsky,Phys.Lett.$\underline{155B}$,299(1985)
26. P.O.Lipas and K.Helimäki,Phys.Lett.$\underline{165B}$,244(1985)
27. M.Zimmermann and J.Dobes, Phys.Lett.$\underline{156B}$,7(1985)
28. A.E.L.Dieperink and I.Talmi,Phys.Lett.$\underline{131B}$,1(1983)
29. A.E.L.Dieperink and R.Bijker,Phys.Lett.$\underline{116B}$,77(1982)

Mixed-symmetry

30. K.Heyde et al.,Discussion meeting report,1986
31. P.Van Isacker,K.Heyde,J.Jolie and A.Sevrin,Ann.of Phys.(N.Y)
32. S.A.A.Eid,W.D.Hamilton and J.P.Elliott,Phys.Lett.$\underline{166B}$,267(1986)
33. O.Scholten,K.Heyde,P.Van Isacker and T.Otsuka,Phys.Rev. $\underline{C32}$ 1729(1985)
34. O.Scholten,K.Heyde and P.Van Isacker,Phys.Rev.Lett.$\underline{55}$,1866(1985)

35. D.Bohle et al.,Phys.Rev.Lett.$\underline{55}$, 1661(1985)
36. O.Scholten et al.Nucl.Phys.$\underline{A438}$,41(1985
37. J.N.Ginocchio,Nuclear Shell Models(World Sientific Publ.Co, 1985),p.220
38. T.Otsuka and J.N.Ginocchio,Phys.Rev.Lett.$\underline{54}$,777(1985)
39. A.Richter, Workshop on Nucl.Coll.Motion(Internal report),1984
40. J.Vervier,Proc.of the XXIII Int.Winter Meeting on Nucl.Physics (Bormio,Italy,1985),to be publ.
41. W.D.Hamilton, A.Irbäck and J.P.Elliot,Phys.Rev.Lett.$\underline{53}$,2469(1984)
42. F.Iachello,Phys.Rev.Lett.$\underline{53}$,1427(1984)

F-spin structure

43. A.Frank, Yale preprint YNT 86-04, to be publ.
44. P.Sala et al.,University of Köln preprint,1986,to be publ.
45. A.Frank and P.Van Isacker,Phys.Rev.$\underline{C32}$, 1770(1985)
46. P.von Brentano, A.Gelberg,H.Harter and P.Sala,J.Phys.G $\underline{11}$,L85 (1985)
47. H.Harter,P.von Brentano, A.Gelberg and R.F.Casten,Phys.Rev.$\underline{C32}$ 631(1985)
48. H.Harter, A.Gelberg and P.von Brentano,Phys.Lett.$\underline{157B}$,1(1985)
49. T.Otsuka, A.Arima, F.Iachello and I.Talmi,Phys.Lett.$\underline{76B}$,139 (1978)
50. A.Arima,T.Otsuka,F.Iachello and I.Talmi,Phys.Lett.$\underline{66B}$,205(1977)

Magnetic (M1,M3) properties

51. A.E.L.Dieperink,O.Scholten and D.D.Warner,KVI preprint 535 (1985),to be publ.
52. P.Halse,Drexel University preprint   (1986),to be publ.
53. K.Heyde,P.Van Isacker and J.Jolie,Hyperfine Int.$\underline{22}$,339(1985)
54. B.R.Barrett and P.Halse, Phys.Lett.$\underline{155B}$,133 (1985)
55. S.Pittel,J.Dukelsky, R.P.J.Perazzo and H.M.Sofia, Phys.Lett. $\underline{144B}$,145(1984)
56. O.Scholten, A.E.L.Dieperink, K.Heyde and P.Van Isacker, Phys.Lett.$\underline{149B}$,279(1984)
57. P.Van Isacker et al.,Phys.Lett.$\underline{144B}$,1(1984)
58. P.Van Isacker,K.Heyde,J.Jolie and O.Scholten,Int.Boson-Boson and Boson-Fermion Systems(World Scientific Publ.Co.,1985),334
59. M.Sambataro, O.Scholten, A.E.L.Dieperink and G.Piccitto,Nucl. Phys.$\underline{A423}$,333(1984)

Classical limit

60. A.Van Egmond and K.Allaart,Phys.Lett.$\underline{164B}$,1(1985)
61. N.R.Walet,P.J.Brussaard and A.E.L.Dieperink,Phys.Lett.$\underline{163B}$,1 (1985)
62. R.Bijker, Phys.Rev.$\underline{C32}$,1442(1985)
63. S.Pittel and J.Dukelsky,Phys.Rev.$\underline{C32}$,335(1985)
64. A.B.Balantekin and B.R.Barrett,Phys.Rev.$\underline{C32}$,288(1985)

Effective charges

65. O.Scholten et al., MSU preprint   (1986),to be publ.

66. P.Sala, A.Gelberg and P.von Brentano,Z.Phys.A323,281(1986)
67. J.N.Ginocchio and P.Van Isacker,Phys.Rev.C33,365(1986)
68. K.K.Seth,Nucl.Phys.A434, 287C(1985)
69. R.D.Smith, V.R.Brown and V.A.Madsen,Phys.Rev.C33,847(1986)
70. V.R.Brown,A.M.Bernstein and V.A.Madsen,Phys.Lett.164B,217(1985)
71. V.A.Madsen and V.R.Brown,Phys.Rev.Lett.52,176(1984)
72. A.Saha et al., Phys.Lett.132B,51(1983)
73. V.A.Madsen,T.Suzuki,A.M.Bernstein and V.R.Brown,Phys.Lett. 123B ,13(1983)
74. A.M.Bernstein, V.R.Brown and V.A.Madsen,Phys.Lett.103B,255(1981)
75. M.Matoba,Phys.Lett.88B,249(1979)
76. A.M.Bernstein, V.R.Brown and V.A.Madsen,Phys.Lett.71B,48(1977)
77. D.E.Bainum,Phys.Rev.Lett.39,443(1977)
78. V.R.Brown and V.A.Madsen,Phys.Rev.C11,1298(19

Experimental

79. D.Bohle et al., Technische Hochschule Darmstadt preprint(1986), to be publ.
80. W.Andrejtscheff, University of Sussex preprint   (1986) to be publ.
81. C.Djalali et al.,Phys.Lett.164B,269(1985)
82. J.A.Carr et al.,Phys.Rev.Lett.54,881(1985)
83. D.Bohle et al., Phys.Rev.Lett.55,1661(1985)
84. U.E.P.Berg et al.,Phys.Lett.149B,59(1984)
85. D.Bohle, G.Küchler, A.Richter and W.Steffen,Phys.Lett.148B, 260(1984)
86. D.Bohle et al.,Phys.Lett.137B,27(1984)
87. A.Richter,Inst.Kernphysik Darmstadt preprint IKDA-85/8; to be publ.
88. A.Richter,Inst.Kernphysik Darmstadt preprint IKDA-84/14
89. A.Richter,Inst.Kernphysik Darmstadt preprint IKDA-83/28

Two-fluid description

90. A.Faessler and R.Nojarov,Univ.of Tübingen preprint(1986) to be publ.
91. O.Civitarese, A.Faessler and R.Nojarov,Univ.of Tübingen preprint (1986),to be publ.
92. A.Faessler, R.Nojarov and S.Zubik, Univ.of Tübingen preprint (1986), to be publ.
93. R.Nojarov, Z.Bochnacki and A.Faessler,Univ.of Tübingen preprint (1986), to be publ.
94. A.Faessler, Z.Bochnacki and R.Nojarov,J.Phys.G12,L47(1986)
95. S.G.Rohozinski and W.Greiner,Z.Phys.A322,271(1985)
96. N.Lo Iudice, CERN preprint TH.4337/1985
97. N.Lo Iudice, Theory of Nucl.Structure and Reactions(World Scientific Publ.Co.,1985),190
98. N.Lo Iudice et al.,Phys.Lett.161B,18(1985)
99. F.Palumbo and A.Richter,Phys.Lett.158B,101(1985)
100. G.De Franceschi, F.Palumbo and N.Lo Iudice,Phys.Rev.C29,1496 (1984)
101. N.Lo Iudice and F.Palumbo,Nucl.Phys.A326 ,193(1979)

102. N.Lo Iudice and F.Palumbo,Phys.Rev.Lett.$\underline{41}$,1532(1978)
103. T.Suzuki and D.J.Rowe,Nucl.Phys.$\underline{A289}$,461(1977)
104. M.Maruhn-Rezwami, W.Greiner and J.A.Maruhn,Phys.Lett.$\underline{57B}$,109 (1975)
105. W.Greiner,Nucl.Phys.$\underline{80}$,417(1966)
106. A.Faessler,Nucl.Phys.$\underline{85}$,653(1966)

Shell-model(specific)

107. K.Heyde and J.Sau,Phys.Rev.$\underline{C33}$,1050(1986)
108. K.Heyde and J.Sau,Phys.Rev.$\underline{C30}$,1355(1984)
109. J.Sau,K.Heyde and J.Van Maldeghem,Nucl.Phys.$\underline{A410}$,14(1983)
110. A.Van Egmond, K.Allaart and G.Bonsignori, Nucl.Phys.$\underline{A436}$,458 (1985)
111. L.Zamick, Phys.Lett.$\underline{167B}$, 1(1986)
112. L.Zamick, Phys.Rev.$\underline{C33}$,691(1986)
113. L.Zamick,Phys.Rev.$\underline{C31}$, 1955(1985)
114. J.Mc.Cullen, B.F.Bayman and L.Zamick,Phys.Rev.$\underline{134}$,B515(1964) and Technical Report NYO-9891(1964)

RPA,Sum rule method,microscopic studies

115. O.Scholten et al., private comm.
116. E.Hammarén et al.,Phys.Lett.$\underline{171B}$,347(1986)
117. H.Kurasawa and T.Suzuki,Phys.Lett.$\underline{144B}$,151(1984)
118. S.Iwasaki and K.Hara,Phys.Lett.$\underline{144B}$,9(1984)
119. D.Bes and R.A.Broglia,Phys.Lett.$\underline{137B}$,141(1984)
120. R.R.Hilton, Z.Phys.$\underline{A316}$,121(1984)
121. E.Lipparini and S.Stringari,Phys.Lett.$\underline{130B}$,139(1983)
122. E.Lipparini and A.Richter,Phys.Lett.$\underline{144B}$,13(1984)
123. M.Traini,Phys.Rev.Lett.$\underline{41}$,1535(1978)
124. I.Hamamoto and S.Åberg, Lund preprint MPh-86/07
125. S.Åberg , Lund preprint MPh-85/02
126. I.Hamamoto and S.Åberg,Phys.Lett.$\underline{145B}$,163(1984
127. A.Bohr and B.Mottelson, Nucl.Structure, Vol II(W.A.Benjamin Inc.,1975, N.Y),p.375

MIXED-SYMMETRY STATES IN PROTON-NEUTRON SYSTEMS

Discussion Session
Following Review Lecture
by
K. Heyde

Moderator:

IGAL TALMI
Weizmann Institute of Science
Rehovat, Israel

Morton Brenner: Referring to your first transparency: there is apparently a lack of experimental data. There are more theories and too little experimental facts. That's why I think I would like to offer you one phenomena which is bothering us. We have some results on chromium 50 which probably you can easily explain in a less sophisticated way than you have explained other features. This is proton inelastic scattering on chromium 50 at the very modest energy of 6 MeV. The trouble is that we see very strong, narrow resonances and groups of resonances which we cannot explain. These resonances decay strongly to the 2_2^+ and the 2_3^+ states in chromium 50. Some resonances decay very strongly to 2^+ state while the 2_3^+ state is not related to very strong resonances. The 4_1^+ is strongly populated too. We don't know the nature of these resonances and I very much like your cross vibration nuclei. When the proton approaches chromium 50 you may have it vibrate in such a strange way. The cross vibration is, as I understand, just like this (demonstrated with hands). Maybe you or somebody else can give a very simple explanation of our observations.

(Session break occurred at this point, Ed.)

Moderator: People complain about some deficiencies of the s,d boson model which have to be fixed. People are eager to add g bosons, s' bosons, d' bosons and so on. My own mind works slowly, so I am still impressed by the phenomenological success of IBA-1 and would like to understand why the hell it works so well. The IBA-1 model is closely related to the geometrical model. We had very interesting talks explaining why it is not exactly equivalent to it and indeed it is not. However, for practical purposes, for the low lying part of the spectrum, it is sufficiently equivalent to the collective model. It reproduces vibrational nuclei, gamma unstable nuclei, rotational nuclei, various collective bands, and it indeed can be related to the collective model. I do not want to describe the whole history of it, starting with the d bosons which have been intimately connected with the collective motion, with the U(6) model of Jolos, Janssen and Donau, who I am happy to see here, and then the model independently suggested by Arima and Iachello who by introducing s bosons made the U(6) a much simpler model. But IBA-1 need not, does not, and cannot have a microscopic basis. Maybe I am not a moderator, maybe I am doing more provocation than moderation, but really I'm willing to explain this point in detail for anyone who wants to challenge it. The main point is that IBA-1, like most collective models, does not recognize the fact which is known even to some nuclear theorists, that nuclei are made of protons and neutrons. It is a sad fact perhaps because identical nucleons are much easier to handle, seniority works, everything works. Yet nuclei are made of protons and neutrons. If you do not incorporate these degrees of freedom then certainly you cannot have a microscopic model of IBA-1. In particular, models which consider identical nucleons with pairing plus quadrupole interactions certainly do not describe reality and certainly are not models of IBA-1. It can be considered as a phenomenological model that stands on its own, gives nice agreement with experiment and can be related to the collective model. This is certainly a logical viewpoint and that was how historically it was developed.

We can try to understand this model from the shell model, from the microscopic point of view. This immediately introduced the IBA-2 which is a different model. It has two kinds of s and d bosons, proton bosons and neutron bosons and indeed it is much more complicated and less elegant than IBA-1. In IBA-1 one can write those dynamical symmetries, it is really beautiful theory and the remarkable thing is that it is both beautiful and agrees with experiment. There need not be contradiction between these two concepts but rarely is there a coincidence and this is one case. IBA-2 is much more complicated. Hopefully, it has more truth in it, as somebody said, but it doesn't have the beauty. Of course, one could write down something that will look like an IBA-1 Hamiltonian. If F-spin is a good quantum number then the states with maximum symmetry, with the maximum F-spin, are equivalent to IBA-1 states and there is nothing new in it. But in real nuclei the interaction between protons and neutrons is very different from the interaction between identical nucleons. As a result, a boson model based on this shell model cannot have the same interaction between proton bosons as between proton and neutron bosons. Here is the problem, how you start with an IBA-2 Hamiltonian which does not have the symmetry between proton and neutron boson and still you get good agreement with experiment. How can you extract from this IBA-2, the beautiful scheme of IBA-1? If there is a large Majorana term then the lower states have good symmetry or if there is an induced quadrupole-quadrupole interaction between identical bosons, then again there may be states which are more symmetric. The trouble with those renormalizations is that they are very difficult to calculate. We know that in the shell model people have been trying to do, to obtain the effective interaction between protons and neutrons, from the interaction between free nucleons for the last 35 years and they are still trying. It is very difficult, you have to sum all sorts of diagrams. You can get with some effort, if you know what you are after, good results. If you are ingenious as Olaf Scholten, then you calculate renormalizations, and it was very nice to see how the renormalization effects can yield lower values of epsilon, introduce a Majorana terms and so on. Incidentally, chi-1 and chi-3 terms can have a microscopic basis. It is only the chi-2 term which does not have a microscopic counterpart. The difficulty with such calculations is that they are very difficult and the results are not definite. Therefore, what one could do is try to determine the parameters of IBA-2 from experiment and see whether one indeed needs specific renormalizations in order to get good agreement with experimental energy levels and transition probabilities. Unfortunately, the situation here is very complex. There are different sets of parameters which give equally good agreement and they look rather different. As a conclusive statement in some old papers I should say that "more theoretical and experimental work should be carried out on this problem." We still do not know, we have not done sufficiently detailed and systematic study of nuclei using the IBA-2 to tell definitely which are the correct sets the parameters. Thus, we are trying to work it from both ways, trying to fit the data with certain parameters and trying to calculate those parameters from theory, as much as possible.

Another difficulty is that we do not know enough of the effective interaction between nucleons and therefore it is almost impossible to derive microscopically the parameters of the boson model. In fact, anyone who starts from the simple minded interactions and gets excellent agreement actually performs a miracle. I just cannot understand such agreement because we know that not every interaction gives rise to any given set of energies. There is only one Hamiltonian, which is the correct one. We do not know it, unfortunately. I would like to discuss a very simple boson model for semimagic nuclei which is still instructive. We consider states in a given j shell obtained by operating with an operator creating a pair with $J = 0$. We then look at correlated pairs given as a linear combination of pairs in all orbits in a major shell. We consider states with N such correlated pairs, the condensate of such pairs, to be an exact eigenstate of the shell model Hamiltonian. If we insist that this is the case, that these are indeed exact eigenstates, then there are two conditions that have to be satisfied. One is that the correlated pair state should be an eigenstate, the other one is the double commutator condition. If you assume that then you can easily show that for any N such states are eigenstates and the energy is a linear and quadratic function of N exactly. In particular, if this is the form of configuration mixing in this major shell, there are no breaks at filling of subshells which is a very important point. A few days ago there was a talk about binding energies and you saw how difficult it is to extract any break in a major shell, certainly in semimagic nuclei. That is about the ground state. How about the 2+ states in those nuclei? You can construct a similar operator creating a coherent pair, a D pair, with $J = 2$. Then you ask under which conditions these states will be eigenstates. Again you have two conditions which are very simple and then you find that for any N this is an eigenstate and the eigenvalues have a very simple form. They are equal to the energy of the ground state plus a constant term, $V_2 - V_0$. In other words there is a constant 0 - 2 spacing in all semimagic nuclei. If you look at the tin isotopes, for example, from the beginning of the major shell to the end (they are all measured now) they are fairly constant. This is the shell model and this is microscopic. Now we will see a boson model which describes the same situation. You can write down the boson Hamiltonian which is very simple. It has single boson energies and boson-boson interactions. And we see that these states have the same eigenvalues as in the shell model. This can be obtained by using the Bose commutation relation between the s,d operators and their hermitian conjugates. They obey exactly the Bose commutation relations. Still we obtain the same eigenvalues. And remember that in the shell model problem, the Pauli principle was obeyed strictly. Here is a very simple mapping, a very simple set of boson states, a simple boson Hamiltonian which gives the exact eigenvalues of the shell model Hamiltonian. And there is no renormalization involved in the model. This is because of the states considered are completely decoupled from all other states. This is an experimental fact of life.

Once you go over to protons and neutrons the situation is much more complicated and there one would expect renormalization effects. If one has a Hamiltonian which is supposed to give correct nuclear energies you should first try it on semimagic nuclei and reproduce those experimental facts. A Hamiltonian which will reproduce the data should depend critically on the single particle energies and certainly on the matrix elements of the two-body effective interactions. No schematic interaction, like the surface delta interaction, if the orbits are not degenerate, will obey these conditions. It is not easy to know what is the correct Hamiltonian that gives rise to the boson model in semimagic nuclei. It is ten times more difficult to know what it is and how the renormalization works when you have both protons and neutrons. Let me stop here for I would like to ask Akiva Novoselsky to show us some of the data that he obtained using IBA calculations in which he tried to see whether indeed the renormalization of epsilon is necessary. If you look at this boson model, epsilon is simply V_2-V_0, the 0 - 2 spacing. This is the only parameter that you can read off the nuclear data. The question, is whether it is renormalized. Another question is whether a Majorama term is really needed in order to get good agreement with experiment.

Since Akiva must leave before he will be able to give his talk perhaps we could hear from him now.

Novoselsky: I used in the IBA-2 Hamiltonian only single particle energy, proton-neutron quadrupole-quadrupole interaction and in addition I added a neutron-proton dipole-dipole interaction. What is important in this Hamiltonian is that the epsilon parameter is equal exactly to the 0 - 2 spacing in the appropriate semimagic nucleus. Consequently, the value of the epsilon parameter was kept constant and was equal to 1.3 MeV for Xenon, 1.4 MeV for Barium and 1.06 for the Cerium isotopes. In this approach no Majorama term was used. I added a dipole-dipole interaction to improve the fit, and there were only four free parameters which were the kappa, chi-pi, chi-nu, and lambda. For each of the isotopic series the chi-pi term was small and consequently change in chi-nu was kept constant. So for each nucleus in the isotopic chain these were only 2-3 parameters. The fit that I obtained for the Xenon and Barium isotopes is very good. What we see is that if I take the epsilon parameter which is equal to 1.4 MeV, and 1.3 MeV for Xenon, and Barium, I get all the reduction in the 2+ states. These are the results for the ground state band and for the beta band, and I get also good agreement with the gamma band. In particular, I get good results not only for the ground state band and the gamma band, but also for all the other known low-lying states in these nuclei and especially in the Xenon-128 and Platinum-196 which are 0(6)-like nuclei. In addition, I get good agreement to the E2 transition from the first 0+ to the first 2+ including the transition in the semimagic nuclei in Cesium and Barium. However, because there was no use of the Majorana operator we obtain strong F-mixing in the wave functions. Thus for the Xenon-128 in the ground state band there is aproximately 95% contribution of Fmax, but in the gamma band it is only 70% contribution of Fmax. In the Barium-128 the situation is more or less equal. But in other states, for example, in the 0^+, 2^+, 3^+, 4^+ states this is no longer the case. For example, in the 2_3^+ state in Barium-128 and Xenon-128, the main contribution comes from the Fmax-1 and not from Fmax.

Alaart: A number of years ago a student of mine, his name is Akkermans, did a similar thing for the krypton isotopes. We also found that by using these large values of epsilon we obtained a very beautiful fit to the spectra, but there was one problem which we discovered a little bit later and that's why it was never published When we looked not only at the BE(2)'s from 0+ to 2+ but also higher in the band we got a much too strong cutoff of the BE(2)'s which seems completely unrealistic and our conclusion was then that probably we still needed a renormalization of epsilon and that was the origin of all the work which we did afterwards [see Z. Phys. <u>A304</u> 245 (1982)], so I don't know whether we were on the wrong track.

Novoselsky: I get good agreement not only to the transition from the 0+ to 2+ but also all the known experimental E2 branching ratios.

Ginocchio, Los Alamos: As you know, we did a similar calculation a number of years ago on Samarium isotopes within the fermion model, distinguishing between neutrons and protons. We had similar terms which, in the fermion language would be pairing, plus quadrupole, plus dipole, plus octupole, we kept all parameters constant, and we also were able to fit the samarium isotopes plus the BE(2)'s. So I think it must be similar to your calculation in the basic physics that goes into it.

Moderator: With your calculation there was some difficulty about the quadrupole operator. In the fermion states it was OK but when you moved to the boson picture. You had a very strange quadrupole operator a two-body quadrupole operator.

Ginocchio: If the shell occupancy Omega becomes large this quadrupole operator becomes the usual IBM quadrupole operator.

Moderator: In the examples shown by Novoselsky it is seen that there are no states which have Fmax exactly or Fmax-1 but there are linear combinations of all these states and therefore in a more realistic situation that is what you would expect.
  Yes, Kumar

Kumar: I have a small comment about the general comparison made by Prof. Talmi. The forefathers of the shell model may not like a deformed shell model but it remains a fact that this model is a direct descendant of the shell model. I agree with you that it is a miracle that it works as well as it does. But is it not a miracle that any shell model works at all?

Moderator: It is --- it is indeed. But we are simply used to it so we take it for granted. It is indeed a miracle and whenever you can see good agreement obtained by the shell model it is a miracle Whenever you can see good agreement obtained by the shell model, let many-body people handle it they say you need corrections here and you need corrections there.

Seyfarth, Julich: I have a question concerning two nuclei which Kris was mentioning: Titanium-46 and Chromium-50. From old times one has the phenomena of cross conjugate nuclei --- that means if one has a nucleus with two protons and four neutrons, and if one makes a proper transformation, one should get the cross conjugate nucleus which in this case would be Chromium-50, which has two neutron holes and four proton holes. The spectra and decay properties are similar. One finds now in titanium this mixed 1+ state should this state then also show up in the cross conjugate nucleus Chromium-50?

Moderator: I am sure it should because MBZ have really had this symmetry. In fact, they do not even bother to calculate the conjugate nuclei. They just look at the data for the other nucleus. I am sure the results of the MBZ calculations are the same for both nuclei.

Heyde: I just wanted to comment that the cross conjugate nuclei and the original ones should have the same spectra, I know it's always exactly the same spectra. This is a strict statement within the shell model. But the experiment should clearly decide what's coming out and I'm clearly convinced that even if you have the cross-conjugate situation where protons change into neutrons, particles into holes --- the single particle spacing between the unfilled and filled shells (the $1f_{7/2}$ to $1f_{5/2}$, $2p_{3/2}$, $2p_{1/2}$ orbitals) will be slightly different from Coulomb effects so you could have clearly different effects into that. But in principle within this small valence shell model space everything should be coming out identical.

Dieperink: I have a comment and a question. In Heyde's talk he seemed to imply that the state Faessler calculated 16 years ago at 15 MeV is the same state as we now discuss in IBA. I disagree with that because it seems to me that the state Faessler was discussing was an isovector quadrupole giant resonance which in deformed nucleus has a $K = 0$, $K = 1$, and $K = 2$ component; he has pointed out the $K = 1$ components. But I think that his state is a giant resonance, corresponding to irrotational flow. The phenomenon that we are discussing is the low-energy, the shell model equivalent of the state, so they have different inertial and different restoring force parameters. That's the comment.

Heyde: I think as Faessler pointed out in the paper that when you look to what he did in developing the collective model approach to protons and neutrons that he had also the anti-phase quadrupole vibration. When you look at his 1964 paper its just within the standard hydrodynamic approach.

Dennis Hamilton, Sussex: I remember the first time that I talked to Faessler about this and he clearly claimed or believed that he had calculated the 2+ state and what he thought he got wrong was the amplitude of the vibrations and the number of participating particles. If you change that, he believed it was quite simple that it could come down and that in fact is a follow up paper which came this year.

Moderator: I think that it would have been much better if Faessler had been here, in fact I saw his name on the printed program. I wish he were here and tell us what he did, but in his absence I don't think there is much sense arguing these points without him present.

Dieperink; The question I really have is, to come back to the comment Broglia made this morning about the interesting fact that if you do neutron calculations, particle hole calculations with 1+ states, in restricted space, there's no escape, you always find there is an appreciable $h11/2$ proton contribution because that is the only --- there's no spin orbit part to cancel the spin. So if you analyze the wave function you find an appreciable single particle spin contribution and I think the work of Aberg and Hamamoto in that respect an honest calculation. But this is a real puzzle: Why does the spin contribution completely disappear as has been shown in the proton-proton, (p,p') experiments? Is it that you have to go to huge space, a huge diagonalization, to include $h9/2$ spin in addition to the $h11/2$? I would like to ask some of the experts in the fields what is going on.

Moderator: I'm not an expert, but I will say something nevertheless. First of all I don't think that Copenhagen needs my defense, but I think the question of spin contribution and orbital contribution that Kris Heyde did mention is very important because in the f7/2 half shell certainly if you trust the MBZ wave function to get the energies, then do an experiment and find them, you should also expect the wave function to be reasonably given by MBZ. Incidentally it is more than 80% S and D pairs, but that is not the point. The question is whether that spin contribution does not disappear because of all the effects that have been invoked for the Gamow-Teller transitions. So I agree that that should be investigated. For example, in Calcium-42 the state with 1+ which is the f7/2 f5/2 state is barely seen in the p,n reaction and that is a puzzle. Somehow it loses its strength. Maybe something like that happens also here.

Ring: We did an RPA with a selfconsistent RPA with Skyrme force, and it turned out that in this case we have only orbital contributions to that mode. We did not analyze completely what's happening with this one but my feeling is that it depends on the spurious state that you remove in that problem.

Moderator: The effect is that the major shells are determined by the strong spin orbit interaction and therefore one cannot disregard the shell model for the convenience of some cross section. I think that the problem should be investigated thoroughly. As you may remember when the p-n people came out with the Gamow-Teller transitions they decided that the f7/2 shell does not exist and Calcium 42 is given by the singlet S wave function and Scandium by the triplet S as in the old days of Wigner but this is not true.

Bohle, Darmstadt: I've got a viewgraph (no figure supplied...Ed.) it shows the form factor of the strong 1^+ states measured, not only at Darmstadt, but also higher angular momentum transfer measured in Amsterdam. The form factor of the mixed symmetry 1+ state in Titanium-48 which is actually at 3.74 MeV, is nicely described by using the wave function of Zamick, as well in the second maximum as in the first one, so I think that already gives you an impression that the wave functions that Zamick uses in his calculations are actually very accurate.

Moderator: I am very happy to see that but then the puzzle is even thicker. Why where is the spin contribution and incidentally I would like to ask the expert the question myself. They speak sometimes about spin flip sometimes they speak about spin contribution. What is the story? In the f7/2 shell you don't have any spin flip because you just have f7/2 all alone. To me spin flip means going from f7/2 to f5/2, so what so which contribution is not seen the spin or spin flip?

Scholten: The thing which is not seen is simply the spin contribution. As far as I know, related to RPA calculation, the way way to get rid of it is simply to include in your RPA calculation a sigma dot sigma force, of course, which is a repulsive spin-spin force, to push up the spin contribution to something like 10 MeV where the giant resonance is.

Von Brentano: As we are discussing this question of the spin contribution I want to ask the expert, if you have no spin contribution would the proton g factor be unity?

Scholten: Yes.

Von Brentano: But that may be then a really nice way of getting it experimentally tied down because I am sure we can improve the upper limit for the (p,p') cross section to the 1^+ state considerably. If one can get a much better limit in the (p,p') experiment then one would see that the ratio of orbital to spin contributions is not ≥ 20 as it is now but maybe ≥ 100.

Moderator: The problem is the interpretation of the experiment. The question is, if you don't see it in p,p' what does it really mean? The same thing happens with those Gamow-Teller transitions --- you don't see the strength and the question is, where is it? You may invoke deltas that come down and get mixed with nuclear states or you can shift the strength up to higher energies. There are several mechanisms which claim to explain it, so maybe something like that happens here. You see only the orbital part of these transitions, you do not see the spin part. The question is: What is the mechanism of p,p'? What does it really tell you? I do not know. The point is that before we junk jj-coupling we should better see what we are doing.

Bohle: We see some spin induced strength in ^{46}Ti.

Iachello: My understanding of the experimental situation is that, at least in some of these heavy nuclei, such as Gd-156, that the comparison between e,e' and p,p' is quite clear in the sense that it appears that most of the contribution to that state is the orbital contribution. Now, one could try to make this more quantitative, and that is where it becomes very difficult because the question is whether we understand well enough the mechanism of p,p' reactions to be able to make a very definite statement. But certainly as far as an upper limit I would feel quite comfortable.

Moderator: Let me ask you a question. If you take Calcium-42 and you do p,p' and you get to a state which you know is f7/2 f5/2 coupled to 1^+ and you don't see it, what will you say then?

Iachello: Maybe we don't understand the mechanism of the p,p'.

Moderator: That is my suspicion, yes.

Eugene Henry, LLNL, Livermore: I'd like to change the subject a little bit. We have one single piece of data which might suggest that two nucleon transfer reactions might populate these mixed symmetry states. Could anybody say anything about that --- as a signature?

Iachello: If the structure of that state is related to correlated pairs of protons and neutrons, then you would expect that two-neutron transfer would excite those states. Just think of the situation in spherical nuclei. Now there might be in front some recoupling coefficient which may give a small cross section but as far as exciting it there is no reason why two-neutron transfer reactions should not excite mixed symmetry 1^+ states. Also there is no reason why it should not excite the 2+ antisymmetric state. In fact, if the transfer mechanism is such that it picks up only a proton pair or a neutron pair then both the first 2+ state and the antisymmetric 2+ state should be excited almost equally.

Henry: The evidence is that the 2+ is strongly populated and maybe the 1+ also.

Iachello: That's very good.

Tamura: Many people, including myself, were attending the theoretical session this morning and missed Dr. Bohle's paper. I would like to ask him to explain for us the essence of his experiment.

Bohle: Yes, very briefly, what we actually did was compare the (p,p') at 200 MeV and small scattering angles with (e,e') work. At 200 MeV it is mainly the sigma-squared tau-cubed part of the nucleon-nucleon interaction that dominates, and that means that if the mode is not excited in proton scattering it would be essentially an orbital excitation mode. If it would be excited as it was in Titanium-46 then the wavefunction has also a spin component. In this case we were able to derive a value but only upper limits in case of the rare earth nuclei. We have tested this method using (e,e') and (p,p') to excite the strong neutron f7/2 to f5/2 spin flip transition in Calcium-48 where it's actually sure that it's really, to a very high degree, a pure spin flip transition, and the agreement between these two experiments was excellent. So we do think that the method works, whatever difficulties with the (p,p") excitation mechanism there are in some cases.

Otten: I have a question to Dieperink from this morning's discussion of the g factors. He assured us that he could renormalize the g,p factor by 0.2 nuclear magnetons. This was compensated more or less by a rise g_n from 0 to 0.13. So all together, the isoscalar moment remained almost unchanged in your calculation which seems reasonable. But the experimental fit showed strong renormalization for the g_p down to 0.6 no renormalization for g_n. How can one understand this situation?

Dieperink: Very briefly, a very qualitative way to incorporate some of the effects of the core, which presumably are there in addition to the valence space, but I found that g goes sort of in between N-pi over N and Z over A. But I did get a reduction of the proton g factor and a small enhancement of the neutron g factor.

Moderator: There are complications with these g factors and M1 transitions. An expert said this morning that E2 transitions are collective and therefore it is difficult to learn too much from them, but the trouble with the magnetic transitions, the M1 operator in particular, is that it can be extremely sensitive to details of the wave functions. And the difference between 10^{-2} and 10^{-3} is not so big. If it is determined by one promil of the wave function then it doesn't matter if it is 5 promil or 1 promil. There are difficulties with the M1 transitions throughout the periodic table, even in the shell model things are not very happy with M1 transitions.

Moderator: If there are no more pressing questions let us close this session by thanking all the speakers.

G-BOSON RENORMALIZATIONS AND MIXED SYMMETRY STATES

O. Scholten
Kernfysisch Versneller Instituut
Rijksuniversiteit Groningen,
Zernikelaan 25, 9747 AA Groningen, The Netherlands

1. INTRODUCTION

In the IBA model the low-lying collective states are described in terms of a system of interacting s- and d-bosons. A boson can be interpreted as corresponding to collective J=0 or J=2 fermion pair states[1]. As such the IBA model space can be seen as only a small subsector of the full shell model space. For medium heavy nuclei such a truncation of the model space is necessary to make calculations feasable. As is well known truncations of a model space make it necessary to renormalize the model parameters. In this work some renormalizations of the Hamiltonian and the E2 transition operator will be discussed. Special attention will be given to the implication of these renormalizations for the properties of mixed symmetry states.

The effects of renormalization are obtained by considering the influence of fermion pair states that have been omitted from the model basis. Here we will focus attention on the effect of the low-lying two particle J=4 state, referred to as g-boson or G-pair state. Renormalizations of the d-boson energy, the E2 effective charges, and symmetry force will be discussed.

2. THE QUADRUPOLE OPERATOR

The dominant part of the shell model neutron-proton interaction is the quadrupole-quadrupole interaction,

$$V_{\nu\pi} = F_2 \, Q_\pi^{(F)} \cdot Q_\nu^{(F)} \qquad (1)$$

where the quadrupole operator can be written as[2]

$$Q_\tau^F = \kappa_\tau Q_\tau + \kappa_\tau' Q_\tau' + \kappa_\tau'' Q_\tau'' \qquad \tau = \nu, \pi \qquad (2)$$

In this expression Q_τ represents the part of the quadrupole operator acting within the s-d subspace,

$$Q_\tau = (s^\dagger \tilde{d} + d^\dagger s)_\tau^{(2)} + \chi_\tau (d^\dagger \tilde{d})_\tau^{(2)} \tag{3}$$

Q'_τ represents the part of the operator that has non vanishing matrix elements between the s-d and the non-s-d space. In the sdg model approximation this becomes

$$Q'_\tau = (d^\dagger \tilde{g})_\tau^{(2)} + (g^\dagger \tilde{d})_\tau^{(2)} \tag{4}$$

The last term in eq. (2), Q'', is that part of the operator that acts fully in the non-sd part of the space and will be omitted from the following considerations. Due to the coupling to states outside the s-d space, via Q' term in the interaction (see eq. (1) and (2)) the effective quadrupole operator in the s-d space should be renormalized. In perturbation theory (see fig. 1) the expression for the renormalized proton quadrupole operator is

$$\tilde{Q}_\pi = Q_\pi + Q'_\pi \frac{1}{\Delta E}(Q'_\pi \cdot Q_\nu)$$

where the use of Q' insures that only non s-d states are considered in the summation over intermediate states. This expression can be evaluated in the s-d space yielding to a good approximation

$$\tilde{Q}_\pi = Q_\pi + \alpha_\pi Q_\nu n_{d_\pi} \tag{5}$$

where the constant

$$\alpha_\pi = \frac{F_2}{\Delta E} {\kappa'}^2 \sum_\lambda \sqrt{2\lambda+1} \begin{Bmatrix} 2 & 2 & \lambda \\ 2 & 2 & 4 \end{Bmatrix} \langle\langle (d^\dagger \tilde{d})_\pi^{(\lambda)} \rangle/\langle n_{d_\pi} \rangle$$

is a measure for the coupling to the g-boson state. The expression for $\tilde{Q}_\pi$ has been obtained here by using perturbation theory arguments applicable, strictly speaking, only to vibrational nuclei. A similar expression for the effective quadrupole operator, valid in the SU(3) limit of IBA, has been obtained by Otsuka and Ginocchio[3].

Using the above results the effective (s-d) space E2 transition operator can be written as[4]

$$\tilde{T}(E2) \simeq e_\pi \tilde{Q}_\pi + e_\nu \tilde{Q}_\nu =$$
$$\tag{6a}$$
$$= e_\pi Q_\pi + \alpha_\nu e_\nu n_{d_\nu} Q_\pi + e_\nu Q_\nu + \alpha_\pi e_\pi n_{d_\pi} Q_\nu$$

which can be reduced to

$$\tilde{T}(E2) \simeq \tilde{e}_\pi Q_\pi + \tilde{e}_\nu Q_\nu \tag{7a}$$

where the boson effective charges are introduced,

$$\tilde{e}_\pi = e_\pi + \alpha_\nu e_\nu \langle n_{d_\nu} Q_\pi \rangle / \langle Q_\pi \rangle$$

$$\tilde{e}_\nu = e_\nu + \alpha_\pi e_\pi \langle n_{d_\pi} Q_\nu \rangle / \langle Q_\nu \rangle \tag{7b}$$

The brackets $\langle \rangle$ denote expectation values. These charges are introduced to reduce the effective 2-body transition operator to the usual 1-body operator of the IBA model. While the effective 2-body operator (6) is independent of the structure of the nucleus, its 1-body equivalent is structure dependent and has effective charges that are mass dependent. In table 1 the calculated values, using eq. (7), are given for the Nd isotopes, using $\alpha_\pi = \alpha_\nu = 0.75$ and $e_\nu/e_\pi = 0.8/1.8$ (the shell model value)[5]. For the light, near vibrational isotopes, the expectation value of n_d in the ground-state is small. As a result the second terms on the r.h.s. of eqs. (7b) give a negligable contribution, and the ratio of the boson effective charges are close to the shell model ratio. This asymmetry in the operator induces a relatively large transition strength to states of mixed symmetry character. The heavier isotopes are more deformed with a large value for

Table 1: Calculated values for the boson effective charges and the $2_1^+ \to 0_1^+$ transition materix elements in the N_d isotopes, using the formulas given in the text

A	$\tilde{e}_\nu$	$\tilde{e}_\pi$	M(E2; $2_1^+ \to 0_1^+$)	
			Calc	exp.
142	0.049	0.110	0.52	0.53 (4)
144	0.071	0.118	0.66	0.76 (1)
146	0.088	0.126	0.85	0.88 (1)
148	0.110	0.138	1.13	1.178 (9)
150	0.144	0.157	1.69	1.678 (11)

$\langle n_d \rangle \sim 2/3$ N. The second term in eq. (7b) now gives a considerable contribution to the effective charges. Since $e_\pi > e_\nu$ this enhances the effective neutron boson charge more than that of the proton boson, with the result that in ^{152}Nd, $\tilde{e}_\nu \simeq \tilde{e}_\pi$. For the more deformed nuclei the effect of the g-boson renormalization is thus to strongly increase the neutron boson effective charge and, as a result, to make E2 transition operator more symmetric. Also for this reason one would not expect to excite mixed symmetry 2^+ states in deformed nuclei. The mass dependence of the boson effective charges, calculated here, is very similar to what has been used in some recent calculations by Otsuka and Ginocchio[6]. In table 1 also the calculated $2_1^+ \to 0_1^+$, E2 matrix element is compared with experiment[7]. It should be noted that for the calculation of the E2 strength leading to the higher lying 2^+ levels the 2-body operator of eq. (6) or the 1-body operator of eq. (7) give almost identical results.

3. THE D-BOSON ENERGY

The renormalization procedure discussed in the preceeding section can also be applied to calculate the renormalization of the single boson energies. The procedure followed is very similar to that used in ref.2.

The effective neutron-proton quadrupole interaction acting in the s-d subspace can be written as

$$V^{eff} = F_2 \tilde{Q}_\nu \cdot \tilde{Q}_\pi$$

where $\tilde{Q}$ is given by eq. (5). This interaction contains a three body term of the kind

$$F_2 \left[Q_\nu \alpha_\pi n_{d_\pi} Q_\nu + \alpha_\nu n_{d_\nu} Q_\pi \cdot Q_\pi \right] \qquad (8)$$

This term can be reduced to an effective one and two body interaction by using

$$\langle (Q_\nu \cdot Q_\nu - \langle Q_\nu Q_\nu \rangle) \cdot (n_{d_\pi} - \langle n_{d_\pi} \rangle) \rangle \simeq 0 \qquad (9)$$

eq. (9) is only an approximation which is valid in the SU(5) limit of IBA where $\langle n_d \rangle$ is a good quantum number and in the SU(3) limit where Q·Q is diagonal, but strictly speaking not in the transitional region.

Using eq. (9), the term of eq. (8) can be rewritten to give

$$+ F_2 \alpha_\pi n_{d_\pi} \langle Q_\nu Q_\nu \rangle + F_2 \alpha_\pi Q_\nu Q_\nu \langle n_{d_\pi} \rangle - F_2 \langle n_{d_\pi} \rangle \langle Q_\nu Q_\nu \rangle + P_{\nu\pi}$$

where the last term indicates a similar term with neutron and proton indices interchanged. The third term is a constant which only contributes to binding energies, the second represents an effective quadrupole force between like particles, induced by the coupling to the g-boson. The first term is proportional to n_d and thus renormalizes the d-boson energy. In the SU(5) limit the effective d-boson equals

$$\tilde{\varepsilon}_{d_\pi} = \varepsilon_{\nu_\pi} + 5 \alpha_\pi N_\nu F_2$$

where $\langle Q_\nu Q_\nu \rangle$ has been replaced by its expectation value in the SU(5) limit, $5N_\nu$. Using the values for $\alpha = 0.75$, $F_2 = -.13$ and $N_\nu = 2$ as applicable to ^{146}Nd we obtain $\varepsilon = \varepsilon_0 - 1.0 = 0.6$ MeV for $\varepsilon_0 = 1.5$ MeV. This value is close to what has been used in phenomenologic calculations ($\varepsilon \sim 0.6$ MeV). It should be noted that especially for nuclei where $N_\nu \neq N_\pi$ the reduction of ε_d may be considerably different for neutrons and protons.

4. THE MAJORANA FORCE

In most phenomenological IBA-2 calculations an ad-hoc force, the Majorana force, is introduced to shift the position of levels of mixed symmetry character with respect to symmetric levels. The s-d neutron-proton quadrupole force does give rise to a splitting of the energies, but the coupling to the g-boson via Q·Q force does not give an additional contribution, and thus cannot explain the Majorana force. However, the neutron-proton force also does have a considerable hexadecapole component[8]. In this section it will be shown that the Majorana force in the effective IBA-2 interaction can be explained through the presence of both the quadrupole and the hexadecapole component in the interaction.

In the SU(5) limit we will consider the simplified interaction

$$V_{\nu\pi} = F_2 Q_\nu^{(2)} \cdot Q_\pi^{(2)} + F_4 Q_\nu^{(4)} \cdot Q_\pi^{(4)}$$

with $Q^{(2)} = (s^\dagger \tilde{d} + d^\dagger s)^{(2)} + \chi_2 (d^\dagger \tilde{g} + g^\dagger \tilde{d})^{(2)}$

and $Q^{(4)} = (s^\dagger \tilde{g} + g^\dagger s)^{(4)} + \chi_4 (d^\dagger \tilde{d})^{(4)}$

This simplified form for the multipole operators has been taken since it suffices to make the point. For simplicity of writing, the symmetric $n_d=1$ state is labelled $|s\rangle$ and the mixed symmetry state $|a\rangle$[9],

$$|s\rangle = \frac{1}{\sqrt{N}} (\sqrt{N_\nu} |s_\nu^{N_\nu-1} d_\nu s_\pi^{N_\pi}\rangle + \sqrt{N_\pi} |s_\nu^{N_\nu} s_\pi^{N_\pi-1} d_\pi\rangle)$$

$$|a\rangle = \frac{1}{\sqrt{N}} (\sqrt{N_\pi} |s_\nu^{N_\nu-1} d_\nu s_\pi^{N_\pi}\rangle - \sqrt{N_\nu} |s_\nu^{N_\nu} s_\pi^{N_\pi-1} d_\pi\rangle)$$

It can easily be checked that the matrix elements of the hexadecapole force between these states vanish.

Using perturbation theory the splitting due to the coupling to non s-d states of the symmetric and mixed symmetry 2^+ levels, ΔE, can be investigated,

$$\Delta E = \langle a|V \frac{1-p}{E} V|a\rangle - \langle s|V \frac{1-p}{\varepsilon} V|s\rangle$$

where ε is the energy denominator and $1-p$ is the projection operator on the non s-d space. If the interaction would have been a pure $Q_\nu^{(2)} \cdot Q_\pi^{(2)}$ or a pure $Q_\nu^{(4)} \cdot Q_\pi^{(4)}$ then $\Delta E \alpha \frac{1}{N}(N_\nu - N_\pi)^2$ i.e. the splitting between symmetric and anti-symmetric states vanishes when $N_\nu = N_\pi$, contrary to phenomenology. However, when both the quadrupole and hexadecapole interaction are present an additional contribution, $\Delta E \alpha \frac{1}{N} N_\nu N_\pi$ is obtained. This contribution arises from the graph shown in Fig. 1. Part of the presence of the Majorana-force in the effective

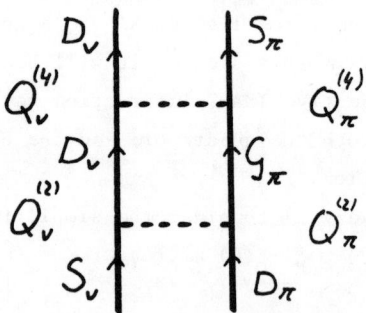

Fig. 1. A first order correction which induces a component of the Majorana force.

interaction is thus shown to arise from g-boson renormalization effects, where it is essential to consider both the quadrupole and the hexadecapole component of the neutron-proton force[8]. The derivation is only valid in the U(5) limit of the IBA-model, it can however be shown, making use of the intrinsic state formalism, that also in deformed nuclei the presence of both the quadrupole and the hexadecapole interaction gives rise to the introduction of a Majorana force in the effective interaction. One of these components by itself does not give rise to a Majorana force.

5. CONCLUSIONS

The effects of g-boson renormalizations have been considered. It is shown that these play a crucial role in the understanding of the reduction of the boson effective d-boson energy in non magic nuclei as well as a variety of phenomena related to mixed symmetry states. g-Boson renormalizations strongly affect the E2 effective boson charges, introducing a strong mass dependence, derived from an indiced 2-body term in the operator. In more deformed nuclei these renormalizations make the operator more symmetric under interchange of neutrons and protons and an excitation of mixed symmetry 2^+ states therefore becomes unlikely, while g-boson renormalizations arising from a quadrupole force do not induce a Majorana force in the effective interaction, this component is induced by the presence of a hexadecapole component in the neutron-proton interaction.

REFERENCES

1) Otsuka, T., Arima, A. and Iachello, F., Nucl. Phys. A309, 1 (1979).
2) Scholten, O., Phys. Lett. 119B, 5 (1982).
3) Otsuka, T. and Ginocchio, J.N., Phys. Rev. Lett. 55, 276 (1985).
4) Scholten, O, et al., to be published.
5) Scholten, O. and Kruse, H., Phys. Lett. 125B, 113 (1983); ibid MSUCL Ann. Rep. 1982 - 1983 p. 66.
6) Otsuka, T. and Ginocchio, J.N., Phys. Rev. Lett. 54, 777 (1985).
7) Endt, P.M., At. Data and Nucl. Data Tables 26, 48 (1981).
8) Scholten, O., Phys. Rev. C28.
9) Scholten, O., Heyde, K., Van Isacker, P., Jolie, J., Moreau, J. and Waroquier, M., Nucl. Phys. A438, 41 (1985).

SOME REMARKS ON MIXED-SYMMETRY STATES IN IBA-2

A.E.L. Dieperink
Kernfysisch Versneller Instituut
Zernikelaan 25, 9747 AA Groningen
The Netherlands

ABSTRACT
Some aspects of the neutron-proton IBA model are discussed, in particular magnetic dipole properties.

1. INTRODUCTION

There are two features which distinguish the Interacting Boson Model from most other microscopic approaches to nuclear collective properties[1-3]. First in the IBA model it is assumed that the collective quadrupole properties of low-lying states are dominated by the pairs of valence nucleons only. Secondly, on the basis of its microscopic foundation it was realized shortly after the introduction of the original phenomenological version (IBA-1) that it would be important to distinguish the neutron and proton building blocks. The interest in the neutron-proton degree of freedom has been stimulated very much when experimental evidence was presented for collective M1 strength in deformed nuclei which had been predicted as a signature of states with a mixed symmetric neutron-proton character.

In this talk I will briefly review some aspects of the neutron-proton degree of freedom in the IBA model. I will begin with a brief discussion about the role of the still somewhat mysterious Majorana interaction (section 2). Next I will discuss some magnetic dipole properties which seem most relevant in connection with mixed symmetry states (section 3).

2. THE MAJORANA INTERACTION IN THE IBA-2 MODEL

A phenomenological successful form of the IBA-2 is given by

$$H = \sum_{\rho=\pi,\nu} \varepsilon_\rho d_\rho^\dagger \cdot \tilde{d}_\rho + \kappa Q_\pi^{(2)} \cdot Q_\nu^{(2)} + \sum_\rho \kappa_\rho Q_\rho^{(2)} \cdot Q_\rho^{(2)} + \lambda_M \hat{M}, \quad (\rho=\pi,\nu) \qquad (2.1)$$

where $Q^{(2)}$ is the general quadrupole operator $Q_\mu^{(2)} = d_\mu^\dagger s + d^\dagger \tilde{d}_\mu + \chi (d^\dagger \tilde{d})_\mu^{(2)}$, and $\hat{M}$ the majorana operator[1-3]

$$\hat{M} = \xi_2 (d_\pi^\dagger s_\nu^\dagger - s_\pi^\dagger d_\nu^\dagger)^{(2)} \cdot (\tilde{d}_\pi s_\nu - s_\pi \tilde{d}_\nu)^{(2)} + \sum_{\lambda=1,3} \xi_\lambda (d_\pi^\dagger d_\nu^\dagger)^{(\lambda)} \cdot (\tilde{d}_\pi \tilde{d}_\nu)^{(\lambda)} , \qquad (2.2)$$

which in the special case of $\xi_1 = \xi_3 = -\frac{1}{2} \xi_2 = 1$ is related to the quadratic Casimir invariant of the group $U^{(\pi+\nu)}(6) \subset U^{(\pi)}(6) \times U^{(\nu)}(6)$

$$\hat{M} = \frac{1}{2} N(N+5) - \frac{1}{2} C_2 [U(6)] . \qquad (2.3)$$

In the limit of a totally symmetric hamiltonian, i.e. for $\kappa_\nu = \kappa_\pi = \frac{1}{2}\kappa$, $\varepsilon_\pi = \varepsilon_\nu$ and $\chi_\pi = \chi_\nu$ the eigenstates of (2.1) can be classified according to the neutron-proton symmetry character (F-spin), $\langle M \rangle = \frac{1}{4}N(N+2) - F(F+1)$ with $F = \frac{1}{2}N, \frac{1}{2}N-1, \ldots \frac{1}{2}|N_\pi - N_\nu|$.

Up to now the majorana interaction has mainly been used as a convenient phenomenological tool to split the various F-spin multiplets. Investigations of the microscopic origin of the interaction parameters $\xi_\lambda (\lambda=1,2,3)$ have been restricted mainly to the U(5) region of the IBA model. Van Egmond and Allaart[4] found that the d-boson conserving part of $\hat{M}$ can be understood as the effect of the truncation of the shell model space to S and D pairs only. Less collective pairs (such as S', D' and G pairs) couple more strongly to the symmetric IBA states than to the mixed-symmetry states. In practice this leads to an effective attractive force between the np d-bosons coupled to even L rather than a repulsion between the bosons coupled to L=1,3.

In addition in the U(5) limit the ξ_2 part of $\hat{M}$, necessary to split the two L=2 states with one d boson, can be simulated to a large extent by the $Q_\pi^{(2)} \cdot Q_\nu^{(2)}$ force, since the $\Delta n_d = 0$ part of this operator is very similar. In the SU(3) limit the situation is more complicated, and no microscopic calculations of the parameters ξ_λ have been reported. Unlike the U(5) case an interaction of the form $Q_\pi^{(2)} \cdot Q_\nu^{(2)}$ or $(Q_\pi - Q_\nu) \cdot (Q_\pi - Q_\nu)$ is not able to split the symmetric and mixed symmetry $(\lambda,\mu) = (2N-4,2)$ β and γ bands, at least not for $N_\pi = N_\nu$. Inclusion of additional degrees of freedom such as the L=4 boson and the hexadecupole interaction hardly improves the situation as can be seen from the results reported in ref. [5]. The reason is that in the SU(3) limit of IBA all intrinsic states are approximate eigenstates of Q_ρ. On the other hand the situation in the Nilsson model is rather different. There one decomposes the hamiltonian $H=H_0+\Delta V$ in an unperturbed part, H_0, that generates the deformed Nilsson field and a residual particle-hole interaction ΔV which can be fine-tuned to generate the appropriate isoscalar and isovector multipole excitations; however, no self-consistency between the mean field and the residual interaction is imposed.

We note that an alternative approach[6,7] to understand the Majorana force is to search for a relation with a similar quantity, namely the symmetry energy that occurs in the mass formula

$$E_s = \langle \int d\tau K \, (\hat{\rho}_\pi - \hat{\rho}_\nu)^2 / (\hat{\rho}_\pi + \hat{\rho}_\nu) \rangle , \qquad (2.4)$$

where K is the symmetry energy strength function, and $\hat{\rho}$ the density operators. This interaction can be thought to originate from the fact that the average the neutron-proton interaction is stronger than the interaction between like fermions. By expanding the density operator $\hat{\rho}_\pi, \hat{\rho}_\nu$ around a spherical value

$$\hat{\rho}(\vec{r}) = \hat{\rho}_0(r) + \sum_{\lambda=0,2,4} \left(\gamma(r) \delta_{0\lambda} \hat{N} + \alpha_\lambda(r)(d^\dagger \tilde{d})^{(\lambda)} + \beta_2(r) \, \delta_{2\lambda}(d^\dagger s + s^\dagger \tilde{d}) \right) \cdot Y_\lambda(\theta, \phi)$$

$$(2.5)$$

one obtains after insertion in (2.4) for E_s in lowest order a multipole expansion, which after recoupling has basically the same form as (2.2). Calculation of the interaction parameters requires detailed knowledge of the transition densities $\alpha_\lambda(r)$ and $\beta_2(r)$ and the density dependence of the symmetry coefficient K.

3. MAGNETIC DIPOLE PROPERTIES

3.1 g-Factors in the Ground State Band

Due to large renormalization effects the effective boson charges e_π, e_ν that appear in the boson quadrupole operator have almost equal values and therefore the E2 matrix elements have predominantly an isoscalar character. On the other hand due to the nature of the M1 operator renormalization effects are expected to be less important for magnetic dipole matrix elements, and therefore the M1 operator is more likely a good probe of mixed-symmetry states.

Let us first discuss magnetic dipole moments. It has been shown that in case F-spin is a good quantum number a simple expression for the g-factor of states with $F = F_{MAX} = \frac{N}{2}$ holds[8]

$$g_R = \frac{1}{N}(g_\pi N_\pi + g_\nu N_\nu), \qquad (3.1)$$

where g_π, g_ν are the boson g-factors.

For comparison we also give the expressions for g factors in the projected Hartree-Fock approximation[9]

$$g_R^{PHF} = \langle M \cdot J \rangle / \langle J_\perp^2 \rangle = \langle J_\perp^2 \rangle_p / \langle J_\perp^2 \rangle + \Delta_{spin}^{PHF} \qquad (3.2)$$

where M is the magnetic dipole operator, and the cranking model[9]

$$g_R^{CR} = \mathcal{J}_p^{CR} / \mathcal{J}^{CR} + \Delta_{spin}^{CR}, \qquad (3.3)$$

where $\mathcal{J}$ is the moment of inertia.

The essential difference with (3.1) is that in the last two cases the contributions to g_R are not restricted to the valence shells only.

In the crudest approximations $g_R^{CR} \sim g_R^{PHF} \sim \frac{Z}{A}$, to be compared with $g_R^{IBA} \sim \frac{N_\pi}{N_\pi + N_\nu}$. Whereas the former value does not describe the experimental g-factors in the rare-earth nuclei at all the IBA result provides a qualitative first approximation[10]. In realistic calculations using density-dependent interactions and pairing effects the methods (3.2) and (3.3) yield satisfactory results; especially in the cranking approach the energy weighting of the quasi-particle contributions near the Fermi-surface plays an important role[9].

It has been noted that a quantitative agreement with experiment over a large mass region can also be obtained by treating the boson g-factors in (3.1) as parameters independent of N_π, N_ν; for a series of Nd-Dy isotopes Wolf et al[11] found $g_\pi \sim 0.65$ and $g_\nu \sim 0.05\ \mu_N$. This result does suggest that the parametrization (3.1) and thus F-spin is valid but also prompts an explanation for the reduction of g_π as compared to the naive value $g_\pi = 1.0\ \mu_N$.

In a microscopic picture the boson g-factors are directly related to the g-factor of the correlated D-pair. Since microscopic calculations indicate that the net spin contribution in the fermionic M1 operator is very small, $\sum_i \langle s_i \rangle = 0$ (as is also the case in the PHF and cranking approaches), within the SD space one finds $g_\pi \sim 1$, $g_\nu \sim 0$. Another possibility is that there is a non-negligible core contribution to g_R. This is also suggested by the fact that the moment of inertia of strongly deformed nuclei cannot fully be explained in a single-shell model space[12]. To see how the core contribution can be incorporated let us assume that g_R can be obtained by projection from an intrinsic pair state and that in addition to the valence orbitals there is a non-negligible core contribution, i.e. that $\langle J_\perp^2 \rangle$ in eq. (3.2) can be decomposed into a valence part $\langle J_\perp^2 \rangle_v$, and a core contribution $\langle J_\perp^2 \rangle_c$, i.e.

$$g_R = (\langle J_\perp^2 \rangle_{p,v} + \langle J_\perp^2 \rangle_{p,c}) / (\langle J_\perp^2 \rangle_v + \langle J_\perp^2 \rangle_c). \quad (3.4)$$

The ratio $\langle J_\perp^2 \rangle_{p,v} / \langle J_\perp^2 \rangle_v$ is given by $\frac{N_\pi}{N}$. If we make the reasonable assumption that $\langle J_\perp^2 \rangle_{p,c} / \langle J_\perp^2 \rangle_c \sim \frac{Z_0}{A_0} = 0.38$ then in terms of the parameter for the fraction of $\langle J_\perp^2 \rangle$ contained in the core, $x = \langle J_\perp^2 \rangle_c / \langle J_\perp^2 \rangle_{tot}$, one obtains

$$g_R = (1-x)\frac{N_\pi}{N} + \frac{Z_0}{A_0}x = \left(1-(1-\frac{Z}{A})x\right)\frac{N_\pi}{N} + \frac{Z_0}{A_0}x\frac{N_\nu}{N}$$

$$= g_\pi^{eff}\frac{N_\pi}{N} + g_\nu^{eff}\frac{N_\nu}{N}. \quad (3.5)$$

Eq. (3.5) with $g_\pi = 1-(1-\frac{Z_0}{A_0})x$ and $g_\nu = \frac{Z_0}{A_0}x$ has the same N_π, N_ν dependence as (3.1); however, the g-factors should now be considered as effective g-factors to be used in the IBA valence model space. For the value of x=0.33 one has $g_\pi = 0.79\ \mu_N$, $g_\nu = 0.13\ \mu_N$, which values give a reasonable description of experimental g-factors throughout the rare-earth region[11].

3.2 g-Factors of Excited Bands

In the IBA-2 approach with the lowest order M1 operators all g-factors of totally symmetric states are the same. Recently accurate measurements of ratios of g-factors have been reported[13] which show for some nuclei large deviations, e.g. $g_{2_2}/g_{2_1} = 1.45 \pm 0.18$ and 0.44 ± 0.15 in ^{188}Os and ^{184}W, respectively, and even more surprisingly $g_{4_1}/g_{2_1} = 1.32 \pm 0.17$ in ^{188}Os.

There are several possible explanations for these variations, i.e. F-spin admixtures, effects from L=4 pairs, or contributions from two-quasiparticle admixtures in these states. One could attempt to investigate the first two effects by introducing a generalized one-boson M1 operator[15] in the sdg space:

$$T(M1) = \sqrt{\frac{3}{4\pi}} \sum_{\rho=\pi,\nu} \left(g_{d,\rho} L^{(1)}_{d,\rho} + g_{g,\rho} L^{(1)}_{g,\rho} \right), \qquad (3.6)$$

which can conveniently be rewritten as ($\vec{L} = \vec{L}_\pi + \vec{L}_\nu$, $\Lambda^{(1)} = \Lambda^{(1)}_\pi + \Lambda^{(1)}_\nu$)

$$T_\mu(M1) = \sqrt{\frac{3}{4\pi}} \left(g_R L^{(1)}_\mu + (g_{R,\pi} - g_{R,\nu}) L^{(1)}_{a,\mu} + \frac{N_\pi h_\pi + N_\nu h_\nu}{N} \Lambda^{(1)}_\mu + \ldots \right) \qquad (3.7)$$

Here we introduced the $\Delta F=1$ M1 operator, $L^{(1)}_a = \frac{1}{N}(N_\nu L^{(1)}_\pi - N_\pi L^{(1)}_\nu)$, and a term associated with the difference of d- and g-boson angular momenta:

$$\Lambda^{(1)}_\rho = \frac{1}{7} \left(4 L^{(1)}_{d,\rho} - 3 L^{(1)}_{g,\rho} \right).$$

The corresponding g-factors are $g_{R,\rho} = \frac{1}{7}(3 g_{d,\rho} + 4 g_{g,\rho})$, $h_\rho = g_{d,\rho} - g_{g,\rho}$, $g_R = \frac{1}{N} \sum_\rho N_\rho g_{R,\rho}$.

The second term in (3.7) contributes to magnetic moments of low-lying states (e.g. the γ-band) through F-spin mixing terms in the hamiltonian, for example $\kappa_\pi = \kappa_\nu \neq \frac{1}{2}\kappa$ or $\chi_\pi \neq \chi_\nu$ in (2.1). Using perturbation theory one can obtain simple analytic results for the perturbed g factors and B(M1) values in limiting cases[14]. However, it seems unlikely that the Os data can be explained in this way.

The third term in (3.7) takes into account that the g-factors of the L=2 and the L=4 pairs could be different, e.g. due to spin contributions of the probably less collective L=4 pairs. In the large N limit of the SU(3) limit the magnetic moments in the γ-band can be expressed as[15]

$$\mu_\gamma = g_R L + g' \frac{4}{1+L} ,$$

where $g' = (N_\pi h_\pi + N_\nu h_\nu) / N$.
Simple microscopic estimates indicate that $|g_g - g_d| \leq 0.1\ \mu_N$ and thus $g' \sim 0.1\ \mu_N$ [15].

ACKNOWLEDGEMENT

This work has been performed as part of the research program of the Stichting voor Fundamenteel Onderzoek der Materie (FOM) with financial support of the Nederlandse Organisatie voor Zuiver Wetenschappelijk Onderzoek (ZWO). I also acknowledge the support by the NATO research grant RG85/0036.

REFERENCES

1. Arima, A. and Iachello, F., Ann. Rev. Part. Sci. 31, 75 (1981).
2. Arima, A. and Iachello, F., Adv. Nucl. Phys. 13, 139 (1984).
3. Dieperink, A.E.L. and Wenes, G., Ann. Rev. Nucl. Part. Sci. 35, 77 (1985).
4. van Egmond, A. and Allaart, K., Nucl. Phys. A425, 275 (1984).
5. Pittel, S. et al., Phys. Lett. 144B, 145 (1984).
6. Faessler, A., Bochnacki, Z. and Nojarov, R., J. Phys. G12, 47 (1986).

7. Faessler, A. and Nojarov, R., Phys. Lett. 166B, 367 (1986);
    Bohle, D. et al., Phys. Lett. 137B, 27 (1984).
8. Sambataro, M. and Dieperink, A.E.L., Phys. Lett. 107B, 249 (1981).
9. Sprung, D. et al., Nucl. Phys. A326, 37 (1979).
10. Sambataro, M. et al., Nucl. Phys. A423, 333 (1984).
11. Wolf, A., Warner, D.D. and Benczer-Köller, N., Phys. Lett. 158B, 7 (1985).
12. Pannert, W., Ring, P. and Gambhir, Y.K., Nucl. Phys. A443, 189 (1985).
13. Stuchberry, A.E. et al., Z. f. Physik A320, 669 (1985).
14. Dieperink, A.E.L., Scholten, O. and Warner, D.D., to be published.
15. Wu, H.C. et al., to be published.

MAGNETIC PROPERTIES OF NUCLEI IN THE COLLECTIVE MODEL [1]

M. S. M. Nour El-Din,[2] S. G. Rohoziński,[3] J. A. Maruhn,[4] and W. Greiner
Institut für Theoretische Physik
Universität Frankfurt
D6000 Frankfurt am Main, West Germany

ABSTRACT

The separation of proton and neutron deformation degrees of freedom is discussed within the geometric collective model. Its implications for the magnetic properties of low-lying states and for the description of the recently discovered quadrupole isovector excitations are presented.

1. INTRODUCTION

The idea that a partial separation of proton and neutron shape degrees of freedom in nuclei might influence the magnetic properties came quite early in the development of the collective model. Faessler and Greiner [1] explained the deviations of g-factors from the constant value Z/A by the difference in pairing strength of protons and neutrons. Then while Greiner[2] extended these ideas to the description of $E2/M1$ mixing ratios, Faessler[3] pursued the idea of excited states based on relative vibrations of the protons with respect to the neutrons, i. e. the quadrupole isovector states. Unfortunately, his estimate for the energy of these vibrations placed them much higher than what seems plausible now on the basis of recent experiments[4,5]. Later, V. Maruhn-Rezwani and coworkers [6] extended the model to link it to the generalized collective model as developed by Gneuss and Greiner[7], and Seiwert[8] applied it to high spin states. Only recently has the experimental evidence for quadrupole isovector states prompted renewed investigation of the description of these states within the model by Rohoziński et al.[9].

2. GENERAL FRAMEWORK

---

[1] Work supported by the Bundesministerium für Forschung und Technologie and by the Gesellschaft für Schwerionenforschung.

[2] Present address: Physics Department, Faculty of Science, Benha University, Benha, Egypt.

[3] Fellow of the Alexander von Humboldt Foundation. Permanent address: Institute for Theoretical Physics, University of Warsaw, Poland.

[4] Invited speaker.

Extending the usual definition of the quadrupole deformation tensor $\alpha_{2\mu}$ to define separate surfaces for protons and neutrons,

$$R_i(\theta,\phi) = R_{0i}\left(1 + \sum_\mu \alpha^i_{2\mu} Y^*_{2\mu}(\theta,\phi)\right), \quad i \in \{p,n\} \tag{1}$$

the collective coordinates α^p and α^n are introduced. Because these deformations should be coupled strongly by the symmetry energy, it is advantageous to introduce instead the average deformation α and the relative separation ξ according to

$$\alpha = \frac{B_p \alpha^p + B_n \alpha^n}{B}, \quad \xi = \alpha^p - \alpha^n, \tag{2}$$

where B_p and B_n are the mass parameters for proton and neutron collective motion, respectively, and $B = B_p + B_n$. The associated transformation of the conjugate momenta π and η (the latter conjugate to ξ) is

$$\pi = \pi^p + \pi^n, \quad \eta = \frac{B_p \pi^n + B_n \pi^p}{B}. \tag{3}$$

The Hamiltonian of the system is decomposed according to

$$H = H_1(\alpha) + H_2(\xi) + H_I(\alpha,\xi). \tag{4}$$

Computing the change in symmetry energy arising from small nonzero values of ξ, it was found[10] that it can be approximated quite well by a parabola, so that a natural assumption for $H_2(\xi)$ appears to be that of a harmonic oscillator:

$$H_2(\xi) = \frac{C_\xi}{2}[\xi \times \xi]^0 + \frac{1}{2B_\xi}[\eta \times \eta]^0. \tag{5}$$

For the first applications of this model it was assumed that, according to Faessler[3], the vibrational energy described by the Hamiltonian (5) would be of the order of $15 MeV$. In this case the eigenstates of the total Hamiltonian (4) consist of well-separated bands distinguished by the vibrational quantum number n_ξ and corresponding essentially to the coupling of the usual low-energy vibrational and rotational states to these ξ-vibrations. The observed low-energy spectrum in particular is then produced by coupling the eigenstates of $H_1(\alpha)$ to the ground state in ξ.

For the interaction energy also in this case the approximation of small vibrations in ξ can be made, leading to

$$H_I(\alpha,\xi) = C_1[\alpha \times \xi]^0 + C_2[[\alpha \times \alpha]^2 \times \xi]^0 + \cdots. \tag{6}$$

If we are interested in transitions between the low-lying states only, this coupling potential can be treated in perturbation theory. It couples those states only to the $n_\xi = 1$-bands, and calculating the matrix elements of the E2 and M1 operators between perturbed states $|\psi>$ based on the coupled states $|J_\alpha>|n_\xi = 0>$ yields for the magnetic moment and transition operator

$$<\psi|M1|\psi'> = i\sqrt{10}\left(\frac{B_p}{B} + \frac{C_1}{C_\xi}\frac{B_n B_p}{2B^2}\right)<J_\alpha|[\alpha \times \pi]^1|J'_\alpha>$$

$$+i\frac{C_2}{C_\xi}\sqrt{\frac{5}{2}}\frac{B_n B_p}{B^2}<J_\alpha|[[\alpha \times \alpha]^2 \times \pi]^1|J'_\alpha> \tag{7}$$

and for the electric transitions

$$<\psi \mid E2 \mid \psi'> = \rho_0 R_0^5 \Big[\Big(1 + \frac{C_1}{C_\xi}\frac{B_n}{2B}\Big) <J_\alpha \mid \alpha \mid J'_\alpha>$$
$$-\Big(\frac{10}{\sqrt{70\pi}} - \frac{C_2}{C_\xi}\frac{B_n}{2B} + \frac{C_2}{C_\xi}\frac{10}{\sqrt{70\pi}}\frac{B_n}{B_p}\Big) <J_\alpha \mid [\alpha \times \alpha]^2 \mid J'_\alpha>\Big]. \qquad (8)$$

So both magnetic and electric operator can be replaced, for transitions between the low-lying levels, by *effective operators* which take into account the presence of the ξ-mode. The magnetic operator of equation (7) is identical to the one originally proposed by Greiner[2].

The first term of eq. (7) contributes to the g-factor, while the second term modifies the g-factor and gives rise to M1-transitions. The theory in this form contains the three parameters C_1/C_ξ, C_2/C_ξ, and B_p/B beyond those contained in the usual low-energy Hamiltonian $H_1(\alpha)$.

The theory, coupled to the Gneuss-Greiner Hamiltonian[7] was applied to some Osmium and Platinum isotopes[6], yielding good agreement with experiment both for the mixing ratios and the g-factors. Here, however, we want to present the new calculations of M. S. M. Nour El-Din, in which the structure of the operators given in eqs. (7) and (8) is used within the rotation-vibration model.

3. APPLICATION IN THE ROTATION-VIBRATION MODEL

In the spirit of the rotation-vibration model it appears appropriate to take into account effects of the ξ-vibrations only to lowest order. Following the treatment of Seiwert et al.[8], we consider the potential energy

$$V(\alpha, \xi) = V(\alpha) + \frac{C_\xi}{2}[\alpha \times \xi]^0 + \frac{C_{\alpha\xi}}{2}[\alpha \times \xi]^0. \qquad (9)$$

The potential takes its minimum for

$$\xi_\mu/\alpha_\mu = -C_{\alpha\xi}/C_\xi \equiv \tan\delta. \qquad (10)$$

Thus the core is forced to oscillate with a constant ratio of ξ to α, and one can introduce a new coordinate u_μ by $\alpha_\mu = u_\mu \cos\delta$, $\xi_\mu = u_\mu \sin\delta$. Now the calculation proceeds as follows: the Hamiltonian is fitted in the normal way to the energy levels, but u replaces α everywhere (which does not change the energies). Then transition probabilities and g-factors are obtained by expressing α^p in terms of u, which will yield expressions similar to those in eqs. (7) and (8), but with the coefficients functions of the parameter δ. The matrix elements of these can then be evaluated between the rotation-vibration eigenstates.

In this procedure the parameters of the Hamiltonian are fitted to the spectra only, and then δ as the *single new parameter* is used to enhance overall agreement with g-factors, mixing ratios, and B(E2)-values. Here we can present a small, but characteristic selection of results only. only.

Table 1 shows a comparison of E2/M1 mixing ratios with experimental data and with other models. Apparently the overall agreement is quite competitive; it is instructive that most models still have trouble with the sign in some cases. In the present model the sign appears in some cases to be in conflict with B(E2) values: the B(E2) values could be improved vastly by accepting a wrong sign in some $\delta(E2/M1)$ value. However, one should

Isotope	Transition	$\delta(E2/M1)$			
		Exp.	Present	PPQ	DNSB
^{152}Sm	$2^+_\gamma \to 2^+_g$	-9.6	-11.9	-24.0	
	$4^+_\gamma \to 4^+_g$	-2.8	-6.0	-10.0	
	$2^+_\beta \to 2^+_g$	-8	-11.3	+11.0	
	$4^+_\beta \to 4^+_g$	+8	-6.1	+4.0	
^{154}Gd	$2^+_\gamma \to 2^+_g$	-9.7	-13.4	-41.0	-16.3
	$4^+_\gamma \to 4^+_g$	-4.1	-6.5	-12.4	-6.2
	$2^+_\beta \to 2^+_g$	+8.3	-14.5	+4.9	-10.1
	$4^+_\beta \to 4^+_g$	+2.9	-7.6	+2.1	+69.6
^{156}Gd	$2^+_\gamma \to 2^+_g$	-17.2	-10.1	-41.0	
	$4^+_\gamma \to 4^+_g$	-4	-9.2	+14.0	
	$2^+_\beta \to 2^+_g$	-14	-18.8	+21.0	

Table 1: Mixing ratios for three isotopes computed in the present model and compared to experiment[11], the dynamic pairing-plus-quadrupole model[12] DPPQ, and the dynamic Nilsson, Strutinsky, and Belyaev model[13] DNSB.

not forget that signs of mixing ratios are notoriously difficult to measure and signs have changed in the past.

Fig. 1 shows some B(E2)-values in ^{152}Sm in comparison with the standard rotation-vibration model and some other theories. The improvement is quite impressive and reproduces the structure of the experimental data faithfully. In general, of course not all isotopes can be described as well, but there is always significant improvement over the standard rotation-vibration model.

4. STRONG COUPLING MODEL AND 1^+ STATES

The discovery[4,5] that the quadrupole isovector excitation in ^{156,158}Gd is located at a much lower energy than expected previously[3] has stimulated interest also in the excited states of ξ-vibrations themselves, not only in their influence on the properties of the low-energy spectrum. Rohozinski and Greiner[9] have investigated the properties of a Hamiltonian like (4) in the strong coupling approximation.

They keep the approximation of a harmonic oscillator for $H_2(\xi)$, but use a much more general interaction Hamiltonian:

$$H_I(\alpha, \xi) = \sum_\mu \left(u^{(1)}_{2\mu}(\alpha) + \sum_{\lambda\lambda'}[w^{(1\lambda')}_\lambda(\alpha) \times [\pi \times \pi]^{\lambda'}]^2_\mu \right) \xi^*_\mu$$
$$+ \sum_{\lambda\mu} \left(u^{(2)}_{\lambda\mu}(\alpha) + \sum_{\lambda'\lambda''}[w^{(2\lambda'')}_{\lambda'}(\alpha) \times [\pi \times \pi]^{\lambda''}]^\lambda_\mu \right) [\xi \times \xi]^{\lambda*}_\mu \quad (11)$$
$$+ \sum_{\lambda\mu} v^{(1)}_{\lambda\mu}(\alpha)[\pi \times \eta]^{\lambda\dagger}_\mu + \sum_{\lambda\mu} v^{(2)}_{\lambda\mu}(\alpha)[\eta \times \eta]^{\lambda\dagger}_\mu + \sum_{\lambda\mu\lambda'}[v^{(11)}_{\lambda'}(\alpha) \times \xi]^\lambda_\mu[\pi \times \eta]^{\lambda\dagger}_\mu + h.c.,$$

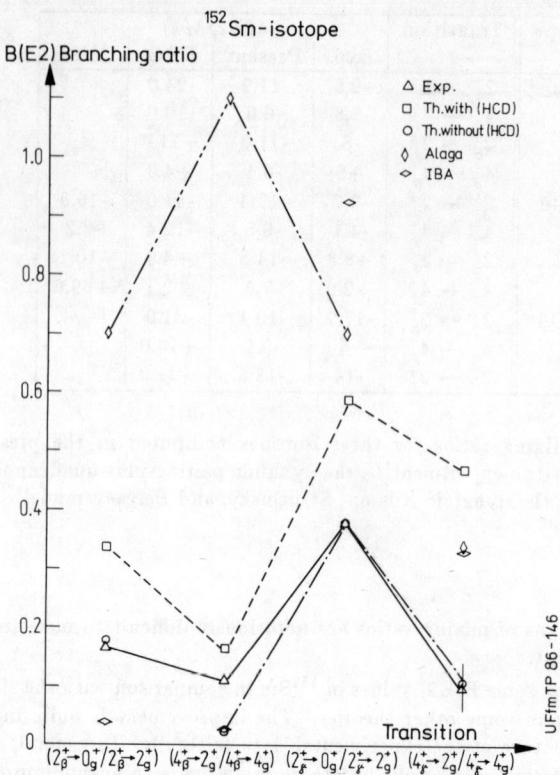

Figure 1: B(E2) branching ratios for ^{152}Sm in experiment[14], the present theory ("without Homogeneous Charge Distribution"), the rotation-vibration model ("with HCD"), the Alaga rule[14], and the IBA[15]. The transitions concerned are indicated on the abscissa.

with all α-dependent terms in turn being expanded in powers of α. For example,

$$u_{2\mu}^{(i)} = \chi_{21}^{(i)} \alpha_\mu + \frac{1}{2}\chi_{22}^{(i)} [\alpha \times \alpha]_{2\mu} + \cdots . \tag{12}$$

This example also indicates the indexing used later on.

The intrinsic system is defined by the average shape parameter α, so that the intrinsic coordinates are the Euler angles ϑ_1, ϑ_2, and ϑ_3 as well as the intrinsic mass deformations $a_0 = \alpha_0'$, $a_2 = \alpha_2'$ and the real and imaginary parts of the intrinsic relative deformations:

$$x_0 = \xi_0', \quad x_k = (\xi_k' + (-1)^k \xi_{-k}')/\sqrt{2}, \quad y_k = (\xi_k' - (-1)^k \xi_{-k}')/\sqrt{2}, \quad k \in \{1,2\} \quad . \tag{13}$$

Now the strong coupling approximation proceeds as in the case of the pure rotation-vibration model[16]. The coordinates a_0 and a_2 are assumed to deviate only slightly from their equilibrium values β and 0, respectively, while the equilibrium value for the x's and y's is determined by the interaction. The purely α-dependent part of the Hamiltonian reduces exactly to the same expression as in the rotation-vibration model, $H_2(\eta)$ retains its harmonic oscillator form, and the coupling part can be written as

$$H_I(x,y) = \sum_{k=0}^{2} H_k(x_k, y_k) \tag{14}$$

where

$$H_0(x_0) = -\frac{\hbar^2}{2B_k}\frac{\partial^2}{\partial x_0^2} + \frac{1}{2}C_0(x_0 - \varsigma)^2 \tag{15}$$

and

$$H_k(x_k, y_k) = -\frac{\hbar^2}{2B_k}\left(\frac{\partial^2}{\partial x_k^2} + \frac{\partial^2}{\partial y_k^2}\right) + \frac{1}{2}C_k(x_k^2 + y_k^2) \, , \, i \in \{1,2\}. \tag{16}$$

It can be shown that the excitations described by eqs.(15) and (16) correspond to angular momentum projection $K = 0$ and $K = k$ respectively. On each vibrational state the low-energy part of the Hamiltonian will build rotational bands with angular momenta $L = K, K+1, K+2, \ldots$ for $K \neq 0$ or $L = 0, 2, \ldots$ for $K = 0$. Thus for the lowest excitations of the proton-neutron relative vibrations we end up with three rotational bands of a definite $K^\pi = 0^+, 1^+, 2^+$ and oscillator quantum numbers $(n_0, n_1, n_2) = (1,0,0), (0,1,0), (0,0,1)$, respectively.

In this approach the interaction is hidden in the parameters B_k, C_k ($k \in \{1,2\}$) and ς, which depend in complicated ways on k and β. In particular, a static axially symmetric relative deformation

$$x_0 = \varsigma = -\left(\beta\chi_{21}^{(1)} - \frac{1}{2}\sqrt{\frac{2}{7}}\beta^2\chi_{22}^{(1)}\right)/C_0 \tag{17}$$

arises mainly owing to the interaction term $[\alpha \times \xi]^0$. The other interaction terms serve to shift the $K^\pi = 0^+, 1^+, 2^+$ bands relative to each other.

It is worth while to mention two characteristics of the 1^+ state in this model. Its excitation energy is equal to

$$E_1^+ = \hbar\sqrt{C_1/B_1} + \frac{\hbar^2}{2\mathcal{J}}, \tag{18}$$

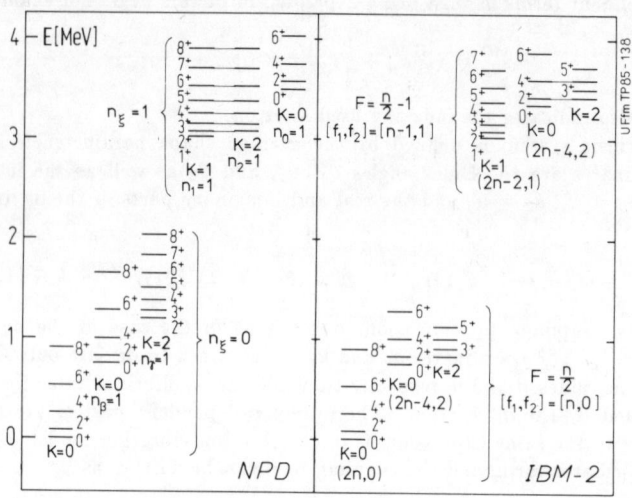

Figure 2: Schematic comparison of the energy spectra in well-deformed nuclei predicted in the strong coupling approximation of the present model ("NPD") and the $SU(3)$ limit of the IBA-2[20] under the assumption of equal excitation energy of the lowest 0^+, 1^+, and 2^+ states in both models. All rotational bands are, in principle, infinite in the NPD, whereas they are finite in the IBA-2.

and the reduced transition probability to the ground state is

$$B(M1, 0^+ \to 1^+) \approx \frac{9}{4\pi} \mu_N^2 g_{rel}^2 \beta^2 \sqrt{\frac{B_1 C_1}{\hbar^2}} \qquad (19)$$

The experimental data for the lowest 1^+ state in ^{156}Gd [5] read $E_{1^+} = 3075 keV$ and $B(M1, 0^+ \to 1^+) = 1.3 \pm 0.2 \mu_N^2$. Taking the values $\beta = 0.307$ for the deformation[17], $J = 33.7 \hbar^2/MeV$ for the moment of inertia[18], and $g_{rel} = 1$ for the gyromagnetic factor, we obtain $C_1 = 58.9 MeV$, $B_1 = 6.3 \hbar^2/MeV$. This should be contrasted with the early calculations of Faessler [3], who obtained $C_\xi \approx 2000-3000 MeV$ from the symmetry energy. Recently, however, Faessler and coworkers [19] have revised this value, the reason for the difference in estimates being that the symmetry energy should be evaluated for densities near the surface and not for bulk density.

Fig. 2 shows a comparison of the level spectra in this model to that of the IBA-2[19] in the $SU(3)$ limit. As usual, the main *practical* difference between the two models is that in the IBA the rotational bands are cut off, while the geometric model in principle allows arbitrary angular momenta. The principal bands are in agreement and finer details like the ordering of the multiplet $K^\pi = 0^+, 1^+, 2^+$ can be adjusted by fitting the interaction terms that can be added in both models.

REFERENCES

[1] Faessler, A. and Greiner, W., Z. Physik 179, 343 (1964).
[2] Greiner, W., Nucl. Phys. 80, 417 (1966).
[3] Faessler, A., Nucl. Phys. 85, 653 (1966).
[4] Richter, A., Proceedings of the International Conference on Nuclear Physics, Florence, Italy, 1983. Blasi, P., Ricci, R.A. (eds.), Vol. II, p. 189. Bologna, Tipografia Compositori 1983.
[5] Bohle, D., Richter, A., Steffen, W., Dieperink, A.E.L., Lo Iudice, L., Palumbo, F., Scholten, O., Phys. Lett. 137B, 27 (1984).
[6] Maruhn-Rezwani, V., Greiner, W., Maruhn, J.A., Phys. Lett. 57B, 109 (1975).
[7] Gneuss, G. and Greiner, W., Nucl. Phys. A171, 449 (1971).
[8] Seiwert, M., Hess, P.O., Maruhn, J.A., and Greiner, W., Phys. Rev. C23, 2335 (1981).
[9] Rohoziński, S.G., Greiner, W., Z. Physik A322, 271 (1985).
[10] Maruhn, V., Ph. D. Thesis, University of Frankfurt, 1975.
[11] Sakai, M., At. Data Nucl. Data Tables 31, 399 (1984).
[12] Lange, J., Kumar, K., Hamilton, J.H., Rev. Mod. Phys. 54, 119 (1982).
[13] Kumar, K., Gupta, J.B., Hamilton, J.H., Aust. J. Phys. 32, 307 (1979).
[14] Konijn, J. et al., Nucl. Phys. A373, 397 (1982).
[15] Scholten, O. et al., Ann. Phys. 115, 325 (1978).
[16] Eisenberg, J.M., Greiner, W., Nuclear Theory, 2nd Edn., Vol. 2. Excitation Mechanisms of the Nucleus. Amsterdam, North Holland 1976.
[17] Löbner, K.E.G., Vetter, M., Hönig, V., Nucl. Data Tables A7, 495 (1970).
[18] Burrows, T.W., Nucl. Data Sheets 18, 553 (1976).
[19] Faessler, A., Bochnacki, Z., Nojarov, R., J. Phys. G12, L47 (1986).
[20] Iachello, F., Nucl. Phys. A358, 89c (1981).

THE NUCLEAR SPECTROSCOPY OF MIXED SYMMETRY STATES

W.D. Hamilton

Physics Division, University of Sussex, Brighton BN1 9QH, U.K.

ABSTRACT

Nuclear spectroscopy experiments provide critical evidence for the identification of mixed symmetry states. Examples of these states in vibrational nuclei are considered with some reference to rotational and γ-unstable nuclei.

The possibility that states of a mixed symmetry, or isovector character, might be found in nuclei was considered twenty years ago by Faessler.[1] He concluded that these states would lie rather high in energy in the region of 10MeV or so on the basis of assumptions about the number of neutrons and protons participating in the out of phase collective motion and the amplitude of this motion. Some ten years later Greiner's group[2] returned to the problem when considering M1 transitions in deformed nuclei but they made no estimates of level energies. Such studies were made within the framework of the Bohr-Mottelson model[3] and no further development took place until the advent of the IBM-2.[4] Indeed such developments do not occur naturally in the usual range of nuclear models which do not distinguish between the possible separate motions of neutrons and protons. The IBM-2 provides a natural framework and the extra πν degree of freedom allows states to occur which are not totally symmetric in the sd space of the model. Of course such states must lie at a higher energy than the normal states which are successfully accounted for by IBM-1.

Thus the problem we are faced with is: How do we identify such states? In this review I shall concentrate for the most part on mixed symmetry states in vibrational-like nuclei and Dr Bohle in the following contribution will discuss the case of rotational nuclei.

In vibrational nuclei the lowest lying mixed symmetry state will contain a d and an s boson and will have spin-parity 2^+ while in rotational nuclei the lowest state will be the 1^+ ground state of the $K^\pi=1^+$ band. A detailed discussion of the mixed symmetry level sequence for the different IBM limits may be found in ref[5]. An example of a 1^+ MS state was first found in ^{156}Gd [6] and subsequently many similar 1^+ states have been identified in other rotation nuclei[7].

It was the lack of correspondence between IBM-1 fitted levels and the experimental ones that led to the identification of the 2^+_{MS} state in the vibrational N=84 isotones[8]. Not only was there no model counterpart to the 2^+ level which occured in ^{140}Ba, ^{142}Ce and ^{144}Nd at ∿2MeV but these 2^+ levels decayed predominantly to the 2^+_1 levels by essentially pure M1 transitions.

Within the U(5) limit of IBM-2 there are two states each with one d boson corresponding to the 2^+_1 and 2^+_{MS} levels. We may write these in terms of the ground state as

$$|2^+_1\rangle = N^{-1/2}(d^\dagger_\nu s_\nu + d^\dagger_\pi s_\pi)|0^+\rangle \qquad (1)$$

$$|2^+_3\rangle = \{(N_\pi/NN_\nu)^{1/2} d^\dagger_\nu s_\nu - (N_\nu/NN_\pi)^{1/2} d^\dagger_\pi s_\pi\}|0^+\rangle, \qquad (2)$$

where N_π and N_ν are the numbers of proton and neutron bosons and $N = N_\pi + N_\nu$. Now from the definition of transition operators we obtain the reduced matrix elements and the two which are relevant to to the $2^+_{MS} - 2^+_1$ transition are

$$\langle 2^+_1 ||T^{E2}||2^+_{MS}\rangle = 5(N_\nu N_\pi/N^2)^{\frac{1}{2}}(e_\nu \chi_\nu - e_\pi \chi_\pi) \qquad (3)$$

$$\langle 2^+_1 ||T^{M1}||2^+_{MS}\rangle = 3(5N_\nu N_\pi/2_\pi N^2)^{\frac{1}{2}}(g_\nu - g_\pi) \qquad (4)$$

Those for other transition may be found in ref[8].

The results of a group of experiments at the ILL provided data on the magnitude and sign of the E2:M1 mixing ratio of the $2^+_{MS}-2^+_1$ transitions and also the $T(2^+_{MS}-0^+_1)/T(2^+_{MS}-2^+_1)$ branching ratios in the N=84 isotones. These together with $B(E2;2^+_1-0^+_1)$ values were used to obtain a set of IBM-2 parameters when the 2^+ levels at ~2MeV were assumed to have a mixed symmetry description in the U(5) limit. Although several simplifying assumptions were made the parameter set was judged to be very satisfactory and it was concluded that the γ-decay properties of these 2^+ levels were well described by those of the lowest mixed symmetry state in the vibrational limit of IBM-2.

The IBM parameters may also be used to make some predictions about other quantities in these nuclei. In particular the quadrupole moment of the 2^+_1 level is given by

$$Q(2^+_1) = 1.7(N_\pi e_\pi \chi_\pi + N_\nu e_\nu \chi_\nu)/N \tag{5}$$

The experimental value for ^{142}Ce is $Q(2^+_1) = -0.12(9)$b while that obtained by substituting derived parameters into eq(5) is $Q(2^+_1)=-0.11$b. A more impressive agreement is found between the predicted and measured transition probabilities in ^{144}Nd.

Table 1 Transition probabilities from the 2^+_{MS} state in ^{144}Nd

Transition Probability	Predicted	Experiment
$B(E2 ; 2^+_{MS} - 0^+_1)$	$0.012 e^2 b^2$	$0.013(3) e^2 b^2$
$B(M1 ; 2^+_{MS} - 2^+_1)$	$0.20 \mu_N^2$	$0.28(7) \mu_N^2$
$B(E2 ; 2^+_{MS} - 2^+_1)$	$0.014 e^2 b^2$	$0.020(5) e^2 b^2$

The experimental values in Table 1 are obtained from the estimated mean-life of the 2.073 MeV level[9] and the measured branching ratio and mixing ratio of the $2^+_{MS} - 2^+_1$ transition[10].

The isovector quadrupole-vibration model has recently been applied to the N=84 isotoms[11]) and it accounts well for the observed properties. In particular the transition probabilities are very sensitive to the number of nucleons contributing to the low-energy isovector vibration and the above values indicate that only nucleons in the last open shells participate as is assumed in the IBM-2 calculations. (This is contrary to the conclusion reached in ref[11] where the transition probability data were not available).

The occurance of states in IBM-2 additional to those obtained in an IBM-1 analyses and for which there appeared to be in general no experimental counterpart was initially resolved by a suitable choice of the Majorana term parameters $\xi_1 \xi_2$ and ξ_3 which placed the additional levels at a sufficiently high energy so that they were no longer an embarrassment. Once having identified the 2^+_{MS} states these may now be satisfactorily accounted for, but a unique set of ξ_i-parameters cannot be obtained with only one level to fit in each isotope. It is usual to choose $\xi_1=\xi_2=\xi_3$ with $\xi_i \simeq 0.12$ MeV being a typical value. This procedure may be used in an attempt to identify other predicted MS states since only the MS states show a strong dependence on the value of ξ_i parameters while the 'regular' fully symmetric states remain unaltered as ξ_i are varied. By this method a 1^+ at 2310keV in ^{140}Ba has been tentatively identified. This decays by an intense (>75% branch) to the 2^+_1 level with a predominantly M1 transition (δ =0.31$\pm$0.05). The situation is less clear in ^{142}Ce where there are more than ten 1^+ levels at approximately the energy of the one phonon excitation above the 2^+_{MS} state[13]).

It is not possible to identify any candidate for a 2^+_{MS} assignment in the N=86 isotones where the level structures become more complex and indeed where we might not expect the vibrational limit to apply because of the proximity of the onset of permanent nuclear deformation.

A further sensitive indication of a mixed symmetry description is given by the F-spin contribution in a state[14]). Now $F=\frac{1}{2}$ for bosons with $F_{max} = (N_\pi + N_\nu)/2 = N/2$ and states with maximal F-spin are symmetric. The states with $F=N/2$ lie lowest in energy followed by $F=N/2-1$ etc. with the separation between states with different F values determined by the Majorana term. Thus in a level fit to IBM-2 the F_{max} contribution is a good indication of the degree of symmetry of a level. This is well illustrated by the 2_3^+ levels in ^{76}Se and ^{78}Se both of which have no counterpart in the IBM-1 level sequence, but they can be successfully fitted by IBM-2 and the F_{max} contribution as indicated in the fig 1 is substantially less than for other 2^+ levels.

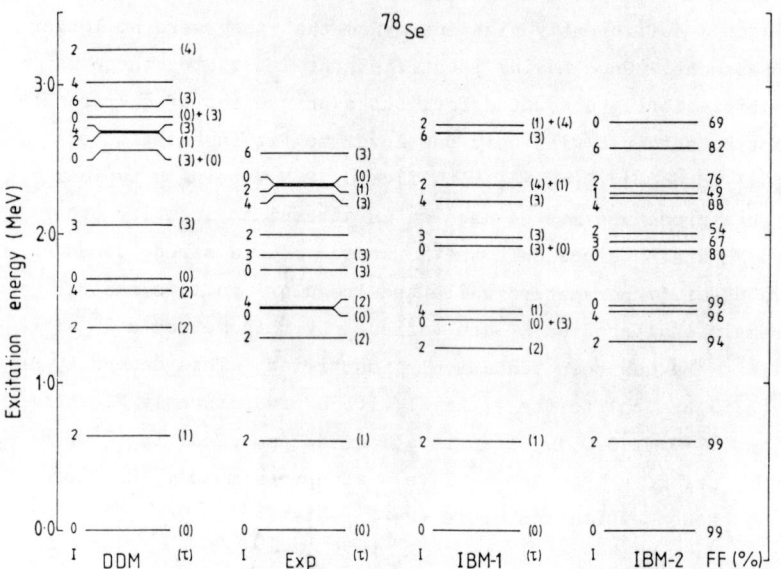

FIG.1. The experimental positive parity states are compared with the predictions of the DDM and IBM-1, which are insensitive to MS states, and IBM-2 in which additional 2^+ and 1^+ levels occur.

Fig.1 shows the experimental levels in ^{78}Se and the IBM-1 and IBM-2 level fits together with the percentage F_{max} contribution for the latter[15]. It would thus appear that the levels at 1.778 MeV (^{76}Se) and 1.996 MeV (^{78}Se) have a mixed symmetry character. If we now examine transition probabilities for the decays of these 2^+_{MS} levels we at first sight appear to find a contradiction.

The $2^+_{MS} - 2^+_1$ decay in ^{78}Se is by a predominantly M1 transition with $\delta=0.44(10)$ but this decay mode has a small M1 transition probability with $B(M1; 2^+_{MS} - 2^+_1) = 4(2) \times 10^{-3} \mu_N^2$. This is also small compared with the competing value of $B(E2; 2^+_3 - 2^+_1) = 6(4) \times 10^{-4} e^2 b^2$ [~0.3W.u.] and it would appear at first sight that $B(M1)$ is too small to warrant a mixed symmetry description of the 1.996 MeV level since on the basis of previous examples we might expect $B(M1) \simeq 0.2 \mu_N^2$. However in these examples it was considered that $g_\pi = 1.0$ and $g_\nu = 0$ and if the g-factors do not have their limiting values then the M1 matrix element will be considerably reduced since in eq(4) it is seen to be proportional to $g_\pi - g_\nu$.

The evaluation of $g_\pi - g_\nu$ is possible from the $2^+_2 - 2^+_1$ transition matrix element since we know $B(E2)$ and δ for this transition. We obtain $g_\pi - g_\nu \simeq 0.2 \mu_N$. This result is consistent with g_π and g_ν values obtained from magnetic moments of 2^+_1 states of nuclei in this region. Thus we may now understand that the low $B(M1)$ value arises from the near equality of g_π and g_ν which effectively reduces the M1 strength by ~25. A similar analysis may be made for the 1.778MeV level in ^{76}Se. This illustrates that a large $B(M1)$ is not necessarily a sufficient signature to indicate an MS state. The F-spin analysis also indicates that there should be a 1^+ level at 2.314MeV which is also an MS state, but this has not been seen.

We now understand why the identification of a mixed symmetry state is not necessarily as straightforward as in the case of the N=84 isotones. A further complication arises when the mixed symmetry description for a particular spin component is not confined to one level.

To some extent this may be true of the 2_4^+ level at 2.335 MeV in fig.1. where we see that its F_{max} contribution is less than maximum.

A good example of sharing of mixed symmetry characteristics is found in ^{56}Fe.[6]) The two 2^+ levels at 2.568 and 2.960 MeV have almost identical properties apart from the signs of the multipole mixing ratios of the 2^+-2^+ transitions, fig.2. Both transitions

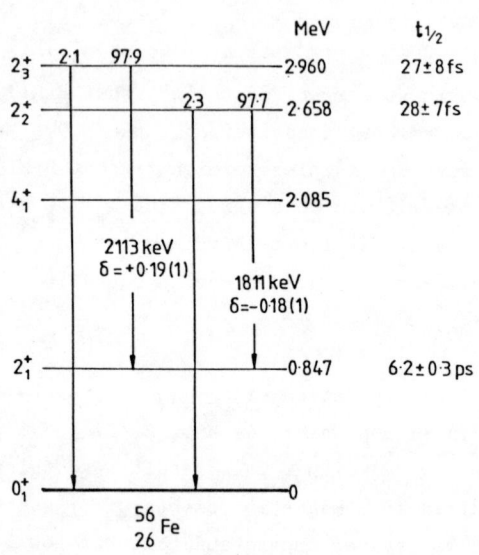

FIG.2. Level energies and half-lives in ^{56}Fe and E2:M1 mixing ratios δ, and transition branching ratios.

have strong M1 components and might be identified as mixed symmetry states. However it is unreasonable for two such states to be so close in energy and it was assumed that a mixed symmetry state $|m\rangle$ coincided with a fully symmetric one $|s\rangle$ such that

$|2_2\rangle = \alpha|s\rangle + \beta|m\rangle$ and $|2_3\rangle = \beta|s\rangle - \alpha|m\rangle$

From the B(M1) values we find that the components α and β of the two contributions in the 2^+ states are approximately equal and that $B(M1, m-2_1^+)=0.38\mu_N^2$ This value is consistant with the expected result for a vibrational nuclius if $|g_\pi-g_\nu|=1.0$ which agrees with g-values obtained for neighbouring one-boson nuclei. While the M1 decay occurs via the $|m>$ component the E2 may decay by either the $|s>$ or $|m>$ components to the fully symmetric 2_1^+ state. The near equality of the δ-values and their opposite signs show that the $|s>$ component dominates.

Other examples in this A=50 to 60 mass region are discussed by Collins[17] in a contribution to these proceedings. He presents data suggesting that a higher lying 2^+ level in ^{56}Fe also contains an MS contribution and that the 4_2^+ level has an MS character. These data also indicate that the energy of the 2^+ MS state falls on approaching a closed shell.

This sharing of an MS character between several levels illustrates another problem in identifying these states since the sharing weakens the MS properties of a particular level. It is this sharing of the MS character over a number of 2^+ levels that has made it so difficult to identify the second member of the $K^\pi=1^+$ band in ^{156}Gd[18]).

A 2^+ level is also expected to be the lowest lying MS state in a γ-unstable nucleus. One of the best examples of such a nucleus is ^{124}Te[19]) and a 2^+ level at 2.09 MeV which could not be fitted by IBM1 and which decays by a predominantly M1 transition to the 2_1^+ level has the required signature of an MS state in IBM-2[20]).

We mentioned previously that it was conventional to choose the three ξ_i parameter to be equal although there appears to be no *a priori* reason for doing so. In the equation describing the Majorana force ξ_1 and ξ_3 play a different role from ξ_2 and it has been shown that in rotational nuclei it is possible to introduce a low lying MS 0^+ level while the position of the $K^\pi = 1^+$ band head remains unaltered[21]). The 0_3^+ level at 1.405 MeV in ^{172}Yb may be an example of such a level. The measured $\rho(E0; 0_3^+ - 0_1^+) = (1.4 \pm 0.2) \times 10^{-2}$ value [22]) is consistent with an MS description of the level but, since we have as yet no knowledge of effective boson monopole changes, no firm conclusion can be drawn. The alternative explanation is that the 0_3^+ state is a quasi particle excitation and this is referred to in the contribution by Andrejtscheff. This explanation has been given some additional support in recent measurements reported on by Subber[23]) which indicates yet another lower lying 0^+ state that might be associated with a quasi-neutron pair while the 1.403 MeV level is due to a quasi-proton pair. If this is true then it would unnecessary to invoke an MS description to account for this low lying 0^+ level.

To summarise; examples of mixed symmetry states have been found in the three broad classes of nuclear types viz. vibrational, rotational and γ-unstable. And while the main characteristic of these states is their strong M1 decay to fully symmetric states the actual values of multipole mixing ratios and B(M1) values will depend sensitively on the values of the effective changes and g-factors entering eqs. 3 and 4. The success of any search for an MS state will be greatly influenced by these parameters for as we have seen the M1 matrix element will tend to zero as the g_ν and g_π - factors approach equality while the E2 matrix element is strongly dependent on e_ν and e_π. The ability to identify an MS state will also be greatly influenced by level density since the MS strength can be spread over neighbouring levels to such an extent that the MS character may be lost. It is thus important to choose relatively simple level schemes and this feature of the sharing of MS characteristics between levels will make it difficult to identify higher energy and hence higher spin MS states. Already we have some indication of the boson number dependence of the MS 2^+

state in lighter nuclei and this region is also important since shell-model calculations[24] can be made here with some confidence.

REFERENCES

1. Faessler A. Nuclear Physics, 85, 653 (1966).
2. Maruhn-Rezwani V., Greiner W. and Maruhn J.A. Phys.Lett., 57B, 109(1975).
3. Bohr A. and Mottelson B.R. *Nuclear Structure* vol.2. (New Yorke:Benjamin) (1975).
4. Arima A. and Iachello F. *Adv.Nucl.Phys.* vol.13. J.W. Negele and E. Vogt(new York: Plenune)(1984).
5. Van Isacker P., Heyde K., Jolie J. and Scholten O., in Interacting Boson-Boson and Boson-Fermion Systems ed.O.Scholten(World Scientific Pub.Co.,Singapore, 1984).
6. Bohle D., Richter.R., Steffen W., Dieperink A.E.L., LO Iudice N., Palumbo F. and Scholten O., Phys. Lett.137B, 27(1984).
7. Bohle D., these proceedings (1986).
8. Hamilton W.D., Irbäck A., and Elliott J.P. Phys.Rev.Lett.53, 2469 (1984).
9. Metzger F.R. Phys.Rev.187, 1700 (1969).
10. Snelling D.M. and Hamilton W.D. J.Phys.,G 9, 763 (1983).
11. Faessler A., and Nojarov R., Phys.Lett., 166B, 367 (1986).

12   Robinson S.J., Hamilton W.D., Hungerford P., Pfeiffer B., Jung G., and Snelling D.M., to be published J.Phys.G, (1986).

13   Michelakakis E., Hamilton W.D., Hungerford P., Jung G., Pfeiffer P. and Scott S.M., J.Phys. G $\underline{8}$,111 (1982).

14   Van Isacker P., Heyde K., Jolie J., and Sevrin A., accepted for publication in Ann.Phys.(N.Y.)

15   Subber A.R.H., Robinson S.J., Hungerford P., Hamilton W.D., Van Isacker P., Kumar K., Park P., Schreckenbach K., and Colvin G., to be submitted to J.Phys.G,(1986).

16   Eid S.A.A., Hamilton W.D., and Elliott J.P., Phys.Lett., $\underline{166B}$, 267 (1986).

17   Collins S.P., Hamada S.A., Hamilton W.D., Hoyler F., and Robinson S.J., a contribution to this Conference (1986).

18   Bohle D., Richter A., Heyde K., Van Isacker P., Moreau J., and Sevrin A., Phys.Rev.Lett., $\underline{55}$, 1661 (1985).

19   Robinson S.J., Hamilton W.D., and Snelling D.M., J.Phys G $\underline{9}$, 961 (1983).
     Subber A.R.H., Park P., Hamilton W.D., Kumar K., Schreckenback K., and Colvin K., accepted for publication in J.Phys.G,(1986).

20   Park P., Subber A.R.H. Hamilton W.D., Elliott J.P., and Kumar K., J.Phys.G $\underline{11}$, L251 (1985).

21   Scholten O., Heyde K., Van Isacker P., and Otsuka T.,

Phys.Rev. C <u>32</u>,1729 (1985).

22 Andrejtscheff A., Petkov P., Protochristow Ch., Koskov L.K., Hamilton W.D., and Hoyler F., accepted for publication in J.Phys.G,(1986).

23 Subber A.R.H., Hamilton W.D., Schreckenbach K.,Colvin G., and Jumar K., a contribution to this Conference (1986).

24 Evans J.A., Elliott J.P., and Szpikowski S., Nucl. Phys. <u>A435</u>, 317, (1985).

INVESTIGATION OF MIXED SYMMETRY $J^\pi=1^+$ AND 2^+ STATES WITH (e,e'), (γ,γ') AND (p,p') TECHNIQUES*

D. Bohle

Institut für Kernphysik, Technische Hochschule Darmstadt,
D-6100 Darmstadt, Germany

ABSTRACT

Mixed symmetric $J^\pi=1^+$ states have been discovered in a large number of nuclei. Three arguments are presented here that the new M1 mode is a rather purely orbital mode: The new mode exhausts a large fraction of sum rule strength, comparison of proton and electron scattering experiments prove the dominance of the orbital over the spin part of the transition matrix element and finally the comparison of (e,e') and (γ,γ') experiments on ^{156}Gd is used to determine the distribution of orbital M1 strength in this nucleus. Furthermore the systematics of excitation energy and transition strength of the new M1 mode are briefly discussed. Candidates for mixed symmetric $J^\pi=2^+$ states in rotational nuclei are presented and the search for the corresponding level in the transitional nucleus ^{110}Pd is described.

1. INTRODUCTION

A new magnetic dipole mode has been discovered[1] in high resolution, inelastic electron scattering in heavy deformed nuclei at 2-3 MeV excitation energy carrying a transition strength of typically 1-2 μ_N^2. This excitation of the nucleus can mainly be regarded as a small amplitude vibration in terms of the angle between the two symmetry axes of the axially deformed neutron and proton bodies. This mode is thus essentially an orbital mode. The main of this talk is to show this in three ways. The first is through the comparison to sum rules, secondly the dominance of the orbital part of the M1 transition operator will be shown through the comparison of (e,e') and (p,p') experiments and, thirdly, it will be demonstrated that also through the comparison of inelastic electron scattering and nuclear resonance fluorescence a distinction between spin and convection current type excitations is possible. Furthermore, a few remarks on the systematics of excitation energy and M1 transition strength will be made. In the last part of this talk possible candidates for mixed symmetric $J^\pi=2^+$ states in rotational and transitional nuclei will be presented.

---

*Work supported by the Deutsche Forschungsgemeinschaft

2. FEW EXAMPLES FOR MIXED SYMMETRIC $J^\pi=1^+$ STATES IN DEFORMED NUCLEI

The systematics of mixed symmetric $J^\pi=1^+$ states in rare earth nuclei has been reported in detail[1,2]. Furthermore, we have identified the states excited by the new mode in two other regions of the periodic table: the $f_{7/2}$-shell nuclei ^{46,48}Ti and the actinide nuclei ^{232}Th and ^{238}U. In figs. 1 and 2 electron scattering on these nuclei are shown. At least one strong peak that we interprete as the excitation of a $J^\pi=1^+$

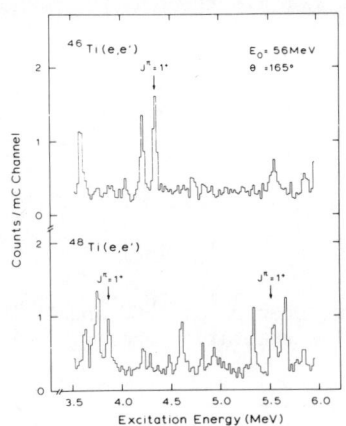
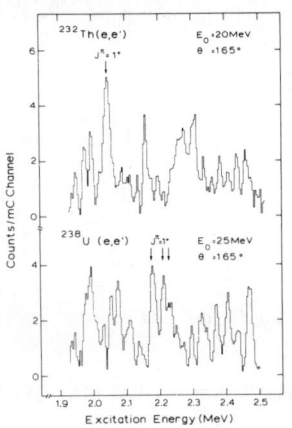

Figure 1
Inelastic electron scattering spectra on ^{46,48}Ti. The new M1 mode excites the $J^\pi=1^+$ states marked by arrows.

Figure 2
Inelastic electron scattering spectra on the actinide nuclei ^{232}Th and ^{238}U. Peaks marked by arrows are interpreted as $J^\pi=1^+$ states due to their appropriate form factor behaviour.

state is marked in each spectrum. That mixed symmetric $J^\pi=1^+$ states have to be expected in $f_{7/2}$ shell nuclei has first been pointed out by Zamick[3]. The search for mixed symmetry $J^\pi=1^+$ states in Th and U has been done in collaboration with U.E.P. Berg and U. Kneissl[4] from Giessen.
Such $J^\pi=1^+$ states have therefore now also been found in the second region of the periodic table where large deformations occur: the actinides. Presently we are extending our measurements also to higher excitation energies to search for weakly excited 1^+ states. First surveys as shown in fig.3 give no indication for further strong $J^\pi=1^+$ states in ^{156}Gd and ^{238}U. We take presently more spectra up to $E_x \approx 10$ MeV and will analyze them with the help of statistical methods developed earlier[5]. Since the relation for the excitation energy of the $J^\pi=1^+$ states, i.e.

$$E_x \simeq 66\delta A^{-1/3} \text{ MeV}, \qquad (1)$$

however, is fulfilled so well in a large number of nuclei it is at present not very likely to assume that the 1⁺ states seen are only the low energy side remnants of a whole distribution centered at higher excitation energy.

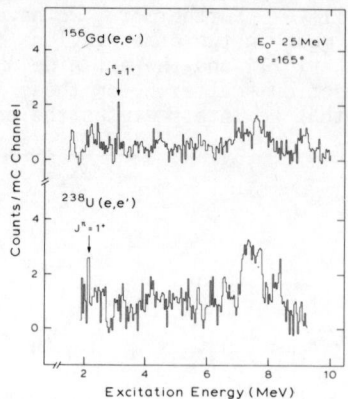

Figure 3
Comparison of inelastic electron scattering spectra on ^{156}Gd and ^{238}U. Strong $J^\pi = 1^+$ states are only observed at low excitation energy.

3. ARGUMENTS FOR AN ORBITAL MODE

3.1 Sum rules

Our first point, if no proof that the new mode is an orbital mode, is made with the help of the sum rules of the interacting boson model for deformed nuclei derived under the assumption that the small amplitude vibration leads to a purely orbital mode. One has[1,6]

$$B(M1)\uparrow = \frac{3}{4\pi} \frac{N_\pi N_\nu}{N_\pi + N_\nu} (g_\pi - g_\nu)^2 \quad [\mu_N^2] \; , \qquad (2)$$

where $N_\pi (N_\nu)$ is the proton (neutron) boson number and $g_\pi (g_\nu)$ are the proton (neutron) boson g-factors. If the g-factors are taken from ref.6, typically 3.5 μ_N^2 are predicted for a deformed rare earth nucleus. If the g-factors of Wolf[7] are taken instead only 1/3 of that number is expected in which case already the strongest excited $J^\pi = 1^+$ state exhausts the sum rule eq.(2).

3.2 Comparison of inelastic electron and proton scattering experiments

No direct proof can be given that the new mode corresponds to the orbital motion of protons about neutrons as the classical picture of Palumbo and Lo Iudice[8] suggests if electron scattering experiments are not contrasted with proton scattering experiments. In (e,e') experiments the transition can be excited through the orbital as well as the spin part of the M1 transition operator. In proton scattering at $E_0 = 200$ MeV and small scattering angles, however, the effective nucleon-nucleon interaction in a nucleon-nucleus collision is dominated by the spin part.

Comparison of (e,e') with (p,p') might thus help to disentangle the orbital and spin contribution.

In electron scattering experiments we measure the quantity

$$B(M1)\uparrow = [\pm\sqrt{B(\ell)} + \sqrt{B(\sigma)}]^2 , \quad (3)$$

where $B(\sigma)$ denotes the spin strength measured directly in inelastic proton scattering experiments at $E_0 = 200$ MeV and $\Theta = 0°$. Assuming constructive interference in eq.(3) we get a lowest estimate of the ratio of spin and orbital strength. If we insert the appropriate g-factors we obtain a ratio of matrix elements

$$\frac{\langle f || \sum_i \ell_i \tau_i^3 || i \rangle}{\langle f || \sum_i \sigma_i \tau_i^3 || i \rangle} = \frac{g_s}{2}\left[\left(\frac{B(M1)\uparrow}{B(\sigma)\uparrow}\right)^{1/2} -1\right] \quad (4)$$

where the quenched[9] spin g-factor g_s enters. For the deformed rare earth nuclei only upper limits were obtained[10], namely $B(\sigma) < 0.2 \mu_N^2$. This already leads to the important fact that the orbital matrix element is at least 3.5 times larger than the spin one. In fig.4 electron and proton scattering spectra on ^{46}Ti are displayed. Note the strongly excited $J^\pi = 1^+$ state at $E_x = 4.319$ MeV in both experiments. In fig.5 the angular distribution of the scattered protons is shown. Since in $f_{7/2}$-shell nuclei protons and neutrons fill the same shell the spin matrix

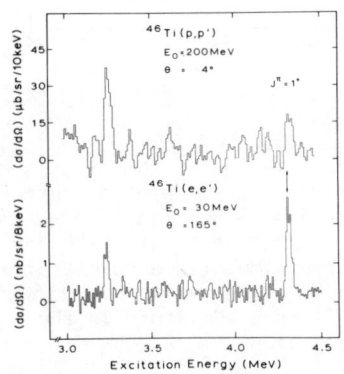

Figure 4
Comparison of inelastic proton and electron scattering spectra. The mixed symmetry $J^\pi = 1^+$ state at E_x 4.319 MeV is excited in both experiments.

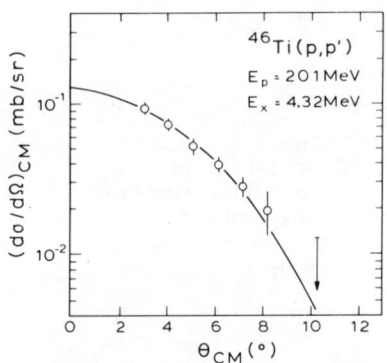

Figure 5
Angular distribution of protons scattered inelastically to the $J^\pi = 1^+$ state at $E_x = 4.319$ MeV.

element is not as much reduced with respect to the orbital matrix element as in the rare earth nuclei. In ^{46}Ti we measure [10] $B(M1)\uparrow = 1.0 \pm 0.2 \mu_N^2$ and $B(\sigma) = 0.13 \pm 0.03 \mu_N^2$ which leads to a ratio of orbital to spin matrix elements of 5 which is a really large number. Zamick[11] predicts a ratio of orbital spin matrix elements that are about three - which is already comforting through not as large as the experimental one - while Civitarese et.al.[12] argued recently that the spin matrix element should be much larger than the orbital matrix element. This is clearly ruled out by experiment. So far the description of Zamick[3,11] is still the best approximation to the experimental results.

With the 200 MeV (p,p') experiment compared to the Darmstadt (e,e') experiment the dominance of the orbital over the spin matrix elements was proven. Because of the high resolution of $\Delta E = 8$ keV obtainable in low energy inelastic proton scattering it was recently feasible[13] to get an even better estimate for the ratio of orbital to spin matrix elements. A comparison of low energy proton and electron spectra taken on ^{156}Gd at

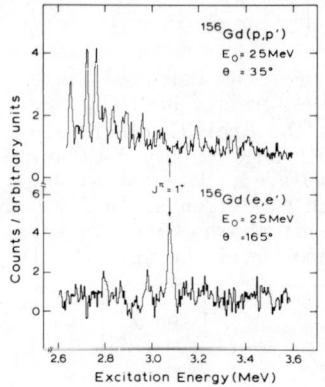
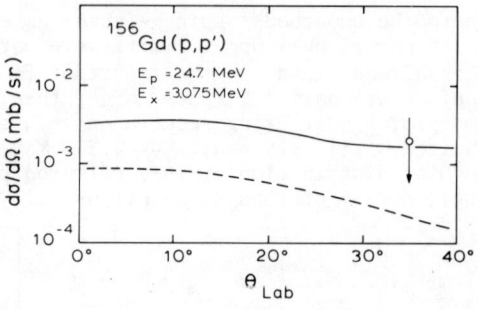

Figure 6
Comparison of inelastic proton and electron scattering spectra both taken at low bombarding energies. The new mode is not excited by protons.

Figure 7
DWBA calculation of the cross section for inelastic proton scattering at $E_0 = 25$ MeV using IBA-2 wave functions (see text). The dashed curve simulates a purely orbital mode, while the full one allows for a 2% spin admixture in the wave function.

low bombarding energy are shown in fig.6. The new M1 mode is only barely, if at all, excited through protons. The theoretical cross sections displayed in fig.7 are the result of a DWBA calculation[13] where a IBA-2 wavefunction was used for the nuclear structure description that had already been sucessfully employed to calculate the electron scattering form factor[1] for ^{156}Gd. The use of the IBA-2 wave function implies that the ratio of orbital to spin matrix elements is 45. If we take the upper limit, the experiment yields as shown in fig.7, that this number is at least 20. The high resolution of low energy proton scattering has hence led to the result that the new M1 mode in rare earth nuclei is a rather purely orbital mode and that the spin admixture of the wave function of the mixed symmetric $J^\pi = 1^+$ state is less than about 4%.

3.3 Comparison with nuclear resonance fluorescence

Through the comparison of inelastic electron scattering and nuclear resonance fluorescence experiments a distinction between spin and convection current type excitations is possible[14]. Even though both experiments rely solely on the electromagnetic interaction this distinction is possible since the transition strength measured directly at the photon point together with the electron scattering form factor specifies the excitation mechanism uniquely. In fig.8 a (γ,γ') spectrum is compared with the sum of all electron scattering spectra on ^{156}Gd taken at $\Theta=165°$. All $J^{\pi}=1^+$ states are marked by arrows. The decomposition of the electron scattering spectrum into individual lines is also shown. Since the electron scattering form factor and the radiative width measured at the photon point have to be described at the same time we have to conclude that all states in fig.8 are indeed excited by M1 transitions.

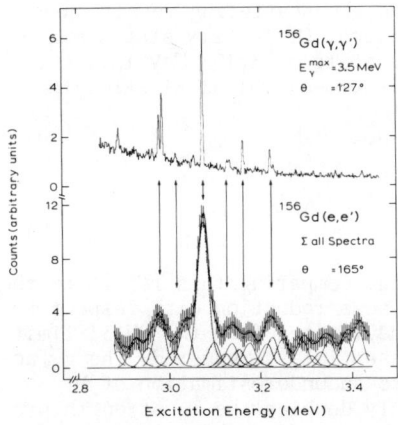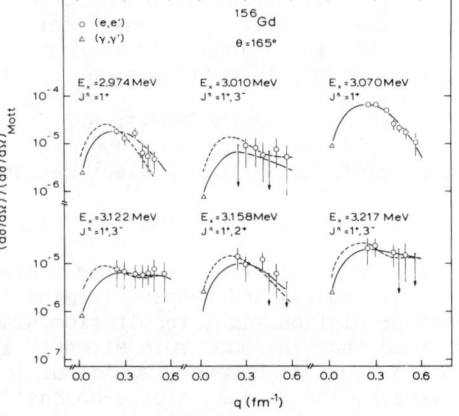

Figure 8
Comparison of (γ,γ') and (e,e') spectra in ^{156}Gd in the region of the new M1 mode where corresponding peaks have been marked by arrows. Note, that peaks from inelastic and elastic scattering occur simultaneously in the gamma ray spectrum. The unfolding of the electron scattering spectrum using the lineshape of the elastically scattered electrons is also shown.

Figure 9
M1 transition form factors including the data from (e,e') and (γ,γ') experiments. The full curve is obtained whenever it is assumed that the $J^{\pi}=1^+$ states are excited by the M1 mode, the dashed curve is obtained when a spin-flip excitation mechanism is assumed instead. Where forward angle measurements indicated electric contributions to the (e,e') cross section the appropriate Tassie model form factor has been added.

Furthermore, since the form factors of spin and convection current excitations are different the $B(M1)\uparrow$ value measured directly at the photon point distinguishes between these possibilities as shown in fig.9.

The forward angle electron scattering data - which are not displayed here - show that within the given energy resolution electric transitions contribute to the respective form factors. Whereever this is the case the appropriate Tassie model form factor has been added in order to describe the (e,e') data. Only the M1 transitions to the states at E_x= 2.974 MeV and 3.070 MeV are separated from transitions of other multipolarities. These transitions shall be discussed here as an example of our method. In the case of the collective transition to the $J^\pi=1^+$ state at 3.070 MeV electron scattering data were sufficient to show [1] that the state is excited by the new mode. The IBA-2 form factor (full drawn curve) gives an excellent description of the data including also the photon point. The dashed line corresponds to a two-quasiparticle spin-flip calculation employed in ref.1. It is compared to the weaker state at E_x= 2.974 MeV. Using this form factor for the extrapolation to the photon point we obtain $B(M1,k)\uparrow = 0.56\ \mu_N^2$ while we get $B(M1,k)\uparrow = 0.34\ \mu_N^2$ in excellent agreement with the photon scattering result when the IBA-2 form factor is used for comparison. Of the six states excited by M1 transitions only the one to the state at 3.158 MeV is identified as a spin-flip transition. The total summed orbital M1 strength amounts to

$$\Sigma B(M1)\uparrow = 2.3 \pm 0.5\ \mu_N^2 \qquad (5a)$$

in case of excitation by electron- and

$$\Sigma B(M1)\uparrow = 2.1 \pm 0.3\ \mu_N^2 \qquad (5b)$$

in case of excitation by photon scattering. Comparing this result to the sum rule eq.(2) and keeping in mind that some reduction with respect to that prediction might result from small deviations from the SU(3) limit we find that the sum rule strength is essentially exhausted. The g-factors of Wolf[7], however, would now lead to an underestimation of M1 strength. The inclusion of g-bosons[15] only doubles the M1 strength predictions which would still be insufficient to redeem this. In case of ^{156}Gd, g-bosons do not seem necessary for our understanding of the summed orbital M1 strength.

Since the orbital M1 strength is spread into five states that are closely centered around the collective state at 3.07 MeV the estimate of the mixing matrix element derived by the assumption that the mixed symmetric $J^\pi=1^+$ state couples strongly to the background of symmetric g-bosons $J^\pi=1^+$ levels[16] has to be modified. It becomes smaller since the orbital M1 strength rests essentially in one level.

4. FEW REMARKS ON SYSTEMATICS OF EXCITATION ENERGY AND TRANSITION STRENGTH

Very briefly, some remarks on recent results[17] on the systematics of excitation energy and transition strength will be made. In the framework of the interacting boson model (IBA) the excitation energy of mixed symmetry $J^\pi=1^+$ states in rotational nuclei can be determined from a schematic Hamiltonian

$$H = \varepsilon_d n_d + \varkappa Q_\pi \cdot Q_\nu + \lambda M \tag{6}$$

where the first term describes the pairing, the second term the quadrupole interaction between protons and neutrons and the third term the Majorana interaction that specifies the energy difference between symmetric and mixed symmetric excitations. Adjusting the parameters of the first two terms of Hamiltonian eq.(6) to the low energy spectrum of the studied nuclei and the Majorana force parameter λ to the excitation energy of the collective $J^\pi=1^+$ states we obtain a systematics of the Majorana force paramter λ. For nuclei of the same mass number A but different deformations δ the same value of the ratio λ/δ is found. The ratio λ/δ which is plotted in fig.10 is well described through the relation

$$\lambda/\delta = 3.7 \ (N_\pi \cdot N_\nu)^{-1/2} \tag{7}$$

The value of the constant on the rigth hand side of eq.(7) depends of course on the structure of the quadrupole interaction chosen in the Hamiltonian. An explanation for the predictive power of eq.(7) is at present still lacking.

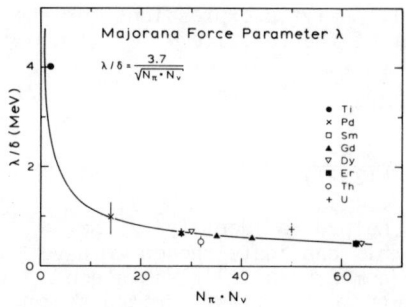
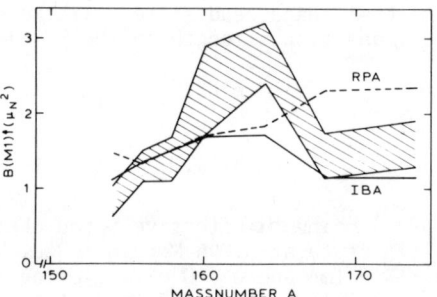

Figure 10
Majorana force paramter λ normalized to the mass deformation δ as a function of $N_\pi \cdot N_\nu$. Compared to the experimental data is the function eq.(7).

Figure 11
M1-strength into the strongest $J^\pi=1^+$ states in rare earth nuclei (dashed area) compared to two theoretical model predictions.

In fig.5 the M1 strength into the strongest $J^\pi=1^+$ states in rare earth nuclei is shown. The dashed area corresponds to the experimental result considering the experimental uncertainty. The theoretical predictions have been adjusted by an additive constant to yield the observed transition strength of $B(M1)\uparrow = 1.3\pm0.2 \ \mu_N^2$ for the collective $J^\pi=1^+$ state at $E_x=$ 3.07 MeV in ^{156}Gd. Normalized this way the curve RPA summarizes all those predictions that follow a

$$B(M1)\uparrow \sim \delta A^{4/3} \tag{8}$$

law. The prediction that the M1 strength should rise with mass number which is also the result of the calculations of Hilton [18] and Nojarov et.al. [19] is not verified by the experiment. The curve labeled IBA shows the trend predicted in the rotational limit of the IBA. As observed experimentally a maximum at A=164 is indeed predicted, however, it fails to reproduce the experimentally observed strength by a factor of 2. This may be a hint on the necessity to include g-bosons in the IBA-description [15] of the M1 strength in these nuclei.

5. MIXED SYMMETRIC $J^\pi=2^+$ CANDIDATES IN ROTATIONAL AND TRANSITIONAL NUCLEI

The lowest mixed symmetric state in rotational nuclei is the $J^\pi=1^+$ state found in a number of deformed nuclei. In vibrational nuclei, however, a $J^\pi=2^+$ state is believed [20] to be lowest. Recent experiments [21] have shown that the third excited 2^+ state in some N=84 isotones has all predicted properties.

Even though the $J^\pi=2^+$ member of the $K^\pi=1^+$ band is not strongly excited in rotational nuclei the identification of the 2^+ member of the band is not too complicated since the excitation energy difference to the $K^\pi=1^+$ band head state is rather small. Analytic expressions for the E2 transition strength in the SU(3) limit of the IBA, i.e.

$$B(E2, 0_1^+ \to 2_1^+)\uparrow = (e_\nu N_\nu + e_\pi N_\pi)^2 (2N+3)/N \qquad (9)$$

and

$$B(E2, 0_1^+ \to 2_M^+)\uparrow = (e_\nu - e_\pi)^2 \frac{3(N-1)}{N(2N-1)} N_\nu N_\pi , \qquad (10)$$

$e_\pi(e_\nu)$ being the effective boson charges, helped to identify [22] the $J^\pi=2_M^+$ state at 3.096 MeV in ^{156}Gd. With this candidate chosen we have $e_\pi \simeq 13$ e fm^2 and $e_\nu \simeq 10$ e fm^2. Using these numbers eq.(9) gives a good approximation to the E2 transition strength to the first excited states in the neighbouring nuclei which would not be case for another choice of candidate.

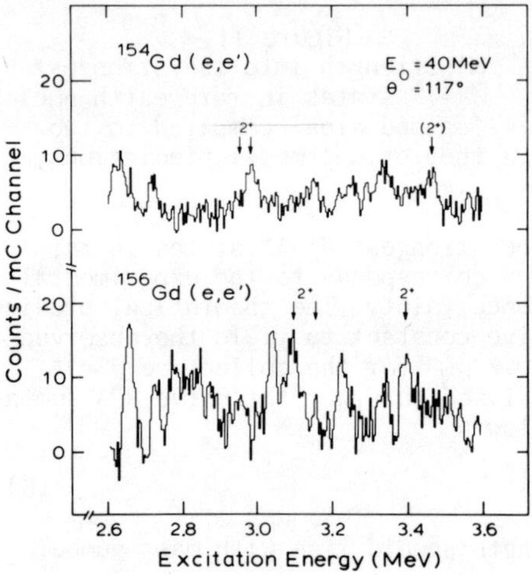

In fig.12 a comparison of electron scattering spectra on ^{154}Gd and ^{156}Gd are shown taken under kinematical conditions favouring E2 transitions. These $J^\pi=2^+$ states that have been marked by arrows have excitation energies higher than the collective $J^\pi=1^+$ state. In ^{154}Gd the two lowest $J^\pi=2^+$ states shown in fig.12 are candidates for mixed symmetric $J^\pi=2^+$ states. Both

Figure 12: Comparison of ^{154,156}Gd(e,e') spectra.

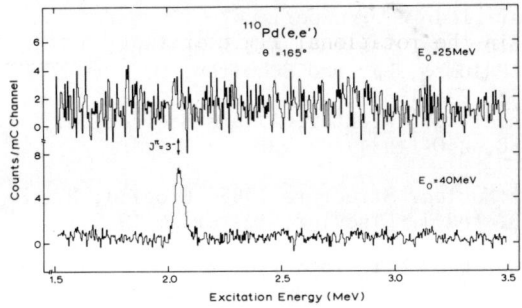

Figure 13

Electron scattering spectra on ^{110}Pd. The search for mixed symmetric states in ^{110}Pd has not led to the discovery of any single strong $J^\pi = 1^+$ and 2^+ state.

are energetically close enough to the $J^\pi=1^+$ state at 2.98 MeV. The lower $J^\pi=2^+$ state is favoured by the estimate of eq.(10) using the effective charges determined [22] from ^{156}Gd. One has to keep in mind, however, that ^{154}Gd is not as good a rotational nucleus as ^{156}Gd. Therefore, the possibility cannot be ruled out that the other state is the mixed symmetric one.

In the transitional nucleus ^{110}Pd the search for mixed symmetric $J^\pi=1^+$ and 2^+ states has not let to their discovery. While strong $J^\pi=1^+$ states are not expected[16], around 3.0 MeV and also around 3.9 MeV weak excitations are found that might correspond to the predicted ones. In fig.13 two spectra taken on ^{110}Pd are shown. The upper one is taken under kinematical conditions where M1 transitions are favoured. The kinematical conditions of the lower one favour E2 transitions. The only strong transition, however, is a transition to a $J^\pi=3^-$ state. Above the excitation energy of the third excited 2^+ state the E2 strength has to be strongly fragmented.

6. CONCLUSION AND OUTLOOK

Combining the information from electron, proton and photon scattering we have shown that the new M1 mode is rather purely a orbital mode, furthermore it was possible to extract a detailed distribution of orbital M1 strength. Mixed symmetric $J^\pi=1^+$ and 2^+ states have been found in a number of nuclei. At present we are extending our efforts to search for mixed symmetric $J^\pi=3^+$ states, for orbital M1 strength at higher excitation energy and for mixed symmetric states in odd-even nuclei.

ACKNOWLEDGMENTS

I would like to thank Th. Guhr, U. Hartmann, K.-D. Hummel, G. Kilgus, U. Milkau and in particular A. Richter for their help and many fruitful discussions.

REFERENCES

1) Bohle, D., Richter, A., Steffen, W., Dieperink, A.E.L.,
   Lo Iudice, N., Palumbo, F., and Scholten, O.,
   Phys. Lett. 137B, 27 (1984),
   Bohle, D., Küchler, G., Richter, A. and Steffen, W.,
   Phys. Lett. 148B, 260 (1984).

2) Richter, A., in:Nuclear Structure 1985. Broglia, R., Hagemann,G.B.,
   and Herskind, B. (ed.). Elsevier, Amsterdam 1985.

3) Zamick, L., Phys. Rev. C31, 1955 (1985).

4) Kneissl, U., contributed paper to this conference.

5) Müller, S., Beck, F., Meuer, D., and Richter, A.,
   Phys. Lett. 113B, 362 (1982).

6) Sambataro, M., Scholten, O., Dieperink, A.E.L., and Piccitto, G.,
   Nucl. Phys. A423, 333 (1984).

7) Wolf, A., Warner, D.D., and Benczer-Koller, N.,
   Phys. Lett. 158B, 7 (1985).

8) Lo Iudice, N. and Palumbo, F., Phys. Rev. Lett. 41, 1532 (1978).

9) Richter, A., Proc. on the Internat. Conf. on Nuclear Physics,
   Florence, Aug. 29-Sep.3, 1983,Edizione Compositori, Florence.

10) Djalali, C., Marty, N., Morlet, M., Willis, A., Jourdain, J.C.,
    Bohle, D., Hartmann, U., Küchler, G., Richter, A., Caskey, G.,
    Crawley, G.M., and Galonsky, A., Phys. Lett. 164B, 269 (1985).

11) Zamick, L., Phys. Lett. 167B , 1 (1986).

12) Civitarese, O., Faessler, A. and Nojarov, R., preprint Tübingen

13) Wesselborg, C., Schiffer, K., Zell, K.O., von Brentano, P.,
    Bohle, D., Richter, A., Berg, G.P.A., Brinkmöller, B.,
    Römer, J.G.M., Osterfeld, F., and Yabe, M., Zeitschr.f.Phys.
    A323, 485 (1986).

14) Bohle, D., Richter, A., Berg, U.E.P., Drexler, J., Heil, R.D.,
    Kneissl, U., Metzger, H., Stock, R., Fischer, B., Hollick, H.,and
    Kollewe, D., Nucl. Phys. A, in press.

15) Barrett, B.R. and Halse, P., Phys. Lett. 155B, 133 (1985).

16) Scholten, O., Heyde, K., van Isacker, P., Jolie, J., Moreau, J.,
    Waroquier, M. and Sau, J., Nucl. Phys. A438, 41 (1985).

17) Hartmann, U., Bohle, D., Guhr, Th., Hummel, K.-D., Kilgus, G., Milkau, U., and Richter, A., to be published.

18) Hilton, R.R., J. Phys. (Paris) Coll. C-6, 255 (1984).

19) Nojarov, R., Bochnacki, Z., and Faessler, A., preprint Tübingen.

20) Iachello, F., Phys. Rev. Lett. 53, 1427 (1984).

21) Hamilton, W.D., Irbäck, A., and Elliott, J.P., Phys. Rev. Lett. 53, 2469 (1984).

22) Bohle, D., Richter, A., Heyde, K., van Isacker, P., Moreau, J., and Sevrin, A., Phys. Rev. Lett. 55, 1661 (1985).

STUDY OF MIXED SYMMETRY STATES WITH PHOTON SCATTERING*

Ulrich Kneissl

Institut für Kernphysik, Justus-Liebig-Universität Giessen
Leihgesterner Weg 217, D-6300 Giessen, Germany

ABSTRACT

Our former study of low energy collective dipole transitions in heavy deformed nuclei has been elaborated by recent nuclear resonance fluorescence (NRF) experiments performed at the bremsstrahlung facilities of the Stuttgart Dynamitron and the Giessen Linac, respectively. Precise excitation energies, reduced transition probabilities, spins and decay branching ratios of numerous new states have been extracted from the energy and angular distributions of the scattered photons measured with three high resolution Ge(HP)-detectors. Results for ^{156,158,160}Gd, ^{160,162,164}Dy, ^{182}W, ^{232}Th and ^{238}U will be presented and discussed.

1. INTRODUCTION

Low lying collective excitations of the nucleus such as vibrations and rotations have been known for a long time. The present interest in the study of low lying collective states is due to predictions of new collective modes. An outstanding example is the isovector magnetic dipole mode in deformed nuclei, which recently has been discovered in (e,e') experiments at Darmstadt[1]. Low lying E1 excitations also have been predicted leading to α-cluster states[2].

The photon scattering technique (nuclear resonance fluorescence (NRF)) represents the most sensitive method to detect these new collective states[3] due to the high selectivity of the excitation of low spin states and the extreme energy resolution achievable in γ-spectroscopy.

*Supported by the Deutsche Forschungsgemeinschaft.
The experiments have been performed in collaboration with:
U.E.P. Berg, C. Bläsing, R.D. Heil, A. Jung, H.H. Pitz, U. Seemann, R. Stock, F.J. Urban (Giessen); C. Wesselborg, P.v. Brentano, K.O. Zell (Köln); B. Fischer, H. Hollick, D. Kollewe (Stuttgart); D. Bohle, A. Richter (Darmstadt)

However, it should be emphasized that the most powerful tool to
distinguish between different excitation modes is a combined analysis[5]
of form factors from (e,e')-work and of the NRF-data.

2. EXPERIMENTAL TECHNIQUE

The experiments have been performed using continuous bremsstrahlung
produced at the high current Stuttgart Dynamitron ($E \leq 4.3$ MeV, $J < 4$ mA;
duty cycle = 100%) and the polarized bremsstrahlung facility of the
Giessen linac. The scattered photons have been detected by three Ge(HP)-
γ-spectrometers. The photon scattering cross sections have been cali-
brated by performing runs with mixed targets containing ^{27}Al, where
states with well known widths were excited. Details of the experimental
techniques will be presented by R.D. Heil during this conference.

3. RESULTS AND DISCUSSION

All states observed in our experiments can be divided into two
groups corresponding to their decay properties. The ratio $R = \Gamma_{2+}/\Gamma_o$ of
the transition width to the first excited 2^+ state over the groundstate
width contains information about the K value of the decaying state.
Within the rotational limit one expects $R=0.5$ or 2 depending upon
whether $\Delta K = 1$ or 0. Both types of states could be detected in our NRF-
experiments. The first part of the discussion deals with $K = 1$ states.
Their excitation should be ascribed to the new collective M1-mode, which
can be described within the IBA-2-model leading to the excitation of
mixed symmetry $J^\pi = 1^+$ ($K = 1$) states.

3.1 Mixed Symmetry States

3.1.1 <u>Gd-Isotopes</u>: Fig.1 shows a (γ,γ')-spectrum for ^{156}Gd in comparison
with the Darmstadt (e,e') data. The marked peaks indicate groundstate
transitions. The corresponding transitions to the first excited 2^+ state
are observed as satellite peaks shifted by 89 keV respectively. In the
case of the most prominent peak at 3070 keV (e,e') data alone enabled
to identify the excitation as due to the new M1-mode. However, for the
weaker transitions only a combined analysis of (e,e') and (γ,γ') experiments succeeded to distinguish between orbital and spin excitations.

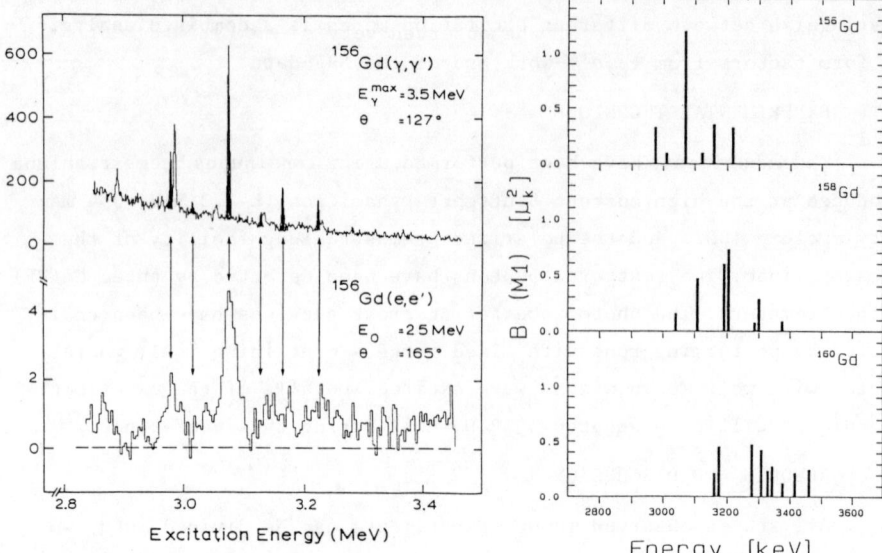

Fig. 1: (γ,γ')- and (e,e')-spectra

Fig. 2: strength distribution in Gd-isotopes (s.text)

Except for the weak transition at 3158 keV all indicated transitions belong to the collective M1 mode. Independently the positive parity of the 3070 keV state has been confirmed by a $(\vec{\gamma},\gamma')$ experiment at the Giessen Linac using linearly polarized bremsstrahlung[4].

The results of the (γ,γ') and (e,e') experiments are in a good agreement. The total orbital M1 strength concentrated in 5 transitions near 3.1 MeV amounts to $(2.06 \pm 0.30)\mu_K^2$ in the (γ,γ') work compared to $(2.33 \pm 0.54)\mu_K^2$ in (e,e'). This strength lies below the value of $\approx 3.8~\mu_K^2$ as predicted by the IBA-2 model in the SU(3) limit.

A more pronounced fragmentation of the strength has been observed for the heavier Gd isotopes ^{158,160}Gd in our NRF-experiments. The energetically shift roughly follows the predicted $66 \cdot \delta \cdot A^{-1/3}$ MeV-dependence. The total strength, assuming all excitations can be ascribed to the orbital M1 mode, amounts to $\approx 2.4~\mu_K^2$ and $\approx 2.7~\mu_K^2$ for ^{158}Gd and ^{160}Gd respectively.

3.1.2 Dy-Isotopes: Most recently we have investigated the isotopes ^{160,162,164}Dy. An example of the observed very clean (γ,γ')-spectra shows fig. 3. As insert the corresponding (e,e')-data[6] are plotted. In contrast to the Gd isotopes in all three Dy-isotopes the strength seems to be concentrated in only 2 or 3 states. The center of the excitation energies is shifted from $\approx$ 2.8 MeV in ^{160}Dy to 3.15 MeV in ^{164}Dy. The total strength observed amounts roughly to $(2.3-3.6)\mu_K^2$ for these isotopes. A detailed analysis is underway[7].

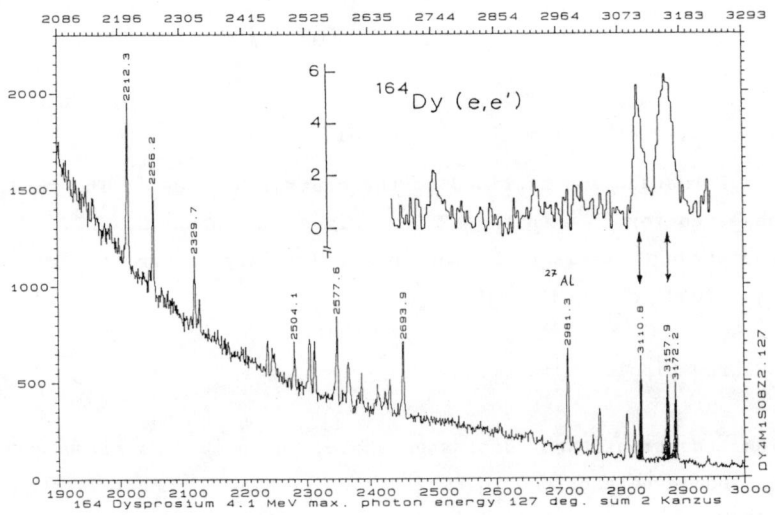

Fig. 3: (γ,γ') and (e,e')-spectra of ^{164}Dy

3.1.3 Actinide Isotopes: It seemed to be of interest to search for the new M1 mode not only in the rare earth region but also in the other island of deformed nuclei: the actinide region. The high atomic number Z and the high level density in these heavy nuclei make more difficult both, (e,e') and (γ,γ') experiments. Nevertheless we succeeded to obtain rather clean (γ,γ')-spectra for ^{238}U and ^{232}Th. An example for ^{238}U is shown in Fig. 4. The 4 most prominent transitions at 2176, 2209, 2245, 2295 keV also could be detected in the corresponding (e,e') experiment

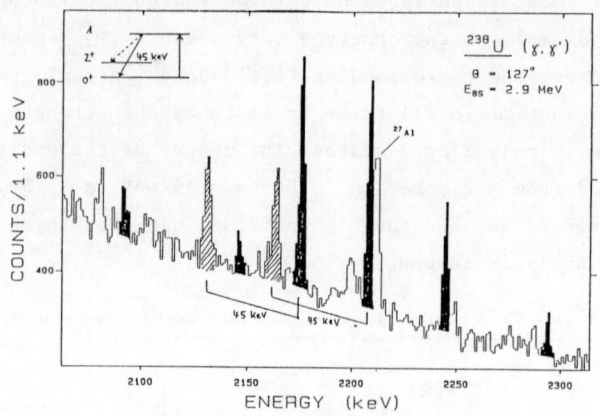

Fig. 4: (γ,γ')-spectrum of ^{238}U (detail)

performed at Darmstadt and ascribed to the orbital M1-mode[9]. The total strength observed for ^{238}U in the NRF experiments is about 2.8 μ_K^2. In ^{232}Th the strength seems to be concentrated mainly in one strong transition at 2043 keV (B(M1) ≈ 1.5 μ_K^2).

3.2 Low Energy Dipole Transitions

Besides the transitions discussed above, which in comparison with (e,e') data should be ascribed to excitations of mixed symmetry states, numerous additional dipole transitions could be observed in the NRF-experiments. Among these transitions, not be evident in (e,e')-data, both $\Delta K = 0$ and 1 decay branching ratios (R ≈ 2 or 0.5) have been observed. The $\Delta K = 0$ transitions then correspond to E1 transitions, the $\Delta K = 1$ transitions can be both of electric or magnetic multipolarity. The $\Delta K = 1$ transitions (fig. 5a) are rather concentrated near 2.8 MeV in the Gd-isotopes, the strength distribution seems to be independent of the deformation in contrast to the orbital magnetic strength distribution (s. fig. 2). The $\Delta K = 0$ transitions are distributed over a larger energy region. The observed groundstate widths of these spin 1-states (up to 70 meV) correspond to B(E1) values which are up to a factor of 6 higher than those to be expected from an extrapolation of the E1-giant dipole resonance (following the assumptions and parametrizations

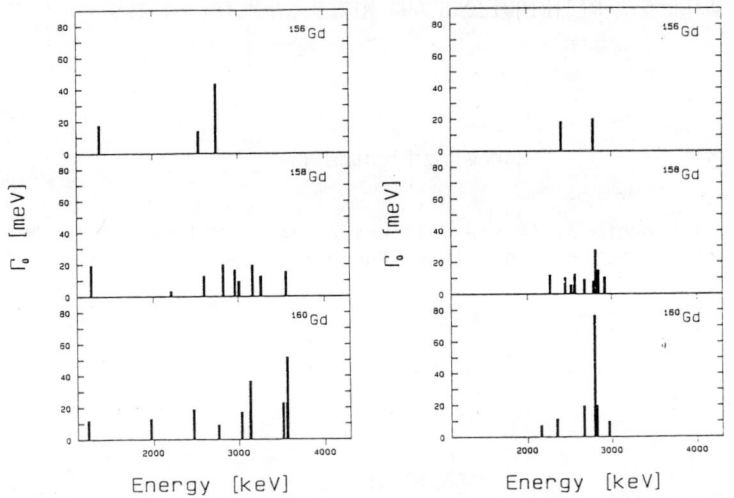

Fig. 5a: ΔK = 1 transitions   Fig. 5b: ΔK = 0 transitions

proposed by Iachello et al$^{2,9)}$. In order to associate these spin 1 states with the predicted α-cluster states more systematic studies over a broader mass region should be performed. In any case the strength of excitations leading to α-cluster states is rather sensitive to the α-cluster admixture amplitudes to the ground state configurations.

REFERENCES

1) D. Bohle et al, Phys. Lett 137B (1984) 27
2) F. Iachello, Phys. Lett. 160B (1985) 1
3) U.E.P. Berg et al, Phys. Lett. 149B (1984) 59
4) U.E.P. Berg, Habilitation Thesis, Giessen 1986, unpublished
5) D. Bohle et al, Nucl. Phys. A, in press
6) D. Bohle et al, Phys. Lett. 148B (1984) 260
7) C. Wesselborg, Thesis in prep.
8) D. Bohle, this conference
9) G. Maino et al, Phys. Rev. C30 (1984) 2101.

STRENGTH DISTRIBUTIONS AND MIXED-SYMMETRY STATES

Marcello Pignanelli

Dipartimento di Fisica dell'Universita', Milano, Italy
Istituto Nazionale di Fisica Nucleare - Sezione di Milano

ABSTRACT

In this contribution we present a search of new classes of collective states by high resolution measurements of proton and deuteron scattering on medium-mass nuclei.

1. INTRODUCTION, METHOD AND RESULTS

In spite of many previous studies on A=90-150 nuclei with various reactions, no detailed information is avalaible on the strength distributions for the different multipolarities. High resolution measurements are therefore important. In the present study the attention is focussed on the E2, E3 and E4 strengths and in particular on the search of 2^+ states, described in the IBA-2 as mixed-symmetry states.[1]

Inelastic differential cross sections have been measured using the AVF cyclotron and the QMG/2 magnetic spectrograph of the Kernfysisch Versneller Instituut, with an energy resolution of the order of 10^{-4}. The nuclei: $^{94,98,100}Mo$, $^{104,110}Pd$, ^{112}Cd, $^{146,150}Nd$ and ^{150}Sm have been studied by 30.7 MeV protons. To have a better information on the structure of the final states, part of these measurements have been repeated using 60 MeV protons and 50.4 MeV deuterons.

About 70 differential cross sections have been obtained for each nucleus. A large part of them belongs to states not observed before. A comparison of the the angular distributions with DWBA or CC predictions, allows the spin-parity assigment. Spin values up to 8 have been obtained. The analysis of the angular distributions gives also important information on the role played by two-step processes.

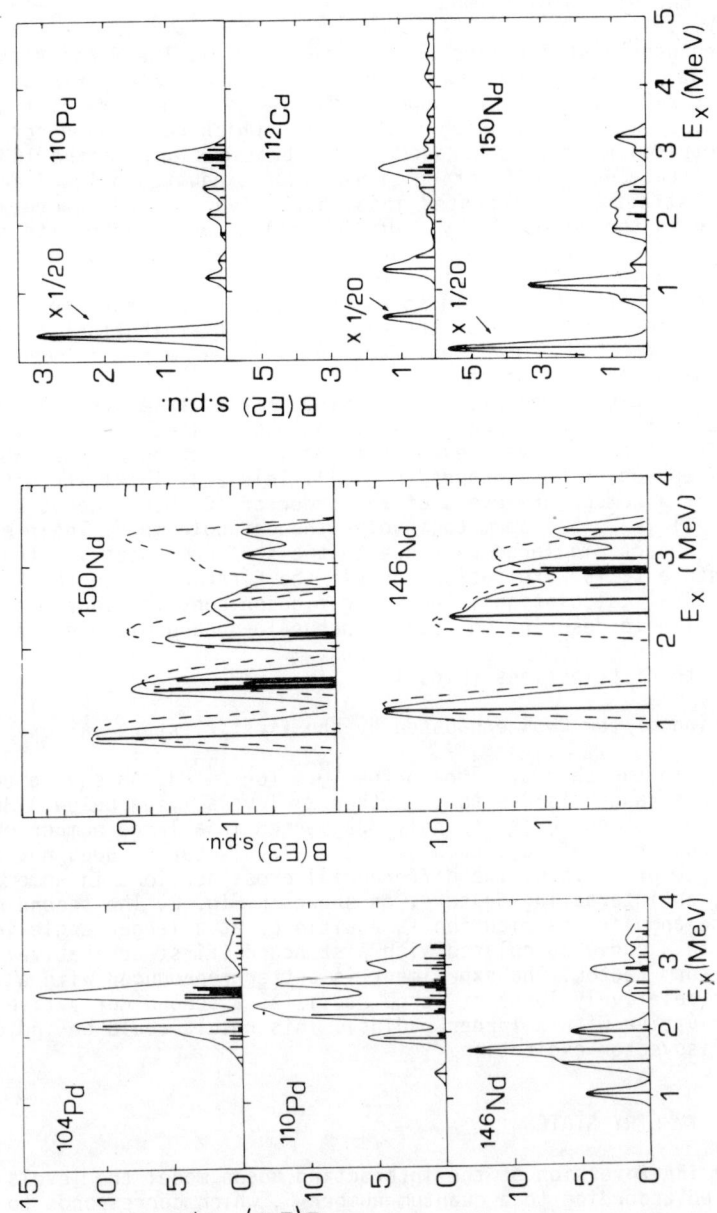

Fig. 1. E4, E3 and E2 strength distributions. The continuous lines are from a smearing of experimental strengths. The dashed lines on Nd data are results from IBA calculations, including f-boson contributions.

2. Eλ STRENGTHS DISTRIBUTIONS

E4 strengths distributions are given in Fig. 1. The first 4^+ state generally displays a relatively small B(E4). Appreciable strengths are observed at 2-3 MeV excitation energy. The largest ones are of the order of 10 s.p.u. These belong to transitions, which could be described, in a geometrical model, as hexadecapole vibrations or, in the IBM, as g-boson excitations. The fraction of the EWSR exhausted below 4 MeV is 4-8 %. This stregth is fragmented into 15-20 levels. The fragmentation is connected to the collectivity of the nucleus and reaches its maximum for a rotational nucleus as ^{150}Nd. The calculated g-boson distribution displays a lower fragmentation with a strength concentrated in few states. These latters in some nucleus, as for instance ^{112}Cd, reproduce the mean structure of the experimental distribution. In other cases the agreement is less satisfactory. Further work is needed to draw conclusions on this point.

The B(E3) value found for the first 3^- state varies from 27, for ^{150}Nd, to 65 s.p.u., for ^{98}Mo. This strength exhaustes the 6-12 % of the EWSR. The other 3^- levels exhaust a smaller fraction of the EWSR, with the exception of Nd isotopes (Fig. 1). This additional strength is fragmented in a number of levels of the order of 10. Other negative parity states (1^-, 3^-, 5^-) seem to involve the octupole mode. Their excitation is, in fact, in large part due to two-step processes via the first 2^+ state (E2+E3 excitation) or via the first 3^- state (E3+E2 excitation). IBA-1 calculations, including f-boson contributions, seem to give a reliable description both to octupole-quadrupole bands as to E3 strength distributions (Fig. 1).[2,3]

The fraction of the EWSR exhausted by the first 2^+ state varies from the 5 % in the case of ^{94}Mo to the 10 % for ^{110}Pd. An equivalent amount of E2 strength is due to the other 2^+ levels lying below 4 MeV. This additional strength is strongly fragmented in a large number of levels (of the order of 20). Each single 2^+ level usually does not reach a strength of 1 s.p.u. The differential cross sections in some cases display an interesting feature, as shown in Fig. 2. The second maximum in the angular distribution is positioned at a larger angle in comparison to a curve calculated with a standard (first derivative) collective form factor. The experiment is better reproduced with a form factor relatively large at large radii (as a second derivative or a first derivative with a larger radius). This result could be indicative of an isovector excitation.

4. MIXED SYMMETRY STATES

In the IBA-2 version of the interacting boson model the levels can be classified according to a quantum number F, which corresponds to the isospin for fermion systems. The levels with F= Fmax= N/2, where N is the total number of bosons, are totally symmetric in respect to proton and neutron bosons. This property is not found when F<Fmax. In this

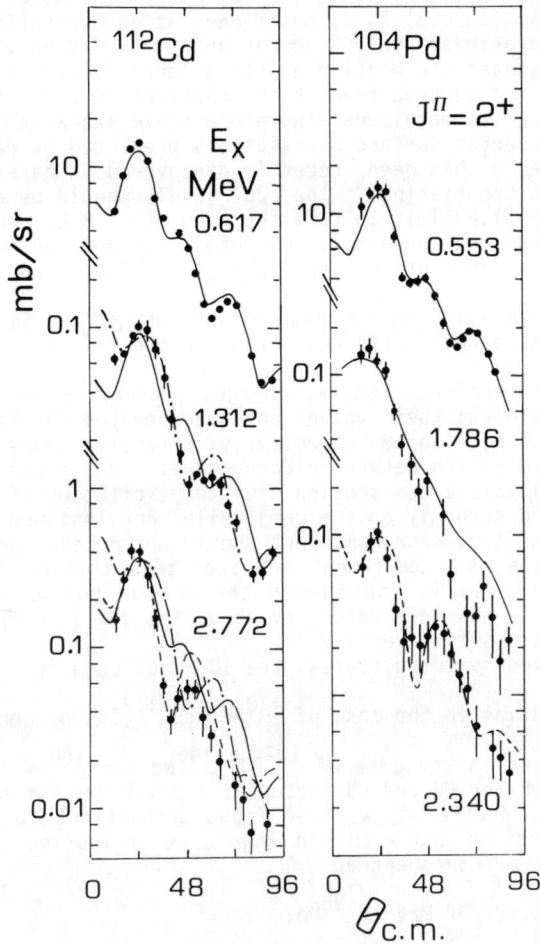

Fig. 2. Sensitivity to transition potentials and reaction mechanism. Solid curves on 0.617, 0.553, 1.312 and 2.772 MeV levels: first derivative. Curve on 1.786 MeV level: microscopic single-particle form factor. Point-dashed curves on 1.312 and 2.772 MeV levels: two-steps Dashed curves on 2.772 and 2.340 MeV levels: second derivative.

case one has a mixed-symmetry state. For F= Fmax-1 the model predicts a triplet of states, ($1a^+, 2a^+, 3a^+$), positioned at an excitation energy of 3-3.5 MeV, in rotational nuclei, and at about 2.5 MeV in vibrational nuclei. For these states the model predicts a large isovector component, being proton and neutron transition densities of opposite sign and of the same order of magnitude. Therefore these states should correspond to the isovector surface oscillations predicted by geometrical models.[4] The state $1a^+$ has been recently discovered in rare-earths nuclei, by electron scattering.[5] The $2a^+$ levels should be easily excited in vibrational nuclei. In fact the predicted B(E2, $0s^+$ $2a^+$) values depend strongly on the effective proton and neutron boson charges, e_p and e_n, being proportional to $(e_p-e_n)^2$. This latter factor

should reach a large value in the case of Pd, Cd, Nd and Sm isotopes. By contrast for these nuclei the transition to the $1a^+$ state has a vanishing strength.

Methods to determine effective charges separately has been outlined by several authors; their values have been quoted for Pd, Nd and Sm isotopes in Ref.s 6, 7 and 8 respectively. Effective charges can be derived from the comparison between electromagnetic and proton scattering data. The inelastic cross sections for the excitation of these 2^+ states should depend strongly on the projectile. For instance a transition with a dominant isovector component should not be excited by an isoscalar projectile as a deuteron. An opposite situation (vanishing proton cross sections) is found when the proton-boson transition density is larger, in absolute value, by about the 30% in comparison to the neutron-boson transition density.

For the 2^+ mixed-symmetry states, the IBA-2 calculations predict a strength fragmented, as in the case of ^{110}Pd and ^{112}Cd, or concentrated in only one level as in the case of ^{104}Pd, ^{146}Nd and ^{150}Sm. The position is around 2 MeV for Pd and Cd isotopes, 1.2-1.5 MeV for Nd and Sm. In proton inelastic scattering we have found transitions to 2^+ levels in these energy regions and with and angular distribution consistent with a large isovector component at:

1.760, 2.207, 2.350, 2.732 MeV in ^{98}Mo;

1.786, 2.341, 2.541, 2.707 and 2.940 MeV in ^{104}Pd;

2.160, 2.494, 2.534, 2.954, 2.996 MeV in ^{110}Pd;

1.470, 2.155, 2.733, and 2.855 Mev in ^{112}Cd;

1.470 MeV in ^{146}Nd; 1.062 MeV in ^{150}Nd and 1.193 MeV in ^{150}Sm.

The deuteron data have been collected in a recent experiment and have been reduced only in part to cross sections. From a first inspection of the results it seems that several transitions listed above, are excited by deuterons exactly as by protons. This result, if confirmed also for the other transitions, should exclude large isovector components and therefore large strengths for asymmetryc states. It must be observed that in the present study the experimental conditions (energy resolution, background and continuum) should allow to detect most of the 2^+ levels with a strength larger than 0.1 s.p.u.

5. CONCLUSIONS

Evidence has been obtained for a systematic presence of strong E4--vibrations, positioned at 2-2.5 MeV excitation energy. In some nuclei here considered, negative-parity bands due to quadrupole-octupole couplings are clearly present. These bands and the E3 strengths distributions are sucessfully described considering f-boson contributions. Some quadrupole transitions, with a strength lower that 1 s.p.u. diplay a characteristic angular distribution that require a collective form factor consistent with an isovector transition. However an interpretation of these transitions in terms mixed-symmetry or isovector components is not confirmed by the comparison between proton and deuteron data.

I would like to acknowledge my colleagues who collaborated on the present study: N. Blasi, W. Borghols, S. Brandenburg, R. De Leo, M.Harakeh, S. Micheletti, M. Schippers and S.Y. van der Werf.

REFERENCES

1) Iachello, F., Phys. Rev. Lett., 53, 1427 (1984).
2) De Leo et al., Phys. Lett., 162B, 1 (1985).
3) Scholten, O., private comunication.
4) Faessler, A., Nucl. Phys., 57B, 653 (1966).
5) Bohle, D. et al., Phys. Rev. Lett., 137B, 27 (1985).
6) Saha, A. et al., Phys. Lett., 132B, (1983).
7) Hamilton, W.D. et al., J.P., Phys. Rev. Lett., 53, 2469 (1984).
8) Otsuka, T. and Ginocchio, J.N., Phys. Rev. Lett., 54, 777 (1985).

SYSTEMATICS OF MIXED-SYMMETRY STATES

H. Harter

(Institut fur Kernphysik Universität Köln)
Zülpicher Str. 77 , 5000 Köln 41
Bundesrepublik DEUTSCHLAND

ABSTRACT

The F-spin purity in ^{128}Xe has been studied. The M1 transition strengths extracted from branching ratios have been used. The purity has been found to be about 98%. However, there exists an additional O(5) selection rule for M1 transitions.

1. INTRODUCTION

In the recent years it has been shown that the low lying levels in collective nuclei can be described by the interacting boson model (IBA-2) as a system of proton and neutron bosons. In the same way as has been done for fermions the isospin of the bosons can be introduced in such a system. This spin is called F-spin to distinguish it from the isospin of the fermions[1]. The idea of the F-spin has recently gained importance owing to the result of a recent experiment by Bohle et al.[2] who observed a number of excited 1$^+$ states in the rare earth nuclei. These can be interpreted as states of mixed proton neutron symmetry in which the total F-spin quantum number differs from the ground state F-spin by one unit. It has also be shown that there exists sets of nuclei in the rare earth which can be interpreted as so called F-spin multiplets. Those multiplets should exist in analogy to isospin multiplets, if the F-spin is an approximately good quantum number and the parameters of the Hamiltonian are constant[3]. As the F-spin is not strongly broken a global fit is suggested, which has been carried out recently[4]. It would therefore be highly interesting to have a reliable estimate of how good the F-spin quantum number is in nuclei, and in particular for the low lying states.

2. M1 TRANSITIONS IN ^{128}XE AND THE PURITY OF F-SPIN[5])

As has been mentioned up to now the F-spin was used as a proper classification scheme of the low lying excitation energies. In the following discussion a test of the purity of the F-spin will be worked out. For this purpose it is necessary to find observables which can give an answer to this question. The magnetic dipole moment is such an operator. In the IBA-2 model the simplest form of the M1 transition operator is the one body operator

$$T(M1) = 3/(8\pi) (g_p-g_n)(L_p - L_n) \tag{1}$$

in nuclear magneton units and where g_p and g_n are the g-factors for the proton and neutron bosons.

This operator has the selection rule $F=0,+1$, because it is an F-vector. In addition transitions between states with maximum F-spin are forbidden[6].

This property allows a test of the F-spin purity when we know the value of g_p-g_n. But the g-factors of the bosons are not well known at the moment. The experimental evidence results from the M1 transition to the mixed symmetry 1^+ state, usually called the isovector state[2], which leads to $0.6 \leq g_p-g_n \leq 1.0$[2]. Similar values are found from systematic fits to the g-factors of the 2_g^+ states which give $g_p=(0.63 \pm 0.04)$ and $g_n=(0.05 \pm 0.05)$[7]. If the magnetic properties of the proton boson and the neutron boson arise from orbital contributions only, a value of $g_p-g_n=1.0$ is obtained. In the following we will use the value $g_p-g_n=1.0$.

So we see that M1 transitions indicate the breaking of the F-spin.

In order to study this question and to determine the F-spin mixing we chose ^{128}Xe as a test case, because there exist particularly detailed data for this nucleus from recent experiments in Köln and Jyvaskyla[8]; moreover, this nucleus belongs to a large class of nuclei in the Xe-Ba region, which have been shown to belong to the O(6) limit of the IBA-1[8]. Thus we have a rather good understanding of the energies and of the E2 transitions of the collective levels of ^{128}Xe and it seems reasonable to start with an IBA-2 Hamiltonian, which projects onto an O(6) IBA-1 Hamiltonian and breaks the F-spin symmetry to the extent required by the M1 transition strengths. The following operator is a Hamiltonian which fulfils this condition

$$H = \varepsilon (n_{dp} + n_{dn}) + k (Q_p \cdot Q_n) + k' (Q_p+Q_n)^2 + \lambda M \tag{2}$$

where $n_{dr} = (d^+_r \cdot \tilde{d}_r)$, $r = p$ or n

$Q_r = (d^+_r s_r + \tilde{d}_r s^+_r)^{(2)} + \chi_r [d^+_r \times \tilde{d}_r]^{(2)}$

and where $M = (F_{max}(F_{max}+1) - F^2)$

This operator projects onto O(6) IBA-1 for any value of the ratio k/k'. Now we will discuss the two limiting cases : the more realistic $(Q_p \cdot Q_n)$ interaction with $k'=0$ and the symmetric quadrupole interaction $(Q_p+Q_n)^2$ with $k=0$, which is an IBA-1 type operator because it is an F-scalar. The O(6) limit requires $\chi_p=0$ and $\chi_n=0$.

The fits to the energies with the two Hamiltonians are shown in fig.1, whereby the Hamiltonian with the $(Q_p \cdot Q_n)$ interaction is denoted by the label A and the one with $(Q_p+Q_n)^2$ by label B. The states having a large amplitude at low F-spin components are denoted by a black dot.

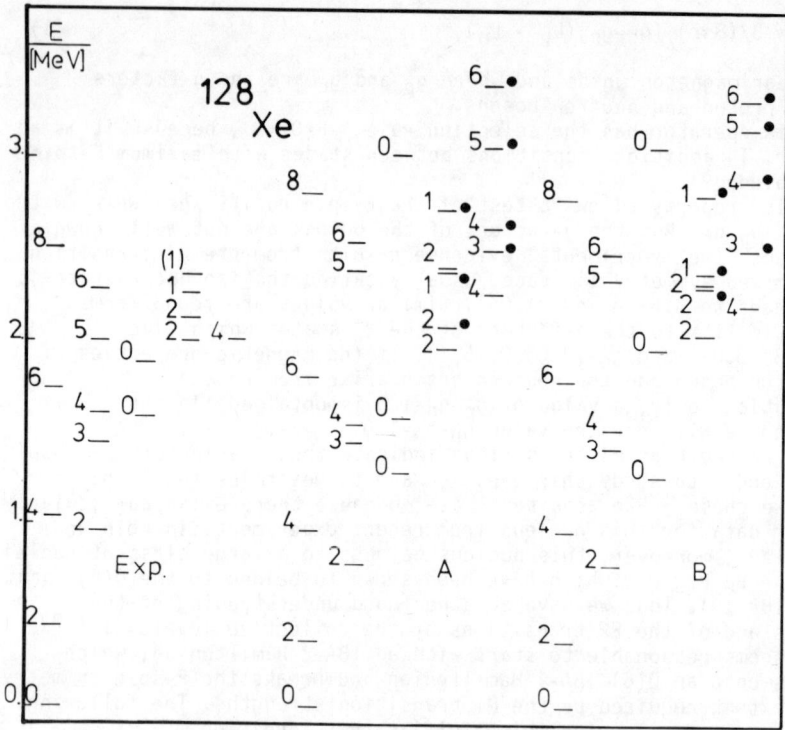

Fig.1. Comparison of the excitation energies of two O(6) type calculations in IBA-2 with the experiment

In order to induce M1 transitions we have chosen to vary the parameter $\chi'=1/2(\chi_p - \chi_n)$ while keeping the parameter $\bar{\chi}=1/2(\chi_p + \chi_n)=0$. Thus an F-vector term is added to the Hamiltonian. The dependence of the M1 matrix elements of two transitions for both types of the Hamiltonian on χ' are shown in fig.2.

As the proper value for the χ' we found 0.220 for the $(Q_p \cdot Q_n)$ and 0.155 for the $(Q_p+Q_n)^2$ interaction.

Actually these values were obtained by fitting the intensity branching ratios. Table 1 show the dependence of our result on the parameter for the Hamiltonian with $(Q_p \cdot Q_n)$.

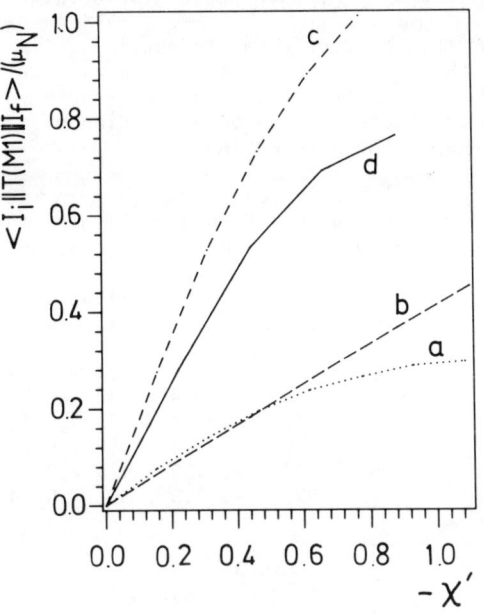

Fig.2
M1 matrixelements
vs. $\chi'=1/2(\chi_p - \chi_n)$
label a and b is the 2_γ to 2_g
label c and d is the 5_γ to 6_g
label a and c for $(Q_p+Q_n)^2$ interaction and
label b and d for $(Q_p \cdot Q_n)$ interaction

Table 1: Relative gamma ray intensity values for ^{128}Xe
Calculation of the theoretical values for $(Q_p \cdot Q_n)$

			Intensities			
I_i	I_f	Exp.	$\chi'= 0.0$	$\chi'=-.22$	$\chi'=-.30$	$\chi'=-1.0$
3_1	4_1	17.8(3)	19.0	21.9	23.9	28.6
3_1	2_2	100	100	100	100	100
4_2	4_1	78.6(15)	53.5	64.1	73.0	104.2
4_2	2_2	100	100	100	100	100
5_1	4_2	13.8(11)	7.4	15.4	21.0	42.8
5_1	6_1	4.1(4)	0.9	4.2	6.0	14.2
5_1	3_1	100	100	100	100	100

Looking at the table one should note that some of the branching ratios depend very strongly on the parameter χ', like e.g. the branching ratio from the 5_1^+ state, whereas other branching ratios such as the branching ratio from the 3_1^+ state depends less strongly on . Therefore as a guide we took the intensities of the 5_1^+ state. This was also motivated by the sensitivity of the M1 matrix elements of the 5_γ to 6_g transition as can be observed in fig.2.

If we consider a deviation between experiment and theory of less than 25% satisfactory, then the values of χ' chosen above give a fair overall agreement.

The reason why we undertook this procedure was to obtain information about the purity of the F-spin. For this purpose a decomposition of the wavefunctions in eigenfunctions of the F-spin was needed. Tables 2 and 3 show the amplitudes of the decomposition of a few wavefunctions of the two types of the interaction.

Table 2
Calculated F-spin amplitudes for the $(Q_p \cdot Q_n)$ interaction ($\chi'=0.220$)

I	E_{exc} (MeV)	Amplitudes		
		F_m	F_m-1	F_m-2
2_1	0.443	-.9827	.1313	-.1306
4_1	1.033	-.9819	.1461	-.1209
2_2	0.969	-.9806	.1547	-.1207
3_1	1.430	-.9786	.1776	-.1049
4_2	1.603	-.9794	.1727	-.1043
5_1	1.997	-.9746	.2090	-.0808

Table 3
Calculated F-spin amplitudes for the $(Q_p+Q_n)^2$ interaction ($\chi'=0.155$)

I	E_{exc} (MeV)	Amplitudes		
		F_m	F_m-1	F_m-2
2_1	0.443	-.9995	.0321	-.0014
4_1	1.033	-.9994	.0343	-.0014
2_2	0.969	-.9982	.0593	-.0015
3_1	1.430	-.9944	.1059	-.0021
4_2	1.603	-.9986	.0524	-.0021
5_1	1.997	-.9915	.1301	-.0034

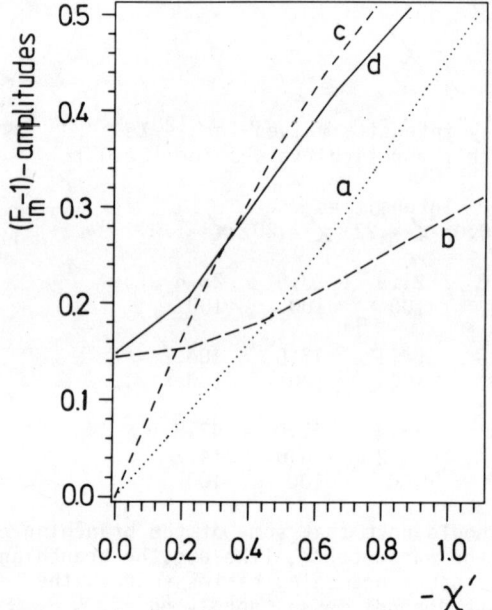

Fig.3
$(F_{max}-1)$ amplitudes vs. $\chi'=1/2(\chi_p-\chi_n)$ for the
a and b 2^+ state
c and d 5^+ state
a and c $(Q_p+Q_n)^2$ interaction
b and d $(Q_p \cdot Q_n)$ interaction

One notes that the F-spin breaking is quite small for the $(Q_p+Q_n)^2$ type Hamiltonian, whereas the situation is more complex for the usual $(Q_p \cdot Q_n)$ IBM-2 Hamiltonian. Due to the F-spin selection rule, in the case of the $(Q_p+Q_n)^2$ type Hamiltonian the M1 transition between the gamma and the ground state band can occur only in the presence of F-spin mixing. To this extend the M1 transitions indicate F-spin mixing. However, in the case of $(Q_p.Q_n)$ the Hamiltonian is a combination of an F-spin scalar and an F-spin tensor. Thus the states considered have an F-spin mixing even at $\chi=0$ although the M1 transition strengths vanish as one should note from fig.3, which shows the dependence of the 5^+_1 and the 2^+_1 of the gamma band on χ'. This unexpected behaviour is due to the symmetry properties of the operator $(Q_p.Q_n)$. This operator still has an O(5) symmetry, which forbids M1 transitions between the gamma and the ground state band[9]. However, as transitions between states with maximum F-spin are rigorously forbidden the F-spin mixing is a necessary condition for the appearance of M1 transition.

3. CONCLUSION

To sum up it has been shown that the F-spin is a very useful quantum number to classify the low lying collecitve levels. It has also allowed to develop an unambiguous way of projecting the IBA-2 Hamiltonian onto the IBA-1 one. Moreover the study of the M1 transitions included in branching ratios is a very sensitive way of determining the Fspin structure of the wavefunction of the IBA-2 Hamiltonian. However, caution is needed in view of special selection rules for M1 transitions, which are independent of the F-spin.

4. REFERENCES

1. Arima, A., Otsuka, T., Iachello, F. and Talmi, T.,
   Phys. Lett. 66B, 205 (1977).
2. Bohle, D., Richter, A., Steffens, W., Dieperink,A.E.L., LoIudice, N., Palumbo, F. and Scholten, O., Phys. Lett. 137B, 27 (1984).
3. Brentano, P. von, Gelberg, A., Harter, H. and Sala, P.,
   Nucl. Phys. 11, 85 (1985).
4. Sala, P., Brentano, P. von and Harter, H.,
   Z.Phys. A 1986 (accepted for publication).
5. Harter, H., Brentano, P. von, Gelberg, A.,
   Phys. Rev.C(1986)(accepted for publication).
6. Isacker, P.v., Heyde, H., Jolie, J. and Severin, A.
   "The F-spin Symmetric Limits of the Neutron-Proton-Interacting Boson Model", unpublished.
7. Wolf, A., Warner, D.D., Benczer/Koller, N.,
   Phys. Lett. 158B, 7 (1985).
8. Casten, R.F. and Brentano, P. von, Phys. Lett. 152B, 22 (1985).
9. Isacker, P.v. , private communication.

PART II

CURRENT AREAS OF STUDY

SECTION IV
SCATTERING REACTION STUDIES

SECTION V
SYMMETRIES IN REACTIONS

SECTION VI
EXPERIMENTAL TECHNIQUES AND STUDIES

SECTION VII
SYSTEMATIC PROPERTIES

SECTION IV

SCATTERING REACTION STUDIES

Session Moderators

N. Hintz
D. Rowe

EXPERIMENTAL STUDIES OF THE PHOTON DECAYS OF GIANT RESONANCES

Alan M. Nathan

Nuclear Physics Laboratory/Department of Physics

23 Stadium Drive

Champaign, IL 61820

USA

ABSTRACT

A review is given of the existing inelastic photon scattering data. These data are discussed in terms of models for the coupling of the giant dipole resonance (GDR) to surface vibrations. The sensitivity of these experiments to the underlying nuclear structure is emphasized.

1. INTRODUCTION

It has long been thought that one of the principal mechanisms for the spreading of collective strength in the giant multipole resonances is through the coupling to low-lying surface vibrations.[1] A major signature for this coupling is in the photon decay modes of the giant resonances to low-lying vibrational states. In fact, these photon decay modes are sensitive indicators of the underlying structure of the low-lying states themselves. The ideal reaction for studying these decay modes is elastic and inelastic photon scattering. Although this was realized many years ago,[2] it has only been in the last few years that one has been able to systematically study this reaction.[3] The purposes of the present paper are to summarize the current experimental situation, to critically assess the degree to which various nuclear structure models can account for the data, and to offer suggestions for further experimental and theoretical work.

We begin with a presentation of the theoretical framework for the interpretation of the data.

2. THEORETICAL FRAMEWORK

The coupling between the GDR and surface degrees of freedom is most easily seen in the context of the hydrodynamic model and its extension, the Dynamic Collective Model (DCM).[4] In these models the coupling arises through the surface boundary condition which requires $E_{GDR} \propto 1/R$, where R is the nuclear radius. This accounts for both the A-dependence of E_{GDR} and for the deformation splitting. If the surface vibrates, then the GDR oscillation is modulated by the slower surface vibrations, resulting in an almost classical problem in the coupling between normal modes. The consequences of this coupling are indicated schematically in the highly simplified 2-state model of Figure 1. The 0-phonon ground state and 1-phonon 1st excited state each have a giant resonance built upon them ($|0\rangle$ and $|1\rangle$, respectively). In the absence of coupling, the states $|0\rangle$ and $|1\rangle$ are connected via the electric dipole operator only to the ground and 1st excited state, respectively. The effect of the coupling, which to lowest order is linear in the surface vibrational amplitude, is to mix the 0-phonon and 1-phonon giant resonances:

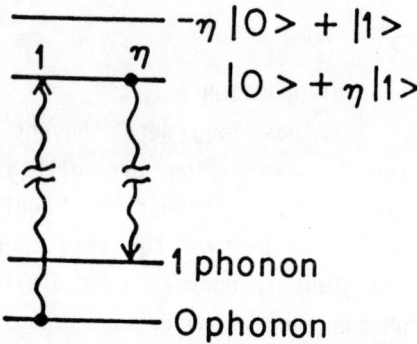

Fig. 1. Schematic 2-state model for the coupling among normal modes.

$$|0'\rangle = |0\rangle + \eta|1\rangle \qquad (1)$$
$$|1'\rangle = -\eta|0\rangle + |1\rangle$$

The interesting physics is contained in the mixing parameter η which affects the properties of the GDR in two ways. First, the ground state dipole strength (or equivalently the elastic scattering amplitude) is split into vibrational satellite peaks, thereby contributing to a general broadening of and substructure within the GDR. Second, the various components of the GDR are connected via the dipole operator to several low-lying states. These photon decay branches, and therefore the mixing parameter η, are directly probed in the inelastic photon scattering reaction. This reaction should be viewed as a resonant 2-step process of photon absorption to the GDR intermediate state followed by photon emission to a low-lying state. In the example of Fig. 1, the state $|0'\rangle$ is populated via the 0-phonon piece of the wave function and decays to the 1st excited state via the 1-phonon piece. Thus the photon scattering reaction is a direct probe of the vibrational components of the giant dipole state.

More generally, the approach of both the DCM and the Interacting Boson Model (IBM) is to write the collective nuclear Hamiltonian as follows:[5,6]

$$H = H_Q + H_D + H_{DQ} \quad \text{(DCM)} \qquad (2)$$
$$= H_{sd} + H_p + H_{psd} \quad \text{(IBM)}$$

where H_D is the part describing the giant dipole modes (the p-boson part), H_Q is the part describing the surface modes (the sd-boson part), and H_{DQ} (H_{psd}) is the coupling interaction between the modes. In the preceeding example, the linear dependence of H_{DQ} on the surface vibrational amplitude leads to[1]

$$\eta = k\beta_0 E_1/E_2, \qquad (3)$$

where β_0 is the zero-point surface vibrational amplitude, E_2 is the energy of the surface vibrations, E_1 is the GDR energy, and k is a model-dependent dimensionless constant of order unity. For the general situation, the mixing is more complicated but we still expect that η will control the degree to which the dipole and surface modes get coupled. Thus we expect that in "soft" vibrational nuclei (β_0 large, E_2 small) the coupling is stronger than in "stiff" vibrational nuclei (β_0 small, E_2 large). In general, a softer nuclear surface results in a greater fractionation of the ground state dipole strength into vibrational satellites and an increased amount of inelastic scattering into the low-lying vibrational levels.

The basic theoretical technique is to diagonalize the Hamiltonian in some convenient basis in order to calculate the properties of the low-lying levels (energies, B(E2)'s, etc.), the energies E_n of the various components of the GDR, and the dipole matrix elements D_{nf} connecting each dipole state n with each low-lying state f. The structure of the photon scattering amplitude connecting initial state i to final state f is given by[2]

$$A_{i \to f} \propto \sum_n D_{in} D_{nf} \left[\frac{1}{E_n - E - i\Gamma_n/2} + \frac{1}{E_n + E' + i\Gamma_n/2} \right], \qquad (4)$$

where $E(E')$ is the energy of the incident (outgoing) photon ($E-E' = E_f - E_i$), and the Γ_n are damping widths. Unfortunately neither model actually calculates the Γ_n. Furthermore, since it is usually the case that the Γ_n are larger than the separation between components of the GDR ($\Gamma_n \sim 3-4$ MeV), much of the interesting structure in the energy dependence of the scattering cross sections is washed out. Fortunately, these widths are not strongly energy dependent nor do they vary much with nucleus, so that there is at least some phenomenological guidance for their selection.

Some comments are in order. First, the parameter η is often large, so that the simple 2-level model described above is not valid. The solution requires a diagonalization of a large matrix involving many vibrational phonons and often resulting in a large fractionation of the ground-state dipole strength and inelastic scattering to many-phonon states. Second, H_Q or H_{sd} can be much more complicated than the harmonic vibrator of our example. This often involves the full Bohr-Mottelson (or full IBM) Hamiltonian, complicated potential energy surfaces, anharmonic mass parameters,[5,6] etc. In fact it is the hope that the complexity of H_Q will be reflected and perhaps even elucidated by the properties of the GDR. Third, the algebraic (or group theoretic) structure of the coupling Hamiltonian is fixed by the symmetries of the problem[5,6] and in particular is independent of the degree of complexity of the low-lying levels. Furthermore, the various coupling coefficients should vary only smoothly (if at all) within a major shell. As yet there is no microscopic theory that will predict these coefficients, although the practitioners of the DCM usually use those coefficients given by the hydrodynamic model.[5] However, it is not clear whether this model is generally applicable. Finally, it is worth emphasizing once again that the damping widths are not specified by the models.

3. EXPERIMENTAL RESULTS
3.1 Spherical Nuclei

The Mo isotopes are essentially spherical vibrational nuclei whose surface progressively gets softer as one proceeds from ^{92}Mo (E_2 = 1.5 MeV, β_0 = 0.10) to ^{96}Mo (E_2 = 0.8 MeV, β_0 = 0.18) to ^{100}Mo (E_2 = 0.5 MeV, β_0 = 0.25). This is reflected in the calculated energy distribution of the ground state dipole strength and in the predicted inelastic scattering, as shown in Fig. 2. For ^{92,96}Mo, the DCM calculation is in remarkable qualitative and quantitative agreement with the data,[3] especially given that the only adjustable parameters are the damping widths and overall normalization of the cross sections. Thus the relationship between the inelastic scattering and the parameter η is well established. A similar study of ^{100}Mo is planned. In that case the coupling is so large that the scattering into the 2-phonon states should be observable. This scattering should be sensitive to anharmonicities in the surface vibrations.

It is apparent from Fig. 2 that there is an excess of elastic scattering above 20 MeV in ^{92}Mo. We believe this is due to the $T_>$ component of the GDR which is split from the more prominent $T_<$ component by the nuclear symmetry energy. Since the isospin degree of freedom is not an inherent ingredient in the DCM, the calculations do not reproduce this feature. For light nuclei, the expected strength of the $T_>$ component should be even larger, as shown in Fig. 3 for ^{60}Ni. The splitting of the scattering cross sections into two major peaks, one at 17 MeV and the other at 20 MeV, is a consequence of isospin splitting and not of the coupling to surface vibrations. The curves show the result of a modified DCM calculation, in which the isospin splitting is inserted in an <u>ad hoc</u> manner.[3] Although the qualitative agreement with the data is reasonable, the inelastic scattering suggests that the coupling to the surface should be

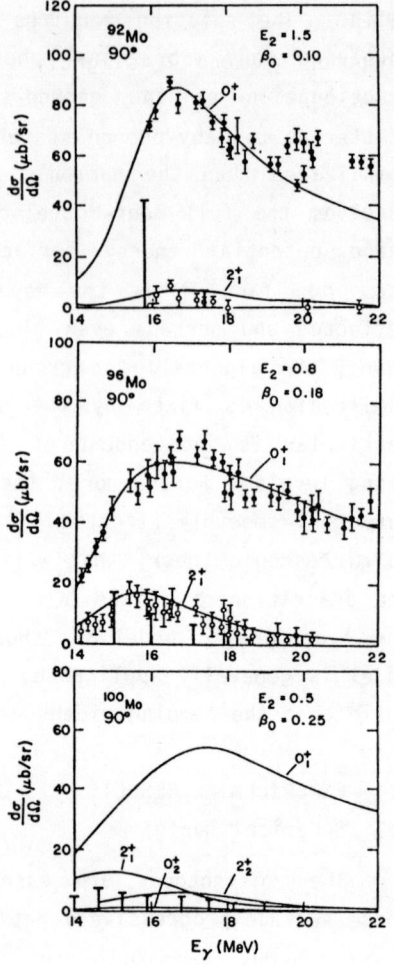

Fig. 2. Photon scattering cross sections and ground-state dipole strength for the Mo isotopes and the DCM predictions.

slightly stronger for the $T_>$ component and slightly weaker for the $T_<$ component. Similar problems are found for other nuclei in this mass region. The absence of a proper accounting for isospin effects remains a major deficiency of the DCM; calculations for this mass region in the framework of IBM-2 would be highly desirable.

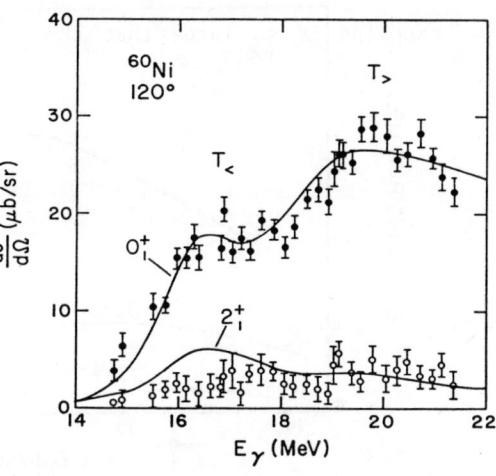

Fig. 3. Isospin splitting in the GDR of ^{60}Ni.

3.2 Deformed Nuclei

In axially-symmetric deformed nuclei, the dominant effect of the coupling to the surface is the splitting of the GDR into a K=0 mode and a two-fold degenerate K=1 mode, corresponding to dipole vibrations along the symmetry axis and transverse axes, respectively. The principal dynamic effect is the coupling to γ-vibrations. Effectively, the nucleus is dynamically triaxial, resulting in further splitting of the K=1 modes and consequentially a substantial photon decay branch to the K=2, 2^+ γ-vibrational band head. The scattering cross section into this level should roughly scale with the mean-square zero-point γ-vibrational amplitude $\beta_0^2 \langle \gamma^2 \rangle$ or, equivalently, $B(E2; 0_g^+ \to 2_\gamma^+)$, and it would be extremely interesting to test this scaling behavior among neighboring nuclei. Thus far, only ^{166}Er has been studied in any detail, and the results are shown in Fig. 4.[7] The only DCM calculations that have been done for ^{166}Er treat the γ-vibrations in the harmonic approximation, for which two parameters are needed: a mass parameter and an energy. The energy is taken from experiment. For the curve DCM-1,[2] the mass parameter is taken

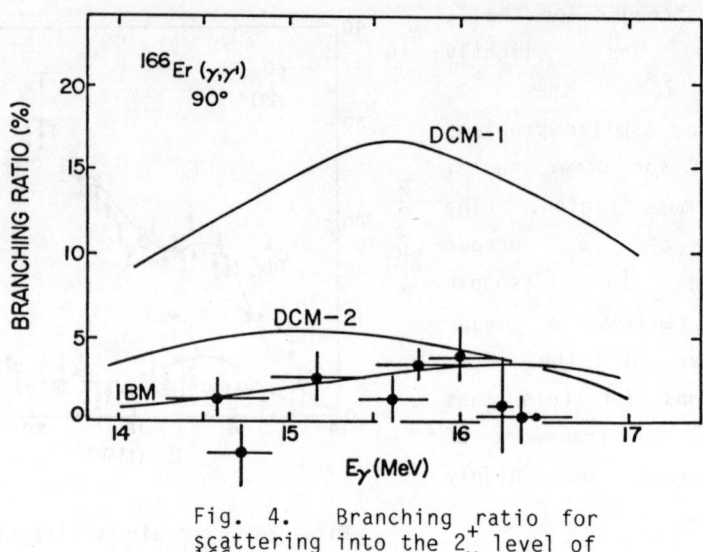

Fig. 4. Branching ratio for scattering into the 2_γ^+ level of ^{166}Er.

from the rotation-vibration model, in which the mass is uniquely specified by the moment of inertia of the ground state band.[8] For the curve DCM-2, the mass parameter is chosen to reproduce B(E2; $0_g^+ \rightarrow 2_\gamma^+$), while the static deformation is reduced from 0.31, which fits the ground state band, to 0.26, which is needed to fit the total photoabsorption cross section.[9] The two calculations differ by a factor of 4 in $\beta_0^2 \langle \gamma^2 \rangle$! Clearly DCM-2 better describes the data. The IBM curve,[10] which was calculated with slightly broken SU(3) symmetry, also describes the data reasonably well.

This example provides an interesting contrast between the two models. For the DCM, the simplest calculation (DCM-1) grossly overpredicts the coupling to the γ-band and it is necessary to make major changes to the simple picture in order to reduce that coupling. For the IBM, the simplest calculation (pure SU(3) symmetry) already does quite well since there is no coupling to γ-vibrations in that symmetry. Only a small deviation from SU(3) symmetry is needed to reproduce the data. It should be emphasized, however, that it is not a question of which model is more correct, but really a question of which model describes the data more naturally. The preceeding example suggests that the coupling to γ-vibrations appears in a more natural way in the IBM than in the DCM.

3.3 Transitional Nuclei

The preceeding examples are those of extreme situations, i.e., spherical vibrators or well-deformed rotors. The real power of the models, however, lies in their ability to describe more general types of collective behavior. Prime examples are the Nd-Sm isotopic chains, which exhibit a transition from spherical to deformed, and the Os-Pt chains, which exhibit a transition from deformed to γ-unstable. In each case, the transition occuring in the low-lying levels is reflected in the structure of the GDR[11,12] and particularly in the photon decay branches. This is illustrated in Fig. 5, which shows some very old DCM calculations[13] of the scattering cross sections for the Nd isotopes. We see a gradual change in structure from a stiff vibrator (^{142}Nd) to a soft anharmonic vibrator (^{146}Nd) to a deformed rotor (^{150}Nd). We are in the process of studying these isotopes at Illinois. Thus far we have completed only ^{146}Nd, which is probably best described as a soft

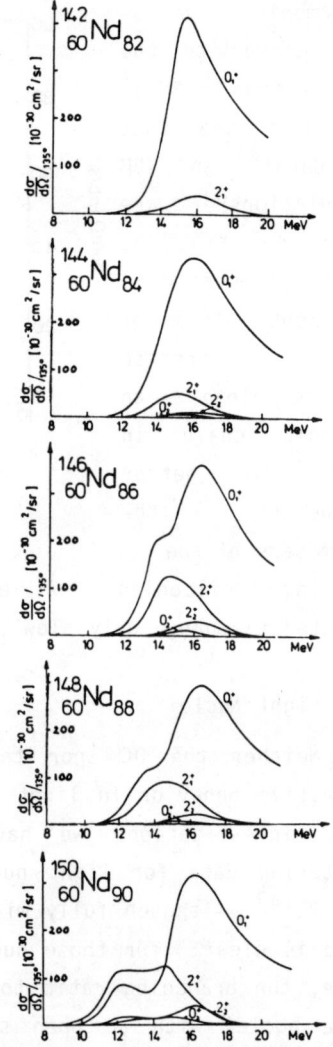

Fig. 5. DCM predictions for photon scattering on the Nd isotopes.

anharmonic vibrator, closer to the U(5) than SU(3) limit of the IBM. The data[14] and IBM calculations[6] are shown in Fig. 6. The agreement is excellent. It is of crucial importance to complete an isotopic chain in order to better define the parameters of the coupling Hamiltonian.

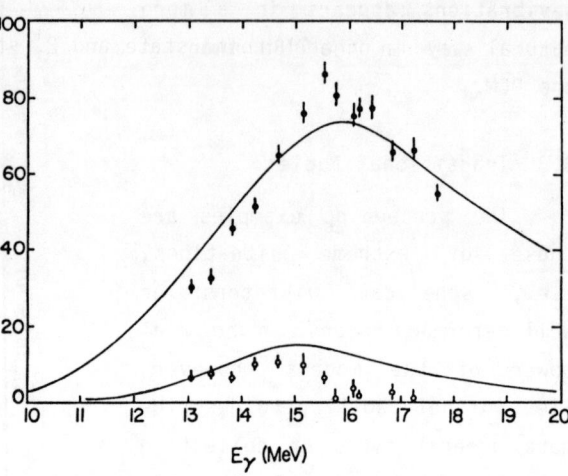

Fig. 6. Scattering cross sections for ^{146}Nd and the IBM predictions.

As emphasized above, these parameters are expected to change only slowly within a major subshell.

3.4 Light Nuclei

Neither the DCM nor the IBM is able to adequately describe collective behavior in light nuclei. Nevertheless, in the course of other investigations we have accumulated considerable inelastic scattering data for light nuclei, and these data are summarized in Fig. 7.[14] Although fully microscopic calculations do not exist, the trend is clear: for those nuclei with closed subshells in the ground state, the branching ratio to the 2_1^+ state is small ($\lesssim 0.3$) while for those nuclei with an open subshell, the branching ratio is large (~1.0). This can be qualitatively understood as follows.[15] For nuclei with a closed subshell, the 2_1^+ level is a 1p-1h excitation so that the GDR built on that level is predominantly a 2p-2h excitation relative to the closed subshell. Thus we expect little overlap with the ground state GDR. For nuclei with an open subshell (e.g. ^{24}Mg), the 2_1^+ state is mainly a rearrangement of nucleons in the open subshell. Thus the possibility exists for a large overlap between the GDR's built on the ground state and 2_1^+ state. A recent shell

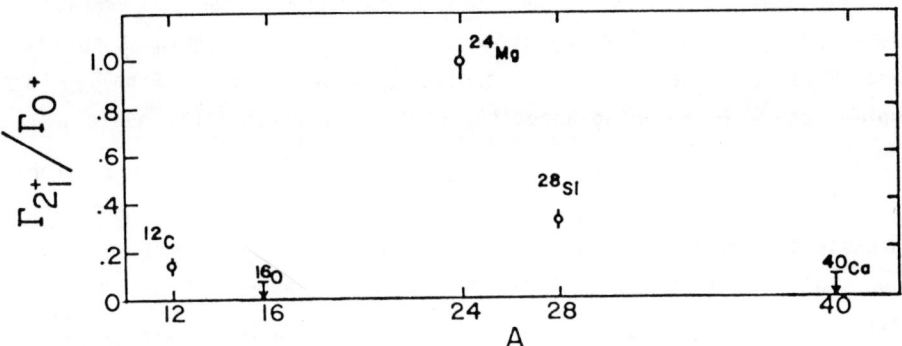

Fig. 7. Measured branching ratios for the photon decay at the peak of the GDR to the first excited state relative to the photon decay to the ground state.

model (TDA) calculation[15] for ^{20}Ne, which should resemble ^{24}Mg, confirms this simple idea. A similar calculation[15] for ^{12}C reproduces the experimental branching ratio to within 10%. Thus we can claim some qualitative and in some cases quantitative understanding of the photon decays in the light nuclei.

4. SUMMARY AND CONCLUSIONS

A review has been given of the experimental photon scattering data as they pertain to the coupling between the GDR and low-lying surface degrees of freedom. Although the need exists for more complete systematic studies in selected regions of the Periodic Table, an interesting picture is already emerging. The data on the Mo isotopes gives credence to the simple scaling arguments based on the parameter η. Data on the f-p shell nuclei, although qualitatively described by a modified DCM calculation, point out the need for a more systematic treatment of isospin effects, perhaps within the context of IBM-2. The ^{166}Er example demonstrates the sensitivity of the inelastic scattering to details about the γ-vibrations and illustrates quite clearly the differences between the IBM and DCM approach. Our best opportunity for a complete understanding of the coupling process lies in the transitional nuclei which have only begun to be studied experimentally. Finally, a simple picture is emerging about the photon decays in light nuclei.

It is a pleasure to acknowledge my many collaborators from Illinois, Argonne, and Beer Sheva, as well as illuminating discussions with Prof. F. Iachello and Dr. Alex Brown. This work was supported by the National Science Foundation under grant PHY 83-11717.

1. Bohr, A. and Mottelson, B., Nuclear Structure (Benjamin, Reading, Mass., 1975), Vol. II, p. 503 ff.
2. Arenhovel, H., et al., Phys. Rev. 157, 1109 (1967).
3. Bowles, T. J., et al., Phys. Rev. C 24, 1940 (1981).
4. Danos, M. and Greiner, W., Phys. Rev. 134, B284 (1964).
5. Rezwani, V. et al., Nucl. Phys. A180, 254 (1972).
6. Maino, G. et al., Phys. Lett. 152B, 17 (1985).
7. Nathan, A. M. and Moreh, R. Phys. Lett. 91B, 38 (1980).
8. Greiner, W. and Eisenberg, J. M. Nuclear Theory (North Holland, Amsterdam, 1975) Vol. I, pp. 127ff.
9. Bergere R. et al., Nucl. Phys. A133, 417 (1969).
10. Schotz, F. G. and Hahne, F. J. W., Phys. Lett. 123B, 147 (1983).
11. Maino, G. et al., Phys. Rev. C 33, 1089 (1986).
12. Sedlmayr, R. et al., Nucl. Phys. A232, 465 (1974).
13. Arenhovel, H., "Photon Scattering by Nuclei", Proceedings of 1973 Asilomar Conference, p. 449.
14. University of Illinois Nuclear Physics Laboratory Annual Report, 1986 (unpublished).
15. Brown, B. A., private communication.

PHOTON SCATTERING IN THE REGION OF GIANT
DIPOLE RESONANCE

S. Turrini

Dipartimento di Fisica dell'Università,
via Irnerio 46, I-40126 Bologna, ITALY

G. Maino and A. Ventura

ENEA, Divisione Fisica e Calcolo Scientifico,
via Mazzini 2, I-40138 Bologna, ITALY

ABSTRACT

Photon scattering by non-magic even-even nuclei is investigated within the framework of the interacting boson model for the excitation of the giant dipole resonance. Elastic scattering includes the Delbrück contribution at the lowest perturbative order, in addition to the classical Thomson term.

1. INTRODUCTION

The interacting boson model (IBM)[1] has been extended in recent years to the treatment of collective particle-hole excitations across a shell closure, particularly the giant dipole resonance (GDR), where calculations have been carried out at the IBM limit symmetries[2-4] and in transitional regions[5-8]. The IBM parameters are so adjusted as to reproduce the total photoabsorption cross section, where the GDR is clearly observed. Inelastic photon scattering[3,7,8] can be evaluated without further adjustments.

As for the elastic scattering, two non-nuclear contributions play an important rôle in this energy region : the classical Thomson scattering by a point-like charge equal to the nuclear charge, Ze, and the radiative correction due to vacuum polarization effects, known as Delbrück scattering[9].

The purpose of the present work is to emphasize the rôle of

non-nuclear effects in elastic scattering and to comment on inelastic scattering as a test of nuclear structure models.

2. THE NUCLEAR MODEL

The excitation of a GDR and its coupling to low-energy collective modes is described in IBM language by means of a P boson, with spin-parity $J^\pi = 1^-$, coupled to the usual s and d bosons responsible for low-lying positive-parity states. The s-d-P Hamiltonian has been discussed by a number of authors[2-8] and is merely reproduced here without further comment :

$$\hat{H} = \hat{H}(s,d) + E_p \hat{n}_p + b^{(0)} \left[(P^+ \times \tilde{P})^{(0)} (d^+ \times \tilde{d})^{(0)} \right]^{(0)} +$$
$$b^{(1)} \left[(P^+ \times \tilde{P})^{(1)} (d^+ \times \tilde{d})^{(1)} \right]^{(0)} + b^{(2)} \left[(P^+ \times \tilde{P})^{(2)} \left[(d^+ \times \tilde{s} + s^+ \times \tilde{d})^{(2)} \right. \right.$$
$$\left. \left. + \chi (d^+ \times \tilde{d})^{(2)} \right] \right]^{(0)}. \tag{1}$$

Its diagonalization by means of the basis states $|\Psi\rangle = (s^+)^{n_s} (d^+)^{n_d} (P^+)^{n_P} |0\rangle$ yields the energies, E_k, of the GDR components. The E1 transition from the m-th low-lying state of spin-parity I_m^+ to the k-th 1^- GDR state is given by the reduced matrix element of the dipole operator, $\hat{T}(E1)$:

$$\langle 1_k^- \| \hat{T}(E1) \| I_m^+ \rangle = \alpha_1 \langle 1_k^- \| (P^+ + P) \| I_m^+ \rangle. \tag{2}$$

The GDR widths, $\Gamma_k = \Gamma(E_k)$, which cannot be evaluated in the frame of the model, are assumed to depend on energy according to the power law $\Gamma(E_k) = \varkappa E_k^\gamma$ [5,7,8]. The P-boson parameters in (1), the effective boson charge, α_1, in (2) and the coefficients of Γ are so adjusted as to reproduce the experimental photoabsorption cross section in the GDR region.

Inelastic scattering and nuclear contribution to the elastic scattering can thus be evaluated without adjustable parameters, thus providing a stringent test of the nuclear structure model, since the Thomson and Delbrück amplitudes have been calculated and coherently added.

3. ELASTIC SCATTERING

The scattering cross sections are computed in the long wavelength limit, $R/\lambdabar \ll 1$, where R is the nuclear radius and $\lambdabar$ the wavelength of incident radiation.
In the elastic case, the well-known formulae[10] containing

Thomson and nuclear contributions are modified so as to include the Delbrück effect. By letting λ (λ') = ±1 be the circular polarization of the incident (scattered) photons and $K_{00}^{\lambda\lambda'}$ the total amplitude for elastic photon scattering by an even-even nucleus, the differential cross section for unpolarized photons can be written as :

$$\frac{d\sigma_{el}}{d\Omega}(\omega,\theta) = \frac{1}{2}\sum_{\lambda\lambda'}\left|K_{00}^{\lambda\lambda'}\right|^2 = \frac{1}{2}\sum_{\lambda\lambda'}\left\{\left|A_{\lambda\lambda'}^{D}(\omega,\vartheta)\right|^2 + \left|A^T + A^N(\omega)\right|^2 \cdot \left[\frac{1}{4}(1+\cos^2\vartheta) + \frac{1}{2}\lambda\lambda'\cos\vartheta\right] + A^{DNT}(\lambda,\lambda',\omega,\vartheta)\right\}. \quad (3)$$

Here, $A_{\lambda\lambda'}^{D}(\omega,\vartheta)$ is the Delbrück amplitude for photons of frequency ω scattered at angle ϑ, evaluated as in ref.[11]. A^T is the usual Thomson amplitude, $A^T = -Z^2 e^2/(AMc^2)$, where AM is the nuclear mass. The nuclear, or Raman amplitude, A^N, reads :

$$A^N(\omega) = \frac{1}{3}\frac{(\hbar\omega)^2}{\hbar^2 c^2}\sum_n \left|<1_n^-\|\hat{T}(E1)\|0_1^+>\right|^2 \cdot \left[\frac{1}{E_n-\hbar\omega-i\Gamma_n/2} + \frac{1}{E_n+\hbar\omega+i\Gamma_n/2}\right]. \quad (4)$$

Finally, A^{DNT} is the interference of Delbrück, Raman and Thomson terms :

$$A^{DNT}(\lambda,\lambda',\omega,\vartheta) = 2\left[\text{Re }A_{\lambda\lambda'}^{D}(\omega,\vartheta)\left[A^T + \text{Re }A^N(\omega)\right] - \text{Im }A_{\lambda\lambda'}^{D}(\omega,\vartheta)\cdot\text{Im }A^N(\omega)\right]d_{-\lambda,-\lambda'}^{1}(0,\vartheta,0). \quad (5)$$

Here, $d_{-\lambda,-\lambda'}^{1}$ is a Wigner rotation matrix, with elements : $d_{1,1}^{1} = d_{-1,-1}^{1} = \cos^2(\vartheta/2)$; $d_{1,-1}^{1} = d_{-1,1}^{1} = \sin^2(\vartheta/2)$.

4. INELASTIC SCATTERING

Let ω' be the frequency of the emitted photon leaving the nucleus in an excited state of spin-parity I_f^π. The differential cross section for unpolarized photons is written in the form[10] :

$$\frac{d\sigma_{in}}{d\Omega}(\omega,\omega',\vartheta) = \frac{\omega'}{\omega}\left|P_J\right|^2 \delta_{JI_f} g_J(\vartheta), \quad (6)$$

where :

$$P_J = \frac{e^2}{[3(2I_f+1)]^{1/2}} \frac{\omega\omega'}{c^2} \sum_n \langle I_f^+ \| \hat{T}(E1) \| 1_n^- \rangle \langle 1_n^- \| \hat{T}(E1) \| 0_1^+ \rangle \times$$
$$\left[\frac{1}{E_n + \hbar\omega' + i\Gamma_n/2} + \frac{(-1)^J}{E_n - \hbar\omega - i\Gamma_n/2} \right] , \qquad (7)$$

is a nuclear polarizability ; the angular distribution, $g_J(\vartheta)$, is given here in the cases $J = 0$ and $J = 2$:

$$g_0(\vartheta) = \frac{1}{6}(1 + \cos^2\vartheta) \quad ; \quad g_2(\vartheta) = \frac{1}{12}(13 + \cos^2\vartheta). \qquad (8)$$

5. RESULTS AND COMMENTS

We present here a few results on photon scattering by ^{168}Er, a strongly deformed nucleus already considered in ref.[3] in the SU(3) limit of the model. The IBM parameters adopted in the present work and listed in Table I allow us to fit the experimental photoabsorption[12] cross section with good accuracy and predict elastic and inelastic scattering cross sections.

Since the elastic cross section has already been evaluated in ref.[3] at large scattering angle, where the Delbrück effect is negligible, we present here calculations at small and intermediate angles. Fig. 1 shows the elastic cross section, its three components and the absolute value of the interference term (whose sign is negative) as functions of scattering angle at fixed photon energy, $\hbar\omega = 9$ MeV, while Fig. 2 gives the same quantities versus photon energy at fixed

TABLE I
IBM parameters

$a_0^{(*)} = 30.0$ keV ;	$a_1^{(*)} = 10.5$ keV ;	$a_2^{(*)} = -8.0$ keV ;
$\chi = -\sqrt{7}/2$;	$E_p = 14.05$ MeV ;	$b^{(0)} = 0.$;
$b^{(1)} = 0.50$ MeV ;	$b^{(2)} = 0.28$ MeV ;	$\alpha_1 = 8.82$ e.fm ;
$\varkappa = 0.045$ MeV$^{1-\gamma}$;	$\gamma = 1.7$.	

(*) Parameters of the Hamiltonian $\hat{H}(s,d)$[1].

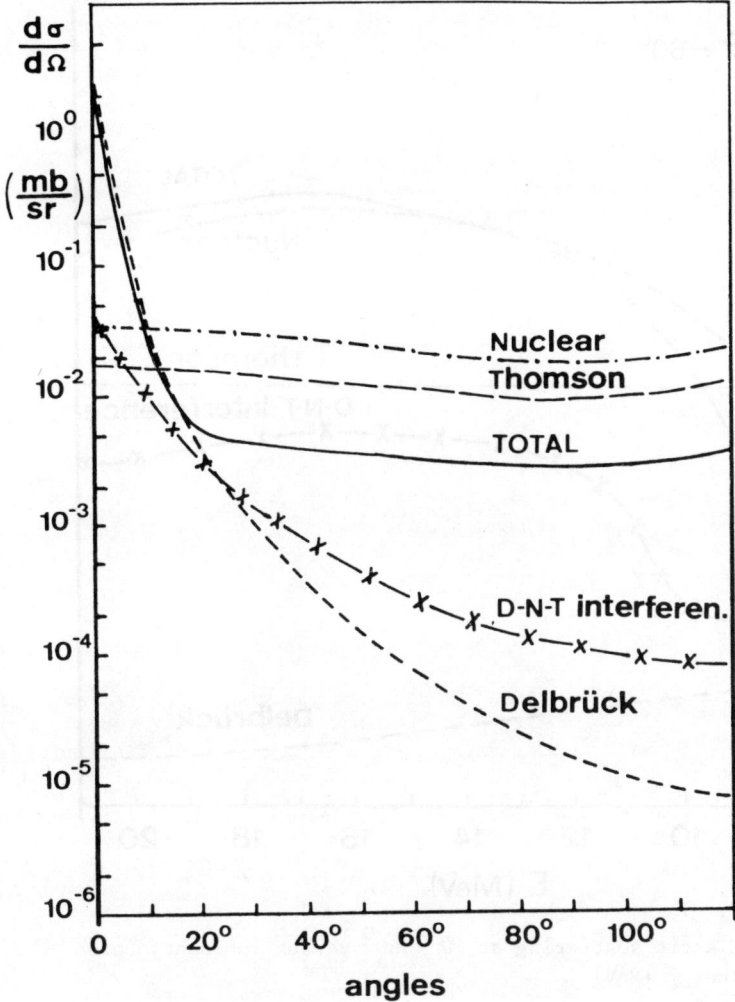

Fig. 1. Elastic scattering of 9 MeV photons by ^{168}Er versus scattering angle.

scattering angle, $\vartheta = 60^0$. In this case the main effect of the Delbrück amplitude is to reduce the total cross section, owing to the interference term.

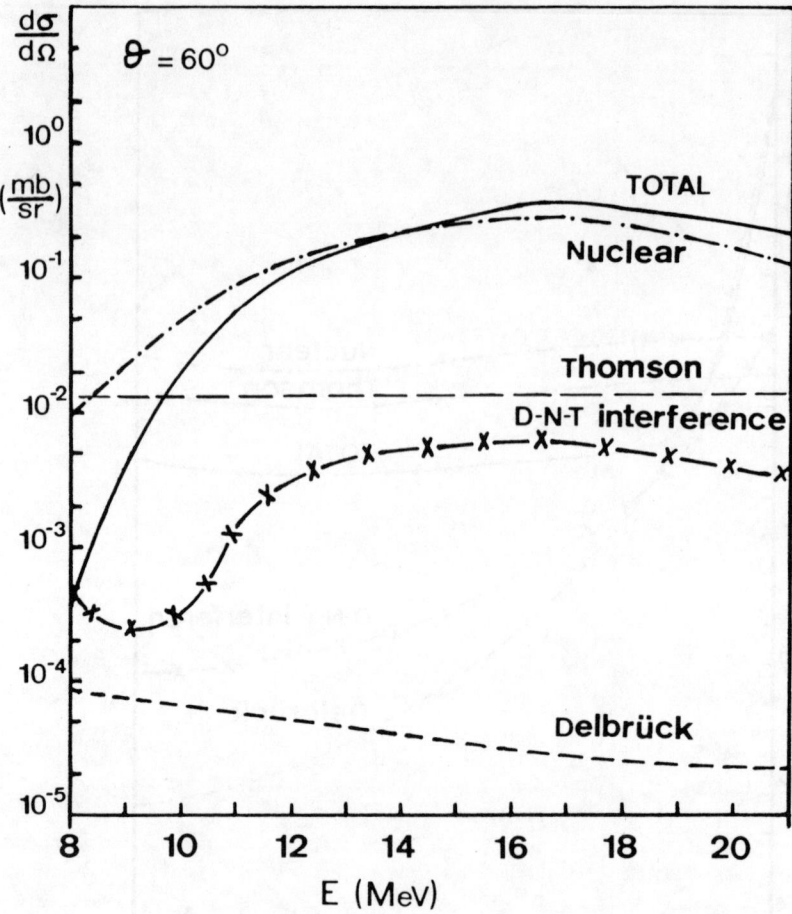

Fig. 2. Elastic scattering at $\vartheta = 60°$ versus incident photon energy (MeV).

As for inelastic scattering, our results are in reasonable agreement with ref.[3], so that it is not necessary to reproduce them here. Comparison with the experimental results of ref.[13] seems to encourage further applications of the IBM to analysis of photon scattering in the GDR region. Accurate calculation of the Delbrück contribution to the elastic channel and particularly of its interference with nuclear and Thomson amplitudes, as described in Section 3., is needed for suitable comparison with experimental information at small and intermediate angles.

Useful conversations with prof. B. De Tollis are gratefully acknowledged.

REFERENCES

1) For a recent review see : Arima,A. and Iachello,F., Adv.Nucl. Phys. 13, 139 (1984).
2) Morrison,I. and Weise,J., J.Phys. G8, 687 (1982).
3) Scholtz,F.G. and Hahne,F.J.W., Phys.Lett. 123B, 147 (1983).
4) Rowe,D.J. and Iachello,F., Phys.Lett. 130B, 231 (1983).
5) Maino,G., Ventura,A., Zuffi,L. and Iachello,F., Phys.Rev. C30, 2101 (1984).
6) Scholtz,F.G., Phys.Lett. 151B, 87 (1985).
7) Maino,G., Ventura,A., Zuffi,L. and Iachello,F., Phys.Lett. 152B, 17 (1985).
8) Maino,G., Ventura,A., Van Isacker,P. and Zuffi,L., Phys.Rev. C33, 1089 (1986).
9) Costantini,V., De Tollis,B. and Pistoni,G., Nuovo Cim. A2, 733 (1971).
10) Eisenberg,J.M. and Greiner,W., Nuclear Theory II : Excitation Mechanism of the Nucleus, North-Holland Publ. Co. (1970), p. 111 ff.
11) De Tollis,B. and Luminari,L., Nuovo Cim. A81, 633 (1984).
12) Gurevich,G.M., Lazareva,L.E., Mazur,V.M., Morkulov,S.Yu., Solodukhov,G.V. and Tyutin,V.A., Nucl.Phys. A351, 257 (1981).
13) Nathan,A.M. and Moreh,R., Phys.Lett. 91B, 38 (1980).

BOSON STRUCTURE FUNCTIONS
FROM INELASTIC ELECTRON SCATTERING

C.W. de Jager

*Nationaal Instituut voor Kernfysica en Hoge Energie Fysica (NIKHEF-K),
1009 AJ Amsterdam, The Netherlands*

ABSTRACT

The even $^{104-110}$Pd isotopes and ^{196}Pt have been investigated at NIKHEF-K by high-resolution inelastic electron scattering. A new IBA-2 calculation has been performed for the Pd isotopes, in which the ratio of the proton and neutron coupling constants is taken from pion scattering. One set of boson structure functions sufficed for the description of the first and second E2-excitations in all Pd isotopes. The data showed no sensitivity for different structure functions for proton and neutron bosons. A preliminary analysis of a number of negative parity states (3⁻, 5⁻ and 7⁻), observed in ^{196}Pt, was performed through the introduction of an f-boson. The first E4-excitation in the palladium isotopes can be reasonably described with a β-structure function, but all other E4-excitations require the introduction of g-boson admixtures.

1. INTRODUCTION

Practically all analyses in the framework of the Interacting Boson Approximation have resorted to static nuclear properties, such as transition rates and quadrupole moments. Inelastic electron scattering offers the opportunity to include radial densities in such studies. Although the electromagnetic interaction is weak, which causes the cross sections to be small, it is well known, so that the nuclear information can be extracted unambiguously. Only direct excitations from the ground state can be studied, which also implies that quite weak transitions can be studied without having to unravel two-step contributions as in hadron scattering. Another implication is that excitations of higher multipolarity can be studied cleanly.

The general form of the one-body transition operator in IBA-2 is:

$$T_\rho^{(l)} = \Sigma_{\rho=\pi,\nu} e_\rho \{\delta_{l2} \alpha_\rho^{(2)}(r) [s^\dagger d + d^\dagger s]_\rho^{(2)} + \beta_\rho^{(l)}(r) [d^\dagger d]_\rho^{(l)}\} \qquad (1)$$

for transitions from a spin-zero ground state to a non-spin-zero excited state. In analyses of static properties the coupling constants α and β entered only as integral properties. The results of electron-scattering experiments now allow the study of the spatial dependence of the coupling of bosons to the external electromagnetic field. In an analogy with the Liquid Drop Model one might naively identify the d-boson with the quadrupole phonon. Then, the α(r) structure function will peak at the surface of the nucleus, thus resembling the first derivative of the ground-state charge distribution, while β(r) - the d-boson conserving term - will behave more like a second derivative.

In Plane Wave Born Approximation the cross section for electron scattering off a spin-zero nucleus is proportional to the square of the so-called form factor, which is a function only of the momentum transfer q. Since this paper concerns only collective excitations, the transverse component of the form factor is neglected. The transition charge density is the Fourier transform of the Coulomb form factor. The Fourier transformation from q-space to coordinate space would require the knowledge of the form factor up to infinite values of the momentum transfer. Systematic observations have, however, shown that the form factor drops off very rapidly once the momentum transfer is larger than twice the Fermi momentum (≈ 3 fm^{-1}). This aspect is used in a model-independent analysis of form-factor data, in which the transition charge density for a transition of multipolarity λ is expanded in a Fourier-Bessel series:

$$\rho_\lambda(r) = \Sigma A_\mu q_\mu^{\lambda-1} j_\lambda(q_\mu^{\lambda-1} r) \qquad r < R_c$$

$$\rho_\lambda(r) = 0 \qquad r > R_c \qquad (2)$$

$$F(q) = F_0 \exp(-\alpha q) \qquad q > q_{max}$$

The even $^{104-110}$Pd isotopes and ^{196}Pt have been studied through inelastic electron scattering at NIKHEF-K. The data cover a momentum-transfer range of 0.4 to 2.5 fm^{-1}. A resolution of 15 to 30 keV was obtained, depending on the primary energy. Form factor data for approximately 30 states in each isotope in the excitation energy region up to 3 MeV were extracted from the spectra through a line-shape fitting routine. The transition density was determined model-independently for a number of collective excitations by the method described above. Details of the experiment and the associated analysis can be found in refs. 1 and 2.

2. PALLADIUM ISOTOPES

The low excitation-energy spectrum of the Pd isotopes exhibits strong vibrator-like characteristics, with a nearly degenerate $(0^+,2^+,4^+)$-triplet at twice the excitation energy of the first 2^+ state. On the other hand, the large value observed for the quadrupole moment and the relatively strong excitation of the second 2^+ state indicate deviations from the pure vibrational limit, necessitating the introduction of anharmonicity. Similarly, Stachel et al.[3] have shown that the Pd isotopes are intermediate between two of the geometric limits of the IBA-1 model, the U(5) and O(6) limits. Van Isacker and Puddu[4] have performed a systematic IBA-2 study of the Pd and Ru isotopes. Although most of the properties studied, such as excitation energies, quadrupole moments, B(E2)-values and branching ratios, could be reproduced quite well, a number of deficiencies induced us to perform a new IBA-study of the Pd isotopes. The excitation energies of the third and fourth 2^+ states were not reproduced, nor the B(E2)-values to the second 2^+ states. Also, equal values were assumed for the coupling constants of proton and neutron bosons e_π and e_ν, while pion-scattering data[5] had shown this assumption to be incorrect. In this new fit to the excitation energies of the 2^+ singlet, the $(0^+,2^+,4^+)$-triplet and the $(0^+,2^+,3^+,4^+,6^+)$-quintuplet the values for ε_d and κ were varied, while the ratio e_ν / e_π was fixed to the value of 0.4 deduced from the pion-scattering data. The value for $(\chi_\pi + \chi_\nu)$ was adjusted to the B(E2)-ratio of the first and the second 2^+ states. With the resulting parameter set all excitation energies studied could be reproduced quite well (as is shown in fig. 1). Also the other deficiencies of the earlier study were remedied. Moreover, the values obtained for ε_d and κ are well within systematics observed over wide ranges of isotopes.

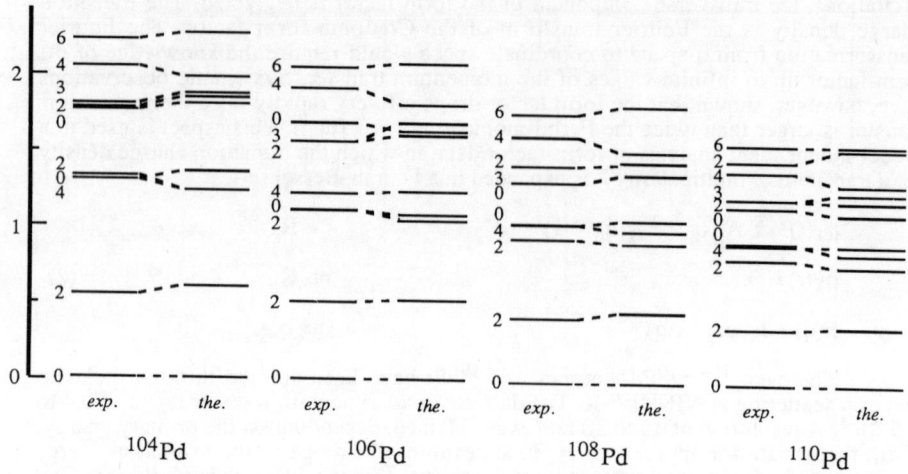

Fig.1 *Excitation energies of collective states in $^{104-110}$Pd below 2 MeV calculated in IBA-2, compared to the experimental values.*

3. QUADRUPOLE EXCITATIONS

From eq. 1 a general form for the quadrupole transition charge density can be derived for a transition from the ground state to the i-th 2^+ state:

$$\rho_i(r) = \Sigma_{\rho=\pi,\nu} e_\rho \{ A_{i\rho} \alpha_{2\rho}(r) + \chi_\rho B_{i\rho} \beta_{2\rho}(r) \} \qquad (3)$$

with the IBA matrix elements:

$$A_{i\rho} = <2_i^+ \| [s^\dagger d + d^\dagger s]^{(2)} \| 0_1^+ > \text{ and } B_{i\rho} = <2_i^+ \| [d^\dagger d]^{(2)} \| 0_1^+ >$$

Care should be taken that the structure functions are correctly normalized, since otherwise the parameter χ would lose its meaning. All quadrupole transitions within one nucleus should be governed by only four boson structure functions $\alpha_{\pi,\nu}(r)$ and $\beta_{\pi,\nu}(r)$. Moreover, if the structure functions are indeed independent of the neutron number, all low-lying 2^+ states in a series of isotopes should be described by the same set. This assumption was tested in a simultaneous analysis of the transition charge densities of the 2_1^+ and 2_2^+ states in all four palladium isotopes studied. The boson coupling constants were assigned values in accordance with the pion-scattering data and the IBA matrix elements were taken from the analysis described in section 2. Since the data showed no sensitivity for different structure functions for proton and neutron bosons, these were assumed to be equal. Indeed, all the transition charge densities mentioned could be reproduced excellently with the two boson structure functions shown in fig. 2. The shapes of $\alpha_2(r)$ and $\beta_2(r)$ resemble that of first- and second-order derivatives of the ground state charge distribution, respectively, in agreement with the naive expectations presented in the introduction.

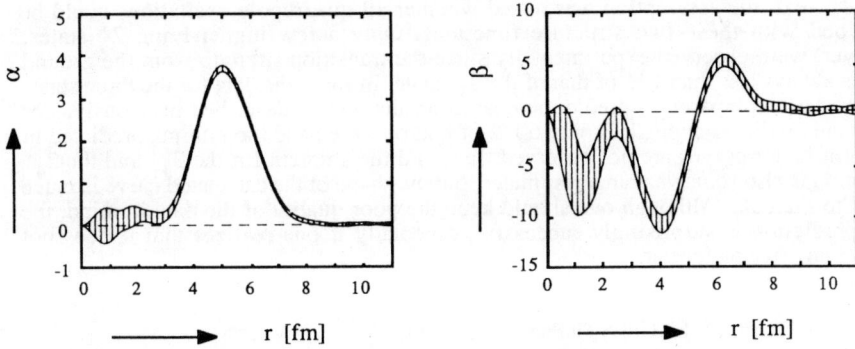

Fig. 2 *Quadrupole boson structure functions deduced from the transition charge densities of the 2_1^+ and 2_2^+ states in $^{104\text{-}110}$Pd.*

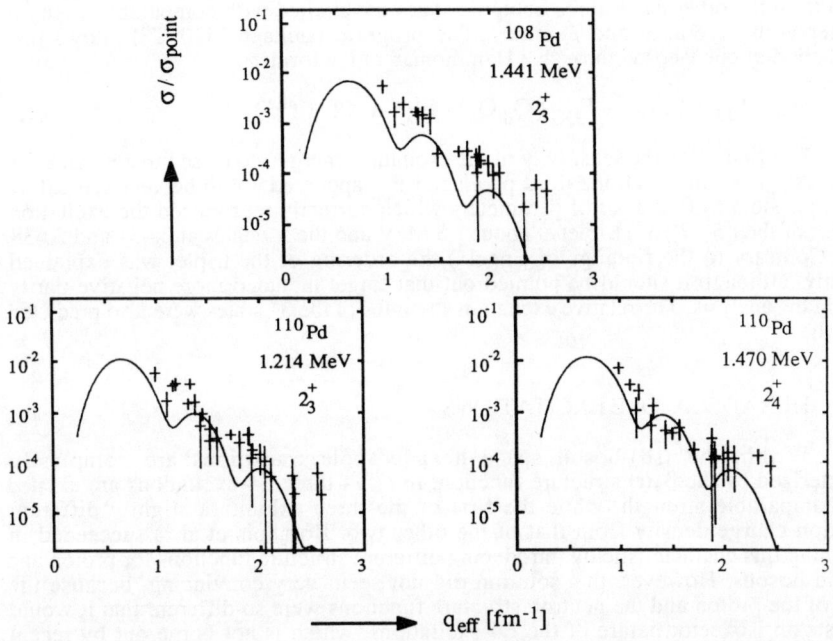

Fig. 3 *Form factor data for the other 2^+ states observed compared to IBA-predictions.*

Next, the assumption was tested whether all quadrupole excitations could be described with these two structure functions. Only a few higher-lying 2^+ states, however, were observed experimentally since the transition strength from the ground state is always less than 1 % of that of the 2_1^+ state. In fig. 3 the data for the three states observed are compared to predictions based on the results described previously. For ^{108}Pd the predicted strength is only 60 % of that observed and the minima predicted in the form factor curves are not observed. In ^{110}Pd the strength for the 2_3^+ and the 2_4^+ excitation is also somewhat underestimated, but the shape of the calculated curve is much closer to the data. Although one should keep the poor quality of the data in mind, the IBA-prediction is surprisingly successful, especially if one realizes that it does not involve any free parameter.

4. OCTUPOLE EXCITATIONS

The study of negative-parity states in IBA requires the introduction of an f-boson. Because several new parameters then have to be determined, such a study requires the observation of several negative-parity states. In the palladium isotopes all octupole strength was observed to be concentrated in one state at approximately 2 MeV. In ^{196}Pt, on the other hand, three octupole states are excited with comparable strength together with several 5^- and 7^- states. The program package PHINT[6]) allows the introduction of one f-boson through a Hamiltonian of the form:

$$H_{df} = F_{LL}(L_d \cdot L_f) + F_{QQ}(Q_d \cdot Q_f) - F_{ex}(d^\dagger f) \cdot (f^\dagger d) \qquad (4)$$

In a first study the sensitivity of the excitation energies to these three parameters was investigated, in which the third parameter F_{ex} appeared not to be very critical. It proved possible to find a set of parameters which correctly reproduced the excitation energies of the ($5^-, 7^-, 3^-$) triplet at about 1.3 MeV and the 3^- states at 2.431 and 2.638 MeV. Contrary to the findings of Engel[7]) the ordering of the triplet was explained correctly, although it should be pointed out that Engel included more negative-parity states in his analysis. The relative excitation strengths of the 3^- states were also predicted correctly.

5. HEXADECAPOLE EXCITATIONS

Within the (sd)-boson space hexadecapole transitions are completely characterized by the $\beta_4(r)$ structure function. In ^{196}Pt three E4-excitations are excited with comparable strength, while the first of the three exhibits a slightly different transition charge density from that of the other two. Borghols et al.[8]) succeeded in explaining this characteristic by introducing different structure functions for proton and neutron bosons. However, this solution did not seem very convincing, because the shape of the proton and the neutron structure functions were so different that it would suggests an isovector nature of the E4-excitations, which is not borne out by recent (α,α')-data[9]).

In the palladium isotopes the first E4-excitations exhibit a transition charge density completely different from that of any other E4-excitation observed. By its node-like structure it resembles a $\beta(r)$ structure function, but not quite (fig. 4). The inner part of the experimental charge densities is much less pronounced and the intersection with the r-axis is located 0.5 fm more to the interior than that of $\beta(r)$. Therefore, a g-boson admixture was included in a pragmatic way.

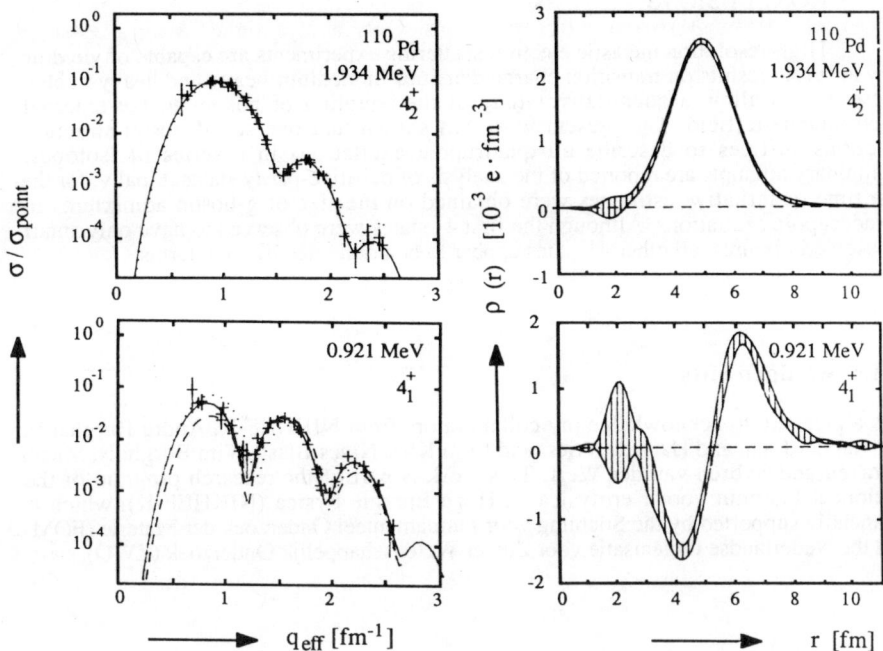

Fig.4 *Form factor data and transition charge densities for 4^+ states in ^{110}Pd.*

In a (sdg)-boson space the hexadecapole transition charge density is given by:

$$\rho_i(r) = e_4 \{ <4_i^+ \| [s^\dagger g + g^\dagger s]^{(4)} \| 0_1^+> \alpha_4(r) + <4_i^+ \| [d^\dagger d]^{(4)} \| 0_1^+> \beta_4(r)$$
$$+ <4_i^+ \| [g^\dagger d + d^\dagger g]^{(4)} \| 0_1^+> \gamma_4(r) + <4_i^+ \| [g^\dagger g]^{(4)} \| 0_1^+> \eta_4(r) \} \quad (5)$$

First the $\beta_4(r)$ structure function is assumed to have the same shape as that of $\beta_2(r)$. Next one assumes the $[s^\dagger g + g^\dagger s]$ term to be the major contribution from the g-boson space. This term, which exhibits a surface-peaked behaviour, is then chosen to have the same shape as the $\alpha_2(r)$ structure function. With only a few percent of this g-boson admixture one obtains a perfect description of the first E4-excitations in all the palladium isotopes, whereas the higher-lying 4^+ states require a substantially larger admixture. Thus, the present electron scattering data allow the first quantitative study of g-boson admixtures, albeit in a rather pragmatic way.

6. CONCLUSIONS

High-resolution inelastic electron scattering experiments are capable of yielding very accurate results on transition charge densities in medium-heavy and heavy nuclei. These results allow a quantitative study of the coupling of bosons to the external electromagnetic field. The present work has shown that one set of boson structure functions suffices to describe all quadrupole excitations in a series of isotopes. Preliminary attempts are reported of the analysis of negative-parity states. Finally, for the first time quantitative estimates were obtained on the size of g-boson admixtures in hexadecapole excitations. Although the first 4^+ states were observed to have only small g-boson admixtures, all other 4^+ states appear to be dominated by such terms.

Acknowledgements

It is a pleasure to acknowledge my collaborators from NIKHEF-K, André Burghardt, Jan van der Laan and Hans de Vries, and from KVI, Nives Blasi, Wim Borghols, Musin Harakeh and Sybren van der Werf. This work is part of the research program of the Nationaal Instituut voor Kernfysica en Hoge-Energie Fysica (NIKHEF-K), which is financially supported by the Stichting voor Fundamenteel Onderzoek der Materie (FOM) and the Nederlandse Organisatie voor Zuiver Wetenschappelijk Onderzoek (ZWO).

References

1. Van der Laan, J.B., Ph. D. Thesis, University of Amsterdam, unpublished (1986).
2. Borghols, W.T.A., Master's Thesis, University of Amsterdam, unpublished (1983).
3. Stachel, J., Van Isacker, P. and Heyde, K., Phys. Rev. C25, 650 (1982).
4. Van Isacker, P. and Puddu, G., Nucl. Phys. A348, 125 (1980).
5. Saha, A. et al., Phys. Lett. 132B, 51 (1983).
6. Scholten, O., The program package PHINT, KVI Internal Report, KVI-63.
7. Engel, J., Phys. Lett. 171B, 148 (1986).
8. Borghols, W.T.A. et al., Phys. Lett. 152B, 330 (1985).
9. Harakeh, M.N., private communication.

DETERMINATION OF BOSON DENSITIES BY ELECTRON SCATTERING.

D. Goutte
Service de Physique Nucléaire - Haute Energie
CEN - Saclay, F 91191 Gif-sur-Yvette Cedex

ABSTRACT

Collective properties of the Os-Pt region and of the Sm isotopes have been explored through the measurements of the transition charge densities of 2^+ and 4^+ states. Their spatial features are interpreted in the framework of the Interacting Boson Model.

1. INTRODUCTION

Electron scattering is an unique tool to provide detailed information on the spatial behavior of the nuclear wave functions. The combination of high resolution and very high momentum transfer allows a very precise (a few percent), model independent determination of the transition densities, even in the inner part of deformed nuclei.

In this contribution I will present a recent measurement done at Saclay on five even-even nuclei of the osmium-platinum region (^{188}Os to ^{196}Pt) and some preliminary results for the samarium isotopes (^{144}Sm to ^{154}Sm).

In both regions (except for the very light Sm isotopes) the Interacting Boson Model (IBM) reproduced with great success integral properties such as excitation energies, transition rates and moments[1]. I will discuss here an interpretation of the radial behavior of the collective nuclear wave functions in terms of IBM.

Originally the model does not take into account any radial degrees of freedom but they can be introduced in a very simple way[2] by writing the transition operator (for an E2 transition for instance) as :

$$T^{(E2)}(r) = \sum_{\rho=\pi,\nu} e_\rho [\alpha_\rho(r)(s^+\tilde{d} + d^+s)^{(2)}_\rho + \chi_\rho \beta_\rho(r)(d^+\tilde{d})^{(2)}_\rho]$$

where $\alpha_\rho(r)$ and $\beta_\rho(r)$ are phenomenological boson functions and e_ρ fixed effective charges. (In fact we shall see later that the choice of this E2 transition operator is not so obvious). A transition charge density can then be written as :

$$\rho_{2^+}(r) = \sum_{\rho=\pi,\nu} e_\rho [\alpha_\rho(r) A^\rho + \chi_\rho \beta_\rho(r) B^\rho]$$

where A^ρ and B^ρ are the matrix element defined by :

$$A^\rho = |<2^+|(s^+\tilde{d} + d^+s)^{(2)}_\rho|0^+>| \qquad B^\rho = |<2^+|(d^+\tilde{d})^{(2)}_\rho|0^+>|$$

As done in most cases we can apply a "pseudo IBM-I" prescription by using only two boson functions :

$$\alpha(r) = \alpha_\pi(r) = \alpha_\nu(r) \qquad\qquad \beta(r) = \beta_\pi(r) = \beta_\nu(r)$$

2. THE OSMIUM-PLATINUM REGION

These nuclei belong to an SU(3) to O(6) transition region. We studied at Saclay ^{188}Os, ^{190}Os, ^{192}Os, ^{194}Pt and ^{196}Pt.

We have first adjusted the parameters of the IBM-2 Hamiltonian in order to fit excitation energies, B(E2)$_\rho$'s and quadrupole moments. The parameters as well as the agreement with the data are close to the one obtained by Bijker et al.[1].

We have measured the transition charge densities of seven 2^+ states in this region (2^+_1 in the five isotopes and 2^+_2 in ^{188}Os and ^{192}Os). Such a large amount of data represent a very strong constraint for any theoretical calculation.

Here we will show that these experimental results allow us to answer important questions about the IBM approach : the first one is of course what quadrupole transition operator do we have to use ? As shown in the introduction, a natural guess is to take a similar expression for $T^{(E2)}$ as the one used for the quadrupole operator $\tilde{Q}$ in the Hamiltonian, but other expressions were proposed [3] and even with this simple one, we don't know which parameter χ_ρ we have to use. An other problem is the determination of the effective charges e_π and e_ν. In fact, in IBM-2, they just appear as normalization factors of the boson densities and do not affect the quality of the fit to the data. If we use a pseudo IBM-1 assumption, then, the ratio e_ν/e_π becomes important and we will see that it has to be adjusted in order to get a good agreement with the data.

By fitting the four boson densities on all the seven 2^+ states the result is almost perfect. By using the same χ_ρ as the one used for the

Hamiltonian or no χ_ρ ($\chi_\rho=1$) we obtained an agreement of the same quality, but very different boson densities (fig.1). $\alpha_\pi(r)$ and $\alpha_\nu(r)$ are quite similar in both cases, where $\beta_\pi(r)$ and $\beta_\nu(r)$ are similar only if χ is used.

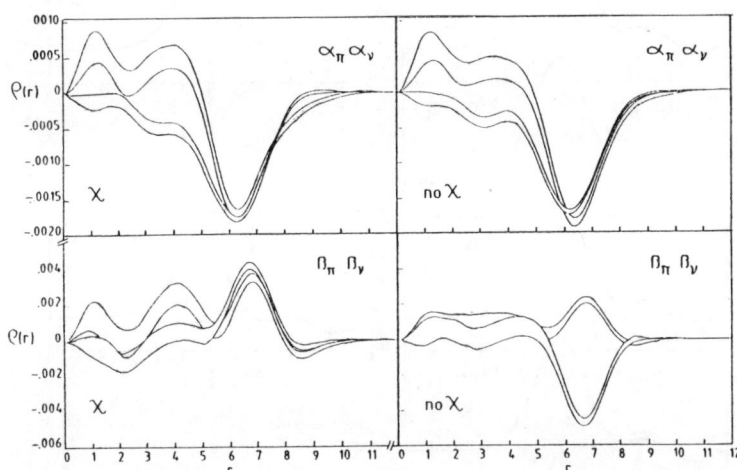

Fig.1 - The four boson densities extracted by using the same value of χ in the $T^{(E2)}$ operator as the one used in the Hamiltonian or using $\chi=1$.

It shows that the "pseudo IBM-1" approximation is valid only if χ is used in the $T^{(E2)}$ operator. Fig.2 shows the comparison to the data within this approximation. The result is almost as good as with IBM-2.

In this latest fit the choice of the ratio of effective charges e_ν/e_π is important. It is determined in the following way. We extracted the two boson densities from the two 2^+ states of ^{188}Os and compared the IBM prediction to the data for the five other levels. The disagreement is then much more sensitive to the change of any parameter than in the case of a fit on all the densities which averages the discrepancies. In fig.3 we show this comparison for the 2^+_2 in ^{192}Os for two different values of e_ν/e_π. The best agreement was obtained with $e_\nu/e_\pi = 0.8$.

With a basis restricted to s and d bosons the only way to excite a 4^+ state is through a $(d^+\tilde{d})^{(4)}$ operator. A transition charge density can then be expressed with only one kind of boson function :

$$\rho_{4^+}(r) = \sum_{\rho=\pi,\nu} \beta_\rho^{(4)}(r) \ |<4^+|(d^+\tilde{d})_\rho^{(4)}|0^+>|$$

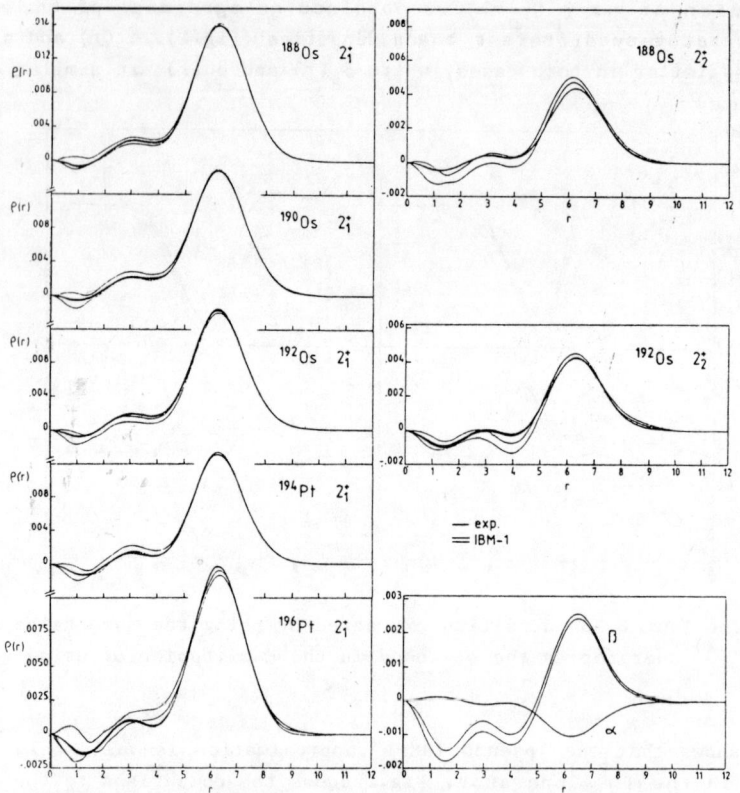

Fig.2 - Comparison between the experimental transition charge densities and the one calculated in the "pseudo IBM-1" framework.

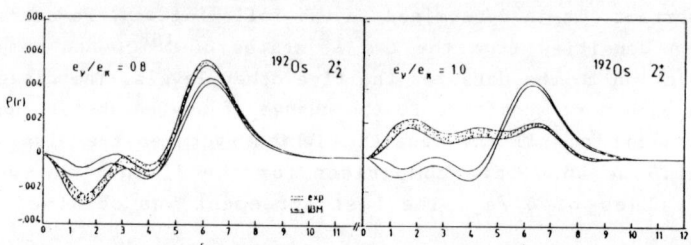

Fig.3 - Comparison between the experimental transition charge density of 2_2^+ in ^{192}Os and the one calculated in IBM-1 for different values of e_ν/e_π.

If we introduce a g boson several other terms will appear in the $T^{(E4)}$ operator such as $(s^+g + g^+s)^{(4)}$ and the associated boson functions will contribute to the transition density.

In this experiment we have determined the transition charge density for five 4^+ states (4_1^+ in ^{188,192}Os ^{194,196}Pt and 4_2^+ in ^{194}Pt). All the 4_1^+ densities have very similar behaviors but the 4_2^+ in ^{194}Pt is peaked at a slightly larger radius than the 4_1^+. That tells us immediatly that an IBM-1 approximation using only one boson function $\beta^{(4)}(r)$ will not be able to reproduce these five densities.

Are the two boson densities $\beta_\nu^{(4)}(r)$ and $\beta_\pi^{(4)}(r)$ sufficient to reproduce these data ? In fig.4 we show the fit to the five 4^+ densities. One can see that the 4_1^+ are well reproduced but we miss the second 4^+ in ^{194}Pt. It is, of course, possible to extract β_π and β_ν from 4_1^+ and 4_2^+ in ^{194}Pt but then the other 4_1^+ densities are not reproduced. This clearly shows that in order to calculate 4^+ states in this region we have to introduce an additional degree of freedom, a g boson for instance.

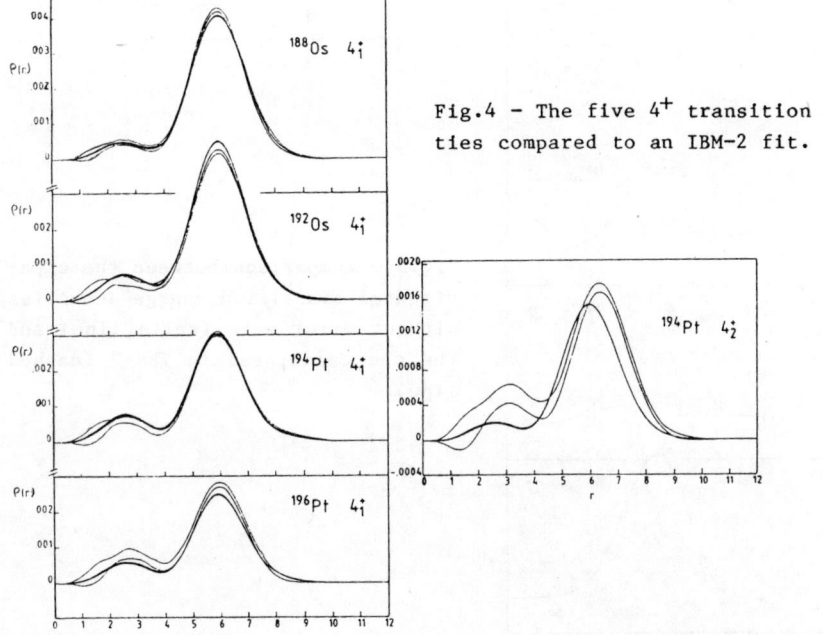

Fig.4 - The five 4^+ transition densities compared to an IBM-2 fit.

3. THE SAMARIUM ISOTOPES

We studied at Saclay five Sm isotopes (^{144,148,150,152}Sm and ^{154}Sm). These nuclei show a clear transition from spherical to deformed shape (U(5) to SU(3) in terms of IBM). The 2_1^+ transition densities measured at Saclay in a previous experiment were used by Moinester et al.[4] in one of the first attempts to reproduce radial behaviors in the IBM framework. The new results presented here are much more accurate because of the larger q-range explored and the better energy resolution reached now at Saclay.

It is now clearly determined that the transition radius of these 2_1^+ states increases slowly with the number of neutrons. That means that a minimum of two different boson functions is needed in order to reproduce all these densities. As $\alpha(r)$ is largely dominant in the expression of the density we can suspect that two different functions $\alpha_\pi(r)$ and $\alpha_\nu(r)$ will be needed.

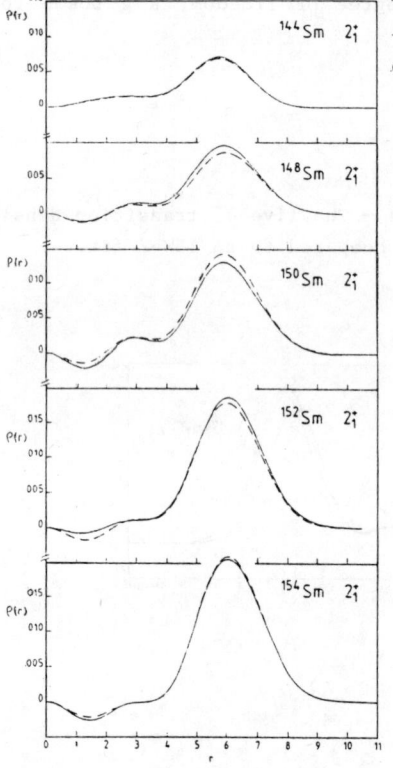
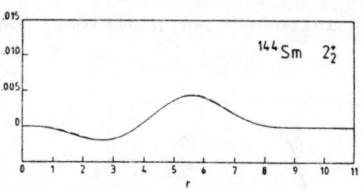

Fig.5 – Comparison between the experimental transition charge densities without error bars (solid line) and the one calculated in IBM-2 (dashed line).

In fact, by fitting in the "pseudo IBM-1" approximation the 2_1^+ and 2_2^+ in ^{144}Sm and all the other 2_1^+ we were not able to reproduce the evolution of the transition radius.

Fig.5 shows a fit to the same data using the four IBM-2 boson densities. The agreement is now quite good for all the levels.

However, this agreement has to be considered carefully : the two functions α_π and α_ν are well determined by the five 2_1^+ transition densities but the β functions are constrained mainly by 2_2^+ states and we have here only one of these in the fit. It means that the agreement between the fit and this 2_2^+ transition density in ^{144}Sm is not really meaningful.

CONCLUSIONS

This analysis of the samarium quadrupole transition charge densities is still preliminary and we need a more detailed study to draw final conclusions. However the explicit distinction between proton- and neutron-boson densities seems to be necessary.

In the Os-Pt region we show that the standard expression of the $T^{(E2)}$ operator and the IBM-1 approximation are well suited to reproduce the quadrupole excitations. But it is also clear that the 4^+ states cannot be reproduced with s and d bosons only. They require the introduction of a g boson.

ACKNOWLEDGEMENTS

The data reduction in the Os-Pt region is due to W. Boeglin as part of his Thesis, and the IBM interpretation of these densities was initiated with the help of J. Engel. I am very grateful to both of them for their very active collaboration.

REFERENCES

1) R. Bijker et al., Nucl. Phys. A344 207 (1980).
   T. Otsuka, PhD Thesis, University of Tokyo, 1979.
   O. Scholten "Interacting Boson in Nuclear Physics", Ed. by
      F. Iachello, Plenum Press, NY 1979.
2) F. Iachello, Nucl. Phys. A358 89 (1981).
3) V. Paar and G. Kynchev, Phys. Lett. 148B 251 (1984).
4) M.A. Moinester et al., Nucl. Phys. A383 264 (1982).

EXPERIMENTAL RESULTS ON THE ELECTRON SCATTERING FROM NUCLEI WITH THE COINCIDENCE REGISTRATION OF CHARGE SECONDARY PARTICLES

S.G.POPOV

Institute of Nuclear Phisics,630090 Novosibirsk, USSR

ABSTRACT

The new generation of experimental results obtained by using continious electron beams and coincidence of the scattered electrons with the secondary particles (SP) are discussed.

The investigation of varius objects by using electron or photon beams was always attractive for experimentalists and theoreticians.The reason is the possibility to use for analysis (i) the well developed theory of the electromagnetic interactions, (ii) the electron "pointlikeness" and (iii) weak interaction of electrons with strongly interacting objects.The interpretation of results obtained by using the electromagnetic probes has no principal theoretical difficulties.However,the small electromagnetic cross section and the necessity to take into account radiation corrections put the high requirements to the electromagnetic experiment luminocity.Here we discuss the nuclear structure investigation by using the electroexcitation of nuclei in the smoll and medium energy region (the excitation energy less than 100 MeV) below the pion production threshold.

In spite of mentioned difficulties the impressive results have been obtained recently in inclusive experiments on electron-nuclei scattering.In particular,that are the elastic scattering experiments for nuclear distribution measurements /1/,the transition dencity measurements by the electroexcitation of isolated levels.I can not stand to mention the example of the 3S-state proton charge density distribution in ^{206}Pb and ^{205}Tl /2/.

In the new generation experiments the reaction product is detected in coincidence with the electron,fixing the momentum transfer and energy of excitation.The additional informational degrees of freedom are the SP type,their energy and the emission direction.Such an experimental performance needs the special technical equipment.The main coincidence experiment demand is to use the continuous electron beam.This is why many projects meeting this demand and high luminocity appear duris last years.The table I shows the list of such facilities.

TABLE 1
THE HIGH DUTY FACTOR FACILITIES

LABORATORY	TYPE OF ACCEL.	ENERGY,MEV	D.F.%	AVERAGE CURR.,MKA
ACTING MACHINES				
1.NOVOSIBIRSK,USSR	Intern.target	100 -500	90	0.5 A
2.STANFORD,USA	Linac	70 -120	75	20
3.ILLINOIS,USA	Racetrac micr.	67	100	2
4.AMSTERDAM,NETH.	Linac	500	2.5	20

5. SACLAY, FRANCE	Linac	600	1.0	1.0	
6. MAINZ, FRG	R.M.	180	100	10	
7. MIT, USA	Linac	700	1.0	0.5	
8. TOCHOKU, JAP.	Stretcher	150	80	0.5	

PROGECTS

9. DARMSTADT, FRG	Linac	130	100	20	
10. NBS, USA	R.M.	185	100	550	
11. LUND, SVEDEN	Stretcher	100 -550	100	10	
12. SASCATOON, CAN.	Linac	300	80		
13. SAO PAULO, BRAS.	LINAC+STR.	17	100	100	
14. SACLAY, FRANCE	Linac+str.	500 -2000	100	100	
15. MAINZ, FRG	R.M.	840	100	100	
16. CEBAF, USA	R.M.	500 -4000	100	200	
17. ILLINOIS, USA	R.M.	450	100	20	
18. MIT, USA	R.M.	250 -1000	100		
19. BONN, FRG	Stretcher	2000	100		
20. MSU, USSR	R.M.	110	100	100	
21. KHARKHOV, USSR	Stretcher	2000	100		
22. NOVOSIBIRSK, USSR	Storage ring	100 -220	90	1.0 A	
23. FRASCATI, ITAL.	Stretcher	500	100	100	
24. MONTREAL, CAN.	R.M.	200	100	300	
25. TSUKUBA, JAP.	Linac	500	2.0	100	

The list of the coincidence experiments with the electroexcitation on nuclei is not complet now, because the new results are appearing from several laboratories.

TABLE II
THE (e,e'X) TYPE EXPERIMENTS

	REACTION	ENERGY (MEV)	EXC.ENERGY (MEV)	LABORATORY
1.	$^{12}C(e,e'p)$	90-126	19-27	STANFORD, USA
2.	$^{16}O, ^{14}N, ^{12}C(e,e'C)$	130	0-70	NOVOSIBIRSK, USSR
3.	$^{1}H(e,e'p)$	110	-	----- " -----
4.	$^{2}D(e,e'd)$	290,400	-	----- " -----
5.	$^{2}D(e,pn)$	180	20-160	----- " -----
6.	$^{12}C, ^{51}V, ^{90}Zr, ^{208}Pb(e,e'p)$	500	0-40	NIKHEF, AMST.
7.	$^{12}C(e,e'\gamma)$	67	4.4	ILLINOIS, USA
8.	$^{208}Pb(e,e'n)$	80	9-16	----- " -----
9.	$^{12}C(e,e'p)$	86		----- " -----
10.	$^{28}Si(e,e'p)$	183		----- " -----
11.	$^{234}U(e,e'f)$	67	5-11.7	----- " -----
12.	$^{238}U(e,e'f)$	170		MIT, USA
13.	$^{238}U(e,e'f)$	180	4-14	MAINZ, GISSEN, FRG
14.	$^{28}Si(e,e'p), (e,e'\alpha)$	180	15-20	MAINZ, FRG
15.	$^{40}Ca(e,e'p), (e,e'\alpha)$	180	10-20	----- " -----
16	$^{2}D(e,e'p)n$		1965-2335	BONN, FRG
17.	$^{3}He(e,e'p)$	560		SACLAY, FRANCE
18.	$^{58}Ni(e,e'p)$	129	15-30	SENDAY, JAP.

Before showing typical examples of the experiments I should stress the main(in my opinion) feature of the coincidence information.This conclusion grounds on the Novosibirsk experiments /3/, which were performed by using the storage ring internal target method with the electroexcitation of light nuclei in broad range of excitation energy (0-100 MeV) and at approximately constant momentum transfer.Fig.1 shows the distribu-

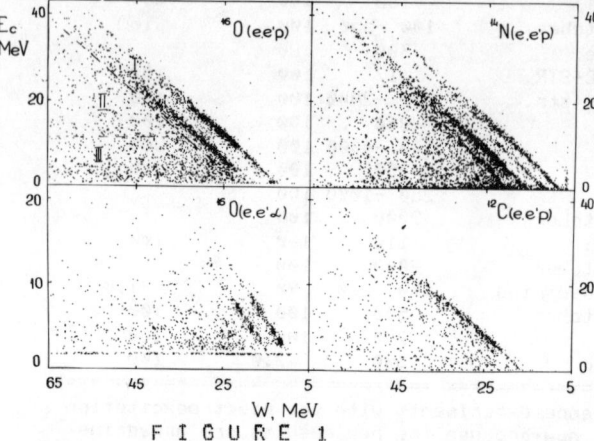

FIGURE 1

tion map of events $A(e,e'C)(A$-nuclei ^{16}O, $^{14}N,^{12}C$; C-charge particles=protons or α) on the plane the of SP energy E (MeV) versus the excitation energy W for the emission direction, which is near to that of the momentum transfer.The analysis of results allows one to separate explicitly three regions, which are connected with various physical processes.The region I contains the "lines" corresponding to the ground and lowlying excited states of a daughter nucleus.In terms of shell structure of the nucleus, the SP appear here as a result of direct knockout from the outer shells.The dense spots on the lines demonstrate the resonance decay channels.The regions II and III have, strictly speaking, a not well defined boundary but also are of a distinguishable phisical nature.Taking into account the angular and energy distributions one can separate the pre-equilibrium (II) and equilibrium events (III).The analysis of the information distributed over regions makes the understanding of nuclear reactions essentially easier.

As an example of such simplified examination I show (Fig.2) the analysis of the "line-p_0" (the case, when after emission the daughter nucleus remains in its ground state).The information is from the coincidence experiment on $^{12}C,^{14}N$ and ^{16}O nuclei(NOVOSIBIRSK /3/).Both the initial and final states are known for this "lines".This is the reason why for this case it is possible to use so-colled multipole analysis of the proton angular distributions at various excitation energies.The presence of the E1 transition and direct processes only in the giant resonance region is assumed.Using an angular distribution, the total cross section (solid line, open circles) and the direct process cross section (closed points)were reconstructed with account of the interference between both processes.The points are:(i)the drastic decreas of the knockout cross section at 22 MeV excitation energy at least for ^{16}O and ^{12}C nuclei; (ii) the account of the interference changes the correlations between the cross sections of both processes.For example,the ^{16}O 21 MeV 1^- resonance cross section changes its value by more then the factor 2 in comparison with the standard procedure.

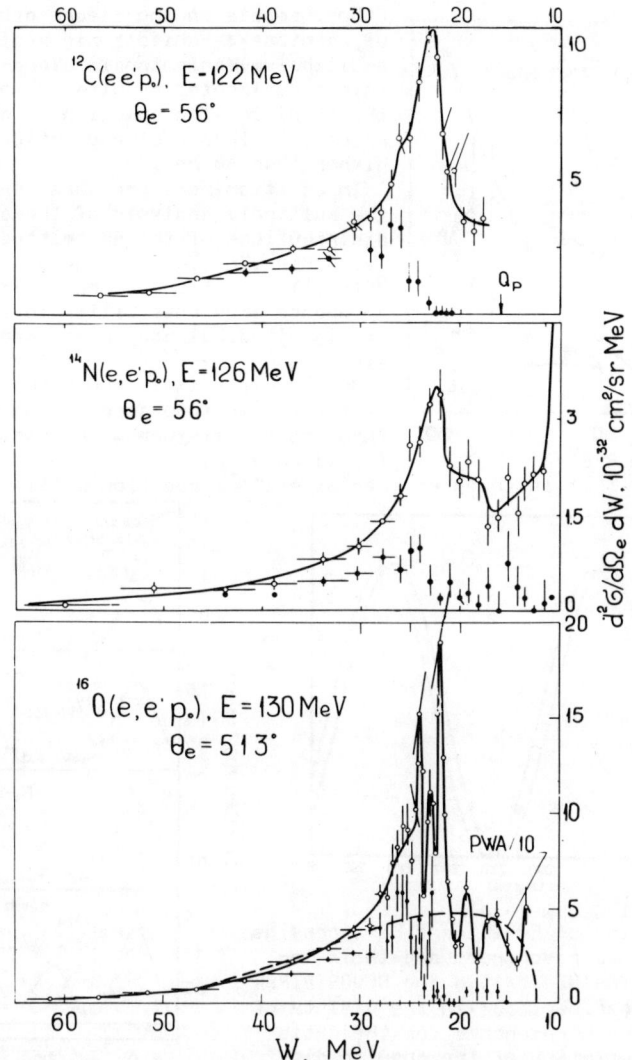

FIGURE 2

As other illustration of the coincidence experiment possibilities let us discuss several results of ILLINOIS, MAINZ, NOVOSIBIRSK laboratories. The quantitative information on the $^{16}O(e,e'C)$ cross sectional structure for the excitation energy range 30-70 MeV is presented in Fig.3 (Novosibirsk). The main contribution comes from the compound nucleus decay with proton (mentioned on the figure o) and alpha- particles

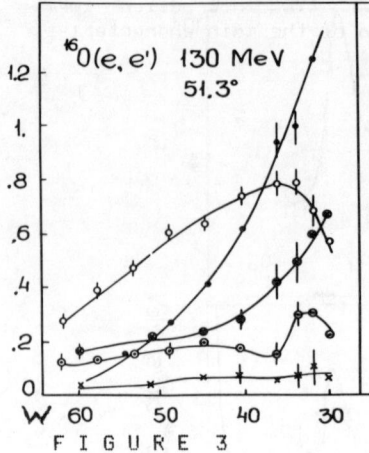

FIGURE 3

Other symbols on the figure are:● -direct of protons, ⊙ -anisotropic protons (pre-equilibrium)✗anisotropic alpha-particles. This experimental results do not support the traditional assumption of a direct process dominance at excitation energy higher than 40 MeV.

In addition, here are three examples of the multipole analysis of the angular distributions of the SP emitted by the resonances. The ILLINOIS measurements $^{12}C(e,e'\gamma)$ reaction on the 4.439 MeV resonance show possibility to separate its longitudinal and transverse matrix elements.

On the Fig.4 it is shown the influence of the taking into account the transverse formfactor admixture with negative phase (solid curve)/4/.

On the Fig.5 it is shown of the $^{28}Si(e,e'\alpha)$ reaction multipole analysis

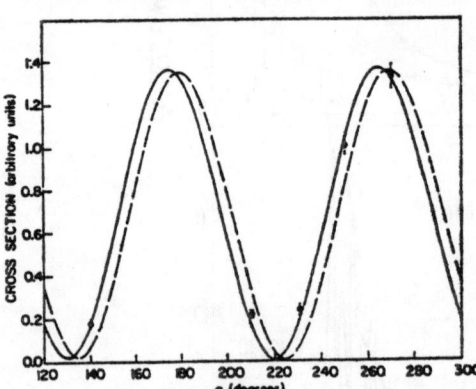

FIGURE 4

for excitation of E0-, E1- and E2-strengths of the several resonanses and their interference (MAINZ /5/). In the NOVOSIBIRSK $^{16}O(e,e'\alpha)$ experiment /3/ in the visinty of the 11.52 MeV 2^+ resonance for the satisfactory description of the angular distribution the excitation of this resonance through the transverse multipole

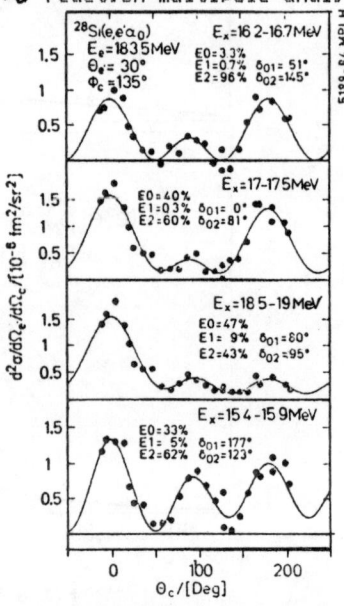

FIGURE 5

operator should be taken into account as well as the presence of the wide 11.26 MeV resonance. By the fitting procedure there were found the longitudinal-transverse ratio -0.087 ± 0.020 and the presence of the 11.26 MeV resonance with the square of the formfactor $(80-60)*10^{-3}$. The confidence level of the fitting is raised by including this parameters from value less than 10^{-3} to 0.15.

In conclusion we can say that the "new generation information" imply the new methods of a reliable determination of the main characteristics of the nucleus structure and of the reaction mechanism,f.e.:
(i) the cross sections of the (e,e'C) reactions;
(ii) properties of the excited nuclear states:
   (a) the spins and parities (without the use of the model dependences of the formfactors on the momentum transfer);
   (b) the decay modes with different emitted particles and various residual nuclear states,
   (c) the longitudinal and transverse parts of the formfactors;
(iii) the mechanismes of cascade processes (through detecting several secondary particles),the widths and nuclear level densities up to the high excitation energies.
In this small revue the simplest examples of the new generation experiment possibilities have been mentioned. The next step should include the experiments with the longitudinal polarized electrons and the polarized targets.

R E F E R E N C E S:

1. B.Frois et al.,Phis.Rev.Lett.,38(1977) 152
2. B.Frois et al.,Nucl.Phys.,A396(1983) 409
3. B.Woitsekhovski et al.,Preprint INP-84-58,Novosibirsk 1984
4. L.S.Cardman Preprint p/85/10/160 Champaign 1985
5. Th.Kihm ,Dissertacion ,Heidelberg 1985

Determination of Transition Matrix Elements and Boson Effective Charges in Pd Isotopes from a Coupled-Channels Analysis of Proton Inelastic Scattering

V. RIECH, E. FRETWURST, G. LINDSTRÖM, K. F. VON REDEN,
R. SCHERWINSKI

Universität Hamburg, FR-Germany

H. P. BLOK

Vrije Universiteit, Amsterdam, The Netherlands

Abstract

From a coupled-channels analysis of high-resolution (p,p') data for ^{104,106,108,110}Pd at $E_p = 15$ and 25 MeV we determined transition matrix elements between the 0_1^+, 2_1^+, 2_2^+ and 4_1^+ states. Calculations with the IBA-2 transition probabilities and using effective charges yield a good description of the data, including those for the 2_3^+ states.

1. INTRODUCTION

Neutron and proton transition matrix elements M_n and M_p for low lying collective nuclear states have been determined for several years by combining the information from different probes[1]. Generally in such analyses the collective model of nuclear shape vibrations was used. The interacting boson model[2] provides a different and valuable basis for the calculation of the various transition moments. In its advanced form, IBA-2[3], neutron and proton bosons are being distinguished. In the framework of this model energy spectra and electromagnetic transition probabilities for the even isotopes of Pd have been calculated by van Isacker and Puddu[4]. In their analysis of $\pi^\pm$ inelastic scattering on the even Pd isotopes Saha et al.[5] obtained transition matrix elements M_p and M_n for the 2_1^+ state excitations separately and determined boson effective charges using the IBA transition matrix elements of van Isacker and Puddu[4]. The extension of such a study to higher lying collective states by proton inelastic scattering is very instructive for various reasons: First, proton inelastic scattering can provide the high energy-resolution necessary to resolve many of the higher lying collective states. Secondly, low-energy protons, like π-particles, interact more strongly with neutrons than with protons, hence in combination with e.m. data the proton and the neutron parts of the transitions can be obtained. Finally, the excitation of low-lying collective states by low-energy protons is ruled by coupled channels (CC) effects, so that information on transition matrix elements not only from the ground state but also from the first excited 2_1^+ state can be expected.

2. EXPERIMENT

Table 1: Excitation energies (in MeV) of low-lying collective states in even Pd isotopes

	^{104}Pd		^{106}Pd		^{108}Pd		^{110}Pd	
	exp.	IBA-2	exp.	IBA-2	exp.	IBA-2	exp.	IBA-2
2(1)[1]	0.556	0.528	0.512	0.477	0.434	0.441	0.373	0.374
2(2)	1.342	1.395	1.127	1.215	0.931	1.034	0.812	0.869
0(2)	1.335	1.717	1.132	1.243	1.052	1.148	0.946	0.989
4(1)	1.324	1.270	1.229	1.171	1.048	1.053	0.921	0.899
2(3)	2.083	2.207	(1.704)	1.899	(1.624)	1.717	1.471	1.477

[1] IBA notation

We performed (p,p') experiments on the even Pd isotopes at 15 and 25 MeV incident energy partly at the cyclotron of the Free University of Amsterdam with a magnetic spectrograph and partly at the cyclotron of the University of Hamburg with solid state detectors. Details of the experimental set-up have been described elsewhere[6][7]. Isotopically enriched targets were employed. The resolution obtained with the spectrograph (9 keV at 15 MeV and 15 keV at 25 MeV) was just sufficient to resolve the narrowly spaced 4^+ 0^+ 2^+ triplet around 1.3 MeV in ^{104}Pd with the help of a spectrum analysis program (s. fig.2 and table 1). Table 1 shows the energies of some of the relevant collective states in the Pd isotopes and the values calculated by IBA-2[2].

3. COUPLED CHANNELS CALCULATIONS

In order to determine transition matrix elements from our inelastic scattering cross sections, we performed CC calculations for the level system $\langle 0_1^+ 2_1^+ 2_2^+ 4_1^+ 2_3^+ \rangle$ (s. also fig.1). The diagonal potentials were taken from a previous CC potential parameter search on the level system $\langle 0_1^+ 2_1^+ \rangle$ using the vibrational model. The transition potentials

$$U_{if} = N\{A_{pp'} \cdot \tilde{\alpha}^{pp'}(r) + B_{pp'} \cdot \tilde{\beta}^{pp'}(r)\} \cdot Y_{\lambda\mu} \quad (\lambda = 2)$$
$$U_{if} = N \cdot C_{pp'} \cdot \tilde{\gamma}^{pp'}(r) \cdot Y_{\lambda\mu}, \quad (\lambda = 4) \quad (1)$$

are made up of the following components:

1. The proton and neutron quadrupole transition densities within the IBA-2 boson model space were taken according to Dieperink[8] as

$$Q_{if}^{x}(r) = A_{if}^{x} \cdot \alpha_{x}(r) + B_{if}^{x} \cdot \beta_{x}(r) \qquad (x = p, n) \qquad (2)$$

with $\quad A_{if}^{x} = \langle f \parallel (s^{\dagger}d + d^{\dagger}s)_{x} \parallel i \rangle \quad$ and $\quad B_{if}^{x} = \langle f \parallel (d^{\dagger}d)_{x}^{\lambda=2} \parallel i \rangle$.

and for the hexadecapole transition

$$H_{if}^{x}(r) = C_{if}^{x} \cdot \gamma_{x}(r) \qquad \text{with} \qquad C_{if}^{x} = \langle f \parallel (d^{\dagger}d)_{x}^{\lambda=4} \parallel i \rangle. \qquad (3)$$

We write $\alpha(r) = \alpha_{x} \cdot \tilde{\alpha}(r)$, $\beta(r) = \beta_{x} \cdot \tilde{\beta}(r)$ and $\gamma(r) = \gamma_{x} \cdot \tilde{\gamma}(r)$ assuming the radial shapes of the boson structure functions $\alpha(r)$, $\beta(r)$ and $\gamma(r)$ to be the same for protons and neutrons due to the strong core polarisation effects. $\tilde{\alpha}(r)$, $\tilde{\beta}(r)$ and $\tilde{\gamma}(r)$ are normalized to $\int \tilde{\alpha}(r) r^4 \, dr = \int \tilde{\beta}(r) r^4 \, dr = 1$ and we also use $\int \tilde{\gamma}(r) r^4 \, dr = 1$. The IBA transition matrix elements can thus be written, skipping the i,f subscripts:

$$Q_{x} = \alpha_{x} \cdot A_{x} + \beta_{x} \cdot B_{x} \qquad \text{and} \qquad H_{x} = \gamma_{x} \cdot C_{x} \cdot K \qquad (x = p, n) \qquad (4)$$

with

$$K = \int \tilde{\gamma}(r) \cdot r^6 \, dr.$$

2. Taking core polarisation into account realistic proton and neutron quadrupole matrix elements M_p and M_n can be written, using the effective charges:

$$M_p = e_p \cdot Q_p + e_n \cdot Q_n \qquad \text{and} \qquad M_n = e_n \cdot Q_p + e_p \cdot Q_n \qquad (5)$$

and similarly for the hexadecapole matrix elements. Two effective charges are sufficient for the self-conjugate Pd-core ($N = Z = 50$).

3. By lack of microscopic calculations the radial shapes of the transition potentials have been taken as the first and second derivatives of an average diagonal optical model potential U(r):

$$\tilde{\alpha}^{pp'}(r) \sim R \cdot \frac{dU}{dr} \qquad \text{and} \qquad \tilde{\beta}^{pp'}(r) = \tilde{\gamma}^{pp'}(r) \sim R^2 \cdot \frac{d^2U}{dr^2} \qquad (6)$$

normalized as explained above. This choice is in accordance with recent results of inelastic electron scattering[9], which show, that with $\rho(r)$ being the ground state charge distribution, $\tilde{\alpha}(r) \sim d\rho/dr$ and $\tilde{\beta}(r) \approx \tilde{\gamma}(r) \sim d^2\rho/dr^2$.

4. The factor $A_{pp'}$ in the transition potentials is written

$$A_{pp'} = v_p \cdot (e_p \alpha_p A_p + e_n \alpha_n A_n) + v_n \cdot (e_n \alpha_p A_p + e_p \alpha_n A_n) \qquad (7)$$

and similarly for $B_{pp'}$ and $C_{pp'}$. In these expressions v_p and v_n denote the relative interaction strengths between the projectile and the nucleons in the target nucleus. The values of v_p and v_n were taken to be 0.25 and 0.75 resp. as follows from the Becchetti-Greenlees[10] parametrisation of the optical model potential.

5. The normalisation factor N has been determined by the requirement, that the cross sections of the 2_1^+ states of all four isotopes have to be reproduced on the average. An attempt to derive the transition potentials by folding an effective interaction into the nuclear density distribution resulted in nearly the same values for N.

From the transition potentials we determine the transition matrix elements as measured in inelastic proton scattering:

$$M_{pp'} = A_{pp'} + B_{pp'} \qquad (8)$$

which according to (7) can be written as $M_{pp'} = v_p M_p + v_n M_n$. To evaluate the transition matrix elements $M_{pp'}$ from our data, we performed CC calculations with the code ECIS-79[11] within the coupling scheme shown in fig.1.

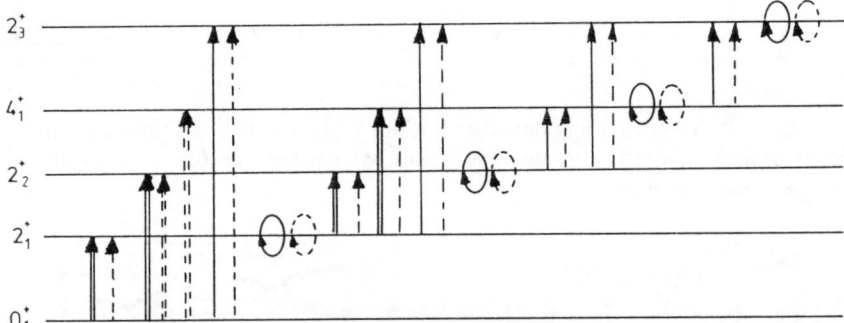

Figure 1: Coupling scheme used in the CC calculations
(Full lines: A_{if}-type transitions ($\lambda = 2$), dashed lines: B_{if}-type transitions)

4. RESULTS

The calculations were started with the boson transition matrix elements A_x, B_x and C_x from van Isacker and Puddu[4] and effective charges from Saha et al.[5], using the same boson coupling constants α_p, α_n as cited there [5) 12)]. The values of β_p and β_n as well as γ_p, γ_n were taken from a preliminary analysis and are discussed below. The results of these calculations, performed on the restricted level system $\langle 0_1^+ 2_1^+ 2_2^+ 4_1^+ \rangle$, are shown by the dashed lines in fig.2 for two selected cases. The description is already rather good.

A search for the most important transition strengths (s. double lines in fig.1), keeping the values of all other couplings on the start values, improved the fits. Since in general small changes in the coupling strengths led already to the best fits, we may conclude that the IBA-2 numbers A_x, B_x and C_x are a good approximation and that our measurements are consistent with the (π, π') data, from which we took the effective charge parameters. We should emphasize, that

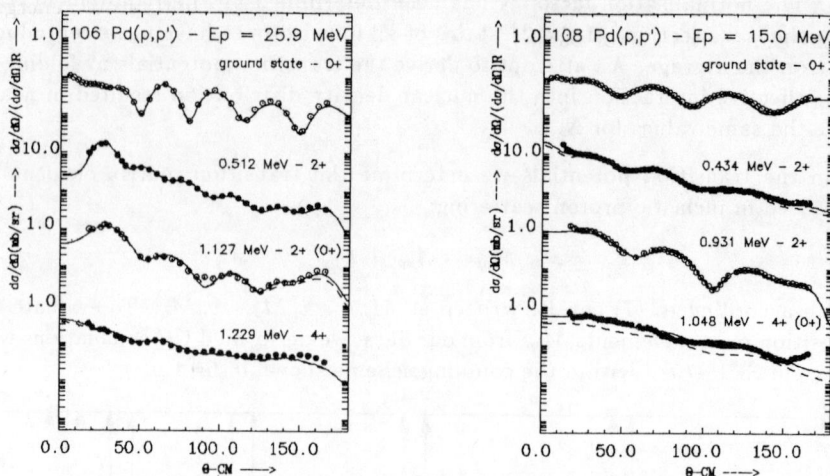

Figure 2: Measured angular distributions with results of CC calculations for two cases. Dashed lines: description with start parameters, full lines: results with adjusted reduced matrix elements.

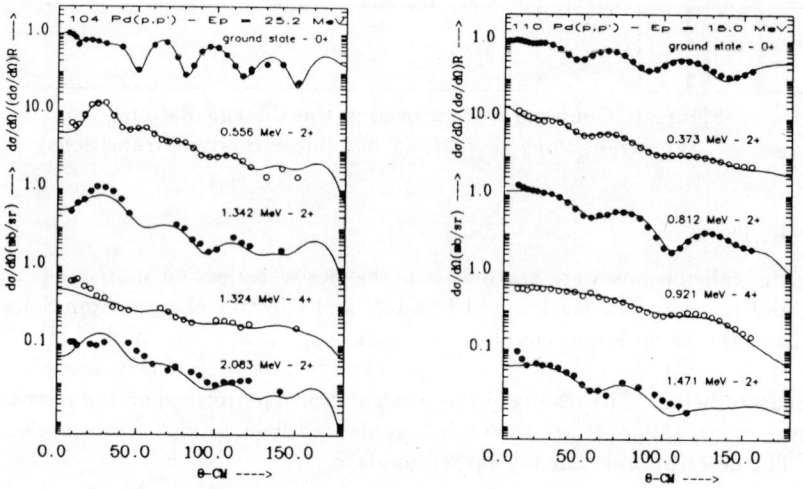

Figure 3: Measured angular distributions including 2_3^+ states with results of CC calculations for two cases.

the searched values of $A_{pp'}$ and $B_{pp'}$ for the same transition at both proton energies are the same within error bars.

Using these values of $A_{pp'}$ and $B_{pp'}$ we determined the matrix elements $M_{pp'}$ (s. table 2). By combining these with previously published M_p data[13] we could derive M_n. Good agreement is achieved with the M_n values of Saha et al.[5] for the $0_1^+ \rightarrow 2_1^+$ transitions. An attempt to find a set of parameters e_p, e_n, α_p, α_n, β_p, β_n, giving the best description of our $M_{pp'}$ and the M_p values, led to our final parameters: Keeping $\alpha_p = 15.2 fm^2$ and $\alpha_n = 16.4 fm^2$ we obtain the effective boson charges $e_p = 1.11 \pm 0.04$ and $e_n = 0.47 \pm 0.03$. Our choice for the β-values $\beta_p = -3.5 fm^2$, $\beta_n = -8.0 fm^2$ is different from that of van Isacker and Puddu[4], who used $\beta_p \approx +2 fm^2$ and $\beta_n \approx -10 fm^2$. It rests on our $B_{pp'}$ best-fit values for the $0_1^+ \rightarrow 2_2^+$ transitions, which are most sensitive to these values, and it agrees in sign with the microscopic predictions of Scholten[12].

We then calculated the matrix elements $M_{pp'}$, M_p and M_n from equation (7), using the boson transition probabilities of ref.[4] and our final parameters. In general they are in good agreement with the experimental values. It should be pointed out, that as a result of the interference between one- and two-step excitation of the levels, also the relative signs of the matrix elements are determined by our analysis and that these signs are also correctly predicted by IBA-2 (see table 2).

Table 2: Measured and calculated transition matrix elements

	$0_1^+ \rightarrow 2_1^+$				$0_1^+ \rightarrow 2_2^+$			
	^{104}Pd	^{106}Pd	^{108}Pd	^{110}Pd	^{104}Pd	^{106}Pd	^{108}Pd	^{110}Pd
$M_{pp'}^{u)}$	99(3)	108(3)	120(4)	134(4)		-16(3)	-13(3)	-20(3)
$M_p^{v)}$	74(2)	83(3)	88(3)	95(3)	-15(1)	-13.5(1)	-13(1)	-11.3(1)
$M_n^{x)}$	107(4)	116(4)	131(5)	147(5)		-17(4)	-13(4)	-23(4)
$M_{pp'} - calc.^{y)}$	97	110	119	133	-12.4	-17.9	-16.4	-18.7
$M_p - calc.^{y)}$	77	82	87	93	-16.5	-18.2	-15.7	-15.8
$M_n - calc.^{y)}$	103	119	130	146	-11.1	-17.8	-16.7	-19.6

	$2_1^+ \rightarrow 2_2^+$				$2_1^+ \rightarrow 4_1^+$			
	^{104}Pd	^{106}Pd	^{108}Pd	^{110}Pd	^{104}Pd	^{106}Pd	^{108}Pd	^{110}Pd
$M_{pp'}^{u)}$		-126(20)	-135(20)	-126(20)	165(15)	185(15)	223(10)	215(10)
$M_p^{v)}$	-67(5)	-83(5)	-116(13)	-95(8)	118(5)	141(9)	159(11)	167(11)
$M_n^{x)}$		-140(30)	-141(30)	-136(30)	181(20)	200(20)	244(15)	231(15)
$M_{pp'} - calc.^{y)}$	-92	-101	-121	-135	155	180	198	221
$M_p - calc.^{y)}$	-73	-74	-86	-92	121	132	141	151
$M_n - calc.^{y)}$	-99	-110	-133	-149	166	196	217	244

All matrix elements are given in fm^2
u) Results of the present analysis
v) From ref.[13]; the signs taken to be the same as for $M_{pp'}$
x) Extracted from $M_{pp'}$ and M_p using $M_{pp'} = 0.25 M_p + 0.75 M_n$.
y) Calculated values using $e_p = 1.11$, $e_n = 0.47$, $\beta_p = -3.5 fm^2$, $\beta_n = -8 fm^2$ together with the IBA-2 transition matrix elements of ref.[4]

5. THE 2_3^+ STATES

A further check on the reliability of our final parameters in combination with the IBA-2 predictions are CC calculations, in which we also included the 2(3) state, as indicated in fig.1. Fig.3 shows for two cases, that the data are nicely described without any fitting.

6. THE HEXADECAPOLE TRANSITION

The description of the cross sections for the 4_1^+ states leads to the following conclusions concerning the $0_1^+ \to 4_1^+$ hexadecapole transition: Unlike β_p and β_n the values of γ_p and γ_n have to be taken positive: $\gamma_p = +4.5 fm^2$, $\gamma_n = +7.0 fm^2$. Taking the same normalisation N as for the $\lambda = 2$ transitions we obtain hexadecapole transition moments of about 650 fm^4.

7. CONCLUSIONS

Our (p,p') cross sections of low lying collective states in even Pd nuclei can be well described by CC calculations using IBA-2 predictions of the boson transition probabilities, if we introduce the effective boson charges $e_p \alpha_p = 17 fm^2$ and $e_n \alpha_n = 7.5 fm^2$. The fit of the separately measured 4_1^+ state leads to information on the $\lambda = 4$ transition matrix elements, of which very little was known.

8. REFERENCES

1. Bernstein,A. et al., Comments Nucl.Part.Phys. 11,203 (1983)
2. Arima,A. and Iachello,F., Ann.Rev.Nucl.Part.Sci. 31,75 (1981)
3. Otsuka,T. et al., Phys.Lett. 76B, 139 (1978)
4. Van Isacker,P. and Puddu,G., Nucl.Phys. A348,125 (1980)
5. Saha,A. et al., Phys.Lett. 132B, 51 (1983)
6. Blok,H.P. et al., Nucl.Phys. A273, 142 (1976)
7. Lindström,G. et al., Jahresbericht 1980/81, I.Inst.f.Exp.-Physik, Universität Hamburg
8. Dieperink,A.E.L., Nucl.Phys. A358, 189c (1981)
9. Van der Laan,J.B. et al., Phys.Lett. 153B, 130 (1985)
10. Becchetti,F.D. and Greenlees,G.W., Phys.Rev. 182, 1190 (1969)
11. Raynal,J., Report IAEA-SMR-918(1972)
12. Scholten,O., Report MSUCL No.416 (1983)
13. Harmatz,B., Nuclear Data Sheets 30, 305 (1980)
    Haese,R.L. et al., Nuclear Data Sheets 37, 289 (1982)
    De Gelder,P. et al., Nuclear Data Sheets 38, 545 (1983)
    Blachot,J. et al., Nuclear Data Sheets 41, 325 (1984)

Fine Structure of LEOR and GQR in ^{48}Ca and ^{208}Pb

Y. Fujita
College of General Education, Osaka Univ., Toyonaka 560, Japan

M. Fujiwara, S. Morinobu, I. Katayama,
T. Yamazaki, T. Itahashi and H. Ikegami
Research Center for Nuclear Physics, Osaka Univ.,
Ibaraki, Osaka 567, Japan

S. I. Hayakawa
Ashikaga Institute of Technology, Ashikaga 326, Japan

I. INTRODUCTION

Various types of giant resonances (GR) have been observed in many nuclei. In a usual experiment, a giant resonance shows a broad bump-like shape with little structure. Among giant resonances, low-energy octupole resonance (LEOR) is an isoscalar E3 resonance built from 1 $\hbar\omega$ excitations and has been observed in many nuclei ($27 < A < 200$) by means of (α, α') experiments mostly as a 1-2 MeV wide broad bump exhausting around 20 % of the energy-weight sum rule (EWSR)[1]. The center of the bump is located at around 30 $A^{-1/3}$ MeV, which is the excitation region where the particle decay is forbidden or at least hindered. The strength of LEOR, therefore, should be found in discrete levels.

As a result of recent high resolution experiments at RCNP Osaka University, fine structures of N=28 nuclei (^{48}Ca, ^{50}Ti, ^{52}Cr and ^{54}Fe), ^{90}Zr and ^{208}Pb are revealed[2,3,4] and the existence of LEOR in ^{208}Pb was confirmed[3]. Recently in ^{48}Ca and ^{208}Pb, each level observed in the spectra was analyzed up to the high excitation region and it became possible to get the information on the tail part of the giant quadrupole resonance (GQR) as well as on the whole range of LEOR excitation.

The fine structure of these excitations are expected to be very important from the view point of investigating damping of nuclear collective vibrations[5]. In this report the features of L=2 and L=3 strengths are discussed putting special emphasis on ^{48}Ca and ^{208}Pb, which are representative doubly closed-shell nuclei.

II. EXPERIMENT AND SUM RULE ANALYSIS

A proton beam from the Research Center for Nuclear Physics (RCNP) cyclotron at Osaka University was used to bombard isotopically enriched target foils with thickness of around 1 mg/cm^2. Inelastically scattered protons were momentum analyzed by a magnetic spectrograph RAIDEN[6] and a focal plane counter system described in Ref.7. Overall resolutions of 10-15 keV were achieved. The obtained spectrum is shown for ^{48}Ca in Fig. 1. The spectra of ^{48}Ca were analyzed up to as high as E_x=13.6 MeV at angles from θ_{lab}=8° to 70° using a peak deconvolution program in order to obtain accurate cross sections for individual levels. The spectra of ^{208}Pb was analyzed up to E_x=7.5 MeV. The angular distributions were analyzed in the framework of DWBA using a collective-model form factor[8] and optical potential parameters of Sakaguchi et al.[9].

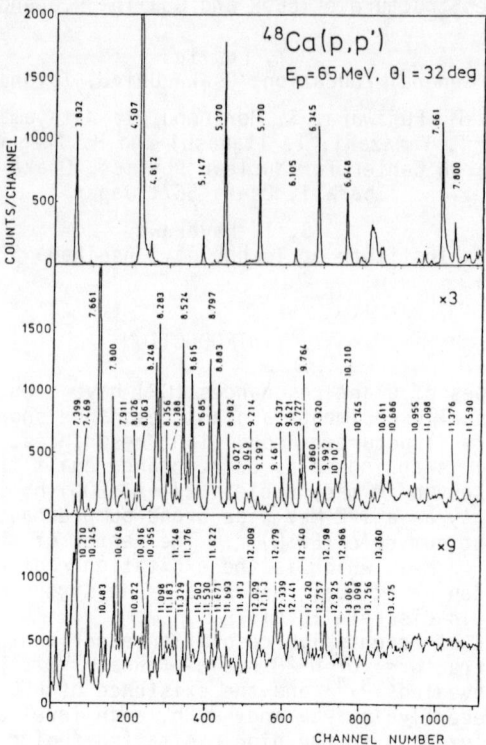

Fig. 1. Spectrum of inelastically scattered protons on the ^{48}Ca target at E_p= 65 MeV.

In order to distinguish natural and unnatural parity states, and also isoscalar and isovector states, one to one correspondence was checked between the levels observed in the ^{48}Ca(p,p') reaction and levels seen in ^{48}Ca(α, α') reaction up to E_x=12 MeV. As a result all of the prominent L=2 and L=3 states were also observed in the (α, α') experiment, which implies negligibly small contribution from either isovector or unnatural parity states. We assume the same situation in ^{208}Pb.

The energy-weight sum rule (EWSR) fractions were derived for these states by using the procedure given in Ref.10. The obtained strength distributions are shown in Figs. 2 and 3 for the states with L=2 and L=3.

III. RESULTS AND DISCUSSIONS

a) L=2 Strength Distribution

It is interesting to note that in ^{48}Ca and also in ^{208}Pb, the L=2 strength may be categorized into three parts, i.e., the strongly excited first 2$^+$ state, several states at around E_x=9 MeV in ^{48}Ca and E_x=5 MeV in ^{208}Pb and many clustering states in the higher excitation region.

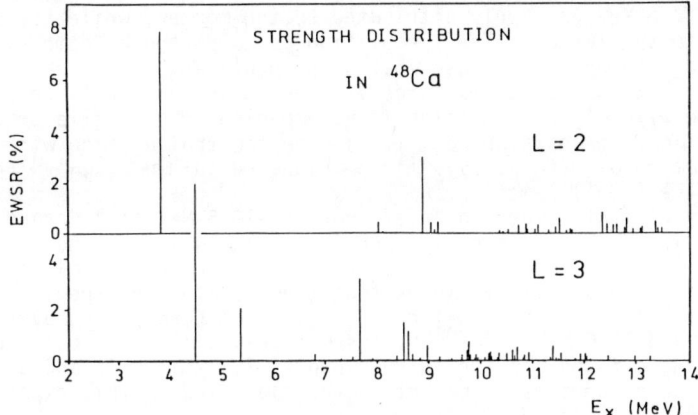

Fig. 2. The distribution of the EWSR percentages for the L=2 and L=3 strengths found in ^{48}Ca.

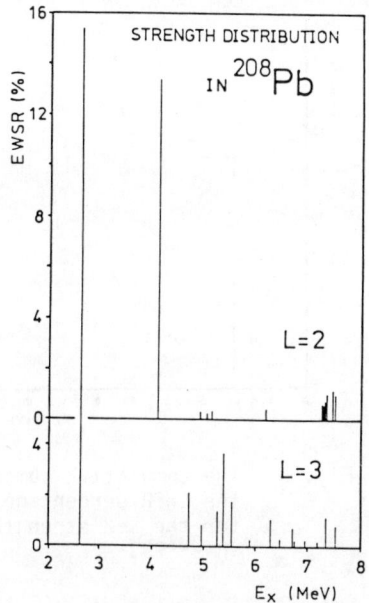

Fig. 3. The distribution of the EWSR percentages for the L=2 and L=3 strengths found in ^{208}Pb.

In ^{48}Ca, natures of the states which belong to the former two categories may be identified by a simple shell model consideration together with the result of comparison between (p,p') and (α,α') reactions. In a simple shell model picture of $f_{7/2}$ on an inert ^{40}Ca core, two configurations, i.e., $\nu(p_{3/2}, f_{7/2}^{-1})$ and $\nu(f_{5/2}, f_{7/2}^{-1})$ with particle-hole (p-h) energies of 2.1 MeV and 8.2 MeV respectively, are possible to make 2^+

states below $E_x=10$ MeV. It is plausible to think that the first 2^+ state at 3.8 MeV is mainly attributed to the former, while the state at 8.9 MeV to the latter. The speculation is supported by comparing the excitation strengths of these states in (p,p') and (α,α') reactions. It is found that the latter is largely hindered in the (α,α') reaction. Since the $f_{7/2} \to f_{5/2}$ transition is accompanied by spin flip process while the $f_{7/2} \to p_{3/2}$ transition not, it is expected that a state with the main configuration of $\nu(f_{5/2}, f_{7/2}^{-1})$ is hindered in the (α,α') reaction than in the (p,p') reaction.

Many 2^+ states were observed above $E_x=10.5$ MeV in ^{48}Ca and above $E_x=7$ MeV in ^{208}Pb. As clearly seen from Figs. 4 and 5, the cumulative sum of L=2 EWSR strength increases rapidly above these excitation energies. It is reasonable to think that these states are the lower tail of the isoscalar GQR, whose center is situated at $E_x=63A^{-1/3}$ MeV (17.3 MeV for ^{48}Ca and 10.6 MeV for ^{208}Pb, respectively).[11] The total sum of the EWSR strength found for the third group is 7.8 % in ^{48}Ca and 5.2 % in ^{208}Pb. The properties of L=2 strengths clarified in this experiment are summarized in Table I.

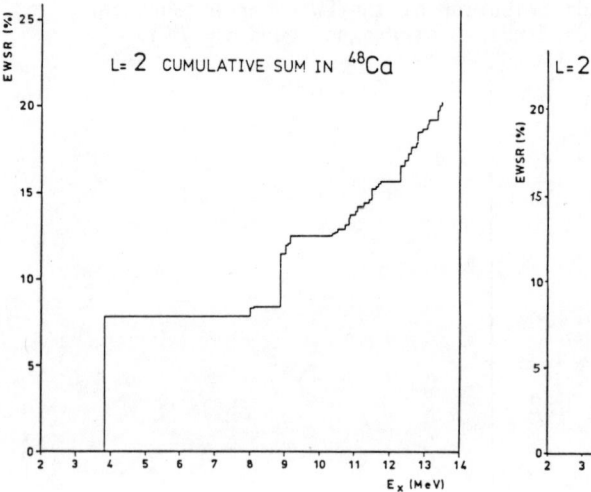
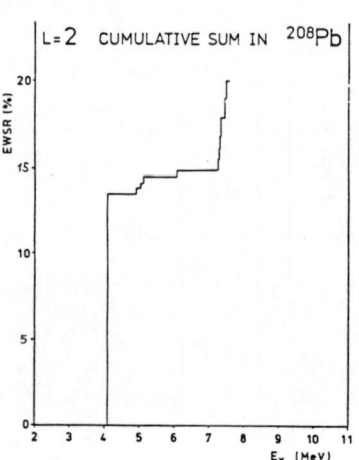

Fig. 4. The cumulative sum of the EWSR percentages for the L=2 strengths in ^{48}Ca.

Fig. 5. The cumulative sum of the EWSR percentages for the L=2 strengths in ^{208}Pb.

b) L=3 Strength Distribution

In the "LEOR region", as seen in Figs. 2 and 3, the L=3 strength is concentrated in each nucleus. The EWSR fractions amount to 15-20 % as summarized in Table II. The envelope of the distribution exhibits a resonance-like shape. The feature is more clearly seen from the plot of cumulative sum of EWSR strength for the observed discrete states (see Figs. 6 and 7). In ^{48}Ca, the increase is very rapid between $E_x=9$ and

Table I Observed properties of quadrupole strengths in ^{48}Ca and ^{208}Pb.

	^{48}Ca	^{208}Pb
E_x of low-lying 2$^+$	3.83 (MeV)	4.09 (MeV)
EWSR of low-lying 2$^+$	7.8 (%)	13.4 (%)
E_x of middle group	8~9 (MeV)	5~6 (MeV)
EWSR of middle group	4.7 (%)	1.4 (%)
E_x of GQR	>10.4 (MeV)	>7.2 (MeV)
EWSR of GQR (tail part)	7.8 (%)	5.2 (%)

Table II Observed properties of octupole strengths in ^{48}Ca and ^{208}Pb.

	^{48}Ca	^{208}Pb
E_x of low-lying 3$^-$	4.51 (MeV) 5.37 (MeV)	2.65 (MeV)
EWSR of low-lying 3$^-$	6.9 (%) 2.0 (%)	20.4 (%)
E_x of LEOR	9.71 (MeV)	5.39 (MeV)
EWSR of LEOR	17.0 (%)	15.2 (%)
Number of L=3 states in LEOR region	45	15
L=3 mean level density	7.6/MeV	5.3/MeV
Γ	2.76 (MeV)	0.69 (MeV)
Γ/E_x of LEOR	0.28	0.13

11 MeV, while it becomes slow above E_x=12 MeV. The same is true in ^{208}Pb, where the increase is very rapid around E_x=6 MeV and slow above 7 MeV. This fact indicates the existence of resonance like structure as a whole. It should be noted that the "degree of fragmentation" is quite different between these two nuclei. As discussed in a previous paper[3] this can be explained by assuming that the fragmentation of LEOR is caused by the coupling of a collective LEOR state with background 2p-2h states containing the 2$^+$ phonon mode as a component.

In order to know the width Γ and the central value of the excitation energy E_0, the cumulative sum was fitted to the integrated Breit-Wigner function

$$Y(E) = Y_0 \{(1/\pi)\tan^{-1}(E-E_0)/(\Gamma/2) + \frac{1}{2}\} ,$$

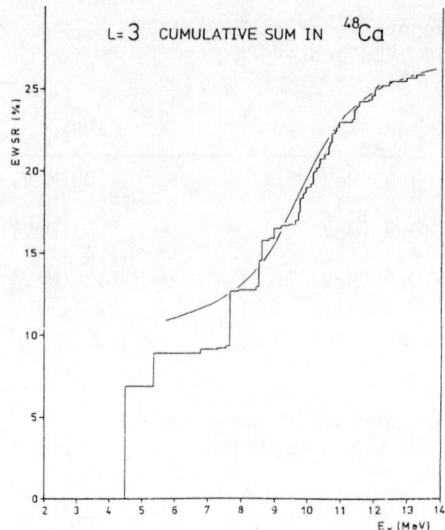
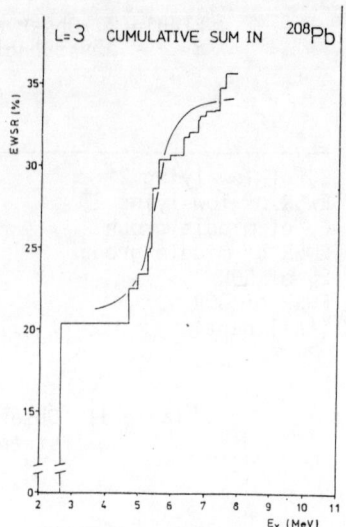

Fig. 6.   The cumulative strength sum of the EWSR percentages for the L=3 strengths in ^{48}Ca. The cumulative strength sum of the LEOR region is fitted to the integrated Breit-Wigner function.

Fig. 7.   The cumulative strength sum of the EWSR percentages for the L=3 strengths in ^{208}Pb. The cumulative strength sum of the LEOR region is fitted to the integrated Breit-Wigner function.

where Y_0 is the integrated strength. The obtained values for E_0 and Γ are shown in Table II. According to Moss et al., the E_0 values are described by a systematics of $E_0 = 30\, A^{-1/3}$. In our experiment, on the other hand, the obtained coefficients are slightly larger than 30 (for ^{48}Ca 35 and for ^{208}Pb 32). It was suggested that the coefficients for other N=28 nuclei, such as ^{50}Ti, are slightly lower (~ 26). Thus, it will be interesting to study theoretically whether the coefficient is large in a doubly magic nucleus or not.

The width Γ is wide in ^{48}Ca, while in ^{208}Pb it is rather narrow. Even if the width Γ is normalized by the excitation energy of LEOR (Γ/E_x of LEOR), the width of ^{48}Ca is twice as large as that of ^{208}Pb. In a schematic view, a GR is described as a strongly coherent 1p-1h excitation and the fragmentation will occur mainly from its coupling to the background states of 2p-2h configurations with the same J^π value as the GR. It is interesting to estimate the strength of interaction V which makes the fragmentation by using the above information and the Fermi's golden rule,

$$\Gamma = 2\pi V^2\, dn/dE \ .$$

The expected value of V is around 100-200 keV. Similar value of V (~140 keV) is estimated for the GQR of ^{208}Pb using an RPA calculation[12].

REFERENCES

1) J.M. Moss, et al., Phys. Rev. C 18 (1978), 741.
2) Y. Fujita, et al., Phys. Lett. 98B (1981), 175.
3) Y. Fujita, et al., Phys. Rev. C 32 (1985), 425.
4) M. Fujiwara, et al., Phys. Rev. C 32 (1985), 830.
5) G.F. Bertsch, et al., Rev. Mod. Phys. 55 (1983), 287; and refs. therein.
6) H. Ikegami, et al., Nucl. Instrum. Methods 175 (1981), 335; Y. Fujita et al., Nucl. Instrum. Methods 225 (1984), 298.
7) Y. Fujita, et al., Nucl. Instrum. Methods 217 (1983), 441.
8) P.D. Kunz. DWUCK, University of Colorado (unpublished).
9) H. Sakaguchi, et al., Phys. Rev. C 26 (1982), 944.
10) G.R. Satchler, Nucl. Phys. A195 (1972), 1.
11) D.H. Youngblood, et al., Phys. Rev. C 13 (1976), 994; and refs. therein.
12) P.F. Bortignon and R.A. Broglia, Phys. Lett. 102B (1981), 303; Nucl. Phys. A371 (1981), 405.

NUCLEAR STRUCTURE WITH PROTONS AND PIONS AT LAMPF

Norton M. Hintz, D. Cook, M. Gazzaly and M. Franey,
University of Minnesota

I. Introduction

We will present here a few examples of nuclear structure information obtained from protons and pions at LAMPF. For extracting charge and current distributions, electron scattering can't be beat. Thus, one of the basic goals in hadron scattering is to obtain neutron matter, spin, and current, static and transition densities. First let us look at the sensitivity of various probes to neutron-proton density differences. If the average projectile-nucleon interaction strengths are g_p and g_n, then we want to maximize the difference, $\Delta\alpha$, for two probes, of the quantity $\alpha = (g_p-g_n)/(g_p+g_n)$ to obtain separate neutron and proton densities. For electron and intermediate energy proton probes, $\Delta\alpha \simeq 1$, as it is for π^+ and π^- near the (3,3) resonance, and low energy neutrons and protons ($g_{np} \simeq 3g_{pp}$). The best combination would be e vs π^- ($\Delta\alpha = 1\frac{1}{2}$). From practical considerations (counting rates, energy resolution, validity of reaction models) the e vs p comparison seems most attractive. The above remarks apply mainly to excitations (including elastic scattering) which are dominated by the central, spin-independent parts of the force. For spin and isospin modes other projectile combinations may be superior. For example, the giant isovector monopole resonance, was first revealed in the (π^+,π^o) charge exchange reaction. To extract transition (or static) densities from experimental data we have been using non-relativistic versions of the KMT or Glauber theories (elastic) or the Distorted Wave Impulse Approximation (DWIA). Sometimes media modified effective forces are used. Recently Dirac relativistic versions of the above have been developed. In the Plane Wave IA (PWIA), the scattering amplitude, $f(q)$ can be written[1] in terms of the sum of terms of the form $t(q)\rho(q)$ where $t(q)$ is the projectile-target nucleon t-matrix and $\rho(q)$ is the appropriate density (matter, spin, current, etc) in q-space. In the IA, $t(r)$ is the object which when Fourier-transformed is equal to the free projectile-nucleon scattering amplitude, aside from

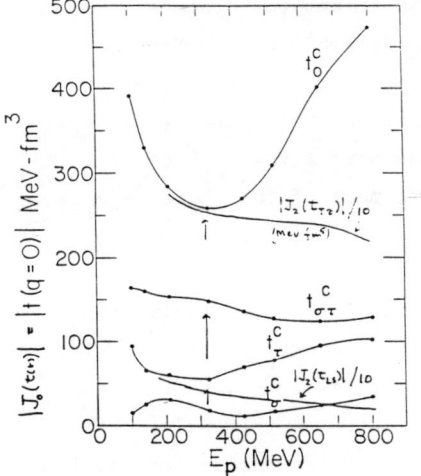

Fig. 1. Energy dependence of N-N t-matrix (σ: spin, τ: i-spin, T: tensor, LS: spin-orbit)(Ref.4)

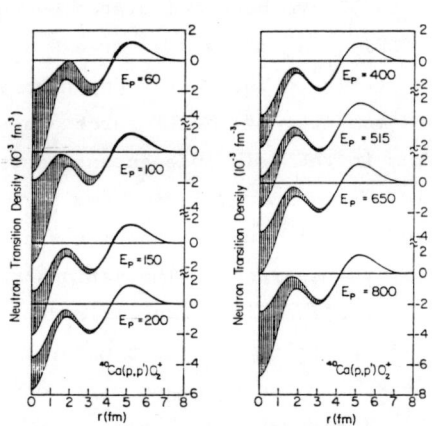

Fig. 2. Error envelopes for ρ_n^{tr} for $^{40}Ca(0_2^+)$ from (p,p') pseudodata (Ref 5).

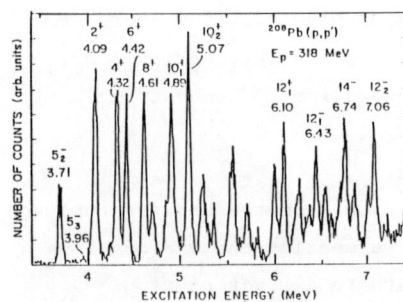

Fig. 3. Spectrum for $^{208}Pb(pp')$ at $\theta_L = 28°$, $\Delta E \approx 40$ KeV.

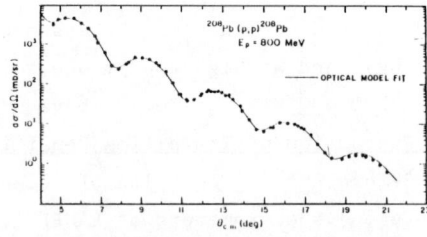

Fig. 4. Elastic cross section for ^{208}Pb with phenomenological optical potential (POP) fit (Ref. 6).

kinematic factors. Medium modified t-matrices for protons ("G-matrices") have been calculated by von Geramb[2] and by Nakayama, et al[3] (GBJ). A widely used free N-N t-matrix is that of Love and Franey (LF)[4]. The energy dependence of the volume integrals (or second moments) of the LF force are shown in Fig. 1. Because of the minimum in the spin-independent central part of the force (t_o^c) at $\simeq$ 300 MeV, this energy is optimum for exciting spin flip modes (including all unnatural parity states) and is the region of greatest nuclear transparency. The various parts of t(q) (central, spin, tensor, etc) couple to specific densities, as defined in Ref. 1. For elastic scattering and most natural parity excitations (in I = 0 targets) only the $t_o^c \rho^J$ and $t_{LS} \rho^J$ contribute significantly, where ρ^J is the Jth moment of the neutron or proton matter density for total angular momentum transfer, J. Transitions to unnatural parity states involve only spin dependent parts of the force (t_σ, t_{LS}, t_T) coupling to spin and orbital densities. In the special case of stretched p-h states (J = j_p + j_h) only the spin density terms survive in both electron and hadron scattering.

An indication of the radial sensitivity to transition densities in (p,p') at various proton energies is given by a model independent analysis of pseudo-data for the 0_2^+ state of ^{40}Ca, shown in Fig. 2 (J. Kelly[5]). The error band near the origin spreads at low energy due to the density dependence of the assumed effective interaction (Pauli blocking), and at high energy due to absorption. The best region is around 200-400 MeV, as noted above.

II. Extraction of Transition Densities

Before showing examples of recent data from the HRS (proton) and EPICS(pion) spectrometers at LAMPF let me mention the various levels of sophistication in the analysis of data to extract neutron transition densities. In most cases we take the proton densities from electron scattering, although this is not necessary in a π^+ - π^- comparison. For protons, either the free (LF) or medium modified (Geramb or GBJ) t-matrix can be used to obtain the elastic distorting potential, and the transition potentials. In many cases

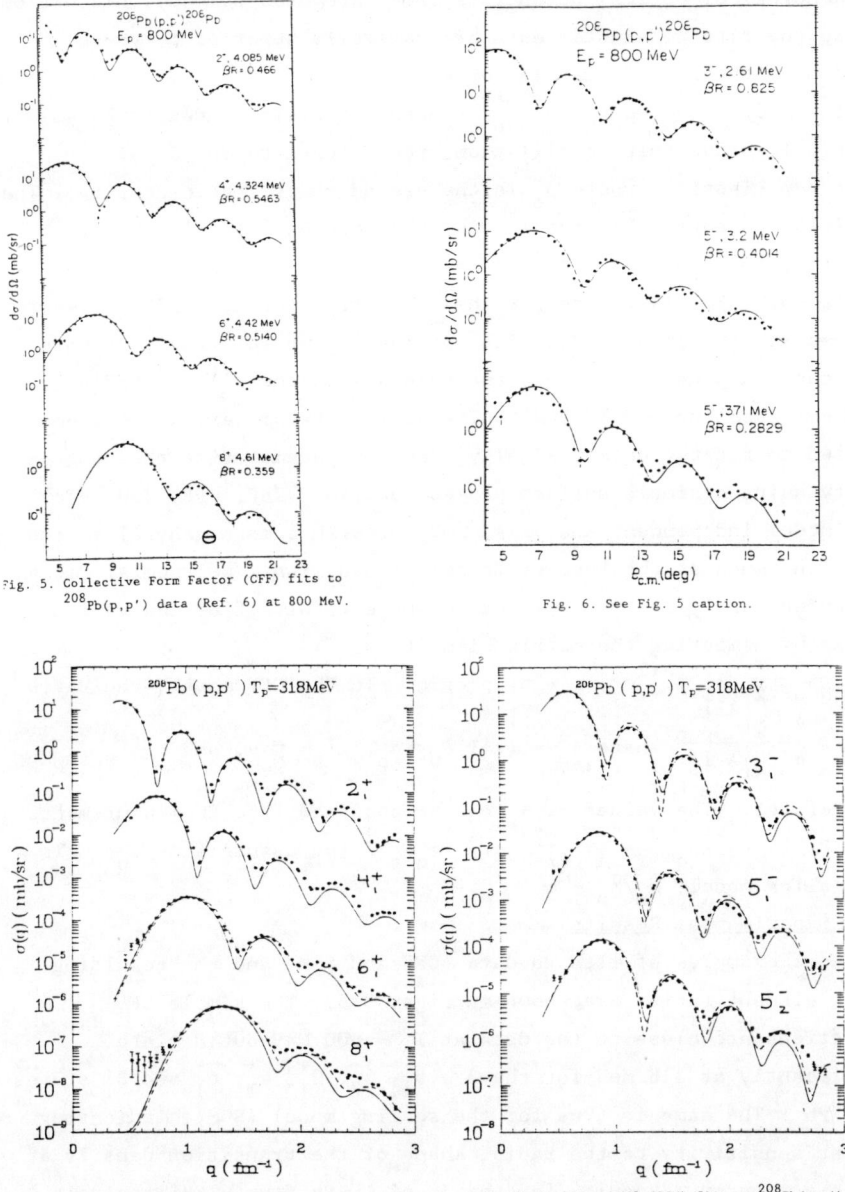

Fig. 5. Collective Form Factor (CFF) fits to ^{208}Pb(p,p') data (Ref. 6) at 800 MeV.

Fig. 6. See Fig. 5 caption.

Fig. 7. CFF fits to ^{208}Pb(p,p') at 318 MeV.

Fig. 8. Scaling model (SCM) fits to ^{208}Pb(p,p') at 318 MeV.

phenomenological optical potentials (POP) are used to model distortion because the fits to elastic data are generally superior to those obtained in the first order IA (or KMT).

The most naive model for ρ_n^{tr} (neutron transition density) is a simple collective shape oscillation proportional to $\partial \rho_o/\partial r$ or $r^{\lambda-1} \partial \rho_o/\partial r$ (Tassie), where ρ_o is the ground state static density. The transition potential, in the IA, is then obtained by folding with the t-matrix ("$t\rho$" approx.). An even simpler prescription (collective form factor, CFF) is to use for the transition potential $\delta_u \frac{\partial U}{\delta r}$, where U is usually the pheonomenological optical potential (POP). A more microscopic approach is to use the DWIA and assume $\rho_n^{tr} = \alpha$ N/Z ρ_p^{tr} (from ee')(scaling model, SCM). The scaling parameter, α, is then adjusted to fit the data. Finally, one can parameterize the neutron density using a simple surface peaked function (2PF, 3PF, 3PG, etc) (semi-model independent analysis, SMI) or with a more general series expansion, such as the Fourier-Bessel or Gaussian x polynomial (model independent analysis, MI). We can compare results from the various methods by computing the matrix element

$$\widetilde{M}_i = \frac{1}{N \text{ or } Z} \int \rho_i^{tr} r^{\lambda+2} dr, \quad i = n, p \text{ or } q.$$ In the CFF model, the ratio

$$\widetilde{M}_n/\widetilde{M}_p = \frac{\delta_n}{\delta_p} \frac{<r^{\lambda-1}>_n}{<r^{\lambda-1}>_p} \text{ where } \delta_n = (\tau+1)\delta_u - \tau\delta_p, \quad \tau = \frac{Z}{N}\nu, \text{ and } \nu \simeq g_n/g_p$$

(see Ref. 6). The values of δ_p can be obtained from EM measurements since $B(E\lambda) = |Z\widetilde{M}_q|^2 = [\frac{Z(\lambda+2)}{4\pi} <r^{\lambda-1}>_q \delta_p]^2$ (we assume $\delta_p = \delta_q$). In the scaling model, $\widetilde{M}_n/\widetilde{M}_p = \alpha$.

III. Experimental Results

Some examples of fits to data (CFF and SCM) and the resulting matrix element ratios are shown in Figs. 3-8. The simple CFF predictions are close to the data at T_p = 800 MeV but deviate significantly at 318 MeV for the 3_1^-, 5_1^-, 5_2^-, 2_1^+, 4_1^+, 6_1^+ and 8_1^+ states of ^{208}Pb. The same is true for the scaling model (SCM), indicating greater sensitivity to the radial shape of the transition density at the lower energy. Results for the 3_1^- of ^{208}Pb from a SMI analysis[7] are shown in Figs. 9 and 10. It is seen that in this model ρ_n^{tr} peaks slightly <u>inside</u> of ρ_p^{tr}.

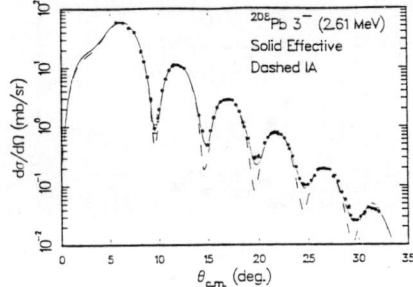

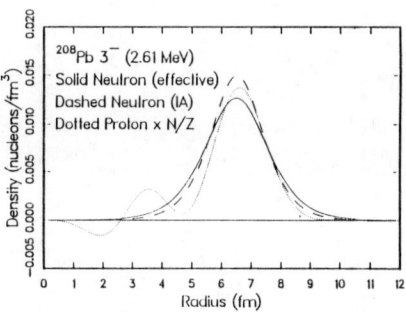

Fig. 10. Neutron transition density from SMI fit of Fig. 9.

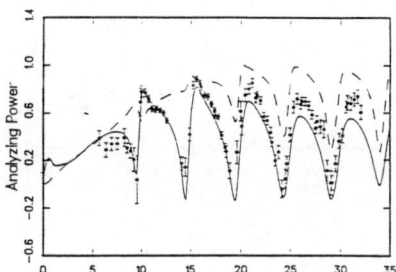

Fig. 9. Semi model Independent (SMI) fits to ^{208}Pb(p,p'), 3_1^- state at 500 MeV (Ref 7).

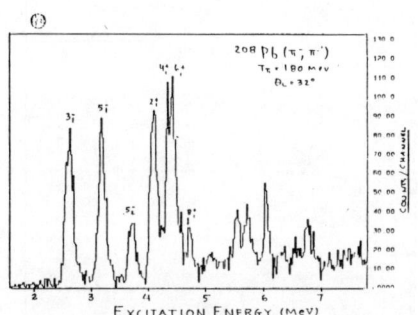

Fig. 11. Spectrum for ^{208}Pb($\pi^-,\pi^{-'}$) at $T_\pi = 180$ MeV.

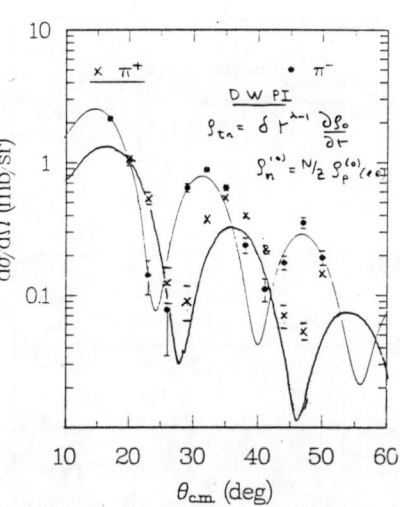

Fig. 12. Tassie transition density fit to ^{208}Pb(π,π'), 2_1^+.

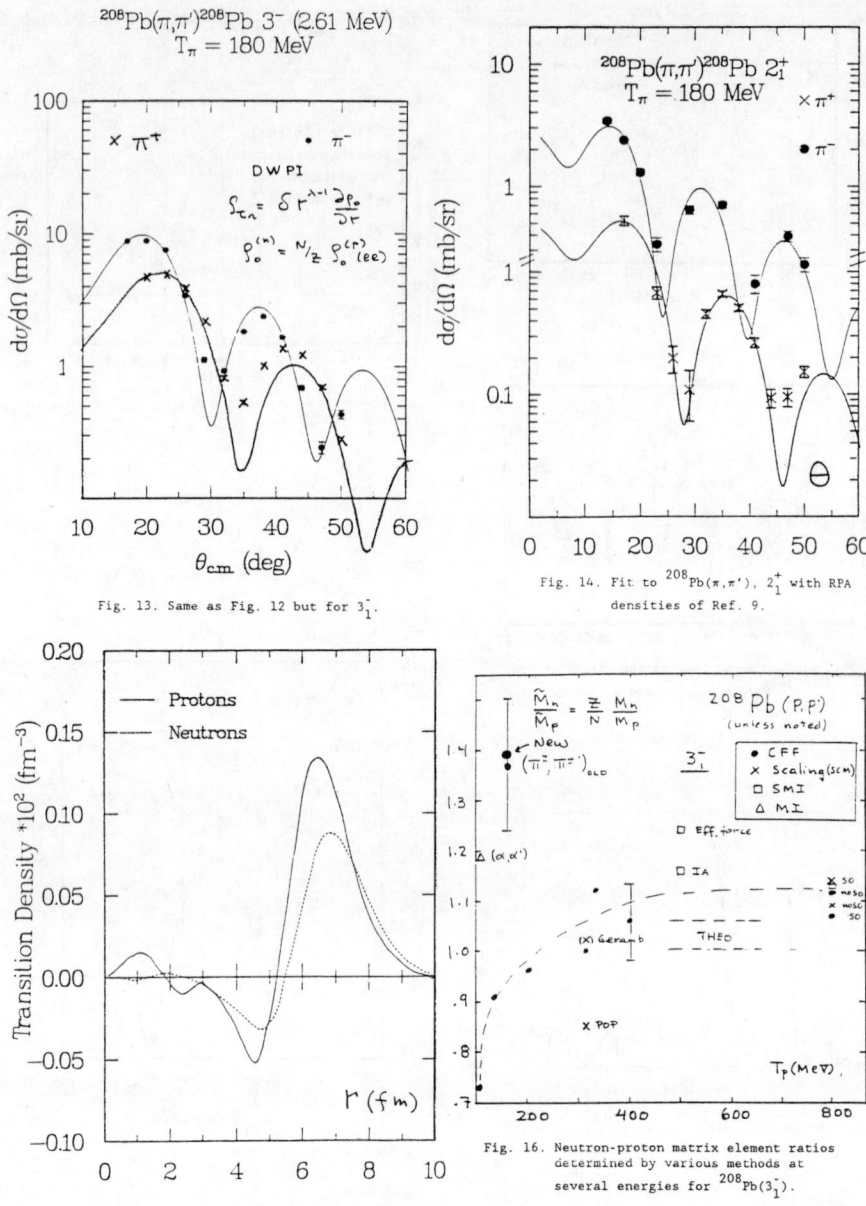

Fig. 13. Same as Fig. 12 but for 3_1^-.

Fig. 14. Fit to ^{208}Pb(π,π'), 2_1^+ with RPA densities of Ref. 9.

Fig. 15. RPA transition densities of Ref. 9.

Fig. 16. Neutron-proton matrix element ratios determined by various methods at several energies for ^{208}Pb(3_1^-).

Some reduced ($\tilde{M}_n/\tilde{M}_p$) neutron-proton matrix element ratios from the 800 MeV CFF analysis are given in Table I for ^{58}Ni (open neutron shell) and ^{90}Zr (open proton). It is seen that the lowest 2^+ has $\tilde{M}_n/\tilde{M}_p > 1$ for the valence neutron case and < 1 for the valence proton nucleus. In each case, however, there are one or more low lying 2^+ "reversed states"[8] with the opposite behavior. This seems to be a rather general situation for open shell nuclei and is an example of states of "mixed symmetry" of intermediate character.

We have also obtained data on $\pi^{\pm}$ inelastic scattering from ^{208}Pb for most of the low lying natural parity states (Figs. 11-15). The simple CFF or Tassie prescriptions with $\rho_n = \alpha$ N/Z ρ_p give relatively poor fits to the data, but for the 2_1^+ state, the RPA densities of Wambach, et al[9] do very well (Figs. 14,15).

The systematics of $\tilde{M}_n/\tilde{M}_p$ ratios determined by various methods (CFF, SCM, SMI, etc) from e-p or $\pi^+ - \pi^-$ comparisons at several (proton) energies are shown in Figs. 16-18. Errors are hard to assign since they are mostly model and not statistical but are at least $\pm$ 10%. It can be seen that a serious energy dependence exists for the derived values below T_p = 500 MeV. We believe this to be due to a deficiency, at higher q, of the central, spin-independent, and possibly spin-orbit parts of the N-N interactions used in the calculations. Even in the CFF analysis, with POP's (U), where the transition potential is $\delta U/\partial r$, this unphysical energy dependence is seen. However, the elastic optical potential depends on t(q) only for $q \leq .2f^{-1}$, whereas for the inelastic states (J = 2-8), q-values in the range $q \simeq .5 - 1.3$ fm^{-1} are involved. The phenomenological effective force of Ref. 7 at 500 MeV does better than the free or von Geramb force below 500 MeV. We conclude that at present no satisfactory theoretical effective interaction exists for $T_p \leq$ 500 MeV. For the present, the 800 MeV $\tilde{M}_n/\tilde{M}_p$ values seem most reliable as they are close to unity, as expected, for N = Z nuclei (^{28}Si, ^{40}Ca) or known isoscalar excitations such as ^{208}Pb(3^-) and ^{58}Ni(6^+) (see below).

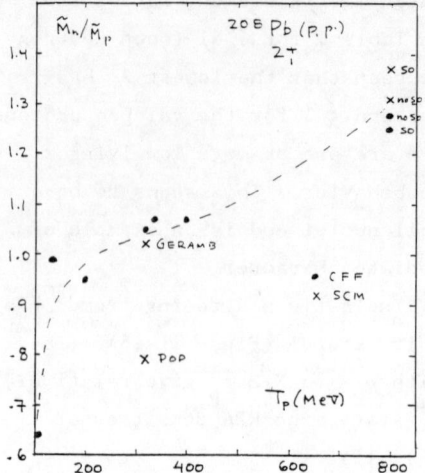

Fig. 17. Same as Fig. 16 but for 2_1^+ state.

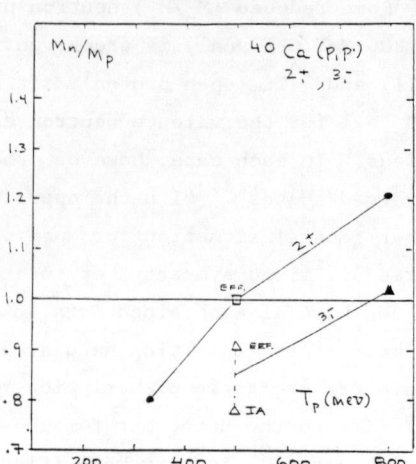

Fig. 18. Same as Fig. 16 but for ^{40}Ca(p,p'), 2_1^+, 3_1^-, states.

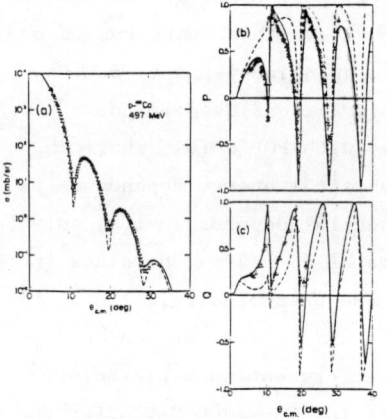

Fig. 19. <u>Dirac</u> and <u>KMT</u> impulse approximation calculations for p + ^{40}Ca at 497 MeV (See Ref. 18).

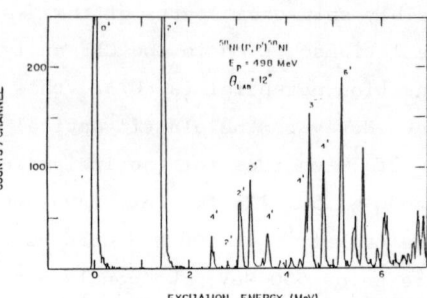

Fig. 20. Spectrum for ^{58}Ni(p,p') at 498 MeV (Ref. 16).

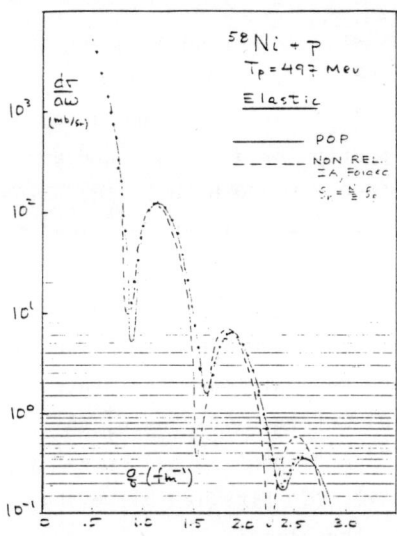

Fig. 21. POP and non relativistic IA (NRIA) predictions for p + ^{58}Ni (elastic cross sections at 497 MeV.

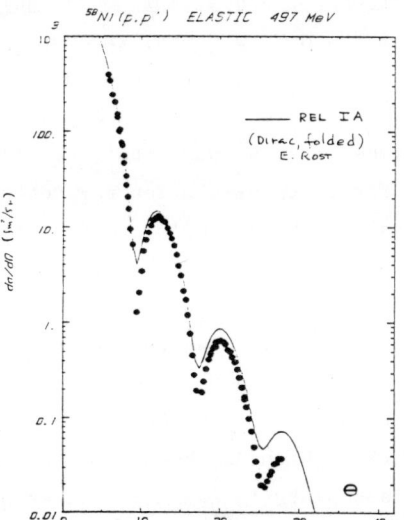

Fig. 22. Dirac IA (DIA) prediction for p + ^{58}Ni elastic scattering.

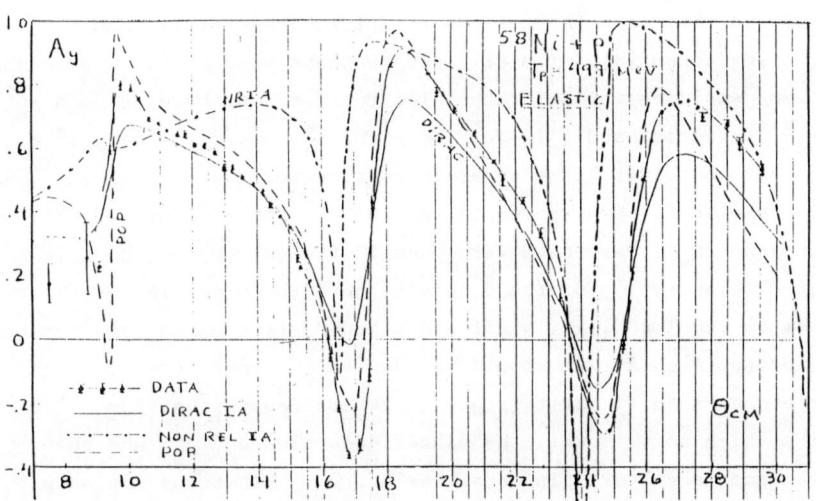

Fig. 23. POP, NRIA and DIA predictions for Ay, p + ^{58}Ni(elastic) at T_p = 497 MeV.

IV Dirac Approaches, Spin Observables

At 800 MeV (protons) the non-relativistic IA (or KMT), using the free N-N interaction has proved very satisfactory in describing elastic cross section and analyzing power (A_y) data. It came as a shock when it was found that the IA did poorly for σ and A_y, and even worse for Q (spin-rotation parameter) at 500 MeV and below. The early successes of Dirac POP's in fitting <u>all</u> of the elastic observables[10] encouraged the development of the Dirac IA for elastic[11,12] (Fig 19) and inelastic[13,14] scattering. To test the Dirac IA for inelastic processes we have measured several cross section and spin observables (σ, A_y, Q, D_{SS} and D_{SL}) for the ground and several low lying states of ^{58}Ni in (pp') at T_p = 500 MeV. Of special interest was the 6_1^+ (5.13 MeV) state which has been seen in (ee')[15] and (p,p')[16] as a nearly pure isoscalar ($f_{7/2}^{-1} f_{5/2}$) excitation (94%). In Figs. 20-31 we show preliminary data on these observables for the elastic and 6_1^+ transitions, along with calculations using a) phenomenological distorting potentials (POP), b) the non relativistic IA (NRIA) (free t-matrix) and the Dirac IA[17] (free relativistic t-matrix, and Dirac IA distorting potential). In all cases we have used proton static and transition densities which fit the electron data and have assumed $\rho_n^{(o)}$ = N/Z $\rho_p^{(o)}$ (elastic) and $\rho_n^{tr} = \rho_p^{tr}$ (for the 6_1^+, $\tau = 0$ state). It can be seen that the Dirac first order IA is superior to the NRIA in fitting the elastic observables (σ, A_y, Q). Of course the POP does best for σ and A_y (these data were used in the search) but does less well for Q (elastic). For the 6_1^+ state, both the Dirac IA and the NRIA seem to require slightly different bound state parameters for the (e,e') and (pp') data. It should be pointed out that for this high spin stretched natural parity state there are four transition densities which contribute: the matter densities ρ_n^J, ρ_p^J; and the isoscalar and isovector spin densities ρ_o^S, ρ_1^S. Electrons see only the ρ_p and ρ_1^S densities. Thus we are free to "fine tune" the ρ_n^J and ρ_o^S densities to optimize agreement with the proton data.

In Fig. 32 we have summarized the normalization (or quenching) factors, $N^2 = \sigma_{exp}/\sigma_{theo}$ for several stretched, or nearly stretched

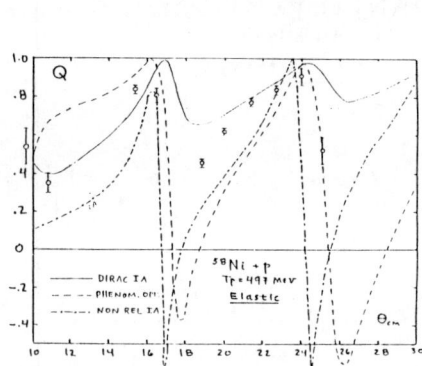

Fig. 24. Same as Fig. 23 but for Q parameter.

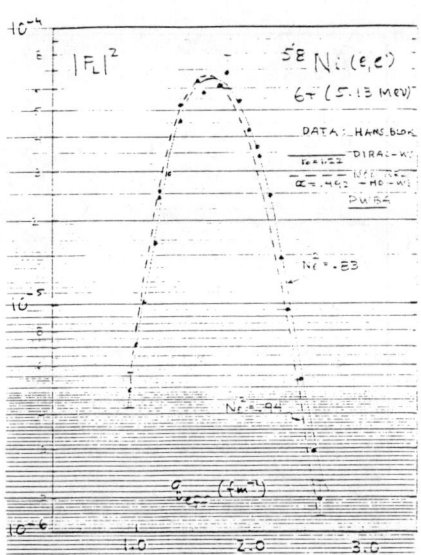

Fig. 25. Dirac and non relativistic fits to ^{58}Ni(e,e') 6_1^+ $|F_L|^2$. Data from Ref. 19.

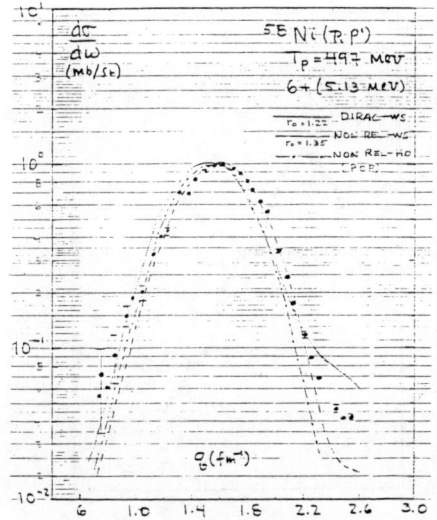

Fig. 26. DIA and NRIA predictions for ^{58}Ni(p,p') 6_1^+ cross section with transition density as in Fig. 25.

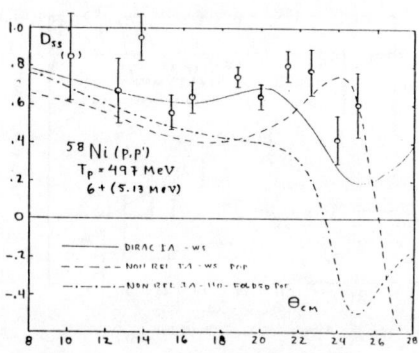

Fig. 28. Same as Fig. 26 but for D_{SS}.

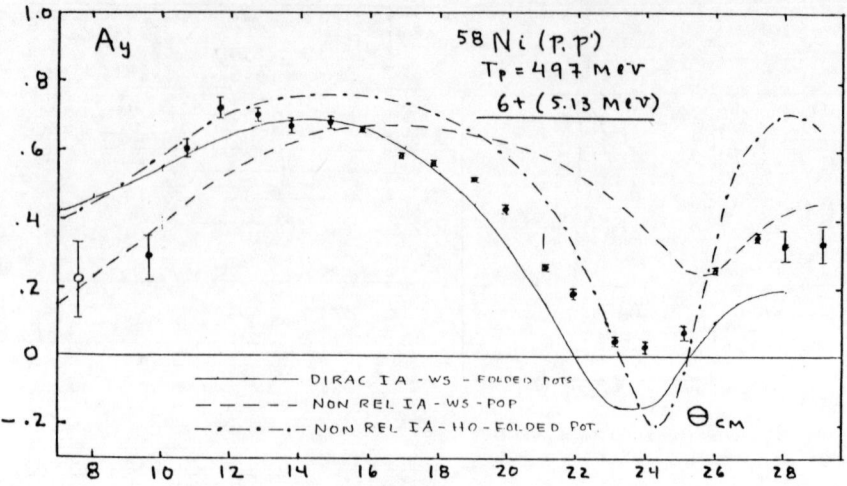

Fig. 27. Same as Fig. 26 but for Ay.

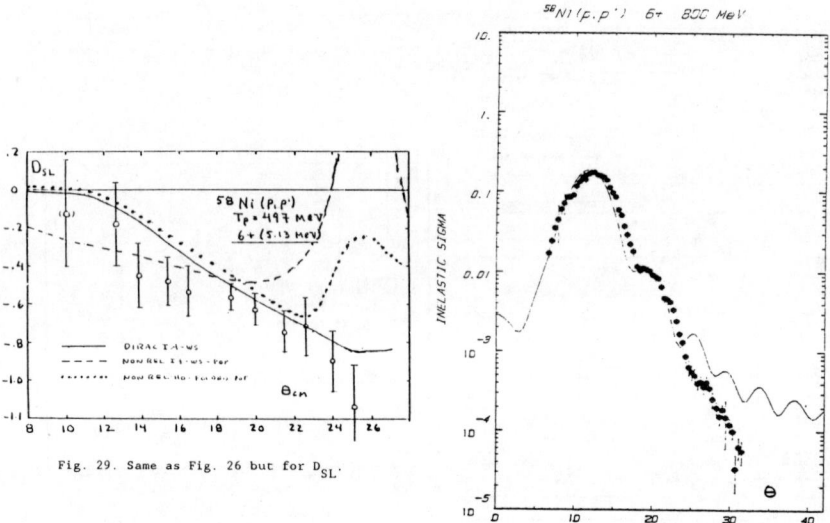

Fig. 29. Same as Fig. 26 but for D_{SL}.

Fig. 30. Dirac IA predictions for ^{58}Ni(p,p')6$^+$ at 800 MeV.

states in ^{28}Si, ^{58}Ni and ^{208}Pb measured by electron, proton and pion scattering. The quantity, σ_{theo}, has been calculated in the DWIA (or Dirac IA), generally assuming a single 1p-1h configuration, such as $(1d_{5/2}^{-1} \, 1f_{7/2})$ for ^{28}Si(6^-).

It can be seen that the (p,p') values at various energies are in reasonable agreement with (e,e') or (π,π') values for the unnatural parity states (dominated by the spin orbit and tensor parts of t(q)) but are not for the natural parity transitions (central and spin-orbit) for $T_p \leq 500$ MeV. Here again we see an energy dependence similar to that for the low spin "collective" transitions discussed in Sec. III. This again suggests the problem in the use of the DWIA lies in the large q behavior of the spin-independent central parts of the free t-matrix at lower proton energies, and is common to both non-relativistic and Dirac approaches.

In conclusion, we believe we are beginning to extract reliable neutron densities from (p,p') and (π,π') data, using for (p,p') the free interaction at 800 MeV or phenomenological effective interactions at lower energy. More theoretical work needs to be done to improve the reaction models.

1. F. Petrovich and W.G. Love, Proceedings of the International Conference on Nuclear Physics, Berkeley, 499C (1980).
2. H.V. von Geramb, private communication.
3. K. Nakagama, S. Krewald, J. Speth and W.G. Love, Nucl. Phys. A431, 419(1984).
4. M.A. Franey and W.G. Love, Phys. Rev. C31, 488(1985).
5. J. Kelley, AIP Conf. Proc. 142, 27(1986).
6. M. Gazzaly, N. Hintz, et al Phys. Rev. C25, 408(1982).
7. L. Ray and G.W. Hoffman, Phys. Rev. C27, 2133(1983) and to be published (with M. Barlett).
8. V.R. Brown, A.M. Bernstein, V.A. Madsen, UCRL-93148(1985).
9. J. Wambach, F. Osterfeld, J. Speth, V. Madsen, Nucl. Phys. A324, 77(1979).
10. B.C. Clark, S. Hana, R.L. Mercer AIP Conf. Proc 97(1983).
11. J.A. McNeil, J.R. Shepard, S.J. Wallace, Phys. Rev. Lett. 50, 1439(1983); Phys. Rev. Lett. 50, 1443(1983).
12. J.A. McNeil, L. Ray, S.J. Wallace, Phys. Rev. C27, 2123(1983).
13. J.R. Shepard, E. Rost, J. Piekarewicz, Phys. Rev. C30, 1604(1984).
14. J.R. Shepard, E. Rost, J.A. McNeil, Phys. Rev. C33, 634(1986).
15. R.A. Lindgren et al, Phys. Rev. Lett. 41, 1705(1978).
16. N. Hintz, et al, Phys. Rev. C30, 1976(1984).

17. We are indebted to E. Rost for the Dirac calculations.
18. See references in S. Wallace, LA-10080-C (Los Alamos Nat Lab) 45(1984).
19. Hans Blok, private communication.

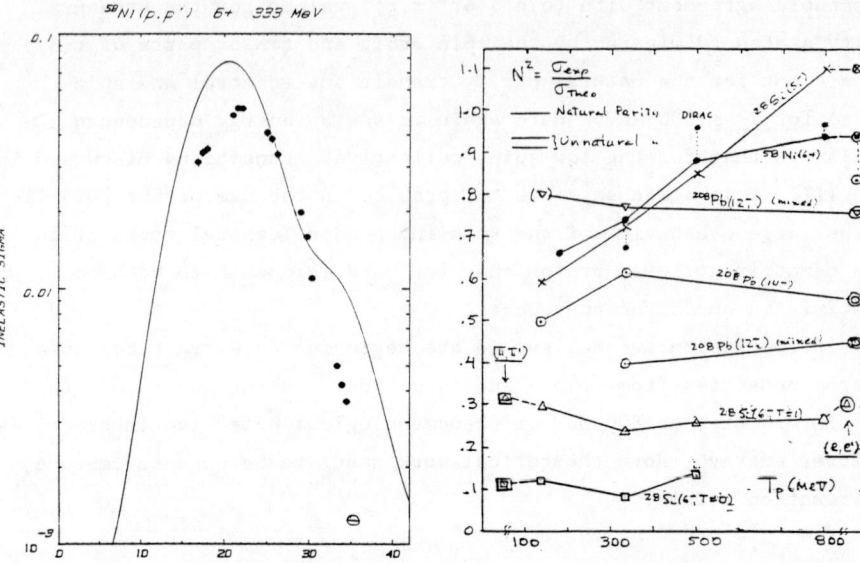

Fig. 31. Same as Fig. 30 but for $T_p = 333$ MeV.

Fig. 32. Energy dependence of normalization factors for inelastic excitation of stretched states.

Table I.   Matrix Element Ratios

Nucleus, J^π(MeV)	$\tilde{M}_n/\tilde{M}_p$	M_1/M_o	ϵ
^{58}Ni, $2_1^+(1.46)$	1.20	.13	
$2_3^+(3.04)$	.88	-.03	.03
$2_4^+(3.26)$	.84	-.05	
^{90}Zr, $2_1^+(2.19)$	.90	.06	
$2_2^+(3.31)$	1.54	.3	.11

$M_1/M_o = (M_n - M_p)/(M_n + M_p)$, $\epsilon = (N-Z)/A$

$\tilde{M}_n/\tilde{M}_p = ZM_n/NM_p$, $M_i = \int \rho_i^{tr}(r) \, r^{\lambda+2} \, dr$

SECTION V

SYMMETRIES IN REACTIONS

F. IACHELLO

Reviewer

L. Biedenharn

Discussion Moderator

Session Moderators

D. Rowe
G. Temmer

A REVIEW OF SYMMETRIES IN NUCLEAR REACTIONS

F. Iachello

A.W. Wright Nuclear Structure Laboratory,
Yale University
New Haven, Connecticut 06511

Abstract

I review the use of group theoretic techniques in the study of nuclear reactions. In particular, I discuss a model suggested recently for the description of elastic scattering between heavy ions.

1. INTRODUCTION

The subject of nuclear reactions is an old one and several approaches have been developed in order to deal with it. The use of symmetry considerations may provide further help. Symmetries can be used here in two ways. First in providing an algebraic description of the bound states of the colliding particles. This aspect is discussed at this Conference by R. Amado [1] and J. Ginocchio [2]. The second way is in providing an algebraic description of the collision process itself. In the last few years, attempts have been made to develop an algebraic theory of scattering [3] which effectively converts the Schrödinger equation into an <u>algebraic</u> equation. In many cases, the algebraic equation can be solved producing S-matrices in closed form. Closed explicit expressions are particularly useful in analyzing data since they require little if no numerical effort at all. Furthemore they provide insights into the problem and quite often lead to new developments. In this talk, I will review only the second of the two aspects mentioned above and refer the reader to the talks of Amado [1] and Ginocchio [2] for a review of the first aspect.

2. PRELIMINARIES

I will concentrate my attention here to elastic

scattering for which most of the results have been obtained so far. In the Schrödinger approach, the cross section is obtained by first solving the differential equation

$$\left[-\frac{\hbar^2}{2\mu}\nabla^2 + V(r)\right]\psi(r) = E\,\psi(r) \quad , \qquad (2.1)$$

with appropriate boundary conditions. For example, for a potential decreasing faster than $1/r$,

$$\psi(r) \underset{r\to\infty}{\to} e^{i\mathbf{k}\cdot\mathbf{r}} + f(\Omega)\frac{e^{ikr}}{r} \quad . \qquad (2.2)$$

This method produces the S-matrix and thus the cross section. For non-relativistic problems it is convenient to introduce a partial-wave expansion of (2.1). The S-matrix for partial wave ℓ can then be written as

$$S_\ell(k) = e^{2i\delta_\ell(k)} \quad , \qquad (2.3)$$

and the scattering amplitude as

$$a(k,\theta) = \frac{1}{2ik}\sum_{\ell=0}^{\infty}(2\ell+1)[S_\ell(k)-1]P_\ell(\cos\theta). \quad (2.4)$$

The cross section is then

$$\frac{d\sigma}{d\Omega} = |a(k,\theta)|^2 \quad . \qquad (2.5)$$

3. S-MATRICES FROM GROUP THEORY

It has been found in the last few years that explicit forms of S-matrices can be obtained from a purely abstract formulation of the scattering problem using group theory. This occurs whenever the problem has a dynamic symmetry described by the group G. The process by means of which the S-matrix can be computed consists in an expansion of the representations of G describing the system in presence of interactions into representations of a group F (called the asymptotic group) describing the system in the absence of interactions, schematically written as

$$\boxed{\text{Representations of } G = \\ = (\text{S-matrix}) \times \text{Representations of } F} \quad . \quad (3.1)$$

It is important to note that, contrary to the case of bound state problems where the knowledge of the dynamic group G is sufficient to determine the properties of the system, for scattering problems we need the knowledge of both the dynamic group G and the asymptotic group F. This is in general a complex mathematical problem, as discussed by Biedenharn [4], but for obtaining the results presented here it is sufficient to consider a simpler form. For non-relativistic problems in a partial wave formalism, the asymptotic group F must contain the Euclidean group E(3). The relation (3.1) then corresponds to the Jost expansion

$$\psi_\ell(k,r,\Omega) \underset{r\to\infty}{\to} \phi^{(-)}(\ell,k)\, e^{ikr}\, Y_{\ell m}(\Omega) +$$
$$+ \phi^{(+)}(\ell,k)\, e^{-ikr}\, Y_{\ell m}(\Omega) \quad , \quad (3.2)$$

as it will become apparent in the sections below.

Group theory provides algebraic relations for the Jost functions $\phi^{(\pm)}(\ell,k)$ which can be solved explicitly.

4. THE COULOMB PROBLEM

The simplest illustration of the group method is provided by the non-relativistic Coulomb problem. Here it has been known for many years [5] that the dynamic group G is SO(3,1), with six generators, the angular momentum, **L**, and the Runge-Lenz vector, **A**. The asymptotic group is E(3) with six generators, the angular momentum, **L**, and the momentum, **P**. The representations of G are labelled by

$$G: \left| \begin{array}{ccc} SO(3,1) & \supset SO(3) & \supset SO(2) \\ \downarrow & \downarrow & \downarrow \\ (\omega,0) & \ell & m \end{array} \right\rangle \quad . \quad (4.1)$$

Those of F by

$$F: \left| \begin{array}{ccc} E(3) & \supset SO(3) & \supset SO(2) \\ \downarrow & \downarrow & \downarrow \\ (\pm k,0) & \ell & m \end{array} \right\rangle \quad . \quad (4.2)$$

In (4.2) I have used the continuous unitary representations of SO(3,1) with

$$\langle C_2 \rangle = \omega(\omega+2) \quad ; \quad \omega = -1+if(k) \quad ;$$

$$\langle C'_2 \rangle = 0 \quad , \tag{4.3}$$

where $C_2 = \mathbf{L}^2 - \mathbf{K}^2$, $C'_2 = \mathbf{L} \cdot \mathbf{K}$, $\mathbf{K} = (\mu\mathbf{A})/k$. The representations of E(3) are characterized by the eigenvalues of $\mathbf{P}^2$,

$$\langle \mathbf{P}^2 \rangle = k^2 \tag{4.4}$$

and labeled by ±k. Also here $\langle \mathbf{P} \cdot \mathbf{L} \rangle = 0$. When written explicitly, Eq.(4.1) reads

$$|\omega,\ell,m\rangle = A_\ell(k) |-k,\ell,m\rangle + B_\ell(k) |+k,\ell,m\rangle \tag{4.5}$$

where $|-k,\ell,m\rangle$ and $|+k,\ell,m\rangle$ are the free incoming and outgoing waves. Eq.(4.5) is identical to the Jost expansion (3.2). The Jost functions $\phi^{(\pm)}(\ell,k)$ are denoted by $A_\ell(k)$ and $B_\ell(k)$ in (4.5). In order to obtain recursion relations for the Jost functions $A_\ell(k)$ and $B_\ell(k)$ one writes the generators of G in terms of those of F. This technique, called <u>Euclidean connection</u>, is the crucial point in the whole derivation and is discussed by Y. Alhassid in his contribution [6]. The Euclidean connection provides recursion relations for $A_\ell(k)$, $B_\ell(k)$ and their ratios $R_\ell(k) = B_\ell(k)/A_\ell(k)$,

$$R_{\ell+1}(k) = -\frac{\ell+1+if(k)}{\ell+1-if(k)} R_\ell(k) \quad . \tag{4.6}$$

The recursion relations can be solved to yield the S-matrix $S_\ell(k) = e^{i(\ell+1)\pi} R_\ell(k)$, i.e.,

$$S_\ell(k) = \frac{\Gamma(\ell+1+if(k))}{\Gamma(\ell+1-if(k))} \quad , \tag{4.7}$$

and the cross section

$$\frac{d\sigma}{d\Omega} = \frac{f^2(k)}{4k^2 \sin^4(\theta/2)} \quad . \tag{4.8}$$

The particular functional form (4.7) is due to the dynamic group SO(3,1). All problems with SO(3,1) dynamic symmetry have S-matrices of this form. The function f(k) is determined either by the relation between the Hamiltonian and the Casimir invariants of SO(3,1), if this relation is known, or can be obtained by fitting the experimental data. For pure Coulomb scattering

$$H = -\frac{\mu\beta^2}{2(C_2+1)} \quad , \quad \beta = Z_1 Z_2 e^2 \quad , \qquad (4.9)$$

gives

$$f(k) = \frac{\mu\beta}{k} \quad . \qquad (4.10)$$

The SO(3,1) scattering amplitude (4.7) can be generalized by making $f(k)$ a function of ℓ. Since ℓ is related to the eigenvalues of the Casimir invariant of SO(3), this generalization maintains the SO(3,1) symmetry but makes the relation between the Hamiltonian H and the Casimir invariants more complex. Making $f(k)$ a function of ℓ, i.e. $f_\ell(k)$, adds slightly to the numerical complexity of the problem. The S-matrix is still given by an explicit expression. The only difficulty now is that the partial wave expansion cannot be summed in closed form as in (4.8). S-matrices of the type

$$S_\ell(k) = \frac{\Gamma(\ell+1+if_\ell(k))}{\Gamma(\ell+1-if_\ell(k))} \qquad (4.11)$$

can be used to analyze scattering by potentials more complex than the Coulomb potential, for example by the Yukawa potential, $V(r) = -\gamma e^{-\nu r}/r$.

5. THE MODIFIED COULOMB PROBLEM

This is the problem of particular interest in nuclear physics since the interaction between two ions is dominated at large distances by their Coulomb repulsion while at short distances this repulsion is modified by the short range nuclear interaction. While, in principle, one could use the SO(3,1) amplitude (4.11) to analyze this case, it is more convenient to use a more general form of scattering amplitudes which separates explicitly the short range from the long range behavior. This is achieved by introducing a more general construct based on the larger dynamic group G≡SO(3,2). The asymptotic group F corresponding to this case is F≡E(3)⊗E(2) composed of a part, E(3), describing the asymptotic space group and a part, E(2), describing the asymptotic internal group. The representations of G are labeled now by

$$G: \left| \begin{array}{c} SO(3,2) \supset SO(3) \times SO(2) \supset SO(2) \times SO(2) \\ \downarrow \quad\quad \downarrow \quad\quad\quad \downarrow \quad\quad\quad \downarrow \\ (\omega,0) \quad\quad \ell \quad\quad\quad m \quad\quad\quad v \end{array} \right\rangle \quad (5.1)$$

while those of F are labeled by

$$F: \left| \begin{array}{c} E(3) \times E(2) \supset SO(3) \times SO(2) \supset SO(2) \times SO(2) \\ \downarrow \quad\quad \downarrow \quad\quad \downarrow \quad\quad\quad \downarrow \quad\quad\quad \downarrow \\ (\pm k,0) \quad (1) \quad \ell \quad\quad\quad m \quad\quad\quad v \end{array} \right\rangle \quad (5.2)$$

where again I have used the continuous unitary representations of SO(3,2) with

$$\langle C_2 \rangle = \omega(\omega+3) \quad , \quad \omega = -\frac{3}{2} + if(k) \quad ,$$

$$\langle C'_2 \rangle = 0 \quad . \quad\quad\quad (5.3)$$

While the quantum numbers $\omega, \ell, m, \pm k$ have the same meaning as in the Coulomb problem, the quantum number v has the meaning of the added interaction as it will become clearer later on.

The Jost expansion is in this case

$$|\omega, \ell, m, v\rangle = A_{\ell v}(k) |-k, \ell, m, v\rangle + B_{\ell v}(k) |+k, \ell, m, v\rangle \quad .$$
$$(5.4)$$

Using the Euclidean connection one can obtain the S-matrix in the form

$$S_\ell(k) = \frac{\Gamma\left(\frac{\ell+v+3/2+if(k)}{2}\right) \Gamma\left(\frac{\ell-v+3/2+if(k)}{2}\right)}{\Gamma\left(\frac{\ell+v+3/2-if(k)}{2}\right) \Gamma\left(\frac{\ell-v+3/2-if(k)}{2}\right)} \times$$

$$\times e^{i(2\ln 2)f(k)} \quad\quad\quad . \quad (5.5)$$

It is interesting to observe that if one lets in (5.5) $v=1/2$ one obtains back the Coulomb S-matrix (4.7) as one can see by using the duplication formula

$$\Gamma(z) \Gamma(z+1/2) = 2^{1-2z} \pi^{1/2} \Gamma(2z) \quad . \quad (5.6)$$

It is thus more convenient to rewrite (5.5) by introducing the quantity $u = v - 1/2$,

$$S_\ell(k) = \frac{\Gamma\left(\frac{\ell+u+2+if(k)}{2}\right) \Gamma\left(\frac{\ell-u+1+if(k)}{2}\right)}{\Gamma\left(\frac{\ell+u+2-if(k)}{2}\right) \Gamma\left(\frac{\ell-u+1-if(k)}{2}\right)} \times$$

$$\times e^{i(2\ln 2)f(k)} \quad , \quad (5.7)$$

where now u has precisely the meaning of the interaction added to the Coulomb interaction.

The amplitude (5.7) can be further generalized by making u and f functions of both quantum numbers k and ℓ. Since in heavy ion collisions we want to have the long range part of the interaction to be purely the Coulomb interaction we shall, for applications to this problem, retain

$$H = -\frac{\mu\beta^2}{2(C_2+1)} \quad , \quad \beta = Z_1 Z_2 e^2 \quad , \quad (5.8)$$

and

$$f(k) = \mu\beta/k \quad . \quad (5.9)$$

The quantity u will instead be a general function of ℓ and k. This function contains all the physical information on the scattering process. The generalized S-matrix which has been suggested [3] should be used to analyze elastic collisions between heavy ions is

$$S_\ell(k) = \frac{\Gamma\left(\frac{\ell+u(\ell,k)+2+i\mu\beta/k}{2}\right) \Gamma\left(\frac{\ell-u(\ell,k)+1+i\mu\beta/k}{2}\right)}{\Gamma\left(\frac{\ell+u(\ell,k)+2-i\mu\beta/k}{2}\right) \Gamma\left(\frac{\ell-u(\ell,k)+1-i\mu\beta/k}{2}\right)} \times$$

$$\times e^{i(2\ln 2)\mu\beta/k} \quad . \quad (5.10)$$

I shall call this the **S-matrix model** of heavy ion collisions.

One can note at this stage that the S-matrix (5.5) is manifestly unitary. However, if one wants to describe elastic collisions between heavy ions in which many channels are open, one needs, in the Schrödinger approach, a complex potential which takes into account absorption into the channels not included. This effect can be taken into account in (5.5) by making v complex. The associated S-matrix is no longer unitary. In order to satisfy the unitarity bound,

$$|S_\ell(k)| \leq 1 \quad , \qquad (5.11)$$

one must have

$$\text{Im}(v^2) \leq 0 \quad . \qquad (5.12)$$

A form of the "potential" u particularly suited for heavy ion collisions is

$$u(\ell,k) = (u_R - iu_I) \frac{1}{1+\exp[(\ell-\ell_0)/\Delta]} \qquad (5.13)$$

with

$$\ell_0 = \ell_0(k) \quad . \qquad (5.14)$$

Fig.1 shows the angular distribution that one obtains in this case. This angular distribution appears to have all the features observed in heavy ion collisions. The quantity ℓ_0 has the meaning of a grazing angular momentum. Its relation to k, Eq.(5.14), will determine the excitation function (i.e. the cross section as a function of energy at a fixed angle). For example, if one assumes the form suggested by semiclassical arguments

$$\ell_0 = \sqrt{R^2(k^2 - k_0^2)} \quad , \qquad (5.15)$$

where R is a measure of the radius at which the nuclear interaction sets in (barrier top) and

$$\frac{k_0^2}{2\mu} = \frac{\beta}{R} = V_c \qquad (5.16)$$

is the Coulomb barrier at r=R, one obtains the excitation function shown in Fig.2. The expression (5.15) is only valid for $k^2 > k_0^2$. In general, a better procedure would be to determine the function $\ell_0(k)$ by fitting the angular distributions at different values of k.

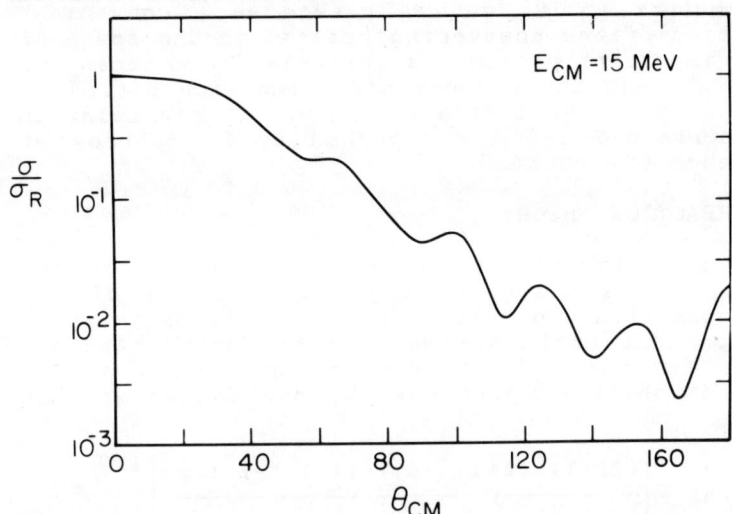

Fig.1. Schematic representation of the angular distribution for the collision of two heavy ions [3].

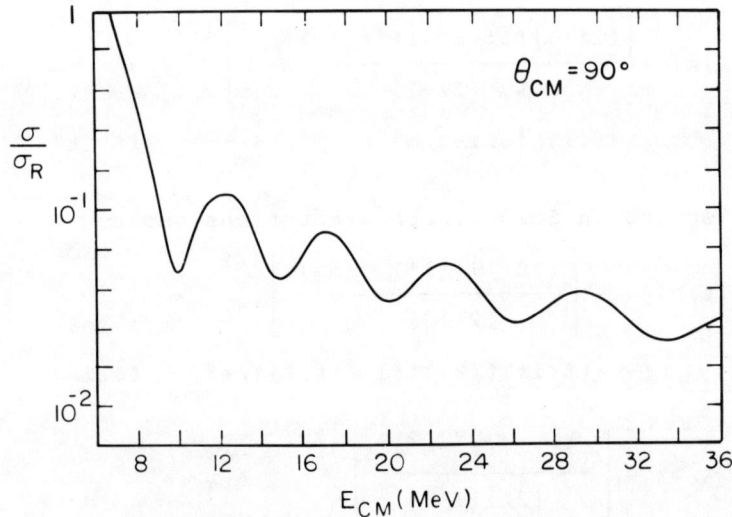

Fig.2. Schematic representation of the excitation function at $\theta_{CM}=90°$ for the collision of two heavy ions.

The S-matrix (5.10) can be viewed as a convenient parametrization of the scattering process in the space of ℓ and k. An important problem is the relation between the function $u(\ell,k)$ and the corresponding function $U(r,E)$ in configuration space. Preliminary studies of this relation have been performed and are reported by D. Sparrow at this Conference [7].

6. REACTION THEORY

The main ingredient in the construction of the analytic S-matrices discussed above is the use of the Euclidean connection. This is a set of relations in which step operators act on both sides of the Jost expansion. For example, for the Coulomb problem, the action of the operators that shift ℓ up or down by one can be written as

$$X_+|\omega,\ell,m\rangle = \left[\frac{(2\ell+1)((\ell+1)^2-m^2)((\ell+1)^2+f^2(k))}{(2\ell+3)}\right]^{1/2} \times$$
$$\times (-i) |\omega,\ell+1,m\rangle , \qquad (6.1a)$$

$$X_-|\omega,\ell,m\rangle = \left[\frac{(2\ell+1)(\ell^2-m^2)(\ell^2+f^2(k))}{(2\ell-1)}\right]^{1/2} \times$$
$$\times (i) |\omega,\ell-1,m\rangle , \qquad (6.1b)$$

whewn acting on $SO(3,1)$ representations and as

$$X_+|\pm k,\ell,m\rangle = -\left[\frac{(2\ell+1)((\ell+1)^2-m^2)}{(2\ell+3)}\right]^{1/2} \times$$
$$\times (\ell+1\pm if(k))(\pm 1)|\pm k,\ell+1,m\rangle, \qquad (6.2a)$$

$$X_-|\pm k,\ell,m\rangle = \left[\frac{(2\ell+1)(\ell^2-m^2)}{(2\ell+1)}\right]^{1/2} \times$$
$$\times (\ell\mp if(k))(\mp 1)|\pm k,\ell-1,m\rangle, \qquad (6.2b)$$

when acting on representations of $E(3)$. These algebraic relations replace the Schrödinger equation since they provide the recursion relations that determine the Jost functions $A_\ell(k)$ and $B_\ell(k)$. For example, comparing (6.1a) and (6.2a) one obtains

$$-i[(\ell+1)^2+f^2(k)]^{1/2} A_{\ell+1}(k)=(\ell+1-if(k)) A_\ell(k)$$
$$-i[(\ell+1)^2+f^2(k)]^{1/2} B_{\ell+1}(k)=(\ell+1+if(k)) B_\ell(k).$$
(6.3)

One can use these relations to construct a purely algebraic theory of scattering. For example, in a multi-channel problem, the Jost functions become an array and the action of the shift operators become a <u>matrix</u>. The Euclidean connection produces matrix recursion relations. If these can be solved in closed form, one can solve the corresponding coupled channel problem in closed form. Otherwise, one can use numerical methods to obtain the solution. If one denotes the Jost arrays by the column vectors

$$\mathbf{A}_\ell(k) = \begin{pmatrix} A_\ell^{(1)}(k) \\ A_\ell^{(2)}(k) \\ \ldots \end{pmatrix} \qquad \mathbf{B}_\ell(k) = \begin{pmatrix} B_\ell^{(1)}(k) \\ B_\ell^{(2)}(k) \\ \ldots \end{pmatrix},$$

(6.4)

the S-matrix is the trasformation matrix from the column vector **A** to the column vector **B**,

$$\mathbf{A} = \mathbf{S}\,\mathbf{B}.$$ (6.5)

A simple example is provided by scattering by a modified Coulomb interaction when the short range potential contains a spin-orbit term. Since this interaction is diagonal in ℓ, s and j, the Euclidean connection matrix is diagonal and the corresponding recursion relations can be solved yielding for spin 1/2,

$$S_{\ell,\,j=\ell\pm 1/2}(k) = \frac{\Gamma\left(\frac{\ell+u^{(\pm)}(\ell,k)+2+i\mu\beta/k}{2}\right)}{\Gamma\left(\frac{\ell+u^{(\pm)}(\ell,k)+2-i\mu\beta/k}{2}\right)} \times$$

$$\times \frac{\Gamma\left(\frac{\ell-u^{(\pm)}(\ell,k)+1+i\mu\beta/k}{2}\right)}{\Gamma\left(\frac{\ell-u^{(\pm)}(\ell,k)+1-i\mu\beta/k}{2}\right)} e^{i(2\ln 2)\mu\beta/k} .$$

(6.6)

Here

$$u^{(+)}(k) = u_c(\ell,k) + \ell u_s(\ell,k) ,$$
$$u^{(-)}(k) = u_c(\ell,k) - (\ell+1)u_s(\ell,k) .$$

(6.7)

Differential cross sections and polarizations can be obtained in the usual way in terms of the amplitudes

$$a(k,\theta) = \frac{1}{2ik} \sum_{\ell=0}^{\infty} \{(\ell+1)[S_{\ell,\ell+1/2}(k)-1] + \ell[S_{\ell,\ell-1/2}(k)-1]\} P_\ell(\cos\theta),$$

(6.8)

$$b(k,\theta) = \frac{1}{2ik} \sum_{\ell=1}^{\infty} \{S_{\ell,\ell+1/2}(k) - S_{\ell,\ell-1/2}(k)\} \times P_\ell^1(\cos\theta) .$$

7. RELATIVISTIC PROBLEMS

The group theory approach to scattering not being directly based on the Schrödinger equation can be easily generalized to <u>relativistic</u> problems. As an example, I quote here the result one obtains for Coulomb scattering of a spin zero particle (such as a $\pi^\pm$). The S-matrix for this process is still given by (4.7) but now

$$f(k) = \beta \frac{\sqrt{k^2+m^2}}{k} .$$

(7.1)

This gives a cross section

$$\frac{d\sigma}{d\Omega} = \frac{\beta^2}{4k^4\sin^4(\theta/2)}(k^2+m^2) \quad , \tag{7.2}$$

where k is the three momentum and m the mass of the particle. With appropriate change of notation, Eq.(7.2) is <u>identical</u> to that obtained by evaluating the Feynman graph of Fig. 3. Similarly, one expects the S-matrix for relativistic scattering of a spin-zero particle from a modified Coulomb interaction to be given by (5.5) but with $f(k)$ given by (7.1). An appropriate choice of the short range interaction $v(\ell,k)$ will thus allow one to construct a realistic model for scattering of $\pi^\pm$ mesons off nuclei. A further generalization to treat <u>relativistic</u> scattering of spin-1/2 particles is much more involved. A preliminary study has been done [8] but further work is needed here before realistic models can be suggested.

Fig.3. Coulomb scattering of π^+ meson.

8. CONCLUSIONS

The algebraic approach appears to provide a powerful technique to analyze scattering processes. It is, in spirit, similar to the S-matrix theory of the early 1960's but it contains one additional ingredient, the concept of dynamic symmetries and groups. The S-matrix theory only made use of analyticity and unitarity concepts and thus was able to provide only dispersion relations. The addition of symmetry ideas provides the functional form of the S-matrix. This idea is similar to that used by Veneziano [9] in constructing dual resonance models. It remains to be seen how useful is this idea in analyzing nuclear collisions especially whether or not the use of algebraic techniques provides simplifications to the solution of coupled channnel problems.

Another area of interest is to the study of <u>relativistic</u> scattering problems. For scattering of particles with no spin, the extension to relativistic problems appears to be straightforward. The situation is more complex for particles with spin. This however is an interesting case since it will allow one to analyze medium energy (1 GeV) proton scattering.

The ultimate goal of the algebraic approach is the

construction of S-matrices in field theories for applications to scattering between elementary particles.

AKNOWLEDGEMENTS

The work reviewed here has been done in collaboration with Y. Alhassid, F. Gürsey and J. Wu. It has also benefited considerably from discussions with R. Amado, L.C. Biedenharn, J. Engel, A. Frank and O. van Roosmalen. It has been performed in part under Department of Energy Contract No. DE-AC-02-76 ER 03074.

REFERENCES

[1] R. Amado, These Proceedings
[2] J.N. Ginocchio, These Proceedings
[3] Y. Alhassid, F. Gürsey and F. Iachello, Ann. Phys. (N.Y.) $\underline{148}$, 356 (1983);
A. Frank and K.B. Wolf, J. Math. Phys. $\underline{26}$, 1973 (1985);
Y. Alhassid, F. Gürsey and F. Iachello, Ann. Phys. (N.Y.) $\underline{167}$, 181 (1986);
J. Wu, F. Iachello and Y. Alhassid, Ann. Phys. (N.Y.), to be published;
Y. Alhassid, F. Gürsey and F. Iachello, Phys. Rev. Lett. $\underline{50}$, 873 (1983);
A. Frank and B. Wolf, Phys. Rev. Lett. $\underline{52}$, 1737 (1984);
Y. Alhassid, J. Engel and J. Wu, Phys. Rev. Lett. $\underline{53}$, 17 (1984);
Y. Alhassid, F. Iachello and J. Wu, Phys. Rev. Lett. $\underline{56}$, 271 (1986)
[4] L.C. Biedenharn, These Proceedings
[5] D. Zwanziger, J. Math. Phys. $\underline{8}$, 1858 (1957)
[6] Y. Alhassid, These Proceedings
[7] D. Sparrow, These Proceedings
[8] J. Wu, C. C. Biedenharn and F. Iachello, J. Phys. A, to be published.
[9] G. Veneziano, Nuovo Cimento $\underline{57A}$, 190 (1968)

SYMMETRIES IN REACTIONS

Discussion Session
Following Review Lecture
by
F. Iachello

Moderator:

L. BIEDENHARN
Department of Physics
Duke University
Durham, North Carolina 27706, U.S.A.

Moshinsky, Mexico: What surprised me here is one thing. You are always using a partial wave expansion in I. Now in your very first problem, the Coulomb problem, you could go to parabolic coordinates. You could get the scattering amplitude directly. And this implies a different subgroup, a subalgebra of 0(3,1). So it seems that there would be a possibility in general of considering differences of sub algebras of your basic algebra and using them in connection with another way of looking at the problem.

Iachello: Yes, this is a good point. It's absolutely right. And in fact what one should do for this problem is perhaps the same type of analysis that has been done in the compact case for IBA, that is to look at the different group chains. Also here there is an interesting point, that is that the case in which one can solve the problem with different group chains corresponds to the fact that the problem is solvable in many coordinate systems as you just mentioned in the Coulomb problem.

Ginocchio, Los Alamos: I just want to make a comment. You pointed out that for example the Coulomb and the Yukawa potential, if you solved exactly you got a ratio of gamma functions. And the algebraic approach gave you a form of ratio of about two gamma functions. But also I want to point out that there are also exactly solvable models in the Schroedinger equation which in fact give you also that form.

Iachello: Yes, this is absolutely right. In fact, Joe Ginocchio himself has developed many of these exactly solvable models. And these models are related to groups. J. Wu in his thesis discussed precisely your potential and related it to that group. It's a slightly more complicated case than that I've discussed here but it is related to a group.

Moderator: I think as moderator I can intervene here because you've hit on a very interesting point. There are some exactly soluble two-dimensional models for which (extending the results of Zamolodehikov) there are exact results for every classical group. Very recently Ojievetsky has shown that one can get the exceptional group G_2 also. The interesting thing is that the scattering matrix is always a ratio of gamma functions.

Iachello: I would like to comment on this. I do not know whether the fact that there are ratios, of gamma functions has a deep significance or not. Maybe all are associated to that fact that in solutions of these problems are hypergeometric functions.

Balantekin, Oak Ridge: I have two comments. One of them is I would like to draw attention to the recent advances in inverse scattering theory. It was rigorously proved that if you had an S matrix for all partial waves and all energies you can get the potential exactly. Sparrow has done this but you have exactly the case here. We have an S matrix for all partial waves and for all energies. Thus one should be able using the powerful inverse scattering techniques to obtain the corresponding potential exactly. Not approximately.

Iachello: Absolutely right. But not for a complex potential. Only a real potential.

Feshbach: I have a number of comments. First of all the reliance on the gamma functions worries me because there are many Schroederinger equations, which do not give you S matrices which are ratios of the gamma functions. In fact I don't know of analytic results for some of them. For example, you have the periodic potential, then you get a Mathieu function. The scattering matrix is not a ratio of gamma functions in that case. You can scatter from elliptical objects and you will get Mathieu functions. And in fact there you can show that these have solutions in terms of continued fraction. So there's a wider class of analytic type functions besides the gamma functions. The gamma functions come in whenever you have either a hyper-geometric or a confluent hyper-geometric equation for those of you who still remember those kinds of equations. And those are related to the geometry because when you compact into spherical coordinates, you introduce a singularity at the origin. You always have a singularity at the infinity because of the fact that you are dealing with a non-compact situation. When you go to elliptic cases you get two singularities at the finite domain and one at infinity and then you don't get the gamma functions any more. Now a physics question. If I had to solve 1842 coupled equations I wouldn't do it by putting it on a computer. I would do it by using the familiar statistical method because I would claim that random approximation is exceedingly good under those circumstances. Now it's better of course to have an exact solution. But on the other hand to say that the only alternative is to solve the coupled equations I think is not correct.

Balantekin, Oak Ridge: There is another way of using algebraic approach to look at scattering problems in the spirit of what Ginocchio and Amado have done. Again if you treat the relative motion semiclassic or classically and the internal motion quantum mechanically, you can go to the particle formalism and if the interaction potential is separable in two coordinates, in the translation of internal coordinates, then the interaction potential plus the internal potential can be written as a linear combination of generator Casimir operators or generators of a group. Then if you go to the pattern they call formalism, exponential of the action becomes an element of the group. That is another way of using the algebraic approach. In that way you can study various approximations. That has been applied to the barrier penetration problems.

Iachello: One comment to what Feshbach was saying. The case of ratio, of gamma functions is associated to the orthogonal groups and to dynamical symmetries. Three or four years ago when we started this project. Alhassid, Gursey and also myself investigated the case in which instead of compactifying to a sphere, you compactify to an ellipse. And that means that instead of having j_x squared plus j_y squared plus j_z squared, we have j_x squared plus alpha, j_y squared plus j_z squared. And that will lead to Lame' type of equations. But that will be a broken symmetry in this language. The S0(3) symmetry will be broken. So that's a very very good point. In fact this is perhaps an idea on how to try to attack problems with periodic potentials. And it could be used in crystal physics.

Broglia, Milan: I'm a little worried about your expression for the phase-shift of the S-matrix Coulomb nuclear problem. In the sense that it is true that I may try to fit Oxygen-16 on Silicon-28 by using potential theory. I may even get the oscillations in the right place if I have enough parameters to do so. Now the question is (and there the physics is) that the incoming projectile which couples to the low lying 2+, 4+, ... rotational band, is cut and then the projectile goes out leaving the nucleus rotating. Now to fit the elastic scattering you may say who cares? I just fit my parameters and I extract something. But you see one of the things I can measure is what the polarization of the outgoing particles. And they would have a very definite polarization depending on the mechanism they went through. And of course if you do your couple channel whether you neglect the Q value or not, you get the polarization essentially right. Now if you put an effective potential the way you do, there is no reason why you should get the polarization right unless you're going to do other things to remedy that. So I would say that solving couple of equations or at least finding out the physics of the problem, would predict different things that can be checked and I wonder further within the formalism you are discussing you can include that within these potential type of scattering.

Iachello: This is certainly a good point. However, what I would like to point out is that there are several mathematical frameworks in which you can describe the same phenomenon. I am not suggesting that the framework that other people use is wrong or incorrect, I am only saying that we would like to use a different framework. What I tried to present here is this framework. You can use it, if you prefer, to add the excitation 0, 2, 4, 6, 8 states instead of studying the optical model and if we would be able to develop this theory far enough to treat this situation, then we would be able to do the same thing.

Broglia: No, in fact, Franco, don't misread me. I think that I would prefer to use something simple instead of something complicated. I'm just asking you to which extent these simple parameterizations can incorporate those features. I'm trying to see to which extent you can bring more physics in.

Iachello: This I simply don't understand. Because in a way our attempt is that of producing simple results. Now you may complain about the fact that the mathematics is difficult and that maybe people do not know group theory and therefore they find this derivation completely difficult. On the other side, the final result is an S-matrix which has an analytic form. There it is and why not use that? It's as simple as any other. I don't see the point there.

Broglia: I completely agree with you on that Franco. I'm just asking: The polarization that one would measure for the outgoing particles in the inelastic channel, I do not believe are predicted by that formula you have.

Iachello: But that is because I was just doing a particular case which is the elastic scattering. I was trying to repeat in this formalism what people do using the optical model.

Broglia: No, but I say, using the optical model with coupled channels, you get the polarization in the elastic. That is what I'm trying to say.

Moderator: Wait a minute. What is happening is Franco's attempt is an **initial** attempt to try at a general problem, not the final, or complete answer.

Alhassid: I want to make a comment in connection with Feshbach's remark and also Franco's remark. You can get a mathieu equation and also relate it to group theory and in fact you can do even more complicated periodic potentials associated with the group SU(2). For example what you can get in this case from group theory is the band structure plus another important quantity case which is that of the scattering matrix and is called the transfer matrix. This matrix is the one which connects the solution from one period of the potential to the next. It seems in although we have not been able to complete this calculation, that this transfer matrix could be written in terms of generalization of the gamma function known as the gamma-q function.

Iachello: It's a three term recursion relation which produces a gamma-q function.

Alhassid: Yes. That's correct. Also another comment I want to make is about the inversion problem. So far what has done is only for a local potential while in our approach the potential is probably a non-local potential. For local potentials it is enough to know all the phase shifts at the given energy for all l plus the poles (bound states) and residues. The potential is found by solving the Gelfand-Levitan equation.

Iachello: For those of you who would like to have some references, there are some references here that perhaps you could copy.

Gilmore, Drexel: I'd like to follow up the statement by Balantekin. A colleague and I have looked at some molecular physics Hamiltonians and used algebraic techniques to solve the S-matrix in this case. We have written down the exponential of the interaction term and then converted this from unitary to a nonunitary representation. We have prepared our results with the benchmark calculations of Johnson et al. and have been able to reproduce them very easily. We've also done this for the two mode calculation for which there are limited amounts of data and have been able to reproduce all of these. The extension to three and higher numbers of modes hasn't been done because in the classical techniques it costs too much money. However this involves integration of 8 x 8 matrices. It's very simple.

Iachello: I know this work. It works very well. Bob has explained it to me previously.

Rohozinski, Warsaw University: I have a question concerning the description of scattering on complex targets. If you describe targets algebraically, then what you should perform is a calculation of the matrix element of exponential function of generators and it is essentially matrix elements of powers of generators. Now if we think that these matrix elements are approximations of fermion matrix elements, then for every power you have different boson pictures so to say. So, in fact, you have renormalization in every power. So just taking algebraic, I mean the IBM quadrupole operator and taking powers, you are not; maybe you can have the wrong results.

Iachello: That's possible if you think of the bosons as fermion pairs. But for example in the case of molecules the bosons are not the fermion pairs. They just represent the dipole degree of freedom and therefore I don't see any problem there. But certainly that might be the case if you have complicated powers of operators. For example, in IBM, operators if you try to compute q cubed or q to the fourth, since the q's do not commute, you will have problems. You would have to add the higher order terms. But then it becomes very difficult.

Moderator: Do you want to make another comment on it or do you agree? Do we have any other questioners that I have not recognized?

Sunko, Zagreb: I would like to share with you something that I have learned in the book by Gilmore. And that is that there are two ways in which you can write the general group element. One is the ordinary one, an exponent of the generator, and it corresponds to exponentiating along a straight line from the origin parameter space to any point. And if you use instead of a straight line a general line parameterized by some parameter d, then the group element with an exponent of an integral of an operator H(t) with a t-ordering operator in front. In other words it has just the form of the S matrix in Dirac's theory which means if you can suppose a dynamical group for your scattering problem you can replace the Dirac general complicated form of the S matrix by a simple form which is just an exponent of the generator. And it seems to me that that is just what you have been doing. Once you assume the dynamical symmetry it is exact.

Iachello: I don't have any comment to that. That's possible. I don't know. I haven't looked into that.

Moderator: Maybe someone else would like to comment on this particular point? Perhaps Bob?

Bob Gilmore, Drexel: I believe that comment is accurate. But I'd like to comment on a slightly different but related topic. That is that in his calculations, Roelof Bijker ran into some problems with numerical conversions. Is Roelof here? Would you like to explain that? (not present)

Iachello: You mean in the first part of my talk?

Gilmore: There were some difficulties in computing the exponentials and by straightforward exponential expansion. Once we disentangled the operators to write them in a particular operator order form the convergence was straightforward and extremely rapid.

Iachello: Very good. Very good remark.

Tamura, University of Texas: I do not want to ask any fundamental question. I am just curious about the numerical behavior. I can visualize in my head the behavior of the S-matrix in general but cannot visualize the behavior of the ratio of products of gamma functions. In any case, my question is whether you might be able to stay with the simple Coulomb term, one gamma function upstairs and another gamma function downstairs, and then parameterize the arguments. You might be able to obtain results similar to what you did with the four gamma functions.

Iachello: First of all about the results. If you take the ratios of gamma functions, this is what you obtain for example for the S matrix as a function of l in the heavy ion collision. You see that the absolute value of S_l has the usual behavior. Also here, there is the result for the real part of the S matrix. This is the Coulomb phase shift and this is the point at which it starts deviating from the Coulomb phase shift. So although the gamma functions may not look attractive, the end result is what usually everybody obtains.

Tamura: I'm not suspecting your numerical calculations. I am simply asking whether you need four gamma functions or can you just start with the well known Coulomb formula and modify it and parameterize in in the form you did? Can you reproduce this way or will you never be able to do that?

Iachello: Yes it has been done.

Moderator: We can carry this on privately. I've just gotten the news that there's coffee ready for us. And unless there's a pressing question, a very small question.

Kyrchev: In your expose you mentioned that one of the problems is that always the relevant noncontact dynamical group is known. But I remember you had such cases when you knew the compact group, $U(6)$ for example, for the description of the structure, then you had the recipe how to find the noncompact group and then to apply this to the scattering problems. The question is: If you know the compact dynamical group relevant for description of structure problems, can you always find which is the non-compact counterpart?

Iachello: Well I'll try to give a short answer because this is an important problem. We have at the moment only analyzed S-matrices which arose from orthogonal groups. But in principle one could derive also S-matrices starting from unitary groups and maybe also starting from exceptional groups. Those will be related to other kind of but at the moment we have not done it. The only thing we have done is the orthogonal group. I would have liked to use unitary groups because my original intention was to start from $U(4)$ and go to $U(3,1)$. Or from $U(6)$ to $U(5,1)$. But so far we have not done that.

Moderator: I think at this is a good point to stop. Thank you very much for your attention.

Medium Energy Scattering from "Algebraic" Targets

R.D. Amado,[a] R. Bijker[a] and D.A. Sparrow[b]
(a) Department of Physics, University of Pennsylvania,
Philadelphia, PA 19104
(b) Department of Physics, Temple University,
Philadelphia, PA 19122

Medium energy scattering involves many partial waves and in strong coupling many states of the target. The high energy of the projectile permits the partial waves to be summed using the eikonal or Glauber approximation. Since that same high energy is normally large compared with the energy scale of the strongly coupled states of the target, they can be treated in the adiabatic approximation and combined with the eikonal expression. If those states can be described algebraically, it is particularly easy to implement this eikonal-adiabatic- algebraic treatment to yield a closed form expression for elastic and inelastic scattering amplitudes that takes channel coupling into account in all orders. We will develop the formalism, emphasizing the basic physics, and illustrate it with examples from nuclear and molecular physics.

The large number of partial waves in medium energy scattering (For example for 800 Mev proton scattering from Pb the grazing L=50, and some 120 waves are needed for a complete description of the scattering.) suggests converting the partial wave sum into an integral. This can be done exactly using a Fourier-Bessel transform or equivalently an impact parameter representation of the scattering amplitude $t(\vec{q}, E)$ ($\vec{q}$ is the momentum transfer) given by

$$t(\vec{q}, E) = \frac{-ik}{2\pi} \int e^{i\vec{q}\cdot\vec{b}} d^2b (S(\vec{b}, E) - 1) \qquad (1)$$

where $S(\vec{b}, E)$ is the S-matrix at impact parameter $\vec{b}$. If S depends only on the magnitude of b, the transform can be written in the more familiar form

$$t(q, E) = -ik \int_0^\infty b\, db\, J_0(qb)(S(b, E) - 1) \qquad (2)$$

The major obstacle to using (1) or (2) is that in contrast to the partial wave S-matrix, we do not generally have a simple way to calculate S(b,E) from the dynamics.

This problem is solved by the eikonal approximation which is a large energy approximation. If we neglect the range of variation of the potential V(r) compared with the projectile wave length, and the potential strength compared with the incident energy, we can write S(b,E) in eikonal form to yield

$$t(\vec{q}, E) = \frac{ik}{2\pi} \int e^{i\vec{q}\cdot\vec{b}} d^2 b (e^{i\chi(\vec{b},E)} - 1) \tag{3}$$

with

$$\chi(\vec{b}, E) = \frac{-m}{k} \int_{-\infty}^{\infty} V(r) dz \tag{4}$$

where z is the beam direction, and m and k are the projectile mass and wave number respectively. It is well known that this eikonal approximation is an excellent approximation for medium energy scattering in forward directions ($\Theta \leq 60°$).

Now suppose the target is not described by a simple potential but rather has a set of low lying states that are strongly coupled and excited by the projectile. How can we generalize the eikonal approach? As an example consider a deformed nucleus characterized by a static deformation oriented at angle Ω. Assume the potential seen by the projectile depends on Ω, $V(r,\Omega)$, while the nuclear Hamiltonian is just the normal collective rotational Hamiltonian H_{rot} and depends on derivatives with respect to Ω. The total Hamiltonian for projectile and target nucleus is then

$$\frac{p^2}{2m} + H_{rot} + V(r, \Omega) \tag{5}$$

where $p^2/2m$ is the projectile kinetic energy. The essence of the eikonal approximation is that that kinetic energy is the leading term in the Hamiltonian. It is certainly large compared with the energies of the collective states described by H_{rot} (eg 800Mev > 500kev). Hence we can neglect H_{rot}. We can then carry out the eikonal derivation just as before and will get exactly Eqn(3) with $V(r,\Omega)$ in χ of Eqn(4). The t-matrix now depends on

Ω as a parameter. This is the essential feature of the adiabatic approximation. In neglecting H_{rot} we assume that the motion of the nucleus is slow (adiabatic) compared with the interaction time of the rapidly passing projectile so that the deformed nucleus is essentially fixed in space during that interaction. Alternately one can call this the sudden approximation as viewed from the projectile. The t-matrix that emerges from using $V(r, \Omega)$ in Eqn(3) is diagonal in Ω which is not the representation we want. Using the collective wave functions of H_{rot}, we can easily project that t-matrix onto states of definite J and M.

The t-matrix calculated in this way is equivalent to a full coupled channel partial wave t-matrix but is far simpler to calculate. Fig. 1 shows such a calculation [AMS84] for 800 Mev proton scattering from ^{154}Sm with a deformed Woods-Saxon potential and appropriate parameters for this nucleus compared with the data of Barlett et al. [MLB80] We show elastic scattering and the cross section for excitation of the 2$^+$ and 4$^+$ collective states. For eleastic scattering we also show (dotted line) the scattering neglecting the virtual effects of channel coupling, from which we see the importance of that coupling.

Figure 1: Calculation of $800 MeV$ $p-^{154}$Sm elastic and inelastic scattering compared with the data of Barlett et al [MLB80]. The dashed curve is for elastic scattering without channel coupling.

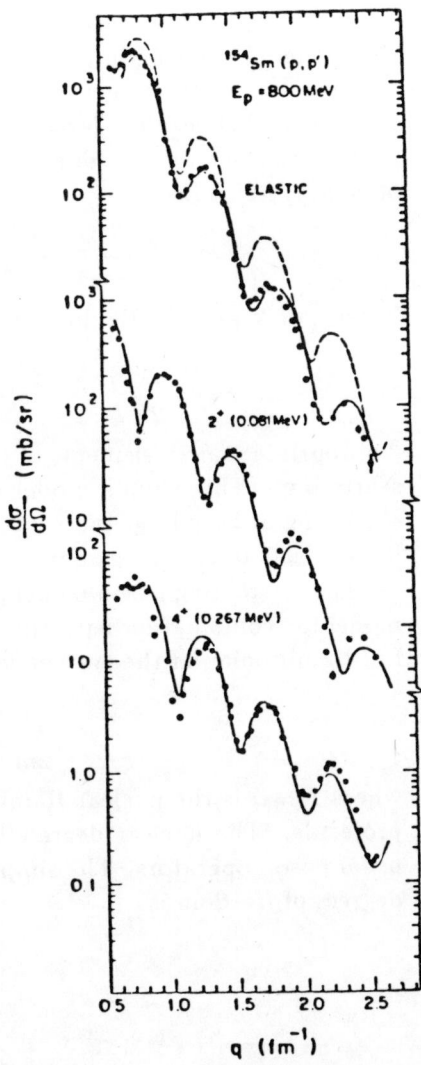

As the next application of the adiabatic-eikonal method, we turn to an algebraic model of a two state system. Consider a nucleus with two important coupled states separated by energy Δ. Suppose the diagonal scattering of the projectile between these states is represented by a potential V and the off diagonal projectile nucleus potential by W. The Hamiltonian is a 2x2 matrix in the target nucleus space that can be represented by introducing the usual Pauli matrices, that is SU(2). The Hamiltonian becomes

$$\frac{p^2}{2m} + V + W\sigma_x + \frac{1}{2}(1 - \sigma_z)\Delta \quad (6)$$

(Note there is no spin in this problem, the Pauli matrices simply carry the 2x2 structure of the target dynamics). Again at high energy we can neglect the spacing between the levels, Δ, compared with the kinetic energy of the projectile. Once we have done this, we can make the eikonal approximation of Eqn(3) but now

$$\chi = \chi_v + \sigma_x \chi_w \quad (7)$$

in term of V and W. Exploiting the SU(2) algebra, the t-matrix becomes

$$t = \frac{ik}{2\pi} \int e^{i\vec{q}\cdot\vec{b}} d^2 b (e^{i\chi_v}(\cos\chi_w + i\sigma_x \sin\chi_w) - 1) \quad (8)$$

Appropriate matrix elements of the t-matrix then give elastic and inelastic scattering. This simple model can be applied to 800 Mev proton^{208}Pb scattering at very large momentum transfer to account for the surprisingly large effects of channel coupling in this problem. [ADS84]

The interacting boson model provides a far more sophisticated algebraic model for combination with the eikonal approximation. For this case the full Hamiltonian of the system can be written

$$\frac{p^2}{2m} + H_{IBM} + V_c \quad (9)$$

where H_{IBM} is the nuclear Hamiltonian and V_c couples the nucleus to the projectile. The nuclear degrees of freedom are described in terms of the usual boson operators. The simplest form for the coupling involving these degrees of freedom is

$$V_c = f(r)(Q_2 Y_2 (\hat{r})_o)$$

where Q^2 is the IBM quadrupole operator. Making the adiabatic assumption, $(H_{IBM} \ll p^2/2m)$ we can exponentiate V_C in eikonal form so the χ will be linear in Q_2. Again this will yield a t-matrix in a "wrong" representation. The correct representation is found by taking matrix elements of $\exp(i\,\chi)$ between states of H_{IBM}. Since χ is linear in Q_2 this simply yields a representation matrix that is straightforward to calculate particularly in one of the dynamical symmetries of the nuclear IBM. [GIN86] Applications of these methods to medium energy scattering from Gd and Sm isotopes are being presented at this conference by Ginocchio and Wenes.

Finally we turn to a non-nuclear example–electron molecule scattering. The principal degrees of freedom of a diatomic molecule are rotation and vibration. These can be well described algebraically in terms of four bosons, the three components of a dipole boson, $\vec{p}$ and a scalar boson, s. This so called vibron model gives a good account of the molecular rotational and vibrational spectra. [FIA81], [FIL82] It is a U(4) model and has two dynamical symmetries. The first, $U(4) \supset O(4) \supset O(3) \supset O(2)$, is related to the Morse oscillator, while the other, $U(4) \supset U(3) \supset O(3) \supset O(2)$, to the three dimensional harmonic oscillator. The principal feature of a non-homopolar molecule is its dipole moment D represented algebraically by $\vec{p}^\dagger s + \tilde{p} s^\dagger$. The scattering electron couples primarily to that dipole moment. The total Hamiltonian is

$$\frac{p^2}{2m} + H_{vib} + f(r)\vec{D} \cdot \vec{r} \qquad (10)$$

where again $\frac{p^2}{2m}$ is the projectile kinetic energy and H_{vib} is the internal Hamiltonian of the molecule expressed in terms of vibrons. The third term is the coupling term between the molecule and the projectile with coordinate r. The function f(r) carries the strength and the r dependence of the electron molecule dipole coupling. For large r, $f \sim 1/r^3$, while it should be cut-off for small r to represent the internal structure of the molecule. Since the molecular spectrum and dipole moment are known all the parameters in (11) are fixed except the nature of the short range cut-off in f(r).

The adiabatic approximation now consists of neglecting H_{vib}. The procedure is as before and now the eikonal profile function, χ, will be linear in the dipole operator D. We have to calculate matrix elements of the exponentiated dipole operator between states of H_{vib}. These are just elements of the representation matrix of D, which is a generator of the vibron algebra. These matrix elements are particularly easy to calculate in one of the dynamical symmetries of the vibron model. [BAS86]

Medium energy for this problem corresponds to 5-10 ev. The molecular rotational states are strongly coupled and are so close in energy, that unfortunately, true eleastic scattering is not easily observed. In Fig.2a we show a calculation for electron scattering from LiF at 5.44ev. [BJA86] The data are normalized to the calculation since absolute cross sections are not measured. The curves show the vibrationally elastic data (averaged over initial rotational states and summed over final) and the cross section for exciting the first vibrational band. Agreement with experiment is remarkably good. At 20 ev, Fig.(2b), agreement is not so good at the larger angles, but there we are beginning to probe the molecular inside for which our model is inadequate. We are working to repair this aspect of the problem. Similar calculations with comparable agreement exist for KI.

It is unfortunate that the experiments cannot resolve the rotational states since as we see in Fig. (2c) the individual cross sections to particular rotational levels have much more structure than the average. Also shown in Fig. (2c) is the Born approximation. We see that it is not valid except at the very smallest angles which are dominated by the very peripheral parts of the long range dipole interaction. We also see that the excitation of the 2^+ and 3^- states, excitations that cannot go in one step for a dipole interaction, are large at large angles. This fact and the failure of the Born approximation both point to the importance of channel coupling or close coupling as it is called in molecular physics. Of course the algebraic-eikonal approach has that coupling in all orders.

Similar calculations can be done for electron scattering from triatomic molecules. Here the algebraic treatment requires two sets of vibrons or $U(4) \times U(4)$. The scattering calculations can be done in a straightforward way exploiting the simplicity of the algebraic-eikonal method. For triatomic molecules these calculations are very much faster than those using any standard method from molecular physics and therefore give promise of wide utility. Finally we should remark that though the vibron model may not

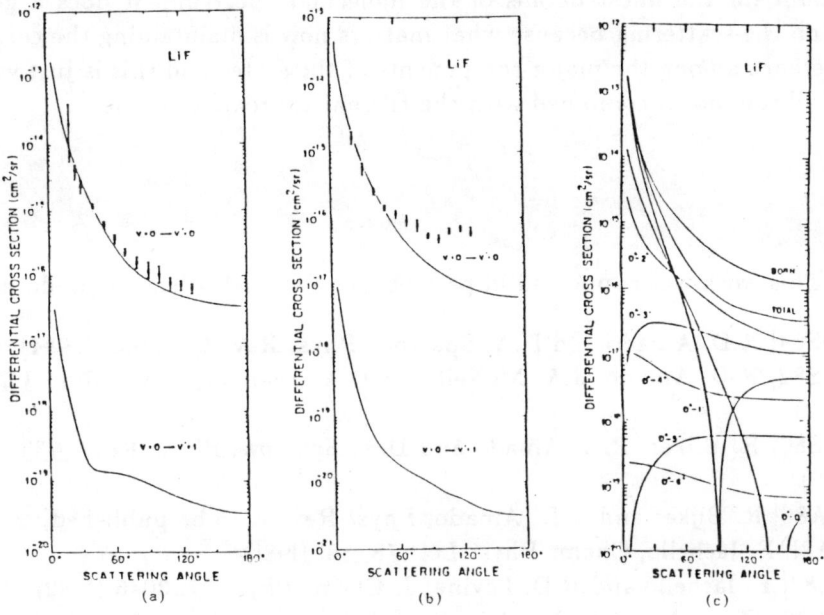

Figure 2: Differential cross sections for electron LiF scattering. a) vibrationlly elastic ($v = 0 \rightarrow v' = 0$) and inelastic ($v = 0 \rightarrow v' = 1$) at 5.44eV. b) at 20.0eV c)at 5.44 eV showing the excitation of the individual rotational states as well as the first Born approximation.

account for the finest details of the molecular spectrum, it does a good job on the scattering because what matters now is maintaining the correct coherence among the major components of the states and this is just what the vibron model combined with the eikonal approximation does.

This work was supported in part by the National Science Foundation.

[ADS84] R.D. Amado and D.A. Sparrow, Phys. Rev. C29, 932 (1984)
[OBS82] R.D. Amado, J.A. McNeil and D.A. Sparrow, Phys. Rev. C25 1 (1982)
[BAS86] R. Bijker, R.A. Amado and D.A. Sparrow, Phys. Rev. A33, 871 (1986)
[BJA86] R. Bijker and R.D. Amado, Phys. Rev. A to be published
[FIA81] F. Iachello, Chem. Phys. Lett. 78, 581 (1981)
[FIL82] F. Iachello and R.D. Levine, J. Chem. Phys. 77, 3046 (1982)
[GIN86] Ginocchio, Otsuka, Amado and Sparrow, Phys. Rev. C33, 247 (1986)
[MLB80] M.L. Barlett, et al, Phys. Rev. C22, 1168 (1980)

THE INTERACTING BOSON MODEL AND MEDIUM ENERGY PROBES

J. N. Ginocchio and G. Wenes

Theoretical Division
Los Alamos National Laboratory
Los Alamos, NM 87545 U.S.A.

ABSTRACT

Medium energy electron and proton scattering are used to determine the quadrupole form factors for the Interacting Boson Model of Nuclei.

1. INTRODUCTION

The interacting boson model (IBM) has been successful in giving a unified description of energy spectra and transition rates of collective nuclei[1]. However, there is now available very high quality electron scattering data which can determine the static and transition charge densities of nuclei down to very small distances[2]. There are also high quality medium energy proton scattering data available on heavy nuclei, which, in conjunction with the electron scattering, can determine the contribution of the neutron and proton degrees of freedom[3]. Can the IBM describe these transition densities as well? This is the topic that we shall discuss in this paper.

2. MEDIUM ENERGY PROTON SCATTERING AND IBM

In order to address this question at all, we need to be satisfied that medium energy proton scattering is handled correctly. Electron scattering from nuclei can be analyzed with distorted wave Born approximation (DWBA). However for medium energy proton scattering coupled channel effects may be important[3,4], especially for collective nuclei. Recently we have been able to calculate these coupled channel effects to all orders in a very elegant way[4,5]. In the Glauber approximation the scattering amplitude for scattering a medium energy proton with momentum k from a nucleus in an initial state i to a final state f and transferring momentum $\vec{q}$ is given by

$$<f|F|i> = \frac{k}{2\pi i} \int d^2 b\, e^{i\vec{q}\cdot\vec{b}} (e^{i\phi(b)} U_{fi}(\vec{b}) - \delta_{fi}) \tag{1a}$$

$$U_{fi}(\vec{b}) = <f|e^{i\hat{\Psi}(\vec{b})}|i> \tag{1b}$$

where $\vec{b}$ is the impact parameter, and $\phi(b)$ is the distorted wave. The transition operator $\hat{\Psi}(\vec{b})$ is a linear combination of the boson operators, and the quadrupole term is given by

$$\hat{\Psi}(\vec{b}) = \left[g(b)\, P + g_2(b)\, T\right] \cdot Y^{(2)}(\hat{b}) \tag{2a}$$

$$P_\mu = (s^\dagger \tilde{d} + d^\dagger s)^{(2)}_\mu \tag{2b}$$

$$T_\mu = (d^\dagger \tilde{d})^{(2)}_\mu \tag{2c}$$

$$g(b) = \frac{2\pi}{k} f \int_{-\infty}^{\infty} dz\, \alpha\left(\sqrt{b^2 + z^2}\right) \tag{2d}$$

$$g_2(b) = \frac{2\pi}{k} f \int_{-\infty}^{\infty} dz\, \beta\left(\sqrt{b^2 + z^2}\right) \tag{2e}$$

where P, T are boson operators acting on the monopole ($s^\dagger$) and quadrupole ($d^\dagger$) bosons in the nucleus, f is the average nucleon-nucleon scattering amplitude, and $\alpha(r)$, $\beta(r)$ are the IBM form factors to be discussed later. The salient point is the following. Since $\hat{\Psi}$ is a linear combination of the boson operators which are generators of the SU(6) dynamical symmetry[1] of the IBM, the operator $e^{i\hat{\Psi}(\vec{b})}$ produces a unitary transformation of the six bosons, $s^\dagger$, $d^\dagger_\mu$, $\mu = -2, -1, \ldots 2$ among themselves, and hence also among all the bosons in the nucleus. The transition matrix $U_{fi}(\vec{b})$ is then the SU_6 group representation matrix where g, g_2 are like rotation angles. That is, the transition matrix is a generalization of the Wigner D-function which is the group representation matrix of SU_2. Therefore this matrix can be calculated in closed form. For special limits of the quadrupole oscillator, the γ-unstable rotor, and the axially symmetric rotor, this transition matrix can be given in analytic form in terms of a hypergeometric power series[4]. For the more general case, the transition matrix can be calculated numerically[5]. Hence all coupled channels within the IBM model space are included in the scattering in a unified manner.

3. THE IBM ELECTROMAGNETIC QUADRUPOLE TRANSITION OPERATOR

Electron scattering from nuclei measures the static and transition densities of protons alone since neutrons carry no charge and the magnetic interactions are very small and are generally neglected. However the IBM is a truncation of the nuclear many-body problem and hence the IBM form factors will consist of both a direct part due to scattering from the <u>included</u> degrees of freedom, and an effective part due to the <u>excluded</u> degrees of freedom. Because neutrons and protons are probed differently in electron scattering, neutron and proton degrees of freedom should be distinguished as in IBM-2[1]. The electromagnetic quadrupole density operator is then given as

$$\hat{\rho}_{em,\mu}(r) = e \left[\alpha_{\pi\pi}(r) \hat{P}_{\pi,\mu} + \beta_{\pi\pi}(r) \hat{T}_{\pi,\mu} + \tilde{\alpha}_{\pi\nu}(r) \hat{P}_{\nu,\mu} + \tilde{\beta}_{\pi\nu}(r) \hat{T}_{\nu,\mu} \right] \quad (3)$$

The form factors $\alpha_{\pi\pi}$ and $\beta_{\pi\pi}$ include both direct and effective interactions of the electron with the valence protons, while $\tilde{\alpha}_{\pi\nu}$ and $\tilde{\beta}_{\pi\nu}$, of course, include only the effective interaction with the valence neutrons. We call these form factors generically IBM proton form factors since they are probed by electrons.

We use an IBM-2 Hamiltonian which reproduces the energy spectra and B(E2)'s of ^{154}Gd,[6]

$$H = \epsilon \hat{N} - \kappa \hat{Q}_\pi \cdot \hat{Q}_\nu + \sum_L \xi_L A_L^\dagger \cdot A_L \quad (4)$$

where $\epsilon = 0.52$ MeV, $\kappa = 0.075$ MeV, $\xi_1 = \xi_3 = 1.2$ MeV and $\xi_2 = 0.03$ MeV, where $A^\dagger_{LM} = (d^\dagger_\nu d^\dagger_\pi)^{(L)}_M$, $L = 1, 3$, and $A^\dagger_{LM} = (d^\dagger_\nu s^\dagger_\pi - d^\dagger_\pi s^\dagger_\nu)_M/\sqrt{2}$, $L = 2$. This last term pushes up the antisymmetric neutron-proton states. The quadrupole operators are $\hat{Q}_{\sigma\mu} = \left[\hat{P}_{\sigma\mu} + \chi \hat{T}_{\sigma\mu}\right]$, $\sigma = \pi, \nu$, where $\chi = -0.9$.

Using the eigenfunctions of this Hamiltonian we can then calculate the matrix elements of the quadrupole density operator given in (3) from the ground state to the measured excited states $J^\pi_i = 2^+_i$, $i = 1, 2, 3$. However since only three transition densities have been measured and we need to determine four form factors, we have assumed that the symmetric form factor $\alpha_{\pi\pi}(r) + \tilde{\alpha}_{\pi\nu}(r)$ has a Tassie form since this symmetric form factor dominates in the excitation of the first 2^+ state in deformed nuclei. This gives the fourth equation, and hence we can solve for the four form factors. The results for $\alpha_{\pi\pi}(r)$ and $\tilde{\alpha}_{\pi\nu}(r)$ are shown in Fig. 1. The form factor $\tilde{\alpha}_{\pi\nu}(r)$ is smaller than $\alpha_{\pi\pi}$ and peaks further out on the surface.

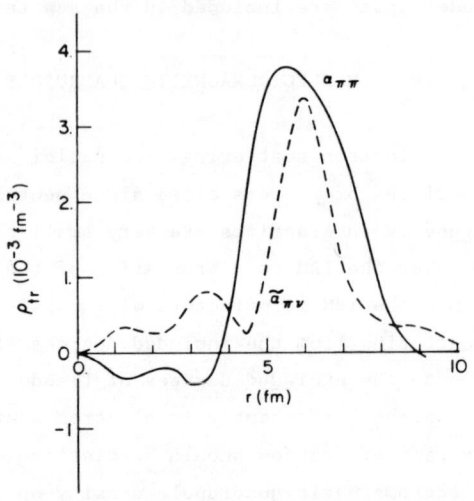

Figure 1. $\alpha_{\pi\pi}$ (solid line) and $\tilde{\alpha}_{\pi\nu}$ as a function of nuclear radius.

This behavior is expected since $\tilde{\alpha}_{\pi\nu}(r)$ is entirely an effective form factor.

4. THE IBM HADRONIC QUADRUPOLE TRANSITION OPERATOR

Since hadronic probes are not elementary particles as is the electron, the hadronic form factors need not necessarily be the same as the electromagnetic form factors.[7] However we ignore this difference in this discussion for simplicity of exposition only.[7] Most importantly there are additional form factors because the proton does interact directly with the valence neutrons whereas electrons do not. For medium energy protons, in fact, the interaction with proton and neutrons is almost equal. For this reason IBM-1 may be adequate to describe proton scattering, but, of course, including coupled channel effects as outlined in section 2. The effective IBM-1 hadronic form factors to be used in Equation (2) will then be[7]

$$\alpha(r) = \alpha_{\pi\pi}(r) + \tilde{\alpha}_{\pi\nu}(r) - \Delta\alpha(r) \tag{5a}$$

$$\beta(r) = \beta_{\pi\pi}(r) + \tilde{\beta}_{\pi\nu}(r) - \Delta\beta(r) \tag{5b}$$

$$\Delta\alpha(r) = \left[N_\nu\left(\alpha_{\pi\pi}(r) - \alpha_{\nu\nu}(r)\right) + N_\pi\left(\tilde{\alpha}_{\pi\nu}(r) - \tilde{\alpha}_{\nu\pi}(r)\right)\right]/N \tag{5c}$$

$$\Delta\beta(r) = \left[N_\nu\left(\beta_{\pi\pi}(r) - \beta_{\nu\nu}(r)\right) + N_\pi\left(\tilde{\beta}_{\pi\nu}(r) - \tilde{\beta}_{\nu\pi}(r)\right)\right]/N \tag{5d}$$

where $N_{\nu(\pi)}$ is the number of neutron (proton) bosons and $N = N_\nu + N_\pi$. The neutron form factors $\alpha_{\nu\nu}$, $\tilde{\alpha}_{\nu\pi}$, $\beta_{\nu\nu}$, $\tilde{\beta}_{\nu\pi}$ are analogous to the form factors given in (3), but can only be probed in hadron scattering and can <u>never</u> be determined in electron scattering. Thus $\Delta\alpha(r)$ and $\Delta\beta(r)$ represent the average difference between proton and neutron boson form factors which can be determined by hadron scattering in IBM-1. As a first step we assume that $\Delta\alpha = \Delta\beta = 0$, and then α and β are determined from electron scattering and are shown in Fig. 2. We note that $\beta(r)$ is similar to the derivative of $\alpha(r)$. We take the IBM-1 Hamiltonian of Ref. 8 since the matrix elements of $\hat{P}$, $\hat{T}$ are almost equal to the matrix elements of $\hat{P}_\pi + \hat{P}_\nu$, $\hat{T}_\pi + \hat{T}_\nu$, of the IBM-2 Hamiltonian in (4), better than those of the IBM-1 Hamiltonian given by the prescription of Ref. 9 for reducing an IBM-2 Hamiltonian to an IBM-1 Hamiltonian.

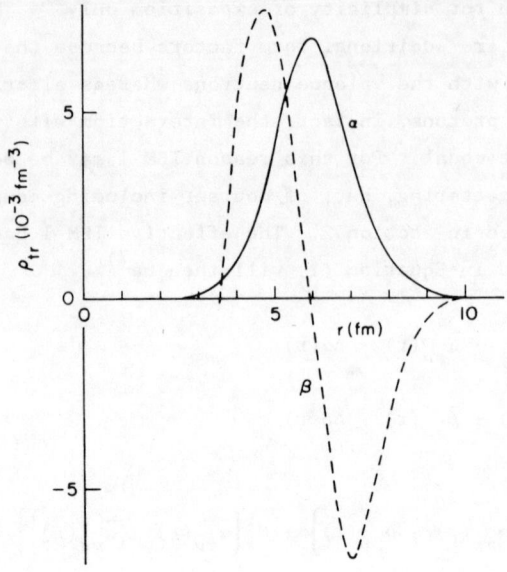

Figure 2. α and β as a function of nuclear radius for $\Delta\alpha, \Delta\beta = 0$.

5. RESULTS AND CONCLUSIONS

In Fig. 3 the differential cross sections for 650 MeV proton scattering on ^{154}Gd with $\Delta\alpha = \Delta\beta = 0$ are compared with the experimental cross sections measured at LAMPF.[10] For elastic scattering, the calculated cross section (solid line) agrees very well. For inelastic scattering the calculated cross section (dashed line) is about 20% too high for scattering to the first 2_1^+ state, but otherwise the minima and maxima are given very well. This result indicates that the proton scattering may be sensitive to the difference in neutron and proton degrees of freedom (i.e., $\Delta\alpha$ and $\Delta\beta$) which we ignored in going from electron scattering to proton scattering. Hence if on the average the proton form factors are larger than the neutron form factors ($\Delta\alpha, \Delta\beta > 0$), there will be a decrease in the cross section to the first 2_1^+ state.

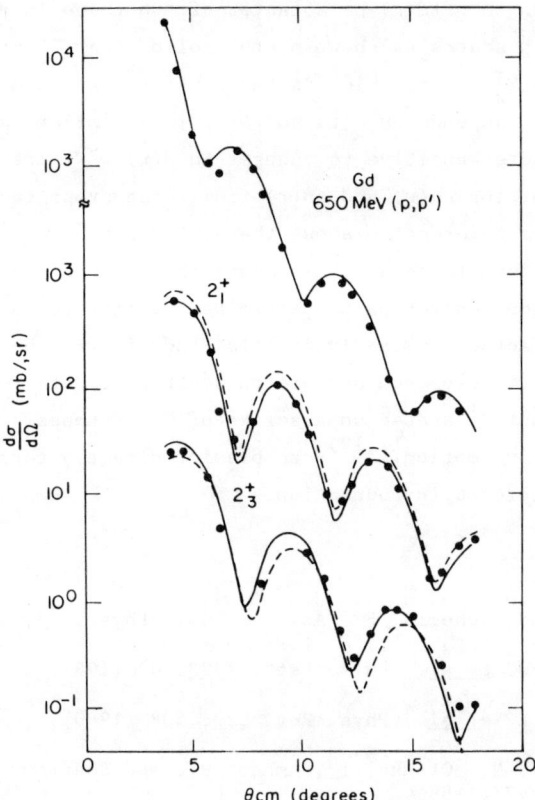

Figure 3. Elastic and Inelastic Scattering Proton Scattering in ^{154}Gd. For inelastic scattering dashed line is with $\Delta\alpha$, $\Delta\beta=0$ and full line with $\Delta\alpha = 0$, $\Delta\beta = 0.5\beta$. For elastic scattering solid line is for both.

For scattering to the 2_3^+ state the calculated cross section (dashed line) is about 20% too <u>low</u> and shifted forward about 0.4° in angle. However this state depends differently on $\alpha(r)$ and $\beta(r)$. A decrease in $\alpha(r)$ will decrease its cross section just as for the 2_1^+, but a drecrease in $\beta(r)$ will increase its cross section. In fact if we decrease the strength of $\beta(r)$ by a factor of two there is very good agreement for both states as shown by the solid line. Furthermore the scattering to the 2_2^+ state, which is the weakest state and is still being analyzed[10], depends on $\alpha(r)$ and $\beta(r)$ in a similar manner as the 2_3^+, but is even more sensitive to changes in $\beta(r)$. We are now in the process of determining $\Delta\alpha(r)$ and $\Delta\beta(r)$ from these cross sections which shall give us some information about the difference between the neutron and proton boson form factors. This means that we can successfully use electron and hadron scattering to determine the difference in neutron and proton form factors. Clearly in this study it will be helpful to have more systematic electron and proton scattering data to both the yrast and non-yrast 2^+ states on a series of Gd isotopes. Also fermion models of collective motion[11,12] can predict directly these form factors from a microscopic foundation.

REFERENCES

1. Arima, A. and Iachello, F., Adv. in Nucl. Phys., 13, 139 (1984).

2. Hersman, F. W. et al., Phys. Lett. 132B, 47 (1983).

3. Barlett, M. L. et al., Phys. Rev. C22, 408 (1980).

4. Ginocchio, J. N., Otsuka, T., Amado, R., and Sparrow, D., Phys. Rev. C33, 247 (1986).

5. Wenes, G., Ginocchio, J. N., Dieperink, A., Van der Cammen, B., Los Alamos Preprint LA-UR-86-510, to appear in Nucl.Phys.A (1986).

6. Otsuka, T., Hersman, F. W., Ginocchio, J. N., and Wenes, G., Japan Atomic Energy Research Institute Preprint (1986).

7. Ginocchio, J. N. and Wenes, G., Los Alamos Preprint LA-UR-86-1024, to appear in Phys. Rev. C (1986).

8. Lipas, P.O., et al., Phys. Lett. 155B, 295 (1985).

9. Harter, H. et al., Phys. Lett. 157B, 1 (1985).

10. Hintz, N., et al., Univ. of Minnesota Progress Report 1984-1985.

11. Ginocchio, J. N., Ann. of Phys. 126, 234 (1980).

12. Wu, C. L., et al., Phys. Lett. 168B, 313 (1986).

GROUP THEORY APPROACH TO SCATTERING AND ITS APPLICATION TO HEAVY ION COLLISIONS

Y. ALHASSID
A.W. Wright Nuclear Structure Laboratory
Yale University, New Haven, Connecticut 06511, U.S.A.

Abstract The dynamical group approach, useful in structure problems is extended to scattering processes. A new way to analyze heavy ion collisions, based on the dynamical group SO(3,2) is suggested. This group is a generalization of the Coulomb Scattering group SO(3,1) and describes scattering from a short-range potential in the presence of a Coulomb field. Applications to elastic heavy ion scattering are shown as well as comparison with other methods.

1. INTRODUCTION

Heavy ion collision is a complicated process involving a many-body system and the presence of many open reaction channels. Various methods have been devised to attack such a problem. The elastic scattering have been successfully described by an effective complex interaction known as an optical potential [1]. Direct reactions can be treated via a distorted wave Born approximation (DWBA) [2] and in the limit of extremely large number of open channels, the statistical model of Hauser and Feschbach [3] is useful. However, in intermediate situations it is usually necessary to solve the complicated coupled-channel problem which is a very difficult numerical job and does not necessarily deepen our physical insight of the problem.

We suggest a new method in which the S-matrix is constructed directly based on appropriate dynamical groups and not via the solution of a Schrödinger equation where the interaction is described by a potential. For simple Schrödinger problems which are connected to dynamical groups our method yields the same S-matrix as the conventional approach.

To motivate our approach we note that for instance the data of heavy ion elastic scattering in the vicinity of the Coulomb barrier is quite systematic. It is customary to plot the cross-section σ relative to the Rutherford cross-section σ_{Ruth} versus the center of mass energy of the colliding ions. At energies below the Coulomb barrier the above ratio is very close to 1 since most of the scattering is affected by the Coulomb field. Around and above the Coulomb barrier many inelastic and reaction channels become open to the system resulting in a strong absorption and a rapid drop of σ/σ_{Ruth} to a level of few percents, where it shows wide oscillations (several MeV's in width) around a constant level. Fig. 1 shows a typical excitation function for the ^{13}C+^{13}C elastic scattering at various scattering angles. In some cases we see also fine structure of narrower resonances superimposed on the above gross structure. Similar behaviour is observed in the angular distributions at a given collision energy.

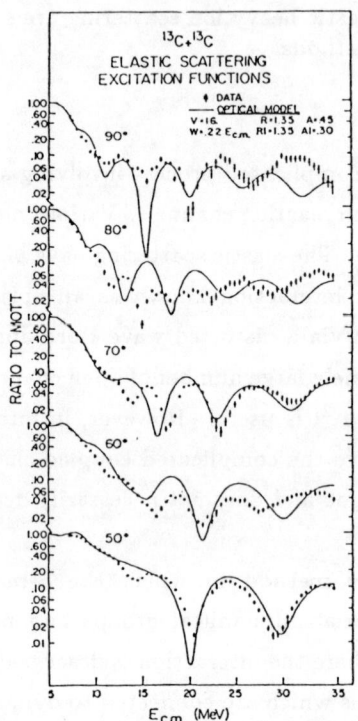

FIGURE 1: Excitation function of the ^{13}C+^{13}C reaction at several scattering angles. σ/σ_{Ruth} is plotted on a logaritmic scale vs. the initial center of mass energy. The full circles are the experimental results (S. Korotky, K. Erb, and D.A. Bromley) and the solid lines are typical optical fits. Taken from the thesis of S. Korotky.

The systematics shown by the data suggest that it should be possible to construct relatively simple S-matrices for the problem. In a recent series of investigations [4-6] we have been able to make the first steps in generalizing the techniques of dynamical groups, useful in the analysis of nuclear structure (Interacting Boson Models [7]) to problems involving the continuum. In particular, when dynamical symmetry occurs in the problem one can construct the S-matrix in a closed form. The content of my presentation will follow the major steps in the development of the approach. In Section 2 the need to use non-compact groups [4] for the description of scattering problem will be clarified. The method to calculate the S-matrix [5] associated with a certain dynamical symmetry is sketched in Section 3 and used in Section 4 to derive an S-matrix for modified Coulomb potentials [6]. Finally, in Section 5 we suggest an S-matrix for elastic heavy ion scattering and in Section 6 an analysis of data and comparison with other methods are presented [8]. A detailed analysis of the properties of the above S-matrix and a connection to the traditional potential approach are described in a paper by R.D. Amado and D. Sparrow [9].

2. SCATTERING AND NON-COMPACT GROUPS

Most potentials of interest in physics have finite number of bound states, so that the groups used in the description of structure problems are usually compact groups whose unitary representations are all finite-dimensional. To describe systems with a continuous spectrum we must use *non-compact* algebras. They possess unitary representations which are *continuous* and infinite-dimensional and thus form a natural candidates for the description of scattering states. To illustrate the mathematical concept of a non-compact group we use the group $SU(1,1)$ (isomorphic to $SO(1,2)$) which is an analytic continuation of the familiar compact $SU(2)$ group. The commutation relations of the $SU(1,1)$ generators $J_i(i=x,y,z)$ differ from the corresponding $SU(2)$ commutation relations by a single minus sign;

$$[J_x, J_y] = -iJ_z$$
$$[J_y, J_z] = iJ_x, \qquad [J_z, J_x] = iJ_y \qquad (2.1)$$

The topological properties of the group are however very different. The space left invariant by $SU(2)$ is the sphere in three dimensions while that left invariant by

SU(1,1) is the hyperboloid which is unbounded and hence the non-compactness of the group.

The SU(1,1) unitary representations are characterized by the eigenvalue $j(j+1)$ of the Casimir invariant

$$C = -J_x^2 - J_y^2 + J_z^2 \quad . \tag{2.2}$$

They are [10]:

(i) Discrete representations, $j = -\frac{1}{2}, -1, -\frac{3}{2}, \ldots$ describing the bound states of the system.

(ii) Continuous principal series representations, $j = -\frac{1}{2} + ik \, (k > 0)$ describing scattering states with momentum k.

(iii) Supplementary series $-\frac{1}{2} > j > 0$ which are missing in the physical applications.

The situation is summarized in Fig. 2 showing the allowed SU(1,1) representations in the complex j-plane. The group SU(1,1) is useful in the analysis

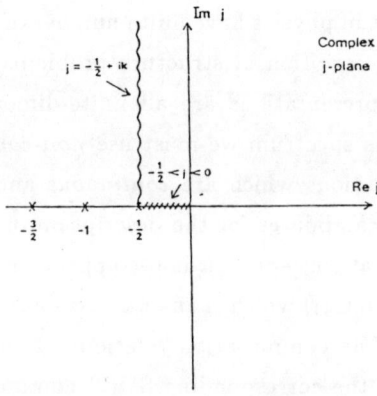

FIGURE 2: The values of j which correspond to unitary representations of SU(1,1) in the complex j plane (see text).

of a variety of one-dimensional problems with solvable potentials. For such problems the group approach offers just an alternative (though more elegant) way to solve them and they are discussed elsewhere [4-5,11,12].

3. THE CONSTRUCTION OF THE S-MATRIX

The major problem in the group approach is the construction of an S-matrix starting from a dynamical group G. In the conventional approach the S-matrix is found from the asymptotic form of the wave function when the separation between the fragments is large. In an algebraic approach where the scattering state is described by an abstract vector in a representation it was not known how to take the asymptotic limit. Our solution was to introduce another group F, which describes the asymptotic problem in the *absence* of interactions [5]. In a very schematic way we can say that the S-matrix appears as a coefficient which connects the representations of F with those of G.

$$\text{Representations of G} = (\text{S-matrix}) \text{ Representations of F} \quad . \quad (3.1)$$

To sketch the method of computation it is best to use a simple and familiar example, Coulomb scattering in three dimensions. A group-theoretical description of this problem and its S-matrix has been the subject of many investigations by various authors [13,14]. We present here our algebraic method which has the advantage that it can be generalized to more complicated problems.

The symmetry group of the Coulomb Scattering in known to be SO(3,1) composed of the angular momentum $\vec{L}$ and a normalized Runge-Lenz vector $\vec{K}$

$$\vec{K} = \frac{\mu}{k}\left[\frac{1}{2\mu}(\vec{p}\times\vec{L} - \vec{L}\times\vec{p}) + \alpha Z_1 Z_2 \hbar c \hat{r}\right], \quad (3.2)$$

where μ and k are the reduced mass and the wave number, respectively, of the colliding charges Z_1 and Z_2 and α is the fine structure constant. The Coulomb representations of this dynamical group G=SO(3,1) are described by the pair $(\omega, 0)$ where $\omega(\omega + 2)$ and 0 are the eigenvalues of the two quadratic Casimir operators of SO(3,1)

$$\begin{aligned}C_2 &= \vec{L}^2 - \vec{K}^2 \\ C_2' &= \vec{L}\cdot\vec{K}\end{aligned} \quad . \quad (3.3)$$

In a scattering representation ω is a continuous function of k and in the Coulomb problem

$$\omega = -1 + i\eta(k), \tag{3.4}$$

where $\eta(k)$ is the Sommerfeld parameter

$$\eta(k) = \alpha \frac{Z_1 Z_2 \mu c}{\hbar k} \quad . \tag{3.5}$$

A scattering state is now characterized by the chain

$$\left| \begin{matrix} SO(3,1) \supset SO(3) \supset SO(2) \\ (\omega, 0), & \ell, & m \end{matrix} \right\rangle, \tag{3.6}$$

where l and m are the angular momentum and its projection.

The asymptotic symmetries are that of a free particle (we set $\alpha=0$) which include in addition to rotations the translations. Thus the corresponding symmetry algebra includes the angular momentum $\vec{L}$ and the linear momentum $\vec{P}$. This is known as the Euclidean group in three dimensions E(3). In our notation we have F=E(3). The incoming$(-k)$ and outgoing$(+k)$ free waves are now characterized by representations "$\pm$k" of E(3) [15]:

$$\left| \begin{matrix} E(3) \supset SO(3) \supset SO(2) \\ (\pm k, 0), & \ell, & m \end{matrix} \right\rangle, \tag{3.7}$$

where k^2 and 0 are respectively the eigenvalues of the two E(3) Casimir invariants

$$\begin{aligned} C_2 &= \vec{P}^2 \\ C_2' &= \vec{P} \cdot \vec{L} \end{aligned} \quad . \tag{3.8}$$

The S-matrix in now defined *algebraically* through the expansion of the SO(3,1) representation in the E(3) representations:

$$| \omega, l, m \rangle = A_\ell(k) | -k, \ell, m \rangle + B_\ell(k) | +k, l, m \rangle, \tag{3.9}$$

where A_ℓ, B_ℓ are k-dependent coefficients. The partial S-matrix at angular momentum ℓ is given by

$$S_\ell(k) = (-)^{\ell+1} B_\ell(k) / A_\ell(k) \quad . \tag{3.10}$$

The calculation of the S-matrix in (3.10) is accomplished by deriving recursion relations in ℓ through the use of the so-called "Euclidean connection" [5] in which the generators of G are expressed as non-linear functions of the generators of F. Such a relation is found by the group theoretical method of expansion [16]. For the Coulomb problem it is given in the $\pm k$ representations by

$$K_3(\pm k) = \frac{i}{k}\left[\frac{1}{2}(P_+L_- - P_-L_+) + (-1 \mp i\eta(k))P_3\right] \quad . \tag{3.11}$$

Applying (3.11) to both sides of Eq. (3.9) we obtain a recursion relation for $R_\ell(k) = B_\ell(k)/A_\ell(k)$:

$$R_{\ell+1}(k) = -\frac{\ell + 1 + i\eta(k)}{\ell + 1 - i\eta(k)} R_\ell(k), \tag{3.12}$$

whose solution yields the familiar partial Coulomb phase shifts:

$$S_\ell^{coul} = \frac{\Gamma(\ell + 1 + i\eta)}{\Gamma(\ell + 1 - i\eta)} \quad . \tag{3.13}$$

4. S-MATRIX FOR A MODIFIED COULOMB PROBLEM

We now discuss the S-matrix associated with a dynamical group SO(3,2) which generalizes that of the Coulomb problem [6]. The additional degree of freedom in SO(3,2) accounts for an additional interaction, added to the Coulomb one.

The scattering states are now characterized by

$$\left|\begin{array}{cccc} SO(3,2) \supset SO(3) \times SO(2) \supset SO(2) \times SO(2) \\ (\omega, \circ), & \ell & m & v \end{array}\right\rangle, \tag{4.1}$$

where

$$\omega = -\frac{3}{2} + i\eta(k) \tag{4.2}$$

and $\omega(\omega + 3)$ characterized the quadratic Casimir operator of SO(3,2). the additional quantum number v characterizes the strength of the interaction which is added to the Coulomb potential. The asymptotic group is now F=E(3)xE(2), where E(3) describes the spatial symmetries and E(2) the "internal" symmetries of the interaction. The Euclidean connection and its use in the S-matrix evaluation are discussed elsewhere [6]. We just note that by using these techniques one obtains a set of four recursion relations in ℓ and v whose solution is the following partial S-matrix [6]

$$S_\ell(k) = \frac{\Gamma\left(\frac{\ell+v+3/2+i\eta(k)}{2}\right)\Gamma\left(\frac{\ell-v+3/2+i\eta(k)}{2}\right)}{\Gamma\left(\frac{\ell+v+3/2-i\eta(k)}{2}\right)\Gamma\left(\frac{\ell-v+3/2-i\eta(k)}{2}\right)} e^{i2\ell n 2\eta(k)}. \quad (4.3)$$

This S-matrix has the following properties:

(i) For $v=1/2$ it reduces to the Coulomb S-matrix, S_ℓ^{coul}, as is easily shown by using the doubling formula for the Γ-function

$$\Gamma(z)\Gamma(z+1/2) = 2^{1-2z}\pi^{1/2}\Gamma(2z) \quad . \quad (4.4)$$

(ii) If v is bounded then for any given k

$$\begin{aligned} S_\ell(k) &\to S_\ell^{coul}(k), \\ l &\to \infty \end{aligned} \quad (4.5)$$

So that the scattering is mostly Coulomb at large impact parameters.

(iii) S_ℓ depends on v^2. For v^2 real the S-matrix is unitary.

(iv) To get absorption we need to have v complex. The unitarity bound is satisfied as follows:

$$\mid S_\ell(k) \mid \leq 1 \quad \text{for} \quad Im(v^2) \leq 0. \quad (4.6)$$

It is important to note that the dynamical symmetry determines the functional form of the S-matrix. However, the parameter v is in principle still a function of ℓ and k. This function determines the specific form of the interaction and depends on the process under consideration. In a specific application it can be determined from theoretical models, or better yet from the experimental data. We define

$$w = v - 1/2, \quad (4.7)$$

so that $w=0$ corresponds to a pure Coulomb interaction.

5. ELASTIC SCATTERING OF HEAVY IONS

Elastic scattering is an important process which accompanies any reaction. A variety of methods have been used in the past to analyze elastic heavy ion scattering such as the optical model [1] and diffraction models [17]. Other parametrized models of the S-matrix were introduced by Blair [18], McIntyre [19], Ericson [20], and Frahn and Venter [21]. We suggest a new method based on the S-matrix associated with the SO(3,2) dynamical group given by Eq. (4.3). This form of the S-matrix is of particular interest in such application as it describes scattering from a short-range interaction in the background of a Coulomb field. A particular useful form that we suggest is a Woods-Saxon (or a Fermi) form in ℓ:

$$w = \frac{w_\circ}{1 + exp(\ell - \ell_\circ)/\Delta}, \qquad (5.1)$$

where $w_\circ$ is a complex number, $\ell_\circ$ is a "grazing" angular momentum and Δ is a diffuseness parameter. The above form is motivated by our physical knowledge that the most important processes in the elastic heavy ion scattering are the peripheral ones due to the short-ranged nuclear interaction and the strong absorption at angular momenta smaller than the grazing one. Furthermore, as shown by Amado and Sparrow [9] using inversion of the S-matrix in the Eikonal approximation, the tail of the potential in the radial coordinate r decreases to zero faster than an exponential (by a certain power) as $r \to \infty$. This is in accord with microscopic calculations of folded Yukawa potentials.

The amplitude $f(\theta)$ and the cross-section are calculated as usual by summing up the partial wave series

$$f(\theta) = \frac{1}{2ik} \sum_\ell (S_\ell(k) - 1) P_\ell(cos\theta) \qquad (5.2)$$
$$\sigma(\theta) = |f(\theta)|^2 .$$

The series (5.2) suffers from the same convergence problem as does the corresponding series for the long-range Coulomb potential. In practice we subtract and add the Coulomb amplitude so that

$$f(\theta) = \frac{1}{2ik} \sum_\ell \left(S_\ell(k) - S_\ell^{coul}(k)\right) P_\ell(cos\theta) + f_{coul}(\theta), \qquad (5.3)$$

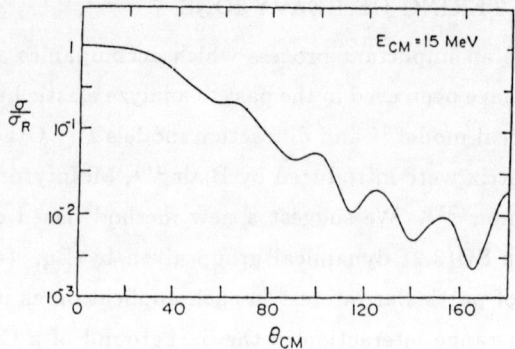

FIGURE 3: Angular distribution for the collision of two heavy ions with $A_1 = 12, A_2 = 14$, and $Z_1 = Z_2 = 6$. The ratio σ/σ_{Ruth} is calculated using the S-matrix (4.3) and (5.1) with $w_o = 1.5 - 2.5i, \ell_o = 15.4, \Delta = 0.5$ at $E_{c.m.} = 15$ MeV.

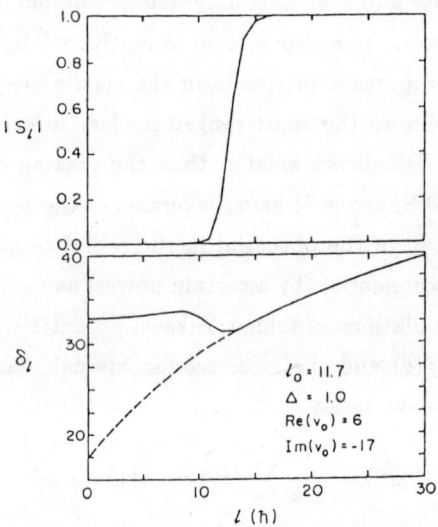

FIGURE 4: The amplitude $|S_\ell|$ (top) and phase δ_ℓ (bottom) of the S-matrix (4.3) and (5.1) versus ℓ. The parameters in (5.1) are $w_o = 6 - 17i, \ell_o = 11.7$, and $\Delta = 1$. The Sommerfeld parameter is $\eta = 3.6$.

where we have used the known analytic summation of the partial Coulomb series $f_{coul}(\theta) = -\eta exp[-2i\eta \ln \sin(\theta/2)]S_o^{coul}/2k\sin^2(\theta/2)$. The series in (5.3) converges rapidly as soon as ℓ is several Δ's above ℓ_o. A typical angular distribution calculated from (5.3),(4.3) and (5.1) using $w_o=1.5$-$2.5i$, $\ell_o=15.4$ and $\Delta=0.5$ is shown in Fig. 3. Writing the partial amplitude as

$$S_\ell = |S_\ell| e^{2i\delta_\ell}, \tag{5.4}$$

we show in Fig. 4 a typical behaviour of $|S_\ell|$ and δ_ℓ vs. ℓ. As expected we get strong absorption for $\ell < \ell_o (|S_\ell| \ll 1)$. The phase δ_ℓ deviate strongly from the Coulomb phase only in the vicinity and below ℓ_o.

To calculate an excitation function versus energy, we need to know the dependence of $\ell_o, w_o,$ and Δ on k. It is found that the most important

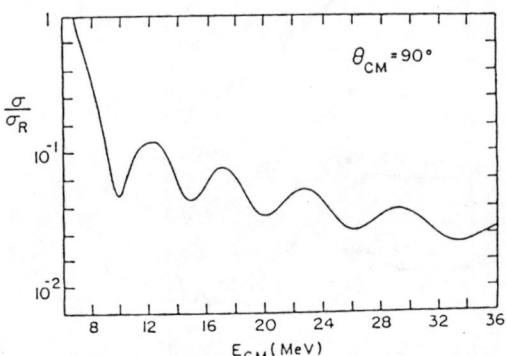

FIGURE 5: Excitation function at 90° for the same reaction as in Fig. 3, and where the energy dependence of ℓ_o is given by (5.5) with R=9.5 fm and V_c=6.5 MeV.

k-dependence is that of ℓ_o. Taking a semi-classical expression for ℓ_o (the incoming energy equal the total centrifugal and Coulomb barrier at the touching distance R of the two ions)

$$\ell_o = \sqrt{R^2(k^2 - k_o^2)}, \tag{5.5}$$

where $k_o^2/2\mu = V_c$ is the Coulomb barrier, we obtain a typical excitation function shown in Fig. 5. Notice that the cross-sections in Figs. 3 and 5 have all the qualitative features observed in heavy ion elastic scattering.

6. ANALYSIS OF SCATTERING DATA AND COMPARISON WITH OTHER MODELS

Using the S-matrix (4.3) and (5.1) we have analyzed elastic scattering data

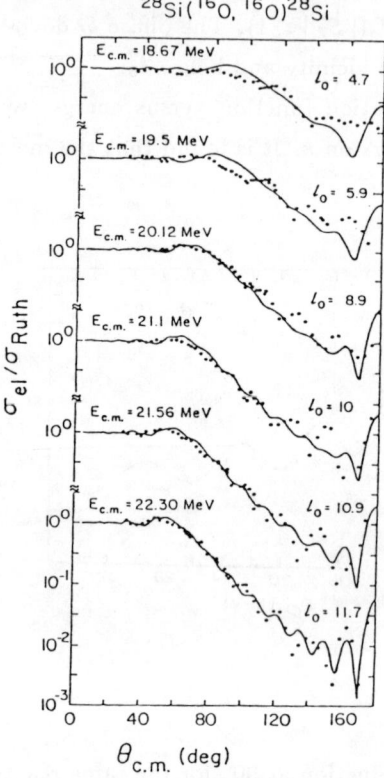

FIGURE 6: Elastic angular distribution for the $^{16}O+^{28}Si$ scattering at several energies above the Coulomb barrier. The full circles are the experimental results of Ref. 22 and the solid lines are calculated by the algebraic model using w_o=6-17i and $\Delta = 1$. The value of ℓ_o is specified in each case.

of the reaction between the spin-0 ions $^{16}O+^{28}Si$. Angular distribution were measured in Saclay by Mermaz et al. [22] in the energy range $E_{c.m.}$=18.6-22.3

MeV just above the Coulomb barrier and in Brookhaven by Braun-Munzinger et al. [23] from 22.7 MeV and above. The calculated angular distribution σ/σ_{Ruth} at $E_{c.m.}=22.3$ MeV was fitted to the measured one. The parameters used were

$$w_o = 6 - 17i$$
$$\Delta = 1 \tag{6.1}$$

and the grazing angular momentum $\ell_o=11.7$. The fit (continuous line) is shown at the bottom of Fig. 6. We have then left w_o and Δ *fixed* and allowed only ℓ_o to change in the fits of all the other distributions shown on Fig. 6. The glory peak around the Coulomb barrier and the drop in the cross-section are well reproduced. The oscillations are reasonably described at the higher energies but less so at the lower energies. However, these fits can be improved if w_o and Δ are also allowed to be functions of k. Optical fits at somewhat higher energies are shown in Fig. 7 [24]. These fits are comparable in quality to the ones reproduced

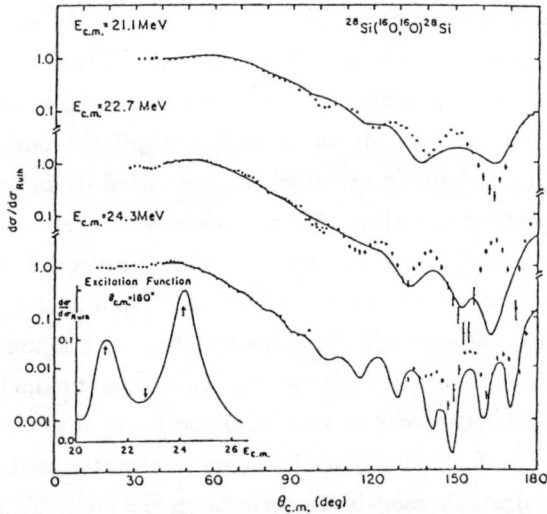

FIGURE 7: Optical fits (solid lines) for the elastic scattering of $^{16}O+^{28}Si$. Resonances are added to the optical amplitude to explain the backward features of the scattering. Taken from Ref. 24.

by the dynamically generated S-matrix (4.3). However, since we produce the S-matrix in closed form the calculations of the cross-sections are straightforward and much simpler. Furthermore, the amplitude (4.3) appears to produce automatically the backward rise observed in the measured angular distributions. In the optical model it is necessary to introduce a signature dependent term $(-)^\ell$ or exchange interaction to produce these features. Also our parametrization uses less parameters than the optical potential and has the Coulomb interaction built intrinsically in the amplitude. While the use of local optical potential is questionable especially at shorter distances where many-body effects are important our approach is free of such problems.

When compared with the Frahn-Venter model our expression has the advantage of being valid in the whole complex plane (being derived by a group theory approach) while the forms of Frahn and Venter are restricted to the physical regime of positive k.

7. CONCLUSIONS

It appears that dynamical groups, useful in the analysis of structure data can be useful in the analysis of scattering data [25]. We remark that the main advantage of the new method introduced here is not in reproducing in a simpler way results for elastic scattering data already obtained by the optical model, but rather that in being simple it can be very useful when generalized to inelastic and transfer reactions. The method can then offer a much more attractive framework than the method of coupled channels. Work in that direction is in progress. An especially simple case is for scattering of ions with spin in the presence of spin-orbit interactions. For spin-1/2 projectile colliding with a spin-0 target the S-matrix is diagonal in the $j = \ell \pm 1/2$ basis. In this context of inelastic excitations it is worth mentioning the work of Amado, Ginocchio et al. [26] where the relative motion is treated semi-classically while the internal states are treated quantum-mechanically by algebraic models (IBA,vibron model etc.). Thus the problem is turned into one of a multi-step excitations of a bound system. In the approach presented here the relative motion is treated quantum mechanically and for inelastic scattering one should further introduce the algebraic description of the internal structure. Other generalizations that we have in mind are to relativistic scattering. In the case of fermions where the Dirac equation has to be used one should include properly the algebra of the γ^μ matrices.

ACKNOWLEDGMENTS

I thank my collaborator in this work F. Iachello. I also thank F. Gürsey, J. Wu, J. Engel and A. Frank for their collaboration on various parts of that work. I acknowledge illuminating discussions with R.D. Amado, D. Sparrow, M. Moshinsky, and L.C. Biedenharn. This work was supported in part by the Department of Energy Contract No. DE-ACO2-76ER03074 and in part by an A.P. Sloan Fellowship.

REFERENCES

1. H. Feschbach, Ann. Phys. (N.Y.) 5, 357 (1958); 19, 287 (1962).
2. See for example in N. Austern, Direct Nuclear Reactions, New York, Wiley (1970).
3. W. Hauser and H. Feschbach, Phys. Rev. 87, 366 (1952).
4. Y. Alhassid, F. Gürsey and F. Iachello, Ann of Phys. (N.Y.) 148, 346 (1983); Phys. Rev. Lett. 50, 873 (1983).
5. Y. Alhassid, J. Engel and J. Wu, Phys. Rev. Lett. 53, 17 (1984); Y. Alhassid, F. Gürsey, and F. Iachello, Ann. Phys. (N.Y.) 167, 181 (1986).
6. Y. Alhassid, F. Iachello, and J. Wu, Phys. Rev. Lett. 56, 271 (1986); J. Wu, F. Iachello, and Y. Alhassid, Yale preprint YNT-86-01
7. A. Arima and F. Iachello, Ann. Phys. (N.Y.) 99, 253 (1976); 111, 201 (1978); 123, 468 (1979).
8. Y. Alhassid and F. Iachello, to be published.
9. R.D. Amado and d. Sparrow, these proceedings.
10. G. Lindbland and B. Nagel, Ann. Inst. Henri Poincare, 13, 27 (1970).
11. A. Frank and K.B. Wolf, Phys. Rev. Lett. 52, 1737 (1984); J. Math. Phys. 26, 1973 (1985).
12. Y. Alhassid, in Boson in Nuclei, D. H. Feng, S. Pittel, and M. Vallieres eds. World Scientific, Singapore (1984).
13. For a review see in B.G. Wybourne, Classical Groups for Physicists, Wiley, New York (1974) and references therein.
14. L.C. Biedenharn and P.J. Brussard in Coulomb Excitation, p. 107, Clarendon Press, Oxford (1965); D. Zwanziger, J. Math. Phys. 8, 1858 (1967); A.M. Perelomov and V.S. Popov, Sov. Phys. JETP 27, 967 (1968); A.O. Barut and W. Rasmussen, J. Phys. B6, 1713 (1973).
15. W. Miller, Jr., On Lie Algebras and some Special Functions of Mathematical Physics, American Mathematical Society, Providence, R.I. (1964).

16. A. Frank, Y. Alhassid, and F. Iachello, Phys. Rev. A (1986).
17. W.E. Frahn, Ann. Phys. (N.Y.) <u>72</u>, 524 (1972).
18. J.S. Blair, Phys. Rev. <u>95</u>, 1218 (1954).
19. J.A. McIntyre, S.D. Baker, and T.L. Walts, Phys. Rev. <u>116</u>, 1212 (1959).
20. T.E.O. Ericson, in Preludes in Theoretical Physics, A. De-Shalit, H. Feshbach, and L. Van-Hove eds., North-Holland, Amsterdam (1965), p. 321.
21. W.E. Frahn in <u>Treatise on Heavy-Ion Science</u>, D.A. Bromley, ed., vol. 1, Plenum Press, New York (1984).
22. M.C. Mermaz, E.R. Chavez-Lomeli, J. Barrette, B. Berthier, and A. Greiner, Phys. Rev. <u>C29</u>, 147 (1984).
23. P. Braun-Munzinger and J. Barrette, Phys. Rep. <u>87</u>, 209 (1982).
24. P. Braun-Munzinger, G.M. Berkowitz, M. Gai, J.W. Harris, C.M. Jacheinski, T.R. Renner, and C.O. Uhlhorn, Phys. Rev. <u>C24</u>, 1010 (1981).
25. For a review see F. Iachello, these proceedings.
26. R.D. Amado; J. Ginocchio, these proceedings.

A SURVEY OF THE GROUP THEORETICAL APPROACH TO SCATTERING[+]

L.C. Biedenharn[++] and A. Stahlhofen

Institut für Theoretische Kernphysik der Universität Bonn
Nußallee 14-16, D-5300 Bonn, West Germany

ABSTRACT

The group-theoretic approach to scattering is reviewed critically, and the concept of the factor space of scattering states introduced. Nonstandard matrix realizations in scattering space lead to a novel view of phase-shifts.

1. INTRODUCTION

The use of symmetry in the analysis of physical problems is well--known to be both fundamental and useful. The mathematics of symmetry is, globally, group theory and, locally, Lie algebra; the use of these disciplines for *bound state* problems is very familiar, with the ($\mathbb{R}^3$) harmonic oscillator (SU3), and the bound Coulomb problem (SO4) the best known examples. Applications of group theoretic techniques to *scattering* (continuum states) is, by contrast, relatively undeveloped. A group theoretic interpretation for Coulomb scattering was given early[1,2], but it was only recently, for example, that the Pöschl-Teller and Morse potentials, long used in molecular and nuclear physics, were related[3,4] for scattering, to SU(1,1) symmetry.

By and large the discussion in the literature has been a mixture of group theory and analysis, and the question as to whether a purely group-theoretic treatment of scattering can be given has only recently been addressed[5]. The present paper attempts a critical analysis of this question.

We begin by reviewing the group-theoretical treatment of Coulomb scattering. We then turn to the general question posed above, develop-

---

[+]Supported in part by the National Science Foundation
[++]Alexander von Humboldt Stiftung Re-Invitation Program
 on leave from Duke University, Durham, NC 27706 USA

ing our answer by an explicit example. The final section states our conclusions.

2. REVIEW OF THE COULOMB SCATTERING EXAMPLE[1,6]

The NR Coulomb problem in $\mathbb{R}^3$, with Hamiltonian: $H = \frac{p^2}{2m} + \frac{\alpha}{r}$, ($\hbar=c=1$), has two vector constants of the motion: the angular momentum $\vec{L} = \vec{r} \times \vec{p}$ and the Lenz-Runge-Laplace operator: $\vec{A} = \frac{1}{2m}(\vec{p} \times \vec{L} - \vec{L} \times \vec{p}) + \alpha\hat{r}$, obeying the commutation relations: $\vec{L} \times \vec{L} = i\vec{L}$; $\vec{L} \times \vec{A} + \vec{A} \times \vec{L} = 2i\vec{A}$; $\vec{A} \times \vec{A} = i(-2H/m)\vec{L}$. We renormalize: $\vec{\tilde{A}} \equiv |2H/m|^{-1/2}\vec{A}$ to obtain:

E < 0: $\vec{\tilde{A}} \times \vec{\tilde{A}} = i\vec{L}$, symmetry group SO(4), bound states,

E = 0: $\vec{\tilde{A}} \times \vec{\tilde{A}} = 0$, $\mathbb{R}^3$ Euclidean group E(3),

E > 0: $\vec{\tilde{A}} \times \vec{\tilde{A}} = -i\vec{L}$, symmetry group SO(3,1), scattering.

To discuss Coulomb scattering one constructs the Temple-Gordon solution[1], a continuum eigenfunction which, asymptotically (beyond a cutoff for the potential) becomes a plane wave e^{ikz} plus outgoing spherical waves. Specializing to $\vec{k}_{inc} \parallel \hat{z}$, one has:

$$\psi(\vec{k},r) = \sum [4\pi(2L+1)]^{1/2} i^L e^{i\sigma_L} \frac{F_{L,\eta}(kr)}{kr} Y_L^0(\hat{r}) , \qquad (1)$$

where the $F_{L,\eta}$ are defined to be regular at r=0 and asymptotically $F_{L,\eta} \simeq \sin(kr - \frac{L\pi}{2} - \eta\log 2kr + \sigma_L)$. (Note the infamous logarithmic term). The Coulomb phase shifts are: $\sigma_L \equiv \arg(\Gamma(L+1+i\eta))$, with $\eta = \alpha/v_\infty$ being the Sommerfeld number.

The Temple-Gordon solution has a direct group theoretic significance. Consider the compact (bound state) case; there are two group chains:

a) $SO4 \supset SO(3) \supset SO2$ with quantum numbers: $(N,O);L;M$.

b) $SO4 \stackrel{loc}{=} SU2 \times SU2 \supset SO2 \times SO2$ with: $(N,O); (j_1 m_1)(j_2 m_2)$. (Since $\vec{L} \cdot \vec{A}=0$, we must have $j_1 = j_2 = \frac{N-1}{2}$.)

The eigenkets of these two chains are related by the SU2 vector addition coefficients, that is:

$$|(N,O); (\frac{N-1}{2},m_1)(\frac{N-1}{2},m_2)\rangle = \sum_{L,M} C_{m_1 \; m_2 \; M}^{\frac{N-1}{2} \; \frac{N-1}{2} \; L} |(N,O);LM\rangle . \qquad (2)$$

The Temple-Gordon formula is the analytic continuation of this result, with $N \to i\eta$. The Coulomb phase shift is then found to be[1] the phase of the Wigner coefficient in this analytic continuation[7]. Noting that the LHS of eqs. (1) and (2) have A_z sharp, whereas on the RHS A_z acts as a shift operator in L, one finds a three-term recursion relation whose solution yields the Coulomb phase shifts[2]. This procedure generalizes to $\mathbb{R}^N$ yielding the $SO(N,1)$ phase shifts[8]. Scattering for the Dirac-Coulomb problem can also be treated group-theoretically[9].

3. SOME GENERAL REMARKS

The treatment of Coulomb scattering in the above section clearly exploits special features: how should one proceed for a general group-theoretical determination of scattering?

Observe that this problem is not well posed: scattering cannot be a purely group-theoretic construct. Both the Coulomb problem and the interaction-free problem ($\alpha=0$) possess exactly the same symmetry group, $SO(3,1)$, yet the scattering differs completely. Clearly scattering involves the Hamiltonian as well as the symmetry group. Given these data the problem may still not be well-posed, in the sense that the Hamiltonian may be given abstractly (in terms of group invariants). For example, the Morse and Pöschl-Teller problems both involve the same symmetry, and the same formal Hamiltonian, yet differ, because of different embeddings[10].

We see that scattering depends not only on the symmetry group and the form of the Hamiltonian, but also on the specific realization.

Scattering is a matter of asymptotics, since the initial (and final) scattering states are measured at spatial infinity. With this in mind, Wu and Alhassid[11] took the *Euclidean group* as an abstration of the asymptotics and introduced (in $\mathbb{R}^2$) two types of states: $|\pm k, m\rangle$, with $\pm k$ = out/in waves), as irreps of the Euclidean group E2. They defined the *Euclidean connection*: the generators of the original group are written in the enveloping algebra of the Euclidean group generators. From this, they derive the scattering matrix.

This is an interesting suggestion, but it suffers from the technical defect that in- and out-going irreps of E(n), n≥2, do not exist[+].

It is easy to see what goes wrong: the in/out spherical waves are singular at the origin. This observation[++] provides the key; to resolve the problem, *we delete the origin from the space of scattering states*. Topologically when we delete the origin, and use (r,φ) coordinates, the radial coordinate becomes contractible. In the limit the space $\mathbb{R}^2 - \{0\}$ contracts to S^1, the circle at infinity. The radial derivative in the limit yields two values, the remnants of in/out waves.

The net result is this: *the proper space for discussing scattering in $\mathbb{R}^2$ is the factor space $S^1 \times Z_2$*. In other words, we simply factor out the radial motion, but preserve the two element space Z_2 distinguishing in/out motion at infinity.

Now consider the group G, realized on $\mathbb{R}^2$. What happens to Lie (G) as we go to spatial infinity? Nothing whatsoever! The Lie algebra must be obeyed everywhere in $\mathbb{R}^2$, with exactly the same relations. The form of the generators may change, but the algebra cannot. *What changes is the space over which the realization is constructed*. Put in this way, we see that the Euclidean group actually plays no rôle in the discussion of scattering. This is reassuring since quantal Coulomb motion never becomes Euclidean, even at infinity, without a cutoff on the potential.

4. AN ILLUSTRATIVE EXAMPLE

The ideas of the last section become clearer from explicit calculation for SO(2,1) in $\mathbb{R}^2$. The generators are:

$$A_1 = \frac{1}{2m}(L_3 p_2 + p_2 L_3) + \alpha r_1/r; \quad A_2 = -\frac{1}{2m}(L_3 p_1 + p_1 L_3) + \alpha r_2/r$$

[+]This is easily seen from the Mackey-Wigner theorem applied to E2. In the spinless case, the plane waves exhaust the irreps. Using a spherical basis, one sees that in- and out-waves necessarily occur together, not separately.

[++]This is obscured in ref. (5), cf. p. 190, because they have used a similarity transformation singular at the origin.

and $L_3 = (r_1 p_2 - r_2 p_1)$, with the commutation relation: $(A_\pm = A_1 \pm i A_2)$
$[L_3, A_\pm] = \pm A_\pm$; $[A_+, A_-] = -2(\frac{2H}{m}) L_3$.
Define the continuum eigenkets: (k>0)

$$H|k,m\rangle = \frac{k^2}{2m}|k,m\rangle, \quad L_3|k,m\rangle = m|k,m\rangle.$$

We renormalize the non-compact generators, as usual, by:

$$A_\pm \equiv \left|\frac{m}{2H}\right|^{1/2} A_\pm \to A_\pm/k$$

on eigenstates. In this form one obtains the standard SO(2,1) commutation relations:

$$[L_3, A_\pm] = \pm A_\pm, \quad [A_+, A_-] = -2L_3.$$

The essential step for identifying the *scattering space* is to introduce the radial decomposition of the linear momentum, omitting terms of order $1/r$ or higher in $A_\pm$. Define the operator $(ik)^{-1}\frac{\partial}{\partial r}$ to limit to the generator R of the group Z_2. Thus the scattering space$_+$ is $S^1 \times Z_2$. (The energy is sharp on scattering space, and this label is suppressed).

The generators which act on the scattering space have the explicit form ($L_3 \to L_3$):

$$A_+ \to A'_+ = -\frac{i}{2}(e^{i\varphi}L_3 + L_3 e^{i\varphi})R + \eta e^{i\varphi},$$

$$A_- \to A'_- = \frac{i}{2}(e^{-i\varphi}L_3 + L_3 e^{-i\varphi})R + \eta e^{-i\varphi}.$$
(3)

Note that $[R, A'_\pm] = [R, L_3] = 0$, and $R^2 = E$. It is easily verified that the generators A'_1, A'_2, L_3 and R are Hermitian and that they generate the group $SO(2,1) \times Z_2$. Thus the scattering space realizes two distinct copies of the original symmetry group. Introduce now the scattering space eigenkets: $|\varepsilon, m\rangle$ and determine the action of the four generators:

More precisely the asymptotic radial motion is factored out (including logarithmic terms), and one verifies that factoring commutes with the action of the generators on the factor space.

$$R|\varepsilon,m\rangle = \varepsilon|\varepsilon,m\rangle, \quad (\varepsilon = \pm 1)$$

$$L_3|\varepsilon,m\rangle = m|\varepsilon,m\rangle, \quad (m=0,\pm 1,\pm 2,\ldots) \quad (4)$$

$$A'_\pm|\varepsilon,m\rangle = (\mp i\varepsilon(m\pm 1/2)+\eta)|\varepsilon,m\pm 1\rangle.$$

One recognizes at once that action of $A'_\pm$ is unusual! One would expect the *standard action* $(J_\pm \to [(j\mp m)(j\pm m+1)]^{1/2}$ for SU(2) which for SO(2,1), with $j = -1/2+i\eta$, is:

$$A_\pm|m\rangle = [(m\pm 1/2)^2+\eta^2]^{1/2}|m\pm 1\rangle,$$

and, indeed, this is exactly the result one has for the original generators $A_\pm$.

The standard action is *not* canonical since it requires a definite phase and normalization convention[12]. To restore the standard action we introduce a new basis: $|\varepsilon,m\rangle_{std} \equiv N(\varepsilon,m)|\varepsilon,m\rangle$. Orthonormality yields: $|N(\varepsilon,m)|^2 = 1$.

Requiring the standard action implies that:

$$A'_+|\varepsilon,m\rangle_{std} = [(m+1/2)^2+\eta^2]^{1/2}|\varepsilon,m+1\rangle_{std}$$

$$= [(m+1/2)^2+\eta^2]^{1/2}N(\varepsilon,m+1)|\varepsilon,m+1\rangle$$

$$= N(\varepsilon,m)(-i\varepsilon(m+1/2)+\eta)|\varepsilon,m+1\rangle, \text{ using } (4).$$

This (and A'_- also) implies the two-term recursion relation:

$$(-i\varepsilon(m+1/2)+\eta)N(\varepsilon,m) = N(\varepsilon,m+1)\cdot[(m+1/2)^2+\eta^2]^{1/2}. \quad (5)$$

It is useful to consider at this point the Euclidean case of no interaction. The recursion relation becomes: $N_{Eucl}(\varepsilon,m+1) = N_{Eucl}(\varepsilon,m)\cdot(-i\varepsilon)$, with the solution: $N_{Eucl}(\varepsilon,m) = (-i\varepsilon)^{m-m_0}N_{Eucl}(\varepsilon,m_0)$.

The general solution to eq. (5) has the form:

$$N(\varepsilon,m) = N_{Eucl}(\varepsilon,m)\cdot\left[\frac{\Gamma(m+1/2+i\eta\varepsilon)}{\Gamma(m+1/2-i\eta\varepsilon)}\cdot\frac{\Gamma(m_0+1/2-i\eta\varepsilon)}{\Gamma(m_0+1/2+i\eta\varepsilon)}\right]^{1/2}, \quad (6)$$

where $m_0=0$ or $1/2$ corresponding to the integer or halfinteger principal series of $\overline{SO(2,1)}$, with $j = -1/2+i\eta$.

To interpret these results, recall that the action of the generators on the original space must go smoothly into the action of the asymptotic generators on the scattering space. Since the original generators had the standard action, it follows that the asymptotic generators must also have the standard action. Hence:

$$|k,m\rangle \simeq (\text{constant})(|+1,m\rangle_{std} + |-1,m\rangle_{std}),$$
$$= (\text{constant})(N(+1,m)|+1,m\rangle + N(-1,m)|-1,m\rangle). \qquad (7)$$

Using the definition of the S-matrix, defined relative to the Euclidean phases, one finds:

$$S \to S_m \equiv e^{2i\sigma_m} = \frac{N(\varepsilon=+1,m)}{N_{Eucl}(\varepsilon=1,m)} \cdot \frac{N_{Eucl}(\varepsilon=-1,m)}{N(\varepsilon=-1,m)}. \qquad (8)$$

Thus: $e^{2i\sigma_m} = \frac{\Gamma(m+1/2+i\eta)}{\Gamma(m+1/2-i\eta)}$, (omitting an overall constant).

This result shows an interesting new view of the origin of the scattering phase shifts.

5. CONCLUSIONS

We abstract from the example above the relevant concepts for a group-theoretic determination of scattering. The essential concept is that of *the factor space of scattering states*: $R^2 \to S^1 \times Z_2$, with Z_2 generated by R. The generators, restricted to the factor space, have two distinct realizations, distinguished by R. The action on these two realizations is non-standard, and linear in L_3. The relative phase, defined by standardizing the two non-standard actions, yields the scattering phase-shifts. *Each of these concepts appears to be group-theoretic* (within the context of a given Hamiltonian, symmetry group and spatial realization). For example, the doubling of the generators is clearly essential, relating to in/out waves, which in turn is related to the sign of $i = \sqrt{-1}$. But scattering requires a continuum and hence a non-compact group G. This in turn defines a Cartan involution, which is a formal imaginary unit. The present work is preliminary, and it will be interesting to see how far these ideas can be carried.

Although we have been critical of details, our results actually validate the insights into a group-theoretic view of scattering developed by the Yale group. We wish to thank Professor Francesco Iachello for helpful discussions.

6. REFERENCES

1) Biedenharn, L.C., Brussaard, P.J., "Coulomb Excitation", Clarendon Press, Oxford (1965)
2) Zwanziger, D., J.Math.Phys. $\underline{8}$, 1858 (1967)
3) Basu, D., Wolf, K.B., J.Math.Phys. $\underline{24}$, 478 (1983); Frank, A., Wolf, K.B., Phys.Rev.Lett. $\underline{52}$, 1737 (1984)
4) Alhassid, Y., Gürsey, F., Iachello, F., Ann.Phys. $\underline{148}$, 346 (1983)
5) ibid. $\underline{167}$, 181 (1986)
6) Wybourne, B.G., "Classical Groups for Physicists", Wiley, New York (1974)
7) Basu, D., Majumdar, S.D., J.Phys. $\underline{6A}$, 1097 (1973)
8) Rasmussen, W.O., Salamo, S., J.Math.Phys. $\underline{20}$, 1067 (1979)
9) Wu, J., Biedenharn, L.C., Iachello, F., Duke Univ., preprint (1986), submitted for publication
10) Gürsey, F., "Group Theoretical Methods in Physics", Springer Lect. Notes in Physics $\underline{180}$, 106 (1983)
11) Alhassid, Y. and Wu, J., Chem.Phys.Lett. $\underline{109}$, 81 (1984)
12) Biedenharn, L.C., Louck, J.D., "Angular Momentum in Quantum Physics", Addison-Wesley, Reading Mass., (1981)

GROUP DEFORMATIONS AND THE ALGEBRAIC SCHEME FOR SCATTERING

A. FRANK[†*], Y. ALHASSID[†] and F. IACHELLO[†]
[†] A.W. Wright Nuclear Structure Laboratory, Yale University
New Haven, CT 06511 USA

and

[*] Centro de Estudios Nucleares, UNAM
Apdo. Postal 70-543, Circuito Exterior C.U.
Mexico D.F. Mexico

ABSTRACT

The algebraic scheme for scattering is shown to be closely related to the concept of group contactions and expansions. The deformations describe the distortion of physical states when evolving from one kind of symmetry to another.This formulation can be applied to the systematic study of dynamical symmetries in scattering systems.The method is illustrated for scattering models in one dimension of the type SO(2,1), which already contain much of the structure needed for the study of higher dimensional models of the form SO(3,2), which characterize modified Coulomb interactions and which seem to be relevant for the description of heavy ion reactions.

1   INTRODUCTION

The use of algebraic methods for the derivation of analytic S-matrices [1-4] has recently been shown to be of practical interest when considering heavy ion scattering through an $SO(3,2)$ dynamical symmetry which describes modified Coulomb interactions [5]. The evaluation of the S-matrix is carried out by connecting the dynamical algebra associated to the scattering problem with an Euclidean algebra which describes the symmetry of the undistorted waves in the asymptotic regime [4]. As has been shown in these studies, the basic requirement of the theory is to be able to express the generators of the scattering algebra in terms of those of the asymptotic one, which leads to the possibility of evaluating recurrence relations for the S-matrix [3,4]. For the cases studied so far, this "connection formula" was derived by proposing a general bilinear form in the Euclidean generators and requiring that the appropriate commutation rules be satisfied [4]. However, the relation between these algebras, as well as the general structure of such connection formulas was not fully understood. We have recently been able to recast this problem in a general group-theoretical framework,that of group contractions and expansions [6], that permits a systematic investigation of these scattering theories and the construction of connection formulas for them[7]. In what follows, we analize the basic concepts involved through a simple example which already displays the basic structure needed for the analysis of higher dimensional models. In the conclusions we suggest some possible future applications.

2 CONTRACTION AND EXPANSION OF SO(2,1) → E(2).

Consider the group SO(2,1), whose elements are the 3x3 matrices that leave invariant the cuadratic form $(x^2+y^2-z^2)$. These transformations thus leave invariant a hyperboloid, whose surface is not bound, hence the non-compact character of the group. The group generators satisfy the commutation relations

$$[J_1, J_2]=-iJ_3, \quad [J_2, J_3]=iJ_1, \quad [J_3, J_1]=iJ_2, \quad (1)$$

with the sign change in the first commutator giving rise to important differences with respect to the well known case of SO(3). The invariant (or Casimir) operator turns out to be

$$J^2 = J_3^2 - J_1^2 - J_2^2 \quad (2)$$

and the analysis of the representations[8,9] is more complicated than the corresponding one for SO(3). We refer the reader to the analysis in Refs.(3,8) for a description of the characteristics of the SO(2,1) group[9] and its application to scattering by the Pöschl-Teller potential. In those references a particular coordinate realization of eqs.(1) in terms of hyperbolic coordinates is introduced, which connects the invariant (2) with the scattering Pöschl-Teller Hamiltonian and, by means of a limiting procedure in the radial coordinate, gives rise to recurrence relations for the S-matrix of this system. Shortly afterwards, it was recognized[4] that this mechanism can be formulated in a completely algebraic fashion, with no reference to any coordinate realization, by noting that the asymptotic generators of SO(2,1) may in fact be expressed in terms of those of E(2), the Euclidean group in two dimensions, which corresponds to the symmetry of free waves, far away from the scattering potential. Through this idea, it was also possible to construct general SO(2,1) S-matrices, with no reference to a particular potential. We shall now return to this simple example and re-examine it in the light of group contraction and expansion. Formally, one can carry out the contraction of SO(2,1) onto E(2) by defining a scale transformation in the generators (1)

$$P_1 = \varepsilon J_1, \quad P_2 = \varepsilon J_2, \quad P_3 = J_3, \quad (3)$$

from which one finds

$$[P_1, P_2] = -i\varepsilon^2 J_3, \quad [P_2, J_3] = i P_1, \quad [J_3, P_1] = i P_2, \quad (4)$$

and in the limit when $\varepsilon \to 0$ the last two relations in (4) are not modified, while the first gives simply

$$[P_1, P_2] = 0 \ . \tag{5}$$

The last two relations of (4) plus (5) can be readily identified with the commutation relations of the group E(2), which corresponds to the symmetry operations that leave invariant a plane to which the hyperboloid "contracts" in the limit. The SO(2,1) eigenstates $|j\ m\rangle$ satisfy the relations[3,4]

$$J^2 |j\ m\rangle = j(j+1)|j\ m\rangle, \quad J_3 |j\ m\rangle = m|j\ m\rangle, \tag{6}$$

with $j = -1/2 + if(k)$ in the continuous representation, with $f(k)$ being a function of the real number k associated in the physical applications[3,4] to the energy k^2. The E(2) eigenstates are in turn characterized by the equations

$$P^2 |\pm k\ m\rangle = k^2 |\pm k\ m\rangle \ , \tag{7a}$$

$$J_3 |\pm k\ m\rangle = m |\pm k\ m\rangle \ , \tag{7b}$$

where $P^2 = P_1^2 + P_2^2$ is the E(2) invariant, corresponding to the energy of the free wave. The contracted (asymptotic) SO(2,1) states can be expressed in terms of the E(2) eigenstates (7) through

$$|j\ m\rangle^{\varepsilon=0} = A_m |-k\ m\rangle + B_m |+k\ m\rangle \ , \tag{8}$$

which corresponds to the usual incoming and outgoing terms in a scattering system. The S-matrix can be determined if we can find the ratio B_m / A_m. If we apply rising and lowering operators $J_\pm$ to the left of (8), we find from the properties of the SO(2,1) algebra that[3,4]

$$(J_\pm |j\ m\rangle)^{\varepsilon=0} = \sqrt{(m\mp j)(m\pm j+1)} |j\ m\pm 1\rangle^{\varepsilon=0} = C_m |j\ m\pm 1\rangle^{\varepsilon=0}, \tag{9}$$

where we have introduced the rising and lowering operators $J_\pm = J_1 \pm iJ_2$ and taken the contraction limit $\varepsilon \to 0$. Since, from (9) we also have

$$|j\ m\pm 1\rangle^{\varepsilon=0} = A_{m\pm 1}|-k\ m\pm 1\rangle + B_{m\pm 1}|+k\ m\pm 1\rangle, \tag{10}$$

we can obtain recurrence relations for the A_m and B_m if we can deduce the action of $J_\pm$ on the E(2) wavefunctions $|\pm k\ m\rangle$. This can be accomplished if we are able to express these generators in terms of those of E(2), for which the action is well known[9], i.e.,

$$J_3|\pm k\ m\rangle = m|\pm k\ m\rangle, \ P_\pm|+k\ m\rangle = k|+k\ m\pm 1\rangle \ , \ P_\pm|-k\ m\rangle = -k|-k\ m\pm 1\rangle. \tag{11}$$

The problem of expressing the SO(2,1) generators in terms of those of E(2) corresponds to the inverse of the contraction mechanism and is known as the expansion of the group. In Refs.(4,5) it was called the "Euclidean connection" and found by constructing bilinear forms in the E(2) generators and requiring them to satisfy the SO(2,1) commutation rules. In fact, this mechanism is a general group-theoretical procedure which can be carried out for any pair of groups, G and G',whenever the contraction of G onto G' can be defined[6]. The general problem has not been solved, but many cases have been studied and a good overview can be found in Ref.(6).

The Casimir invariant J_3^2 of the compact subalgebra SO(2) (which is not modified in the contraction process) disappears in the contracted E(2) invariant P^2. We should therefore use it to reconstruct the original scattering algebra. Since J_3^2 is a scalar under SO(2) rotations and P_μ is a vector, we can easily construct a new SO(2) vector which is non-linear in the E(2) generators by means of the algorithm[7,10]

$$\bar{J}_\mu = 1/(2i)[J_3^2, P_\mu] + \rho P_\mu \quad (\rho \text{ real}) \quad \mu=1,2. \tag{12}$$

Since $\bar{J}_\mu$ is an SO(2) vector it automatically satisfies the correct SO(2,1) commutation relations with J_3. Consider now the commutator $[\bar{J}_\mu, \bar{J}_\nu] = [\bar{J}_{\mu 3}, \bar{J}_{3\nu}]$ (where we have introduced a notation that indicates the plane on which the operator generates a rotation). It corresponds to an antisymmetric tensor with respect to the indices μ and ν. Since it is at most quadratic in P and linear in $J_3 = J_{12}$, the only possible antisymmetric tensor to this power is $J_{12}P^2$. Since $P^2 = k^2$ for the states (7), it follows that the $J_\mu = \bar{J}_\mu/k$ must close under commutation with J_3. In fact, we get precisely the SO(2,1) algebra (1). By calculating the Casimir invariant (2) we readily find

$$\rho = \pm f(k) . \tag{13}$$

The sign in (13) depends on whether we work in the +k or -k representation of E(2). By acting with $J_\mu = \bar{J}_\mu/k$ of (12) on the r.h.s. of (8) and by means of relations (11) one finds

$$C_m|j\ m+1\rangle^{\varepsilon=0} = A_m/(-k)[(-1/2+if(k))(-k) + (m+1)(-k)]|-k\ m+1\rangle$$
$$+ B_m/k[(-1/2-if(k))k + (m+1)k]|+k\ m+1\rangle \tag{14}$$

and since

$$|j\ m+1\rangle^{\varepsilon=0} = A_{m+1}|-k\ m+1\rangle + B_{m+1}|+k\ m+1\rangle \quad , \tag{15}$$

comparison of (14) and (15) leads to recurrence relations for the A_m and B_m which can be solved for the reflection amplitude $R_m = B_m/A_m$[4].

3   CONCLUSIONS

In this paper we have presented a mathematical framework for the algebraic approach to scattering which rests on the notion of group deformations. By providing the theory with a general abstract formulation, it permits the systematic study of dynamical symmetries in scattering problems and their associated S-matrices. Although we have presented here only a simple one-dimensional example, the generalization to higher dimensions in the orthogonal groups does not involve any additional complications, and expansion formulas such as (12) can be derived for the dynamical symmetries associated to the groups $SO(m,n)$, $SU(m,n)$ and $Sp(m,n)$ with arbitrary m,n [7,10], for example. A systematic study of these and other symmetries is required if one wishes to include in the theory other degrees of freedom, different reaction channels or asymptotic symmetries which do not correspond to free particles, but describe distorted waves in some potential for which the problem is solvable. In the latter case the calculation of the additional phase shift caused by including an additional potential can then be obtained (in principle) by a deformation of group G onto another group G'. In particular, relativistic generalizations of these techniques may provide a useful framework for scattering processes in elementary particle physics.

This work was supported in part by CONACYT, Mexico, through project PCCBCEU-020061 and the Department of Energy Contract No. DE-AC02-76 ER 03074 USA.

4 REFERENCES

1. Alhassid Y., Gürsey F. and Iachello F., Phys.Rev.Lett.**50**,2515 (1983).
2. Alhassid Y., Gürsey F. and Iachello F., Ann. Phys.(N.Y.)**148**,346(1983).
3. Frank A. and Wolf K.B., Phys.Rev.Lett.**52**,1737(1984);J.Math.Phys. **26**,1979 (1985).
4. Alhassid Y., Engel J. and Wu J., Phys Rev.Lett.**53**,17(1984);Alhassid Y., Gürsey F. and Iachello F., Ann.Phys. (N.Y.), in press (1986).
5. Alhassid Y., Iachello F. and Wu J., Phys.Rev.Lett.**56**,271 (1986).
6. Gilmore R.," Lie Groups, Lie Algebras and Some of Their Applications", Chapter 10, (Wiley, N.Y. 1974) and references therein.
7. Frank A., Alhassid Y. and Iachello F., Phys.Rev. A , in press (1986).
8. Frank A., in "Interacting Boson-Boson and Boson-Fermion Systems", O.Scholten Ed. (World Scientific, 1984) 318.
9. Vilenkin N.J., "Special Functions and the Theory of Group Representations",(American Mathematical Society,Providence,1968).
10. Wolf K.B. and Boyer C.P., J.Math.Phys. **15**,2096 (1974).

MOLECULAR RESONANCES AND SYMMETRIES*

P. Kramer, R. Bader, Z. Papadopolos and W. Schweizer
Institut für Theoretische Physik der Universität Tübingen
Auf der Morgenstelle 14, D-7400 Tübingen, F.R.G.

ABSTRACT

The channels in a nuclear reaction are characterized by unitary and symplectic quantum numbers. For partners with $s^4p..$ configuration, unitary and symplectic principles exclude the formation of compound states and lead to the occurrence of molecular resonances.

1. INTRODUCTION

In 1985 two of the authors[1] proposed a U(3) principle for the many-body explanation of nuclear molecular resonances. For a review of the experimental work we refer to Cindro[2] and for the theoretical approaches to sources quoted in [1]. In the present contribution we describe the main features of this unitary scheme. Then we extend the analysis to the symplectic group and show that channels may be labeled by selected representations of this group. Orthogonality to the compound states is proposed as a salient feature for molecular resonances in these reaction channels.

The branching of symplectic compound states into reaction channels has been considered for $^8Be \to {}^4He + {}^4He$ by Arickx[3] and by Vasilewski and Filippov[4] and for heavier systems by Hecht and Braunschweig[5] and by Suzuki[6]. We take the opposite view, decompose channel states under the symplectic group, and explore the physics of these channels. Our general analysis is based on the methods described in [7,8].

---

* Work supported by the Deutsche Forschungsgemeinschaft

2. COUPLING RULES FOR NUCLEAR REACTION CHANNELS

Consider two nuclear reaction partners 1,2 and their relative motion (r.m.). If $\underline{J}^1$, $\underline{J}^2$ and $\underline{J}^{r.m.}$ denote the corresponding angular momenta, the total angular momentum is given by

$$\underline{J} = \underline{J}^1 + \underline{J}^2 + \underline{J}^{r.m.} \qquad (1)$$

The corresponding eigenstates of a reaction channel are characterized, with respect to angular momentum, by the coupling

$$j' \times j'' \times j^{r.m.} \rightarrow j \qquad (2)$$

Suppose now that we can characterize the reaction partners by oscillator shell model states labeled by irreducible representations or partitions $h' = [h_1' h_2' h_3']$ and $h'' = [h_1'' h_2'' h_3'']$ repectively of the corresponding unitary groups U(3). For an excitation N = 0,1,2.. the relative motion carries with respect to U(3) a partition [N] = [N00]. A U(3) channel state may now be characterized by the eigenstates of the summed U(3) generators and by the coupling

$$h' \times h'' \times [N] \rightarrow h = [h_1 \; h_2 \; h_3]. \qquad (3)$$

This coupling is governed by Littlewood's rules. To describe a general state of the relative motion we must admit any N in this coupling. If in a channel the overall partition h has been fixed we speak of a U(3) channel.

3. U(3) SYMMETRY IN REACTIONS WITH TWO PARTNERS

The coupling rules for U(3) channels depend only on the Wigner-Racah algebra of this group, but the occurrence of possible partitions h',h" and h is related to the underlying system of A = n-1 nucleons. The relations can be inferred from a comparison of the U(3) channel states of partition $h = [h_1 \; h_2 \; h_3]$ with the compound nucleus states of partition $h^c = [h_1^c \; h_2^c \; h_3^c]$. One obtains the following types of configurations[1]:

Type	partner 1	partner 2	compound system
1a	s...	s...	s...
b			$s^4p...$
2a	s...	$s^4p...$	$s^4p...$
b			$s^4p^{12}(sd)...$
3a	$s^4p...$	$s^4p...$	$s^4p...$
b			$s^4p^{12}(sd)...$

An analysis of the U(3) channels shows the following features:
(a) Only for collisions of type 3b, where two sp partners are colliding, the compound states $s^4p^{12}(sd)...$ are strictly excluded in the U(3) channels. More precisely, for

$$h = [h_1 \, h_2 \, h_3]: \quad h_1 - h_3 > 12 \tag{4}$$

the lowest shell configuration cannot be formed.
(b) We assume without loss of generality $h_1' - h_2' > h_1'' - h_2''$ and denote the quantum excitation of the compound nucleus by m. For

$$h_1' + h_2'' + 2(h_3' + h_3'') > 16 + 2m \tag{5}$$

the formation of a compound state with a quantum excitation m is forbidden.
(c) If n is the lowest value of m which does not fulfill the inequality, the formation of an excited compound state requires the transfer of n nucleons from the sd to the pf shell.

Now we apply the following U(3) principle for nuclear molecular resonances: Nuclear molecular resonances occur in U(3) channels which do not match the lowest shell model compound levels. A detailed account of the fragment configurations and of the numbers n is given in Table 2 of [1].

4. THE EXAMPLE OF THE CHANNELS $^{16}O + ^{12+b}X$, $0 \leq b \leq 4$.

A complete analysis of these channels is given in [1]. The orbital partitions f in these channels combine according to

$$f' \times f'' \to f$$
$$[4^4] \times [4^3 b] \to [4^7 b] \tag{6}$$

With respect to the U(3) partitions, we have

$$[4\ 4\ 4] \times [4\ 4\ b] \times [N\ 0\ 0] = [8\ 8\ 4+b] \times [N\ 0\ 0]$$
$$= \sum_{W=b}^{4} [8+N+W-4\ \ 8\ \ 8+b-W] \tag{7}$$

We call W the mode quantum number. The compound nucleus configurations admit those U(3) partitions which would correspond to $N \geq 16+b$. In contrast, the U(3) channel states require $N \geq 16+2b$. This means that for $4 > b > 1$ the U(3) channels cannot match the compound states for fragment partitions $h' = [4\ 4\ 4]$, $h'' = [4\ 4\ b]$. Moreover it turns out that for $b < 3$ and $N = 16+2b$, only the U(3) coupling mode $W=b$ with the most compact partition is open while all other modes are forbidden. Generally one finds that the mode W only exists if the relative motion is excited by at least $n=W$ additional oscillator quanta. These features then single out candidates for nuclear molecular resonances.

The U(3) channel states may be written as

$$|[h_1\ h_2\ h_3]f\rangle = c^f |[h' \times h'' \times [N]]_h \rangle \tag{8}$$

where c^f is the Young operator for the symmetric group $S(n-1)$ and the square brackets indicate U(3) coupling of fragment and relative motion states. The numbers

$$\eta = \langle [h_1\ h_2\ h_3]f | [h_1\ h_2\ h_3]f \rangle \neq 0 \tag{9}$$

are obtained in [1] as the eigenvalues of the normalization operator [7,8] from the analytic expressions in Bargmann space. For $^{16}O + ^{16}O$, these numbers are non-zero only for the even values $N = 24, 26..$ and hence display the extreme even-odd effect. This effect extends qualitatively to the whole configuration. In the U(3) theory, the partition $h = [h_1\ h_2\ h_3]$ is often interpreted in terms of deformation. In all cases considered, we have $h_1 - h_2 \gg 1$, indicating a strong deformation of the system in one space dimension, in accordance with the simple quasi-molecular picture.

The interaction in the U(3) channel configuration has for $^{16}O + ^{16}O$ been studied in [1] with Brink-Boeker, Volkov and Coulomb forces by analytic methods. The model space is chosen with $24 \leq N \leq 40$ and $0 \leq L \leq 40$. It is found that a good fit to the experimentally observed resonances is obtained only upon inclusion of many U(3) channel states. The good reproduction of the experimental resonances in the low angular momentum region by a quasi-bound calculation suggests that a quasi-molecule is formed and shows that the phenomenon of molecular resonances can be linked to states in the entrance channel which are highly selective with respect to the U(3) and supermultiplet quantum numbers.

5. SYMPLECTIC PROPERTIES OF U(3) CHANNEL STATES

A full analysis of the A-body system may be based on the following group-subgroup diagram:[9]

$$
\begin{array}{ccc}
& Sp(6n,\mathbb{R}) & \\
Sp(6,\mathbb{R}) & \cdots\cdots\cdots & Sp(2n,\mathbb{R}) \\
\downarrow & & \downarrow \\
U(3) & \cdots\cdots\cdots & U(n) \\
\downarrow & & \downarrow \\
O(3,\mathbb{R}) & \cdots\cdots\cdots & O(n,\mathbb{R}) \\
& & \downarrow \\
& & S(n-1)
\end{array} \quad (10)
$$

collective   intrinsic

In this scheme of collective and intrinsic groups, the irreducible representations of pairs of groups connected by broken lines are linked by complementarity relations. We now extend the analysis of channel states form the collective group U(3) to the collective group symplectic Sp(6,$\mathbb{R}$) which is generally linked to collective motion.[10] In addition to the generators of the unitary group U(3), the symplectic group has the generators

$$K_{ij,+} = \frac{1}{2} \sum_{s=1}^{n} (a_i^s)^+ (a_j^s)^+ \qquad i,j = 1,2,3 \qquad (11)$$

$$K_{ij,-} = \frac{1}{2} \sum_{s=1}^{n} a_i^s a_j^s \qquad (12)$$

in terms of oscillator creation and anihilation operators. An irreducible representation of Sp(6,$\mathbb{R}$) possesses a lowest weight state characterized by the lowest representation $\{j_1 j_2 j_3\}$ of its U(3) subgroup. This state is characterized by

$$|\{j_1 j_2 j_3\}>: K_{ij,-} |\{j_1 j_2 j_3\}> = 0 \qquad i,j = 1,2,3 \qquad (13)$$

A general state extremal in U(3) may then be written as

$$|\{j_1\ j_2\ j_3\}\ [h_1\ h_2\ h_3]> \tag{14}$$

in the Sp(6,$\mathbf{R}$) > U(3) chain of groups. We have studied the symplectic decomposition of U(3) channel states wrt the symplectic group. We define the normalized channel state

$$|[h_1\ h_2\ h_3]\ f> = |[h_1\ h_2\ h_3]\ f)\ \ \eta^{-1/2} \tag{15}$$

and consider for fixed f the expansion

$$|[h_1\ h_2\ h_3]\ f> = \sum_{\{j_1 j_2 j_3\}} \Big[|\{j_1\ j_2\ j_3\}\ [h_1\ h_2\ h_3]\ f >$$

$$\alpha(\{j_1\ j_2\ j_3\},\ [h_1\ h_2\ h_3])\Big] \tag{16}$$

The coefficients α have been obtained by solving recursion relations, and examples are given in Tables 1,2, for the configurations discussed already in section 4. We describe here the main features of the expansion eq.(16):

(a) the U(3) channel states of lowest excitation belong to a single symplectic representation identical to this U(3) representation.
(b) the U(3) channel states of higher excitation contain superpositions of symplectic representations. All these states are orthogonal to the symplectic compound states.

For the physics of these states, the extension from the unitary to the symplectic group has important consequences: Not only the overlap, but also the matrix elements of the symplectic generators between channel and compound states vanish. Among them there are the kinetic energy, the collective potential energy and the (mass) quadrupole operator. In the U(3) analysis, a change of frequency between the channel and compound states removes in part the orthogonality due to this group. In the symplectic analysis, a change of frequency cannot mix different irreducible representations, and the orthogonality of channel and compound states persists. The interpretation of the symplectic partition $\{j_1 j_2 j_3\}$ admits an interpretation in terms of intrinsic deformation. In this interpretation, the symplectic channel states are linearly strongly deformed in agreement with the quasimolecular picture. The irreducible representations of the symplectic group admit rotational states of a more general form and involving strong U(3) mixtures.

We therefore propose for nuclear molecular resonances the following symplectic principle: Nuclear molecular resonances occur in symplectic reaction channels which do not match the lowest symplectic shell model compound levels. The results of the calculation on $^{16}O + ^{16}O$ which lead to strong U(3) mixtures support this broader interpretation.

From the point of view of nuclear reaction physics, the channel analysis presented here implies that the channel states built up, for sufficient high mass number, selective symplectic many-body quantum numbers and differ from the usual collective states of the shell model analysis. The dependence of these selective channel states on the mass number should have other relevant consequences for heavier nuclei which should be explored.

REFERENCES

1) Bader, R. and Kramer, P. Nucl.Phys.A441 174(1985)

2) Cindro, N. The Resonant Behaviour of Heavy-Ion Systems, Riv. Nuovo Cimento 4, No. 6 (1981)

3) Arickx, F. Nucl.Phys. A284, 264(1977)

4) Vasilewski, V.S. and Filippov, G.F., Sov.J.Nucl.Phys. 33, 500(1981)

5) Hecht, K.T. and Braunschweig, D. Nucl.Phys. A295, 34(1978)

6) Suzuki, Y. Nucl.Phys. A448, 395 (1986)

7) Kramer, P., John, G. and Schenzle, D., Group Theory and the Interaction of Composite Nucleon Systems, Braunschweig 1980

8) Bader, R., Dissertation, Tübingen, 1985

9) Kramer, P. and Papadopolos, Z., Collective motion, composite particle structure and symplectic groups in nuclei, in Lecture Notes in Physics 135, ed. K.B. Wolf (Springer, Berlin 1980)

10) Kramer, P., Papadopolos, Z. and Schweizer, W., Nucl.Phys. A441, 461 (1985); J.Math.Phys. 27, 24 (1986)

Table 1: Symplectic decomposition in the $^{16}O + ^{16}O$ channel. The unitary partition is [32 + 2k 8 8], k = 0,1,... and the symplectic partition is {32 + 2μ 8 8}, μ = 0,1,...

α^2 μ:	0	1	2	3	4	5	6
k: 6	0.172028091	0.378121150	0.309678119	0.117560667	0.021069092	0.001519136	0.000023744
5	0.223710982	0.413865317	0.273846610	0.078733073	0.009498299	0.000345718	
4	0.294279538	0.439933047	0.220504551	0.042682999	0.002599864		
3	0.391800226	0.443773726	0.149784235	0.014641813			
2	0.528276931	0.403015391	0.068707678				
1	0.721804511	0.278195489					
0	1.000000000						

Table 2: Symplectic decomposition in the $^{16}O + {}^{13}X$ channel. The unitary partition is
a) [29 + 2k 8 5], k = 0,1,.. and the symplectic partition is
{29 + 2μ 8 5}, μ = 0,1,.. (odd channel parity)
b) [30 + 2k 8 5], k = 0,1,.. and the symplectic partition is
{30 + 2μ 8 5} μ = 0,1,.. (even channel parity)

a)

α^2 μ	0	1	2	3	4
k: 4	0.121272128	0.408018088	0.336916458	0.117750836	0.016042490
3	0.189169002	0.487719007	0.274198210	0.048913781	
2	0.306593787	0.538687772	0.154718441		
1	0.526739937	0.473260063			
0	1.000000000				

b)

α^2 μ	0	1	2	3	4
k: 4	0.253613667	0.405889143	0.254414354	0.076633253	0.009449583
3	0.353464135	0.433296099	0.184834287	0.028405479	
2	0.495813026	0.414005513	0.090181461		
1	0.700879165	0.299120835			
0	1.000000000				

PHYSICAL BOUNDARY CONDITIONS FOR NUCLEAR HEAVY ION SCATTERING IN R-MATRIX THEORY*

R. Maass and W. Scheid
Universität Giessen, D-6300 Giessen, F.R. Germany

A. Thiel
Universität Frankfurt, D-6000 Frankfurt, F.R. Germany

ABSTRACT

An R-matrix formalism with approximate physical boundary conditions is developed and applied to the elastic and inelastic scattering of ^{24}Mg on ^{24}Mg.

1. INTRODUCTION

The R-matrix theory in the formulation of Bloch is used for a non-microscopic description of a multi-channel scattering problem in nuclear heavy ion physics. We apply the formalism for the calculation of the elastic and inelastic cross sections of the scattering of ^{24}Mg on ^{24}Mg.

According to the R-matrix theory, the S-matrix elements for the subspace of open channels c and c' with total angular momentum J can be written as follows[1]:

$$S^J_{cc'} = \left(\frac{I_c(a)I_{c'}(a)}{O_c(a)O_{c'}(a)}\right)^{1/2} (\delta_{cc'} + i \sum_{jj'} \Gamma^{1/2}_{jc} G(b^{KP}_c)_{jj'} \Gamma^{1/2}_{j'c'}), \qquad (1)$$

where

$$G(b^{KP}_c)_{jj'} = \langle \tilde{\phi}_j |(H + \mathcal{L}(b^{KP}_c) - E)^{-1}| \phi_{j'} \rangle. \qquad (2)$$

---

*) Work supported by GSI and BMFT

Here, $I_c(a)$ and $O_c(a)$ are Coulomb functions at the matching radius $r=a$, $\Gamma_{jc}^{\frac{1}{2}}$ partial widths, $G_{jj'}$ the matrix elements of the Green function, ϕ_j the basis states orthonormalized over the interaction region $0 \le r \le a$, and $\mathcal{L}(b_c^{KP})$ the Bloch operator with Kapur-Peierls boundary condition parameters.

The basis states ϕ_j have to fulfil arbitrary, but fixed boundary conditions at r=a, which, in general, will not coincide with the boundary conditions of the scattering eigenstate. The latter boundary conditions, denoted as physical boundary conditions (p.b.c.), are unknown at the beginning of the calculation of the S-matrix. Non-physical boundary conditions lead to discontinuities of the slopes of the radial scattering wave functions and, as a consequence, give rise to a slow convergence of the expansion of the scattering state. Therefore, in order to overcome this disadvantage, several methods have been developed[1] which change the arbitrarily chosen initial values for the boundary conditions by iteration procedures into the p.b.c.. In our approach we use approximate p.b.c., avoiding lengthy iteration procedures.

2. APPROXIMATE PHYSICAL BOUNDARY CONDITIONS AND BASIS FUNCTIONS

The parameters B_c^J for the exact physical boundary conditions in the channels c with total angular momentum J are given by

$$B_c^J = a \left[\frac{d}{dr} \ln(I_c(r) \delta_{cc_o} - S_{cc_o}^J O_c(r)) \right]_{r=a} , \qquad (3)$$

where c_o denotes the elastic channel. Without knowledge of the S-matrix the quantities B_c^J for the inelastic channels can be calculated and correspond to Kapur-Peierls boundary conditions for outgoing Coulomb waves. Only the parameter $B_{c_o}^J$ of the elastic channel depends on the S-matrix element of the elastic channel and has to be approximated for the calculation of the basis functions ϕ_j and the S-matrix elements according to Eq.(1).

Besides the use of the physical boundary conditions in the inelastic channels, we approximate the initially unknown parameter $B_{c_o}^J$ by replacing the elastic S-matrix element in Eq.(3) by the corresponding one of the uncoupled scattering problem, which is obtained by means of an optical model code. This choice for the boundary condition of the elastic channel is especially useful if the coupling effects of the inelastic channels are not too strong. For stronger coupling potentials we recommend an iteration procedure, in which the calculated S-matrix elements of the elastic channels are inserted into Eq.(3) and, therefore, the quantities $B_{c_o}^J$ are corrected for a repetition of the calculation.

For the basis functions with total angular momentum J we use the ansatz
$$\phi_j = N_j j_{\ell_{c_j}}(k_j r) |c_j). \quad (4)$$
The complex wave numbers k_j are obtained via the boundary conditions described above. They are selected in such a way that the real parts of the energy expectation values calculated with ϕ_j are distributed symmetrically around $E_{c.m.} = E$. For a fixed J, the number of basis functions in each channel is taken to be the same:
$$N(J) = \mathrm{Max}(N_o - [J/4], 1), \quad (5)$$
where the brackets denote the integer part.

3. APPLICATION TO THE SCATTERING OF ^{24}Mg ON ^{24}Mg

We applied our method to the elastic scattering and single inelastic excitation of the first 2^+-state of ^{24}Mg on ^{24}Mg. For the optical potential we used a Woods-Saxon shape, whose parameters were fitted to four measured elastic angular distributions[2]. For the matching radius we chose the value a=15 fm.

The convergence of our method was found satisfactory for $N_o \geq 21$. Employing one iteration step for the elastic S-matrix element, we improved the convergence markedly. A second iteration did not change the results any more.

In Fig.1 we present a comparison of various calculations with the experimental cross section for the elastic scattering of ^{24}Mg on ^{24}Mg at $E_{c.m.}$ =46.7 MeV. The various calculations are described in the figure caption. The results of the R-matrix calculations for $N_o \geq 21$ are in good agreement with those ones obtained by a coupled channel code. For a better reproduction of the cross section, the optical potential has to be newly investigated, since the coupling potential changes the structure of the angular distribution.

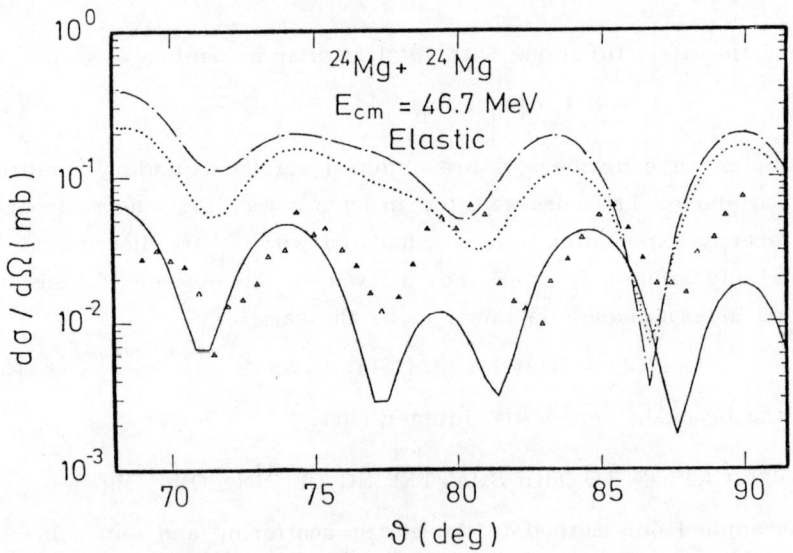

Fig. 1 The differential elastic cross section for ^{24}Mg+^{24}Mg at $E_{c.m.}$=46.7 MeV. Triangles: experimental data[2]; full curve: optical model code; dotted-dashed curve: R-matrix code (N_o=25, one iteration step); dotted curve: coupled channel code.

4. SUMMARY

In summary, we conclude that the R-matrix formalism with approximate physical boundary conditions is a considerable alternative for the description of nuclear heavy ion multi-channel scattering in comparison with coupled channel calculations, especially appropriate for the examination of resonances, since the R-matrix formalism yields the widths of the resonance states directly.

REFERENCES

1. Barrett, R.F. et al., Rev. Mod. Phys. 55, 155 (1983).
2. Zurmühle, R.W. et al., Phys. Lett. 129B, 384 (1983).

DESCRIPTION OF TWO- AND THREE-CLUSTER SYSTEMS OF LIGHT NUCLEI

R. Krivec and M. V. Mihailović,

J. Stefan Institute, University of Ljubljana, Jamova 39,
61111 Ljubljana, Yugoslavia

ABSTRACT

A microscopical theory of reactions based on the variational principle is presented. Scattering states for both binary and tertiary reactions with charged fragments are written in a Generator Coordinate basis which enables one to take into account all symmetries of the system. The calculation of reaction parameters is reduced to the calculation of multi-dimensional integrals.

1. INTRODUCTION

A comprehensive theory is presented which is able to describe two- and three-cluster bound systems and reactions with two and three charged fragments in the channel. The variational principle of Kohn and Kato is used to express the reaction parameters. The emphasis in this presentation is on: (1) How the domain of the functional is defined in the Generator Coordinate (GC) basis in which scattering states are expressed as transforms of the scattering function for point charged particles, and how the latter can be calculated; (2) The form and complexity of the multidimensional integrals needed for evaluating the reaction parameters.

2. THE VARIATIONAL FORMALISM FOR REACTIONS WITH THREE FRAGMENTS IN THE OUTGOING CHANNEL

In order to make the calculation of reaction parameters feasible one has to use the advantage of formulating the problem in such a way that the same matrix elements are used in both the bound (quasibound) problem and the scattering. As compared to the binary reactions, where the solution of the asymptotic Hamiltonian is known in the whole interval from zero to infinity, a major effort is necessary to calculate the solution corresponding to a tertiary reaction.

2.1 Boundary condition in the three fragment channels

Let us introduce the following notation: Let $\underline{X} = (\underline{x}, \underline{y})$ stand for the pair of Jacobi coordinates for the centres of mass of the three-fragment system. Let $\underline{Z} = (\underline{\sigma}, \underline{\tau})$ be the corresponding pair for the centres of the shell model potentials. The hyperspherical coordinates are

$$\underline{X} = (X, A, \hat{\underline{x}}, \hat{\underline{y}}), \quad A = \arctan(y/x)$$
$$\underline{Z} = (Z, G, \hat{\underline{\sigma}}, \hat{\underline{\tau}}), \quad G = \arctan(\tau/\sigma) \qquad (1)$$

Here we are interested in the type of reaction

$$^3He + {}^3He \longrightarrow {}^4He + p + p \qquad (2)$$

The three fragment channel index is specified as $j = (\alpha Klms_1\nu_1 s_2\nu_2 s_3\nu_3)$ where l, m are the total orbital angular momentum, and s_i and ν_i are spins of the fragments. We shall use the notation $\alpha = a$ for the two-fragment and $\alpha = 0$ for the three-fragment channel.

The solution of the Schrödinger equation satisfies the boundary condition at infinity

$$\Psi_i \rightarrow I_i^{(2)}(\underline{r}_a,\underline{\xi}_a) + A_{ii}(\hat{\underline{r}}_a)\, \sigma_i^{(2)}(\underline{r}_a,\underline{\xi}_a) + \sum_j A_{ij}(\hat{\underline{X}})\, \sigma_j^{(3,as)}(\underline{X},\underline{\xi}_0) \qquad (3)$$

where the functions $I_i^{(2)}$ and $\sigma_i^{(2)}$ are the Coulomb in- and outgoing functions for the binary channel. The third term is given by the partial wave expansion of the leading term of the eikonal approximation:[1]

$$\sum_{Kl_1l_2} Y_{Klml_1l_2}(\hat{\underline{X}})(A_{ij}(\hat{\underline{X}})\exp(i\sqrt{E_j}X - \frac{i\zeta(\hat{\underline{X}})}{2\sqrt{E_j}}\ln 2\sqrt{E_j}X) /$$

$$X^{5/2}\Phi^{INT}(\underline{\xi}_0))_{Klml_1l_2} \qquad (4)$$

$$= \sum_{K'l_1'l_2'} A_{ijK'l_1'l_2'}\, \sigma_{jK'l_1'l_2'}^{(3,as)}(X)\, \Phi_j^{INT}(\underline{\xi}_0)$$

where

$$\sigma_{jK'l_1'l_2'}^{(3,as)}(X) = \sum_{Kl_1l_2} Y_{Klml_1l_2}(\hat{\underline{X}})\, \sigma_{jKl_1l_2,K'l_1'l_2'}^{(3,as)}(X),\; \sigma_{jKl_1l_2,K'l_1'l_2'}^{(3,as)}(X)$$

$$= \int d\hat{\underline{X}}\, Y_{Klml_1l_2}(\hat{\underline{X}})\,(\exp(i\sqrt{E_j}X - \frac{i\zeta(\hat{\underline{X}})}{2\sqrt{E_j}}\ln 2\sqrt{E_j}X) / X^{5/2})\, Y_{Klml_1'l_2'}(\hat{\underline{X}}) \qquad (5)$$

The hyperspherical harmonic functions are defined as

$$Y_{Klml_1l_2}(\hat{\underline{X}}) = N_{Kl_1l_2}\, \cos^{l_1}G\, \sin^{l_2}G\, P_{(K-l_1-l_2)/2}^{(l_1+1/2,l_2+1/2)}(-\cos 2G) \qquad (6)$$

where $P_n^{(\alpha,\beta)}(x)$ are Jacobi polynomials.

2.2 The variational approach of the Kohn type

A variational approach of the Kohn type is proposed to calculate the parameters of the above reaction. The following variational equation is used:[2]

$$\langle \delta\Psi_j^{(-)} | H - E | \Psi_i^{(+)} \rangle = 0 \qquad (7)$$

The trial functions are

$$\Psi_i(\omega) = I_i^{(2)}(\underline{r}_a,\underline{\xi}_a) + A_{ii}(\hat{\underline{r}}_a)\, \sigma_i^{(2)}(\underline{r}_a,\underline{\xi}_a) +$$

$$\sum_j A_{ij}(\hat{X}) \sigma_j^{(3,as)}(X,\xi_0) + B_i(\omega), \quad \omega = (x_1, x_2, \ldots, x_{A+B+C}) \tag{8}$$

The square integrable part B should describe the quasibound states.
The third, three fragment, term can be expressed as

$$\sum_{K'l_1'l_2'} A_{ijK'l_1'l_2'} \sigma_{jK'l_1'l_2'}^{(3)}(X) \Phi_j^{INT}(\xi_0) \tag{9}$$

where the radial parts of $\sigma_{jK'l_1'l_2'}^{(3)}$ satisfy the boundary condition

$$\sigma_{jKl_1l_2,K'l_1'l_2'}^{(3)}(X) \to \sigma_{jKl_1l_2,K'l_1'l_2'}^{(3,as)}(X). \tag{10}$$

They are solutions of the Schrödinger equation

$$\left(\frac{d^2}{dX^2} - (K + \tfrac{3}{2})(K + \tfrac{5}{2})/X^2 + E\right) X^{5/2} \sigma_{jKl_1l_2,K'l_1'l_2'}^{(3)}(X) =$$
$$\frac{1}{X} \sum_{K''l_1''l_2''} \zeta_{Klml_1l_2,K''lml_1''l_2''} X^{5/2} \sigma_{jK''l_1''l_2'',K'l_1'l_2'}^{(3)}(X) \tag{11}$$

where ζ denotes the matrix elements of $X \cdot V^{COUL}$ and V^{COUL} is the sum of interfragment Coulomb potentials.

From eq. (7) we have the system of linear equations for coefficients C_p of the quasibound states in B_i, and A_{ij}. We first eliminate the C_p and get the system of equations (with $\lambda = (K, l_1, l_2)$),

$$\langle \sigma_i^{(2)} | \Delta | \sigma_i^{(2)} \rangle A_{ii} + \sum_{\lambda'} \langle \sigma_i^{(2)} | \Delta | \sigma_{j\lambda'}^{(3)} \rangle A_{ij\lambda'} = \langle \sigma_i^{(2)} | \Delta | I_i^{(2)} \rangle,$$
$$\langle \sigma_{j\lambda}^{(3)} | \Delta | \sigma_i^{(2)} \rangle A_{ii} + \sum_{\lambda'} \langle \sigma_{j\lambda}^{(3)} | \Delta | \sigma_{j\lambda'}^{(3)} \rangle A_{ij\lambda'} = \langle \sigma_{j\lambda}^{(3)} | \Delta | I_i^{(2)} \rangle \tag{12}$$

where, with $\mu = (K, l, m, l_1, l_2)$, $R_{i\mu\mu'} = \langle \varphi_{x_i \mu} | H - E | \varphi_{x_i \mu'} \rangle$, φ_{x_i} from B_i,

$$\Delta = H - E - (H - E) \sum_{\mu\mu'} |\varphi_{x_i \mu}\rangle R_{i\mu\mu'}^{-1} \langle \varphi_{x_i \mu'} | (H - E) \tag{13}$$

3. THE GC REPRESENTATION

The basis of the three-centre functions is constructed in a way similar to that of the two-centre ones. Let A (B, C) be the mass numbers of the three fragments and $\psi_i(x_A, s_A)$ ($\psi_i(x_B, s_B)$, $\psi_i(x_C, s_C)$) the single-particle functions of the harmonic oscillator potentials centred at s_A, s_B and s_C respectively. Then we construct the product of A (B, C) functions which corresponds to the cluster A (B, C): $\Phi_{A\alpha}(\omega_A, s_A)$ ($\Phi_{B\beta}(\omega_B, s_B)$, $\Phi_{C\gamma}(\omega_C, s_C)$) with $\omega_A = (x_1, \ldots, x_A)$, etc., and α (β, γ) being the spin configuration of nucleons in the cluster A (B, C). Next we form the set

$$\Phi_{I_i I \nu}^{(3)}(\omega, s_1, s_2) = \{\mathcal{A}[\Phi_{A\alpha}(\omega_A, s_A) \Phi_{B\beta}(\omega_B, s_B) \Phi_{C\gamma}(\omega_C, s_C)]\}_{I_i I \nu} \tag{14}$$

where

$$\underline{s}_1 = \underline{s}_A - \underline{s}_B, \quad \underline{s}_2 = \underline{s}_C - (A\underline{s}_A + B\underline{s}_B)/(A + B), \quad A\underline{s}_A + B\underline{s}_B + C\underline{s}_C = 0 \quad (15)$$

The symbols $I\nu$ denote the total spin; I_i stands for all intermediate quantum numbers which result from coupling the spins of A, B, and C particles to the total spin $I\nu$.

A GC wave function is constructed by superposing functions (14)

$$\Psi^{(3)}(\underline{\omega}) = \sum_{I_i I\nu} \int d^3s_1 \int d^3s_2 \; f^{(3)}_{I_i I\nu}(\underline{s}_1,\underline{s}_2) \; \Phi^{(3)}_{I_i I\nu}(\underline{\omega},s_1,s_2) \quad (16)$$

We first expand in bipolar harmonics

$$f^{(3)}(\underline{s}_1,\underline{s}_2) = \sum_{L_1 L_2 LM} \tilde{f}^{(3)}(L_1 L_2 LM, s_1 s_2) \; B_{L_1 L_2 LM}(\hat{\underline{s}}_1, \hat{\underline{s}}_2)^*$$

$$\Phi^{(3)}(\underline{\omega},\underline{s}_1,\underline{s}_2) = \sum_{L_1 L_2 LM} \tilde{\Phi}^{(3)}(\underline{\omega}, L_1 L_2 LM, s_1 s_2) \; B_{L_1 L_2 LM}(\hat{\underline{s}}_1, \hat{\underline{s}}_2) \quad (17)$$

The desired GC basis is obtained by projecting the functions (16) onto the eigenspace of the angular momentum and parity operators. The angular momentum operator is

$$P^{JM} = \frac{2J + 1}{8\pi^2} \int d\Omega \; \mathcal{D}^{J}_{MK}(\Omega)^* \; R(\Omega) \quad (18)$$

So the eigenspace is spanned by the functions

$$\Psi^{JM\pi}(\underline{\omega}) = \sum_{KI\nu} \int d^3s_1 \int d^3s_2 \; f^{(3)}_{I\nu K}(\underline{s}_1,\underline{s}_2) \; P^{J}_{MK} \; P^{\pi} \; \Phi^{(3)}_{I\nu}(\underline{\omega},\underline{s}_1,\underline{s}_2)$$

$$= (2J + 1)\sum_{KI\nu}\int d^3s_1 \int d^3s_2 \; f^{(3)}_{I\nu K}(\underline{s}_1,\underline{s}_2) \sum_{L_1 L_2 \mathcal{L}\mu\nu'\mu'} \frac{1}{2}(1 + \pi(-1)^{L_1+L_2})$$

$$\tilde{\Phi}^{(3)}_{I\nu}(\underline{\omega}, L_1 L_2 \mathcal{L}\mu, s_1 s_2) \; (-1)^{M+K} \begin{pmatrix} J & I & \mathcal{L} \\ -M & \nu' & \mu' \end{pmatrix}\begin{pmatrix} J & I & \mathcal{L} \\ -K & \nu & \mu' \end{pmatrix} B_{L_1 L_2 \mathcal{L}\mu'}(\hat{\underline{s}}_1, \hat{\underline{s}}_2)^* \quad (19)$$

In the calculation of reaction parameters one has to calculate matrix elements of the terms in the trial function (8). They are expressed by the matrix elements in the hyperspherical basis,

$$\langle \varphi_{x_i} Klml_1 l_2(Z) | H - E | \varphi_{x_j} K'lml_1' l_2'(Z') \rangle$$

$$= \int d\hat{Z} \; Y_{Klml_1 l_2}(\hat{Z})^* \int d\hat{Z}' \; Y_{K'lml_1' l_2'}(\hat{Z}') \langle \varphi_{x_i}(Z) | H - E | \varphi_{x_j}(Z') \rangle$$

$$= \int_0^{\pi/2} \cos^2 G \sin^2 G \; dG \; N_{Kl_1 l_2} \cos^{l_1} G \sin^{l_2} G \; P^{(l_1+1/2, l_2+1/2)}_{(K-l_1-l_2)/2}(-\cos 2G)$$

$$\int_0^{\pi/2} \cos^2 G' \sin^2 G' \; dG' \; N_{Kl_1' l_2'} \cos^{l_1'} G' \sin^{l_2'} G' \; P^{(l_1'+1/2, l_2'+1/2)}_{(K'-l_1'-l_2')/2}(-\cos 2G')$$

$$\bar{H}_{x_i lml_1 l_2, x_j lml_1' l_2'}(Z \cos G, Z \sin G; Z' \cos G', Z' \sin G') \quad (20)$$

where H denotes the matrix elements of H − E in the basis (19).

The functions $\mathcal{O}$ are expanded in the GC basis with amplitudes satisfying the uncoupled integral equations

$$\int_0^\infty dZ \, Z^5 \, (2\pi)^3 \, \exp(-\tfrac{\beta'}{4}(X^2+Z^2)) \, \frac{I_{K'+2}(\beta'XZ/2)}{(\beta'XZ/2)^2} \, f_{j\lambda\lambda'}(Z) = O^{(3)}_{j\lambda\lambda'}(X) \quad (21)$$

4. SOLVING THE INTEGRAL EQUATION FOR THE GC AMPLITUDE

4.1 The radial part of the scattering wave functions

The function $\mathcal{O}^{(3)}(X)$ is the solution of the system of equations (11). In the first step an approximate solution is obtained in the form of an asymptotic series by truncating the system to a finite number, N, of equations, determining N such that the solution provides the necessary accuracy at an appropriate large distance.

If one introduces a matrix Q which transforms the Sommerfeld matrix $n = \zeta/2\sqrt{E}$ into the diagonal form $\hat{n}$,

$$n = Q^T \, \hat{n} \, Q, \quad (22)$$

and substitutes

$$\eta(X,E) = Q^T \mathcal{O}^{(3)}(X,E) = \exp(v/2) \, u(X,E), \quad v = 2i\sqrt{E}X, \quad (23)$$

the system (11) appears as the matrix equation

$$u'' + u' + \left(\frac{i\hat{n}}{X} - \frac{\hat{\mathcal{L}}}{X^2}\right) u = 0. \quad (24)$$

Here $\hat{\mathcal{L}}_{\mu\nu} = \sum_\sigma Q^T_{\mu\sigma} L_\sigma(L_\sigma + 1) Q_{\sigma\nu}$ and the value of L_σ depends on the ordering of the index $\sigma = \lambda_\nu$, $\nu = 1,\ldots,N$. The asymptotic series solution satisfying the proper boundary condition at infinity has the form

$$u(X) = \sum_{k=0}^\infty X^{-k} \sum_{\sigma=1}^N \hat{\eta}^{k\sigma} \, e^{-i\hat{n}_\sigma \ln 2\sqrt{E}X}, \quad (25)$$

with the coefficients $\hat{\eta}$ given by recursion relations.

In the second step the value of the asymptotic series (25) at an appropriate large distance (X = 40 MeV$^{-1/2}$) is used as the starting value for the numerical integration of the system (11) to extend the solution to a small value of the hyperradius X. Testing of the solution shows that the dimension of the truncated system has to be N $\geq$ 49 in order to obtain the radial wave function to an accuracy better than 1 percent.

4.2 Solution of the integral equation for the GC amplitude, eq. (21)

The GC amplitude is expressed in terms of the radial wave functions as follows:

$$f_{jKl_1l_2,K'l_1'l_2'}(Z) = \text{const.} \, (\mathcal{O}^{(3)}_{jKl_1l_2,K'l_1'l_2'}(X) -$$

$$\frac{1}{\rho'\chi} \sum_{K''1''_1 1''_2} \zeta_{jKl_1 l_2, K''l''_1 l''_2} \sigma^{(3)}_{jK''l''_1 l''_2, K'l'_1 l'_2}(\chi)) \qquad (26)$$

This form is obtained as the first approximation of the method of ref. 3)

7. CONCLUSION

1. Scattering states for tertiary reactions. -
It is shown how one can obtain good trial functions which satisfy the required scattering boundary conditions in binary and tertiary channels. The scattering wave functions are expressed in terms of the GC amplitudes which are obtained for different channels from uncoupled integral equations of the first kind. The inhomogeneous terms are expressed in terms of the approximate solution of an infinite coupled system of the "hypergeometrical" differential equations of one variable for scattering of three point charged particles.

2. On the feasibility of the calculation. -
(1) For light nuclei and not too high energies one can satisfy the symmetry requirements exactly using algebraic procedures. The only genuine numerical calculation is the evaluation of two-dimensional integrals for binary reactions, and additionally the integration over hyperangles for tertiary reactions. By employing a theorem[4] one can reduce the amount of work for the two-dimensional integrals for small generator coordinate values.

(2) For heavier nuclei the above procedure is too lengthy. There are two possibilities:
(i) Use other relevant approximate symmetries to evaluate kernels in the three-centre GC basis or even more, to evaluate matrix elements in the basis of scattering functions.
(ii) Use a modern Monte Carlo method to evaluate multidimensional integrals completely numerically (including projection onto the eigenspace of angular momentum). In this case one can use one body functions different from harmonic oscillator functions. The experience in calculations of path integrals in this manner could be valuable.

3. Experience with calculation. -
There are a few stages in the theory which determine the reliability of the reaction parameters up to 1%. The quality of the trial function is mainly determined by the truncation of the system of differential equations for the scattering of three point charges. Dimensions in the range from 49 to 100 provide an accuracy within 1%. The first order in (scattering energy)$^{-1}$ GC amplitude (26) satisfies the integral equation to the same accuracy. Finally, the adequate accuracy of the numerical calculation of integrals in (12) where σ's are expressed in the GC representation using the amplitudes (26) can be reached by a mesh density, in the coordinate Z, of about 1/16 of wavelength.

1. Merkuriev, S. P., Ann. Phys. $\underline{130}$, 395 (1980).
2. M. V. Mihailović, Proc. II La Rabida Intl. Summer School on Theory of Nuclear Structure and Reactions, M. Lozano ed., World Scientific, Singapore 1986, p. 270.
3. M. V. Mihailović and M. Poljšak, Nucl. Phys. $\underline{A388}$, 317 (1982).
4. R. Beck, M.V. Mihailović and M. Poljšak, Nucl. Phys. $\underline{A351}$, 295(1981)

NUCLEAR STRUCTURE IN HEAVY-ION COLLISIONS

K. V. Shitikova
Institute of Nuclear Physics
Moscow State University
Moscow, U.S.S.R.

1. Introduction

A microscopic approach to describing the elastic and inelastic heavy-ion scattering aimed at studying possible manifestations of the nuclear structure in these processes is investigated.

The heavy-ion interaction potential is constructed with the use of nuclear densities calculated in two microscopic models: method of the hyperspherical functions for light ions and quasiparticle-phonon model for heavy ions.

Only one free parameter (renormalization factor of the imaginary part of the optical potential) has been introduced into the calculation.

The elastic and inelastic scattering processes of ^{4}He, ^{6}Li, ^{12}C, ^{16}O on themselves and on targets ^{58,50}Ni, ^{90}Zr, ^{124}Sn, ^{142,144}Nd, ^{208}Pb are considered.

2. Calculational Method

Usually, in the folding calculations there are always two basic ingredients. One is the construction of nuclear densities within the microscopic models and the other is the choice of an appropriate effective N-N interaction between the projectile nucleons and the target nucleons. At first we give a brief description of the nuclear density calculation in the method of hyperspherical functions and the quasiparticle-phonon nuclear model. It should be noted that the knowledge of the wave functions used in the nuclear density calculation enables us to investigate nuclear structure phenomena observed in the heavy-ion scattering.

2.1. Nuclear density distribution in the hyperspherical functions method

The method of hyperspherical functions is specified by that in the 3(A-1)-dimensional space of Jacoby coordinates the spherical coordinates are introduced: the hyperradius ρ and the hyperspherical angles (see Ref. 8-11). These angles $\Theta_1, \Theta_2, ... \Theta_{r-1}$ can be chosen so that the relation between the Jacoby and hyperspherical coordinates has the form:

$$x_1 = \rho \sin\Theta_{n-1} ... \sin\Theta_2 \sin\Theta_1$$
$$x_2 = \rho \sin\Theta_{n-1} ... \sin\Theta_2 \cos\Theta_1$$
$$\vdots$$
$$x_{n-1} = \rho \sin\Theta_{n-1} \cos\Theta_{n-2}$$
$$x_n = \rho \cos\Theta_{n-1}$$
$$\rho^2 = \sum_{i=1}^{n} x_i^2, \quad 0 \leq \rho \leq \infty, \quad 0 \leq \Theta_i < 2\pi, \quad n = 3(A-1) \quad (1)$$

In this way the collective variable ρ is introduced. In the hyperspherical functions method, the total wave function of a nucleus with A nucleons is usually expanded in terms of K-harmonics polinomials $Y_{K\gamma}(\Theta_i)$:

$$\Psi(1,2,...,A) = \rho^{-\frac{1}{2}(3A-4)} \sum_{K\gamma} f_{K\gamma}(\rho) Y_{K\gamma}(\Theta_i) \quad (2)$$

where $\gamma = [f] \in LST$, $\int f_{K\gamma}^2(\rho) d\rho = 1$

The hyperspherical harmonics are eigenfunctions of the angular part of the Laplacian:

$$\Delta_{\Omega_\ell} Y_{K\gamma}(\theta_i) = -K(K+\ell-2) Y_{K\gamma}(\theta_i) \quad (3)$$

Here K is an analogue of the angular momentum for $\ell = 3$ and is called the global moment.

The nuclear Hamiltonian has the form:

$$H = -\frac{\hbar^2}{2m} \frac{1}{\rho^{3A-4}} \frac{d}{d\rho}\left(\rho^{3A-4}\frac{d}{d\rho}\right) - \frac{\hbar^2}{2m\rho^2}\Delta_\Omega + V(\rho) \quad (4)$$

The system of equations for radial eigenfunctions and eigenvalues is written as:

$$\left\{\frac{d^2}{d\rho^2} - \frac{\mathcal{L}_K(\mathcal{L}_K+1)}{\rho^2} - \frac{2m}{\hbar^2}(E + W_{K\gamma}^{K\gamma}(\rho))\right\} f_{K\gamma}(\rho) = \frac{2m}{\hbar^2} \sum_{K\gamma' \neq K\gamma} W_{K\gamma}^{K\gamma'}(\rho) f_{K\gamma'}(\rho) \quad (5)$$

where

$$\mathcal{L}_K = K + \frac{3A-6}{2}, \quad W_{K\gamma}^{K\gamma'}(\rho)$$

are the matrix elements of the potential energy of the nucleon-nucleon interaction

$$V = \sum_{i<j} V(r_{ij}), \quad V(r_{ij}) = f(r_{ij}) W_{\sigma\tau}$$

Then, the radial wave functions $f_{K\gamma}(\rho)$ are used to calculate the matrix elements of the density operator. In the hyperspherical-function method the radial density distribution takes the form

$$n_{ij}(R) = \frac{16}{\sqrt{\pi}} \frac{\Gamma(\frac{5A-11}{2})}{\Gamma(\frac{5A-14}{2})} \int_R^\infty \frac{(\rho^2-R^2)^{\frac{5A-16}{2}}}{\rho^{5A-13}} f_i(\rho) f_j(\rho) d\rho + \quad (6)$$

$$+ \frac{8}{3} \frac{(A-4)}{\sqrt{\pi}} \frac{\Gamma(\frac{5A-11}{2})}{\Gamma(\frac{5A-16}{2})} \int_R^\infty \frac{R^2(\rho^2-R^2)^{\frac{5A-18}{2}}}{\rho^{5A-13}} f_i(\rho) f_j(\rho) d\rho$$

where the density of nuclear states (diagonal matrix elements of the density operator) is normalized according to

$$4\pi \int n_{ii}(R) R^2 dR = A \quad (7)$$

and the rms-radius of each state can be obtained as follows

$$\overline{R_{ii}^2} = \frac{\int n_{ii}(R) R^4 dR}{\int n_{ii}(R) R^2 dR} \quad (8)$$

Figure 1 shows the effective potential

$$V_{eff}(\rho) = \frac{\hbar^2}{2m} \frac{\mathcal{L}_K(\mathcal{L}_K+1)}{\rho^2} + W_{K\gamma}^{K\gamma}(\rho)$$

the energy levels, and the first three radial wave functions found by solution of eq. (5) for the ground and monopole excited states of the nucleus. It can be seen that the wave functions of the excited states have some modes while those of the ground state have none. The hyperspherical functions method provides a convenient basis for a microscopic description of the monopole of "breathing mode" oscillations in nuclei. The point of this method is that a collective variable is being introduced which can be directly associated with the mean-square nuclear radius: $\rho^2 = A<r^2>$, i.e., with the mean nuclear density. The excitations in this variable correspond to the monopole vibrations of the nucleus as a whole; the density being then a dynamical variable.

What is more interesting, the wave functions for the excited states are pushed out towards larger radii. This is illustrated more precisely in Fig. 2 where three wave functions: for ground, first and second excited monopole states in ^{12}C nucleus are shown. It is then evident that the increase in the nuclear size in the excited states is automatically accounted for in the hyperspherical functions method. Fig. 3 shows the dependence of the rms-radii on excitation energy in the hyperspherical-functions method in comparison with that of finite temperature Hartree-Fock-method. The most important outcome which can be reported here is the fact that the present calculation leads to a dependence of the rms-radius on excitation energy so that the nuclear size goes to infinity when the excitation energy of the nucleus reaches its binding energy, while the mean-field theory predicts a finite value [20]. Of course, it is necessary to point out that in such a type of calculation all the energy of the nucleus is concentrated on one degree of freedom, especially the nuclear radius, while in the Thomas-Fermi-Model, by which the temperature has been introduced into nuclear physics, and in the Hartree-Fock approximation, the excitation energy is distributed over a lot of single-particle degrees of freedom. Consequently, one has to expect a stronger energy dependence of the nuclear size than it is given by the mean-field theory.

2.2. Density distribution in the quasiparticle-phonon nuclear model

The general form of the QPM Hamiltonian

$$H = H_{av} + H_{pair} + H_M + H_{sM} \qquad (9)$$

where H_{av} is the average field describing independent single-particle motions; H_{pair} describes the monopole pairing interaction between the neutrons of protons; H_M and H_{sM} are separable multipole and spin-multipole interaction terms generating the nuclear excitations. The one-phonon states with $\lambda^{\pi} = 1^-$, 2^+, 3^-, 4^+,... are generated by the multipole forces, whereas the one-phonon states with $\lambda^{\pi} = 1^+$, 2^-, 3^+,... by the spin-multipole forces. These effective forces include the isoscalar and isovector components:

$$V_\lambda(\bar{\tau}_1,\bar{\tau}_2) = \frac{1}{2}(K_0^{(\lambda)} + K_1^{(\lambda)}\bar{\tau}_1\bar{\tau}_2) r_1^\lambda Y_{\lambda\mu}(\theta_1,\varphi_1) r_2^\lambda Y_{\lambda\mu}(\theta_2,\varphi_2)$$

$$V_L^\sigma(\bar{\tau}_1,\bar{\tau}_2) = \frac{1}{2}(K_0^{(L\lambda)} + K_1^{(L\lambda)}\bar{\tau}_1\bar{\tau}_2)[r_1^\lambda[\vec{\sigma}_1 Y_{\lambda\mu}(\theta_1,\varphi_1)]_{LM} \times r_2^\lambda[\vec{\sigma}_2 Y_{\lambda\mu}(\theta_2,\varphi_2)]_{L-M}$$

(10)

The contributions of H_M and H_{sM} to the ground-state density are negligible, so one has for this case

$$H_0 = H_{av} + H_{pair} = \sum_{jm\tau} E_j a^+_{jm} a_{jm} - \frac{1}{4}\sum_{jj'mm'\tau} G_\tau (-)^{j+j'-m-m'} \times$$

$$\times a^+_{jm} a^+_{j-m} a_{j'-m'} a_{j'm'}$$

(11)

where a^+_{jm} and a_{jm} are the nucleon creation and annihilation operators; $j \equiv (n,l,j)$ is the set of quantum numbers for a single-particle state with the energy E_j and the angular momentum projection m; $\tau = (p,n)$ is the isotopic index; G_n and

G_p are the monopole pairing constants. In the single-particle-basis calculation, the average field is taken as Woods-Saxon potential. After Bogolubov's transformation

$$a_{jm} = u_j \alpha_{jm} + (-)^{j-m} v_j \alpha^+_{j-m} \quad (12)$$

where α^+_{jm} and α_{jm} are the quasiparticle creation and annihilation operators, one can obtain the ground-state density for a heavy (spherical) nucleus in the form:

$$\rho_o(r) = \frac{1}{4\pi} \sum_j (2j+1) |R_j(r)|^2 v_j^2 \quad (13)$$

Here $R_j(r)$ is the radial part of the wave function for a single-particle state $j \equiv (n,l,j)$. Note that we have included into the ground-state-density calculation the effect of monopole pairing interaction between nucleons unlike the shell model densities used in some other folding calculations [2,3]. For the calculation of the wave functions of various excited states in the nucleus-target, the QPM Hamiltonian is transformed into the phonon representation. In contrast with some other microscopic models where the phonons are introduced in a phenomenological way, within QPM the structure of phonons is calculated microscopically, in the random phase approximation (RPA), and the phonons are superpositions of various two-quasiparticle excitations. A short description of this procedure is the following: after the transformation into the space of quasiparticle and phonons, the QPM Hamiltonian can be written in the form

$$H = \sum_{jm} \varepsilon_j \alpha^+_{jm} \alpha_{jm} - \frac{1}{8} \sum_{\lambda i i' \tau} \frac{X_\tau^{\lambda i} + X_\tau^{\lambda i'}}{(2\lambda+1)\sqrt{\mathcal{Y}_\tau(\lambda i) \mathcal{Y}_\tau(\lambda i')}} \times \quad (14)$$

where

$$\times [Q^+_{\lambda\mu i} + (-)^{\lambda-\mu} Q_{\lambda-\mu i}][(-)^{\lambda-\mu} Q^+_{\lambda-\mu i'} + Q_{\lambda\mu i'}] + H_{qph}$$

$$X_\tau^{\lambda i} = \sum_{j_1 j_2} \frac{[f_{j_1 j_2}^\lambda u_{j_1 j_2}^{(\pm)}]^2 \varepsilon_{j_1 j_2}}{\varepsilon_{j_1 j_2}^2 - \omega_{\lambda i}^2}$$

$$\mathcal{Y}_{n,p}(\lambda i) = Y_{n,p}(\lambda i) + Y_{p,n}(\lambda i) \left\{ \frac{K_0^{(\lambda)} - K_1^{(\lambda)}}{2\lambda+1} X_{n,p}^{\lambda i} \middle/ 1 - \frac{K_0^{(\lambda)} + K_1^{(\lambda)}}{2\lambda+1} X_{p,n}^{\lambda i} \right\}^2 \quad (15)$$

$$Y_{n,p}(\lambda i) = \frac{1}{2} \frac{\partial}{\partial \omega} X_{n,p}^{\lambda i}\bigg|_{\omega = \omega_{\lambda i}}$$

Here $f_{j_1 j_2}^\lambda = \langle j_1 \| r^\lambda Y_\lambda(\theta) \| j_2 \rangle$ are the reduced single-particle matrix elements of multipole operator $r^\lambda Y_\lambda(\theta)$ and $u_{j_1 j_2}^{(\pm)} = u_{j_1} v_{j_2} \pm u_{j_2} v_{j_1}$; ε_j and $\varepsilon_{j_1 j_2}$ are the energies of one- and two-quasiparticle states, respectively. The energy $\omega_{\lambda i}$ of one-phonon state $Q^+_{\lambda\mu i} \Psi_0$ is obtained by solving the following RPA equation

$$\frac{K_0^{(\lambda)} + K_1^{(\lambda)}}{2\lambda+1}(X_n^{\lambda i} + X_p^{\lambda i}) - 4\frac{K_0^{(\lambda)} K_1^{(\lambda)}}{(2\lambda+1)^2} X_n^{\lambda i} X_p^{\lambda i} = 1 \quad (16)$$

$$Q^+_{\lambda\mu i} \Psi_0 = \sum_{jj'} \{\psi_{jj'}^{\lambda i}[\alpha^+_{jm} \alpha^+_{j'm'}]_{\lambda\mu} - (-)^{\lambda-\mu} \phi_{jj'}^{\lambda i}[\alpha_{jm'} \alpha_{jm}]_{\lambda-\mu}\} \Psi_0$$

where Ψ_0 is the phonon vacuum. In the one-phonon approximation, i.e in the RPA, the contribution from H_{qph} which describes the quasiparticle-phonon interaction is absent, and one has for an excited state in the nucleus-target: $|\lambda i\rangle = Q^+_{\lambda\mu i} \Psi_0$

Futher, if we define the nuclear transition density as $\rho(r) = \langle f | \sum_k \delta(r - r_k) | i \rangle$

with the multipole expansion of the form

$$\rho(\tau) = \sum_{\lambda\mu} C_\lambda \langle J_i M_i \lambda \mu | J_f M_f \rangle \rho_\lambda(\tau) Y^*_{\lambda\mu}(\theta,\varphi) \quad (17)$$

where the normalization constant C_λ is that defined in ref. [3] we get

$$\rho_\lambda(\tau) = \langle J_f \| \sum_k \tau_k^{-2} \delta(\tau-\tau_k) i^\lambda Y_\lambda(\theta_k,\varphi_k) \| J_i \rangle \quad (18)$$

In the case of single excitation of an even-even nucleus target ($J_f = \lambda$ and $J_i = 0$), after some transformations one obtains

$$\rho_\lambda(\tau) = \frac{1}{\sqrt{4\pi}} \sum_{j_1 j_2} \frac{i^{\ell_2+\lambda-\ell_1}(-)^{j_2+\lambda+\frac{3}{2}}}{2(1+\delta_{j_1 j_2})} \hat{j}_1 \hat{j}_2 \begin{pmatrix} j_1 & j_2 & \lambda \\ \frac{1}{2} & -\frac{1}{2} & 0 \end{pmatrix}$$
$$\times R^*_{j_1}(\tau) R_{j_2}(\tau) u^{(+)\lambda i}_{j_1 j_2} (\psi^{\lambda i}_{j_1 j_2} + \phi^{\lambda i}_{j_1 j_2})[1+(-)^{\ell_2-\ell_1+\lambda}] \quad (19)$$

It should be noted that in some other folding calculations, preference has been given to the macroscopic Tassic model in the nuclear transition density calculation. Since it is just the tails of the transition densities which contribute most to the folding integral for the potential near the strong absorption radius, and those microscopic models which use an oscillator basis do not give the densities accurate enough for such large distances. In the QPM, the single-particle basis is calculated by a numerical method using the code REDMEL and the calculated radial wave functions describe nuclear asymptotics correctly, so the nuclear transition densities calculated within QPM are quite adequate for the use in the folding model.

2.3. The double-folding model

The interaction potential between two nuclei or ions is not completely determined and, therefore, the optical model was used to construct the potential for a system of two nuclei a + A [2,3]. The complete wave function of the system a + A is expanded in internal wave eigenfunctions of individual nuclei

$$\Psi = \sum \Phi_{ai} \Phi_{Aj} \chi_{ij}(R) \quad (20)$$

where $\chi_{ij}(R)$ described the relative motion of the system a + A when nucleus a is in state i and nucleus A is in state j. Elastic scattering corresponds to the wave function $\chi_{00}(R)$. If one neglects the effects of antisymmetrization between two nuclei whose wave functions are separately antisymmetrized, the effective potential of optical model will take the form

$$U_{opt} = V_{00} + \sum_{\alpha\alpha'} V_{0\alpha} \left(\frac{1}{E-H+i\varepsilon}\right)_{\alpha\alpha'} V_{\alpha'0} = U_F + \Delta U \quad (21)$$

where V is the interaction between a and A; the summation runs over all the excited states of one nucleus or both. The first term is the real double-folding potential

$$U_F(R) = V_{00} \equiv (\Phi_{a0} \Phi_{A0} |V| \Phi_{a0} \Phi_{A0}) \quad (22)$$

The integration in (22) is over all internal coordinates of the two nuclei. The remaining term ΔU which contains the coupling to the inelastic channels, is of the dynamic nature, so that one has to know the total excitation spectrum of the colliding nuclei in order to construct it. In phenomenological approaches, U... is approximated by the local complex potential U(R), which is used for example in the Woods-Saxon form. The parameters of the real and imaginary part of OP are often chosen to be independent. Therefore, the radius of the imaginary part of OP is in most cases larger than the radius of the real part.

When $U_F(R)$ is calculated by the method of double folding, the densities of the colliding nuclei a and A are assumed to be nondisturbed, and the interaction potential is the mean value of the nucleon-nucleon interaction averaged over two densities.

$$U_{A_1A_2}(\vec{R}) = \iint \rho_{A_1}(\vec{r_1})\rho_{A_2}(\vec{r_2}) V(|\vec{r_1}+\vec{R}-\vec{r_2}|) d\vec{r_1} d\vec{r_2} \quad (23)$$

Such a determination seems to be justified at the ion collision energies considered here ($\lesssim$ 10 MeV/nucleon) because the elastic scattering of ions is only sensitive to the form of the potential at a distance between the ions near the radius of strong absorption

$$R_{crit} = 1.5(A_1^{1/3} + A_2^{1/3}) \quad (24)$$

The overlap of the densities of colliding nuclei is small in this region, so that one can assume their distortion to be negligible in this case.

Using the Skyrme interaction with δ-forces as nucleon-nucleon interaction, we obtain the following double folding interaction potential 13 for the nuclei a and A:

$$U_{ij,ke}^{a,A}(R) = \pi \int_0^\pi \sin\theta\, d\theta \left\{ \frac{3}{8} t_0 \left[\iint n_{ij}^a(r) n_{ke}^A(|\vec{R}-\vec{r}|)^2 dr + \int_0^\infty n_{ij}^A(r) n_{ke}^a(|\vec{R}-\vec{r}|) r^2 dr \right] + \right.$$
$$\left. + \frac{t_3}{16} \left[\int (n_{ij}^a(r))^2 n_{ke}^A(|\vec{R}-\vec{r}|) r^2 dr + \int_0^\infty (n_{ij}^A(r))^2 n_{ke}^a(|\vec{R}-\vec{r}|) r^2 dr \right] \right\} \quad (25)$$

where ... $|\vec{R}-\vec{r}| = \sqrt{R^2+r^2-2Rr\cos\theta}$; t_0, t_3 ... are the parameters of the two-particle and three-particle Skyrme interaction, respectively; nij(r) is the density distribution of nuclear matter. In the case of finite range forces, the nucleon interaciton potential is chosen to be of the Gaussian form allowing for a soft core in the effective interaction at small distances [15]:

$$U(r) = \sum_{k=1}^{2} U_k \exp\left(-\frac{r^2}{a_k^2}\right) \quad ; \quad r = |\vec{r_1}+\vec{R}-\vec{r_2}| \quad (26)$$

In this case, the Gaussian form with the parameters found from the condition for the best representation of nuclear densities obtained in terms of MHF is also applied to the nuclear matter density distribution [16]. Thus, using the following form for the densities of colliding nuclei,

$$\rho_A(r) = \rho_{0A}\left[\exp\left(-\frac{r^2}{b_{0A}^2}\right) + C_1 \frac{r^2}{b_{2A}^2} \exp\left(-\frac{r^2}{b_{2A}^2}\right)\right]$$
$$\rho_a(r) = \rho_{0a}\left[\exp\left(-\frac{r^2}{b_{0a}^2}\right) + C_2 \frac{r^2}{b_{2a}^2} \exp\left(-\frac{r^2}{b_{2a}^2}\right)\right] \quad (27)$$

we get for the nuclear interaction potential

$$U^{a,A}(R) = \sum_K B_K \left\{ f_{K0a,0A}(R) \exp\left(-\frac{R^2}{\tilde{\delta}^2_{K0a,0A}}\right) + f_{K0a,2A}(R) \exp\left(-\frac{R^2}{\tilde{\delta}^2_{K0a,2A}}\right) \right. \quad (28)$$

$$\left. + f_{K2a,0A}(R) \exp\left(-\frac{R^2}{\tilde{\delta}^2_{K2a,0A}}\right) + f_{K2a,2A}(R) \exp\left(-\frac{R^2}{\tilde{\delta}^2_{K2a,2A}}\right) \right\}$$

where

$$f_{K0a,2A} = C_1 \frac{b_{0A}^3 b_{2A}^3}{\tilde{\delta}^3_{K0a,2A}} \left(\frac{3}{2} \frac{\mu_{K0A}^2}{\tilde{\delta}^2_{K0a,2A}} + \frac{b_{2A}^2}{\tilde{\delta}^2_{K0a,2A}} - \frac{R^2}{\tilde{\delta}^2_{K0a,2A}} \right), \quad \tilde{\delta}_{Kma,nA} = \sqrt{a_K^2 + b_{ma}^2 + b_{nA}^2}$$

$$f_{K2a,0A} = C_2 \frac{b_{2A}^3 b_{0A}^3}{\tilde{\delta}^3_{K2a,0A}} \left(\frac{3}{2} \frac{\mu_{K0A}^2}{\tilde{\delta}^2_{K2a,0A}} + \frac{b_{2a}^2}{\tilde{\delta}^2_{K2a,0A}} \cdot \frac{R^2}{\tilde{\delta}^2_{K2a,0A}} \right), \quad B_K = \pi^3 \rho_{0a} \rho_{0A} V_K a_K^3,$$

$$\mu_{Kns} = \sqrt{a_K^2 + b_{ns}^2}$$

$$f_{K2a,2A} = C_1 C_2 \frac{b_{2A}^3 b_{2A}^3}{\tilde{\delta}^3_{K2a,2A}} \left\{ \frac{9}{4} \frac{a_K^2}{\tilde{\delta}^2_{K2a,2A}} \left(1 + \frac{5}{3} \frac{b_{2a}^2 b_{2A}^2}{a_K^2 \tilde{\delta}^2_{K2a,2A}} \right) + \right. \quad (29)$$

$$\left. + \left[\frac{3}{2} \frac{b_{2a}^2}{\mu_{K2a}^2} \left(1 + \frac{a_K^2 b_{2A}^2}{b_{2a}^2 \tilde{\delta}^2_{K2a,2A}} \right) - 5 \frac{b_{2a}^2 b_{2A}^2}{\tilde{\delta}^4_{K2a,2A}} \right] \cdot \frac{R^2}{\tilde{\delta}^2_{K2a,2A}} + \frac{b_{2a}^2 b_{2A}^2}{\tilde{\delta}^8_{K2a,2A}} R^4 \right\} \quad f_{K0a,0A} = \frac{b_{0a}^3 b_{0A}^3}{\tilde{\delta}^3_{K0a,0A}}$$

In the case of heavy ion interaction we use the zero-range pseudopotential

$$U_\delta(r) = d E \delta(r)$$

in order to take into account the Pauli principle effectively. This correction term depends on the mass number of the system and the incident energy in the following way

$$d(E) = -276 \left(1 - 0.005 \frac{E}{A} \right) MeV \, fm^3 \quad (30)$$

The M3Y interaction is [19,21,22]

$$V(r) = 7999 \frac{\exp(-4r)}{4r} - 2134 \frac{\exp(-2,5r)}{2,5r} - 262 \delta(\vec{r}_{12})$$

The calculation of six dimensional integral is very complicated in the coordinate space but if we work in momentum space, this integral is reduced to a product of three one-dimensional integrals [3] with the multipole expansion (17) one obtains, in the case of single excitation of a spin-zero target, the following expression for double-folded potential

$$U_F(\vec{R}) = C_\lambda \mathcal{U}_\lambda(R) Y^*_{\lambda\mu}(\theta_R, \varphi_R), \quad (31)$$

where

$$\mathcal{U}_\lambda(R) = \frac{1}{2\pi^2} \int k^2 dk \, j_\lambda(kR) \tilde{V}(k) \tilde{\rho}_\lambda^{(a)}(k) \tilde{\rho}_0^{(2)}(k) \quad (32)$$

$$\tilde{f}_\lambda(k) = 4\pi \int r^2 dr \, j_\lambda(kr) f_\lambda(r) \quad (33)$$

The elastic scattering corresponds to $\mathcal{L} = 0$ in these formulae. Further, $U_K(R)$ is taken as a real part of the heavy ion potential into the cross-section calculation. Usually, the imaginary optical potential is supposed to have a Woods-Saxon form

$$W(r) = -\frac{W_V}{1+\exp[(r-R_V)/a]} \quad (34)$$

where $R_V = r_V(A_1^{1/3} + A_2^{1/3})$; W_V, r_V and a_V are defined from the best fit to the elastic scattering data. In this case the imaginary transition potential is defined by deforming the Woods-Saxon potential (34), i.e.

$$W_\lambda(r) = -\beta_\lambda^{(I)} R \frac{dW(r)}{dr} \quad (35)$$

The deformation parameter $\beta_\lambda^{(I)}$ is obtained from the B(Eλ) values scaled according to BR = const. However, we have followed another way supposing that the imaginary potential should have a similar form to the real one.

The scaling factor in elastic channel was searched upon by fitting the theoretical differential cross sections to experimental data.

2.4. Cross sections of elastic scattering

The angular distributions for elastic scattering of ions were calculated using programs TUFO and CHUCK. It was assumed that the forms of the real and imaginary potentials were the same, so that

$$U = U_{00,00}(1+i\alpha) \quad (36)$$

The factor α in the elastic channel was found by fitting theoretical differential cross sections to experimental data. The criterion of fitting is the usual one, namely, we minimized the value

$$\chi^2 = \sum_i \left[\frac{\sigma_{exp}(\theta_i) - \sigma_{theor}(\theta_i)}{\Delta\sigma_{exp}(\theta_i)}\right]^2 \quad (37)$$

where $\sigma_{theor}(\theta_i)$ are the calculated differential cross sections; $\sigma_{exp}(\theta_i)$ are the measured cross sections; $\Delta\sigma_{exp}(\theta_i)$ are the experimental errors.

3. Calculated Results and Discussion

3.1. Nuclear density distribution

The calculations for the light nuclei have been carried out with Brink-Boeker and Volkov potentials in the hyperspherical functions method [17].

Figure 4 shows the potentials which give the best agreement in our calculations for the whole set of experimental data. The full curve presents a sufficiently narrow ($\sim$2 fm) potential B4 (Brink and Boeker) with a very strong ($\sim$700 MeV) core and a deep ($\sim$350 MeV) well. The chain curves (with one and two dots) present, respectively, the potentials B1 and C1 with a core of about 200 MeV, well depth $\sim$50 MeV and width $\sim$3 fm. The potentials differ only in the exchange variant of forces, namely the contribution to C1 from the odd states exceeds that from even states. (The potentials for the odd states are shown by these lines.) Finally, the broken curve presents one of the Volkov potentials, V7, with low core (20 MeV), shallow ($\sim$25 MeV) but wide (4-5 fm) well, and the contribution from the odd states $V_{12}^{odd} - 0.2\, V^{even}$ different from that in the potentials of Brink and Boeker. The comparison with experiment has been made in the case of the nuclei ^{4}He, ^{6}Li, ^{12}C and ^{16}O. The fact is that for these nuclei the experimental data on elastic form factors are known, whereas for the ^{4}He nucleus the position of the monopole 0^+ state and inelastic form factor are also known.

The ^{12}C nucleus: Figure 5 shows the results of calculations of the binding energy, excitation energy of monopole resonance, RMC radius and position of minimum in the form factor for various nucleon-nucleon potentials. The first column of Table 4 lists the relevant experimental data.

The upper part of the figure (a) presents the densities of the ground state and the transition densities with excitation of the monopole resonance. Parts (b) and (c) give the results of calculations of the elastic and inelastic form factors, respectively. The full circles show the experimental results. The theory gives good agreement with experiment for the elastic form factor over a broad range of the transferred momenta with the potential B4. A density depression appears at the centre of the nucleus due to the shell effect accounted for by the increase in the contribution from the 1p state. As for the binding energy and the RMS radius, the potentials examined give a more-or-less underestimated binding energy for the ground state and an overestimated value of the RMS radius. Obviously this result is due to an inadequate selection of the nucleon-nucleon potential.

The ^{16}O nucleus: Figure 6 and Table 3 show the results of the calculation and the corresponding experimental values for the nucleus ^{16}O. It can be seen that the theory with the potential B1 can properly describe the experimental value of the form factor over a broad range of the transferred momenta up to the second minimum ($q \approx 9$ fm^{-1}). In the nucleus-target all calculations have been performed in the distorted-wave Born approximation (DWBA) [21,22]. The phonon amplitudes ψ and ϕ have been calculated by the code RPAS, which performs the RPA calculations within QPM. The structure of the low-lying excited states in the nuclei-targets has been calculated with the sets of constants ($K_0^{(\ell)}$, $K_1^{(\ell)}$) which reproduce the excitation energies and the reduced transition probabilities shown in the Table 1.

In the description of various excitations in the nucleus-target from the point of view of QPM, it is important to know how ($K_0^{(\ell)}$, $K_1^{(\ell)}$) influence the calculated inelastic cross-sections. Theoretical study of the same nuclear states excited in different reactions enables us to choose the most appopriate set of ($K_0^{(\ell)}$, $K_1^{(\ell)}$) for the calculation of the wave functions of excited states with spin ℓ and parity π (see Table 1).

3.2 The Folding-Potentials We shall construct the folding-potentials for the systems with the ^{4}He, ^{6}Li, ^{12}C and ^{16}O ions for various nucleon-nucleon potentials. The results are displayed in Fig. 7. Figure 8 shows more detailed results for the ^{4}He + ^{12}C system. The folding-potentials are deep; their values for all the examined systems at R = 0 are determined by the expression $A_1 T_{00}' \rho(R=0)$ (where T_{00} is the volume integral of effective two-body forces) to within a factor of 2. The radial dependence of the potentials differs from that of the Woods-Saxon potential. In the surface region, the shape of the two curves is the same, although the strength of the folding-potential is somewhat in excess of the strength of the optical potential. From Table 2 follows that the half fall-off radii and the volume integrals of the theoretical and phenomenological potentials differ by 9% and 15% respectively. The folding-potentials with the Skyrme force were calculated in [16]. Figure 8 represents the result obtained for ^{4}He + ^{12}C system. The curves corresponding to the Skyrme force and to the finite-range forces are similar throughout the nuclear region. As it should be expected, however, the half-fall-off radius of the Skyrme potential has proved to be somewhat smaller and the volume integral of the potential is smaller by 20% in this case. Figure 9 shows the results for the ^{16}O + ^{16}O NIP computed in the energy-density formalism (dots), by the double-folding method with the Skyrme forces (the solid line corresponds to NIP calculated with the densities obtained in MHF with the set of parameters B1, and the dashed line with B4), and with finite-range forces allowing effectively for (d $\neq$ 0, the dot-dash line) and disregarding (d = 0), two dots-dash line) anti-symmetrization [18]. The NIP calculated in the double-folding method with finite range forces is characterized by a stronger interaction at high R compared with the result obtained using the . .-force. In particular, the values of the NIP at the critical radius are V(Rcrit) = -1.05 MeV for the parameter set B1 and Skyrme-forces (version (a), and V(R$_{crit}$)

= −2.74 MeV in the case of finite-range forces. The consideration of the antisymmetrization effect in the calculations with the finite-range forces results in a deeper potential at all values of R.

The potentials $V_{11,11}$, $V_{11,12}$ and $V_{11,12}$ for the ^{3}He − ^{12}C system with the Skyrme force are shown in Fig. 10 (the convention is the following: the first pair of indices corresponds to projectile and the second pair of indices to target). The potentials $V_{11,11}$ and $V_{11,12}$ represent the interaction in elastic and inelastic channels, respectively with ^{3}He in ground state and ^{12}C in ground state and in monopole excited state in ^{12}C nucleus.

3.3 Elastic Scattering. We have calculated differential cross section of elastic scattering of ^{6}Li from ^{12}C at laboratory energy of lithium ions T_L = 90.0 MeV in the framework of the optical model. The real part of the optical potential is potential was shown in Fig. 7 and for the imaginary part we have used two forms: Woods-Saxon (WS) and Woods-Saxon squared (WS2). The best fit parameters for SW and WS2 imaginary potentials and volume integrals and mean square radii (rms) for both real and imaginary potentials are shown in Table 3. The results of our calculations are compared with the experimental data in Fig. 11. The solid line represents the best fit differential cross section with WS absorption, and the dashed one with SW2 absorption. As can be seen, the theoretical curves are in qualitative agreement with the experimental data but diffraction structure is reproduced very well. Fig. 12 shows the calculated elastic scattering differential cross sections together with experimental data (points) for ^{3}He − ^{12}C at T (^{3}He) = 108.5 MeV, ^{4}He − ^{12}C at T (^{4}He) = 139.0 MeV/21/, ^{6}Li − ^{12}C at T (^{6}Li) = 90.0 MeV [24] systems. The imaginary potential has the form (36) [14]. The results of the calculations of the cross sections of the 139 MeV α-particle elastic scattering are shown in Fig. 13 together with the experimental data. At α = 0.4 we get a satisfactory agreement with the experimental data up to scattering angles θ = 40°. After that, as the scattering angle increases, the experimental cross section decreases much more rapidly than its theoretical value. Quite a number of reasons may account for such a disagreement between theory and experiment. First, as is noted above, the Woods-Saxon form may be a more realistic form of the absorpton potential. Second, the folding-potential may be corrected for some factors, viz., the density dependence of the effective interaction the contribution of the second-order processes, etc.

Shown for comparison in Fig. 13 is also the cross section calculated with the Skyrme force. It can be seen that the differences in the angular distributions calculated for the finite-range forces and the Skyrme forces are small. Since the Skyrme force includes the density term, while the realistic interaction is independent of the nuclear matter distribution, we cannot draw in any way definite conclusions concerning the role of the finite range of the effective forces from the displayed calculations [15].

The differential cross sections for elastic scattering $^{16}O + ^{16}O$ were calculated at E_{lab} = 49.63 MeV, where the parameter of the imaginary part was α = 0.3. From Fig. 14 it follows that the better description is obtainable at a higher energy [18].

Inelastic scattering: Scattering of ^{6}Li [22]. The scattering of ^{6}Li is of special interest because ^{6}Li appears to be "anomalous" within the framework of the folding model, where the folded potentials must be reduced in strength by a factor of about two [2,3] in order to fit the data. In the MHF, calculation with the N-N potential V7 reproduces the mean squared radius RMS = 2.325 fm and minimum in the form factor of elastic electron scattering q = 5 fm^{-1} compared to the experimental data (RMS$_{expt}$ = 2.353 fm and q_{expt} = 5 fm^{-1}) for ^{6}Li nucleus [17]. We consider the elastic and inelastic scattering of ^{6}Li from ^{90}Zr and ^{58}Ni at energies 34 MeV and 71 MeV, respectively. The structure of excited states in ^{90}Zr and ^{58}Ni-targets shown in Table 1 has been calculated within the RPA. Our calculation shows (see Fig. 15) that elastic and inelastic data for system ^{6}Li + ^{90}Zr at 34 MeV can be fitted rather well without introducing a renormalization coefficient N. However, the data for system ^{6}Li + ^{58}Ni at 71 MeV can be described only with introducing the coefficient N (see Fig. 16). The fit to the elastic data gives ReN = 0.53, and ImN = 0.48 in the case when the shape of the imaginary potential is the same as for the real one, if a volume Woods-Saxon imaginary potential is used, χ^2 is reduced by several times, and a best fit to the elastic data is obtained with N = 0.51, W_v = 17.57 MeV, r_v = 1.198 fm, a_v = 0.832. Our calculations have shown that inelastic data can be reproduced only with introducing the same coefficient N for the transition folded potential as for the optical folded potential. This is consistent with the conclusion made by Satchler in Ref. [2,3].

Scattering of ^{16}O [22]. The results of calculations and elastic and inelastic data for systems ^{16}O + ^{60}Ni at energies 61.4 MeV and 114 MeV, ^{16}O + ^{208}Pb at 104 MeV are shown in Figs. 17-19. As one may see from these figures, our calculations give a good qualitative description for these reactions except for the case of 3_1^- excitation in ^{60}Ni induced by ion ^{16}O at 114 MeV (see Fig. 18). (In this case the calculated cross-section overestimates the data by a factor of about two and it is not understandable from the folding-model point of view.) It is of interest for calculations with the nuclear densities from the MHF to study the influence of different types of the N-N potentials on the calculated cross-sections. An example of this is shown in Fig. 19, where the cross-sections are calculated with two different densities for ^{16}O. The ground-state density obtained with the N-N potential B1 for ^{16}O gives a better fit to the data (the solid curves) than the density obtained with potential B4 (the dashed curve).

Scattering of ^{12}C [21] We consider the elastic and inelastic scattering of ^{12}C ions from some heavy spherical nuclei at different bombarding energies. The nuclear ground-state density for ^{12}C has been calculated in the MHF with the N-N potential V7. This calculation gives the mean squared radius RMS = 2.325 fm, the binding energy E. = 83.1 MeV and minimum in the form factor of elastic electron scattering q = 4.1 fm^{-1} compared to the experimental data (RMS (exp) = 2.294 fm, Eg. (exp) = 92.2 MeV and q(exp) = 4.1 fm^{-1}) for ^{12}C nucleus. The structure of the low-lying excited states in the nuclei-targets has been calculated with the sets of constants ($K_0^{(\ell)}$, $K_1^{(\ell)}$) which reproduce the excitation energies and the reduced transition probabilities shown in the Table 1. The results of the microscopic DWBA calculation for ^{12}C + ^{144}Nd at 70.4 MeV with different sets of ($K_0^{(\ell)}$, $K_1^{(\ell)}$) are shown in Fig. 20. The dash-dotted curves correspond to the K_0 and K_1 which generate the lowest 2^+ and 3^- one-phonon states with the energies ... = 0.696 MeV, ... = 1.51 MeV and the reduced transition probabilities B(E2; $0^+ \to 2_1^+$) = 0.975 e^2b^2, B(E3; $0^+ \to 3_1^-$) = 0.382 e^2b^3, respectively. The dashed curves correspond to ω_{2^+} = 1.25 MeV, ω_{3^-} = 1.9 MeV and B(E2↑) = 0.411 e^2b^2, B(E3↑) = 0.257 e^2b^3; and the solid curves W_{2^+} = 1.15 MeV, W_{3^-} = 1.81 MeV and B(E2↑) = 0.479 e^2b^2, B(E3.) = 0.282 e^2b^3. From various compilations of the experimental data for nucleus ^{144}Nd one has $W_{2^+}^{exp}$ (exp.) = 0.696 MeV, W_{3^-}(exp.) = 1.151 MeV and B(E2.) = 0.40 e^2b^2; 0.51 e^2b^2; (B(E3. .) = 0.26 e^2b^3. As one can see, good agreement of the calculated cross-sections with inelastic data is provided by the sets of constants ($K_0^{(\ell)}$, $K_1^{(\ell)}$) giving the calculated B(E. .)-values close to the measured ones, but the calculated excitation energies $\omega_{\lambda\pi}$ are somewhat higher than the experimental data (see also the Table I). Numerous calculations within QPM have shown that good agreement with the data for both the excitation energy and B(E. .)-value can be reached by including into the nuclear wave function more complicated two-phonon components which allow for anharmonic effects in the collective excitation.

Further we consider the elastic and inelastic ^{12}C scattering on ^{208}Pb at 98 MeV and 116.4 MeV, ^{90}Zr at 98 MeV and Nd isotopes at 70.4 MeV. On the whole, our calculation gives a good description for these data (see Figs. 21, 22) the only exception is the case of 2^+ excitation in ^{208}Pb induced by ion ^{12}C at 98 MeV when the calculated inelastic cross-section underestimates the data by a factor of about two (see Fig. 24b). The calculated elastic cross-sections shown in Fig. 24a for system $^{12}C + ^{208}Pb$ at bombarding energies 96 MeV and 116.4 MeV/37/point to some advantage of the nuclear ground-state densities used in this work compared to the shell model densities used in some other folding calculations [2]. (It is impossible to fill the observed maxima in the elastic cross-sections with the shell model densities not renormalizing the strength of the interaction - the dashed lines in Fig. 21a.)

<u>Inelastic Scattering of Light Ions [14]</u> A first attempt to describe microscopically the angular distribution of inelastic nucleus-nucleus scattering leading to the high excitation of one nucleus was undertaken for the systems $^{12}C(^3He,^3He')^{12}C^{0+}$, $^{12}C(^3He,^3He'^+)^{12}C$, $E_{0^+} = 20.3$ MeV at the incident energy $E_{3He} = 108$ MeV, and $E_{4^+} 0^+ = 20.1$ MeV at $E_{3He} = 65$ MeV. The results shown in Figs. 23 and 24 are in good accordance with the experimental data. We deal with the ^{12}C nucleus for which a 0^+ state at 20.3 MeV was recently found in inelastic scattering of ^3He ions the width of this state being 1.1 MeV. A few years ago in the framework of the hyperspherical functions method the existence of 0^+ states in very light nuclei (A ≤ 16) has been predicted [17]. These states would exhaust a great amount of the EWSR and they would have very small widths. We intend to show that using the transition densities as calculated in the hyperspherical functions method we are able to explain the measured 0^+ state in ^{12}C as being a collective one, in contradiction to the original interpretation that this state exhausts only 2.6% of the monopole EWSR.

Experimental data for inelastic scattering to $0^+(E_x = 20.3$ MeV) state in ^{12}C were analyzed using the coupled channel code CHUCK 2 with two channels (elastic and 0^+ inelastic) explicity included.

<u>The Monopole Energy Weighted Sum Rules (EWSR)</u> The EWSR were estimated according to the formula

$$\sum (E_n - E_0)|M_{no}|^2 = \frac{\hbar^2 A}{2m} \langle 0|r^2|0\rangle \quad (38)$$

where $M_{no} = \langle n|^{1/2} \sum_{i=1}^{A} r_i^2|0\rangle$

$|0>$ and $|n>$ are the wave functions of the ground and excited states, respectively. n enumerates all the 0^+ states, m is the nucleon mass and A is the mass number of the nucleus.

The calculations were carried out for nuclei with A = 4,6, 12,16 and it was shown for all theses nuclei the first 0^+ state exhausts 80-90% of the monopole EWSR. One may believe that this result is not very sensitive to this kind of the nucleon-nucleon effective potential used.

<u>The Width of the GMR</u> It is easy to show in the hyperspherical functions method that among the giant resonances of different multipolarity the width of the GMR should be the smallest. It is determined by a particular nature of its

radial wave function. In Fig. 25 we show three wave functions in their potential wells: the wave function of the monopole $\chi(\tfrac{A}{K})$ and dipole $\chi(\tfrac{A}{K_H})$ excited states in A nucleus and the ground state wave function of the A-1 nucleus. The widths for decay and spread of the GMR are determined by overlap integrals and that is why all the integrals involving monopole excited state wave function (that have a node) will be small. In Ref. [12] the widths of the GMR in ^{16}O have been estimated to be 1 MeV what is similar to the width of the 0^+ state recently measured in ^{12}C.

4. Conclusion

The microscopic nuclear models (the method of hyperspherical functions and quasiparticle-phonon nuclear model) are applied to study the elastic and inelastic heavy-ion scattering within the framework of the folding model. Results of calculations reproduce the experimental data in most considered cases and this indicates the validity of further application of the MHF and QPM to the HI scattering.

The influence of the microscopic structure of the wave functions for considered nuclei on the calcuated cross-sections is discussed.

References

1. Shitikova, K.V., Particles and Nuclei, 1985, v. 16, p. 824.
2. Satchler, G. R., Love, W. G., Phys. Repts., 1979, v. 55, p. 183.
3. Satchler, G. R., Direct Nuclear Reactions. N.Y. Oxford Univ. Press, 1983.
4. Soloviev, V. G., Theory of Complex Nuclei, Pergamon Press, Oxford, 1976.
5. Soloviev, V. G., Particles and Nuclei, 1978, v. 9, p. 580.
6. Vdovin, A. I., Soloviev, V. G., Particles and Nuclei, 1983, v. 14, p. 237.
7. Voronov, V. V., Soloviev, V. G., Particles and Nuclei, 1983, v. 14, p. 1381.
8. Simonov, Y. A., Yad. Fiz, 1966, v. 3, p. 630.
9. Smirnov, Y. F., Shitikova, K. V., Particle and Nuclei, 1977, v. 8, p. 847.
10. Shitikova, K. V., Nucl. Phys., 1979, v. A231, p. 365.
11. Kaschiev, M., Shitikova, K. V., Yad. Fiz., 1979, v. 30, p. 1479.
12. Shirokova, A. A., Shitikova, K. V., Izv. Akad. Nauk SSSR (ser. fiz.), 1979, v. 43, p. 2396.
13. Dymarz, R., Molina, J. L., Shitikova, K. V., Z. Phys., 1981, v. A299, p. 245.
14. Dymarz, R., Shitikova, K. V., JINR, E7-81-653, Dubna, 1981.
15. Burov, V. V., Knyazkov, O. M., Shirokova, A. A., Shitikova, K. V., Z. Phys., 1983, v. A313, p. 319.
16. Bogdanova, N., Shirokova, A. A., Shitikova, K. V., JINR, P4-83-547, Dubna, 1983.
17. Burov, V. V., Dostovalov, V. N., Kaschiev, M., Shitikova, K. V., J. Phys. G: Nucl. Phys., 1981, v. 7, p. 137.
18. Nazmitdinov, R. G., Saupe, G., Shitikova, K. V., Yad. Fiz., 1984, v. 39, p. 1415.
19. Dao Tien Khoa, Shitikova, K. V., JINR, E4-84-305, Dubna, 1984.
20. Saupe, G., Shitikova, K. V., JINR, E4-85-44, Dubna, 1985.
21. Dao Tien Khoa, Shitikova, K. V., JINR, E4-85-143, Dubna, 1985.
22. Dao Tien Khoa, Shitikova, K. V., JINR, E4-85-384, Dubna, 1985.

Table 1

The structure of the low-lying excited states in the nuclei-targets calculated in the RPA

Target	λ^π	$\omega_{\lambda\pi}$ (MeV) expt.	calc.	$B(E\lambda\uparrow)$ ($e^2 b^\lambda$) expt.	calc.	RMS (fm)
^{90}Zr	2^+	2.19	2.63	0.06 , 0.072	0.074	
	3^-	2.75	2.72	0.074 , 0.108	0.071	4.313
^{142}Nd	2^+	1.57	1.75	0.34 , 0.47	0.348	
	3^-	2.08	2.07	0.24	0.22	4.932
^{144}Nd	2^+	0.696	1.15	0.40 , 0.51	0.479	
	3^-	1.51	1.81	0.26	0.282	4.946
^{208}Pb	2^+	4.086	4.64	0.30 ± 0.02	0.301	
	3^-	2.61	2.64	0.69 ± 0.05	0.695	5.588
	5^-	3.194	3.2	0.046±0.006	0.054	
^{58}Ni	2^+	1.45	1.75	0.066, 0.099	0.083	3.823
^{60}Ni	2^+	1.33	1.87	0.092, 0.12	0.092	
	3^-	4.04	4.27	0.016, 0.024	0.023	3.836

Table 2

Woods-Saxon potential parameters and pairing constants

A	N,Z	r_0, fm	V_0, MeV	$\mathscr{X}$, fm^2	α, fm^{-1}	$G_{N,Z}$, MeV
59	N=31	1.31	46.2	0.413	1.613	0.280
	Z=27	1.24	53.7	0.308	1.587	0.302
91	N=53	1.29	44.7	0.413	1.613	0.168
	Z=39	1.24	56.9	0.338	1.587	0.194
121	N=71	1.28	43.2	0.413	1.613	0.122
	Z=51	1.24	59.9	0.346	1.587	0.136
141	N=83	1.27	46.0	0.413	1.613	0.116
	Z=59	1.24	57.7	0.349	1.587	0.122
209	N=127	1.26	44.8	0.376	1.587	0.074
	Z=83	1.24	60.3	0.371	1.587	0.080

Table 2

	$U_{SK}(r)$	$U_{real}(r)$	$U_{opt}(r)$
$R_{1/2}$, fm	2.40	2.56	2.80
$J/4A$ MeV·fm³	323	406	353

Table 3

	$-W$ [MeV]	r_W [fm]	a_W [fm]	Volume integral [MeV fm³]	rms [fm]
Real potential				309.6	3.53
Imaginary potential					
SW	29.8	0.915	1.157	178.1	5.18
SW 2	27.4	1.262	1.200	158.2	4.47

Table 4

for ^{16}O.

	Experiment	B4	B1	V7
E_b (MeV)	127.2	122.2	106.31	146.5
$E_x^{0^+}$ (MeV)		29.1	22.9	25.4
$\bar{R}$ (fm)	2.541	2.460	2.66	2.202
$qA^{1/3}$ (fm^{-1})	4	4.4	4	

for ^{12}C.

	Experiment	B1	C1	B4	V7
E_b (MeV)	92.2	62.5	57.5	69.2	83.1
$E_x^{0^+}$ (MeV)		20.3	20.5	25.3	21.5
$\bar{R}$ (fm)	2.294	2.634	2.585	2.460	2.325
$qA^{1/3}$ (fm^{-1})	4.1	3.9	3.9	4.1	

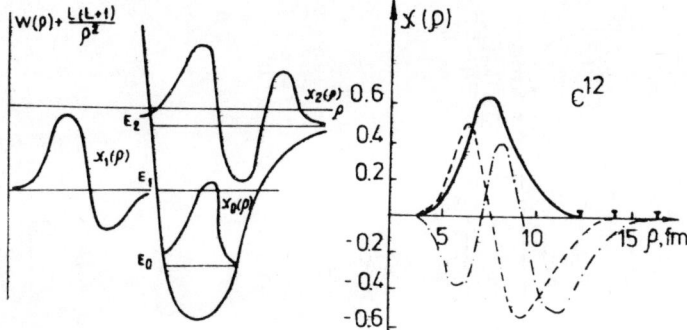

Fig. 1. The radial wave functions for the ground and the two monopole excited states in an effective potential.

Fig. 2. Three wave functions: for ground, first and second excited monopole states in ^{12}C nucleus.

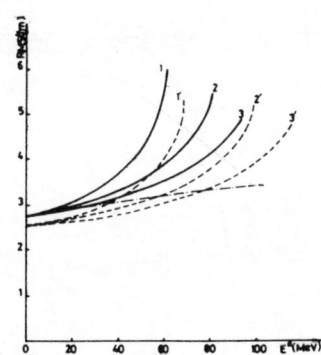

Fig. 3 Dependence of the rms-radii on excitation energy for the nuclei ^{12}C(1,1'), ^{15}O(2,2') and ^{16}O(3,3'). The dash-dotted line represents the result of Pozzolo and Vary.

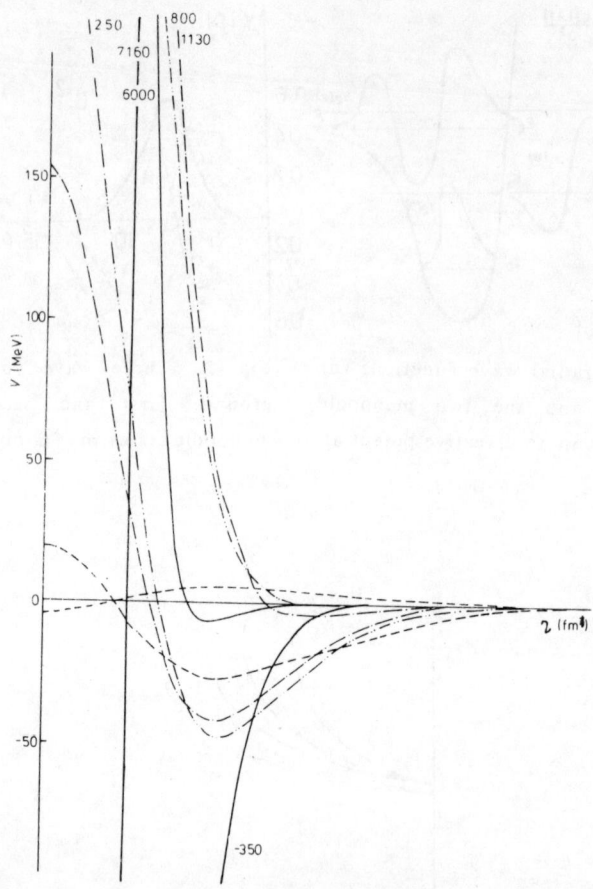

Fig. 4   Central realistic potentials of Brink and Boeker and Volkov. The full curve shows B4; the chain curves with one and two dots show C1 and B1 (Brink and Boeker), respectively; the broken curve shows the potential V7 (Volkov) ; the corresponding potentials for odd states are shown by the thin lines.

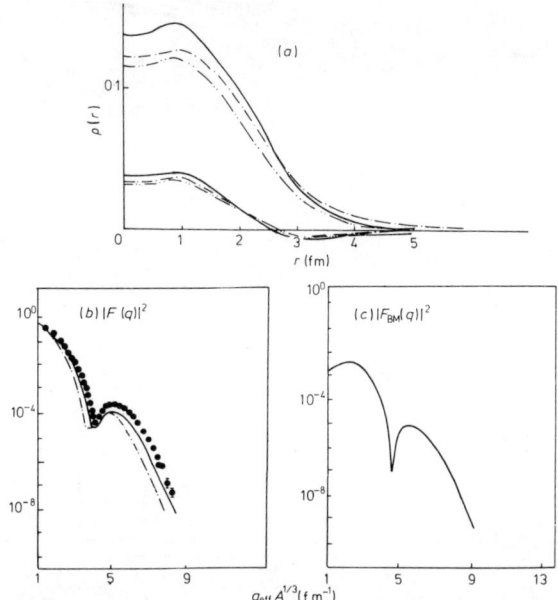

Fig. 5   The results of calculations in the hyperspherical functions method for the ^{12}C nucleus of the binding energy, excitation energy of monopole resonance, RMS radius form factors and densities for various nucleon-nucleon potentials.

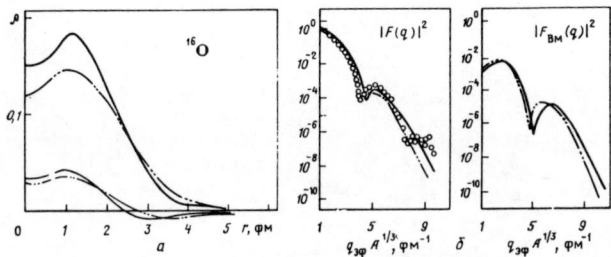

Fig. 6   The same results for the ^{16}O nucleus.

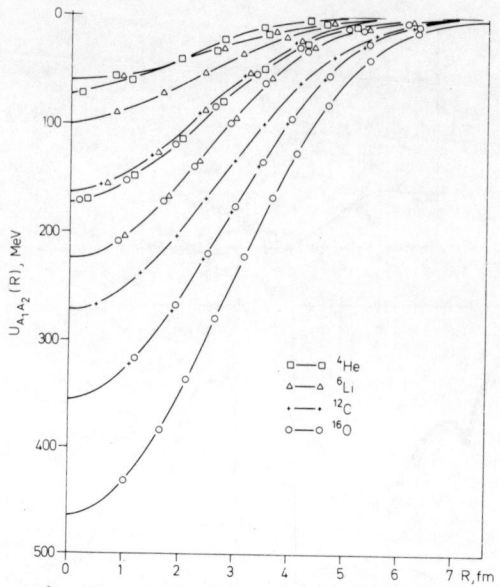

Fig. 7 Folding-potentials for the systems with ions of ^{4}He, ^{6}Li, ^{12}C, ^{16}O.

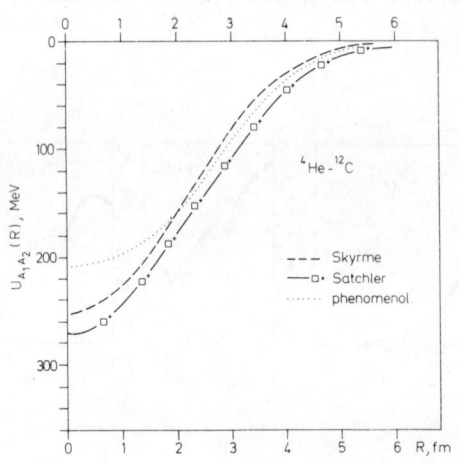

Fig. 8 The same as in Fig. 7 for ^{4}He + ^{12}C.

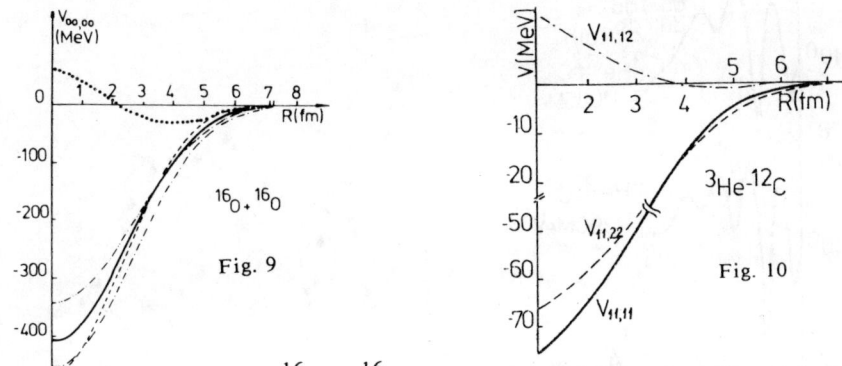

Fig. 9   Potential of $^{16}O + ^{16}O$ in the energy-density formalism (...), in the double-folding method with Skyrme forces: B1 (___), B4 (_ _ _); with Satchler-Love forces including (___ . ___) and disregarding (___ .. ___) antisymmetrization.

Fig. 10   The potentials $V_{11,11}$, $V_{11,12}$ and $V_{11,12}$ for the ^{3}He^{12} system. (The first pair of indices corresponds to projectile and the second pair of indices to target).

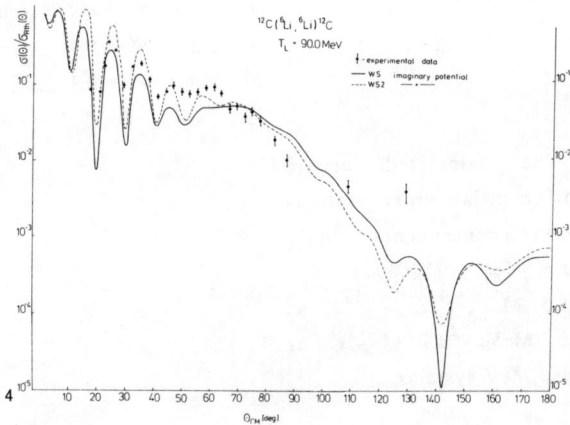

Fig. 11   Ratios of lithium elastic scattering to Rutherford scattering versus the center of mass angle for ^{6}Li - ^{12}C at T_L = 90.0 MeV. The curves are optical model calculations performed with the imaginary potential parameters from the Table 3 and with the real potential in Fig. 7.

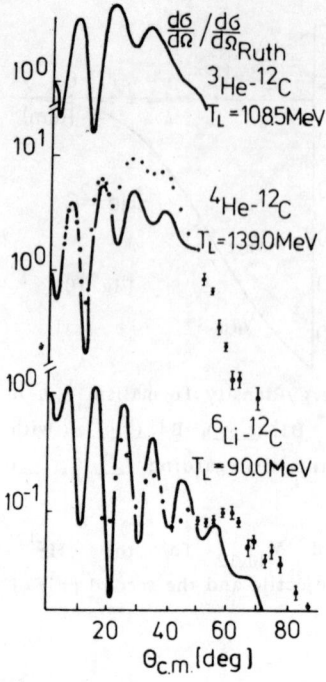

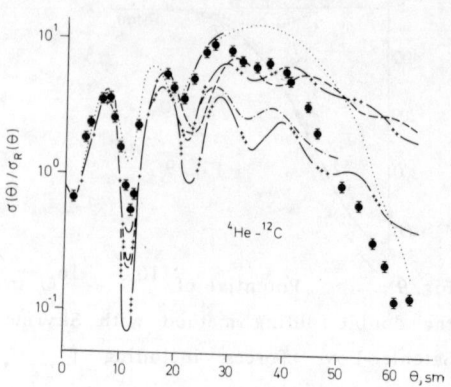

Fig. 13 The calculated elastic scattering cross sections for ^{4}He + ^{12}C at energy 139 MeV and the experimental ones (······ R = 4.2, α = 0.47, Wo = -16.9; Satchler ___ + α = 0.2; ___ ++ α = 0.4; ___ +++ α = 0.6; Skyrme --- α = 0.4; ___ , ___ α = 0.6.

Fig. 12 The calculated elastic scattering differential cross sections together with experimental data (points) for ^{3}He - ^{12}C at T_L (^{3}He) = 108.5 MeV, ^{4}He - ^{12}C at T_L = 139.0 MeV, ^{6}Li - ^{12}C at T_L (^{6}Li) = 90.0 MeV systems.

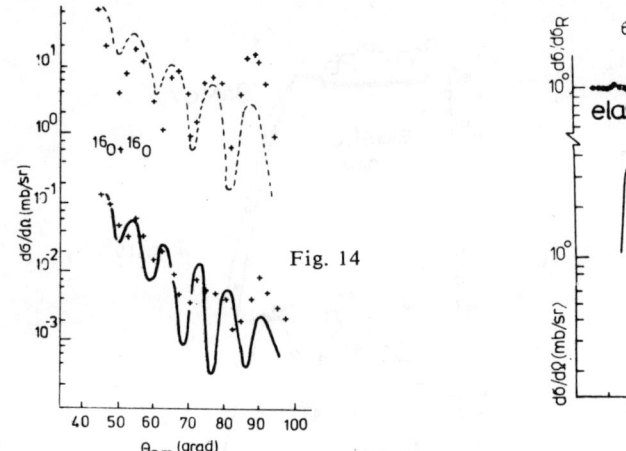
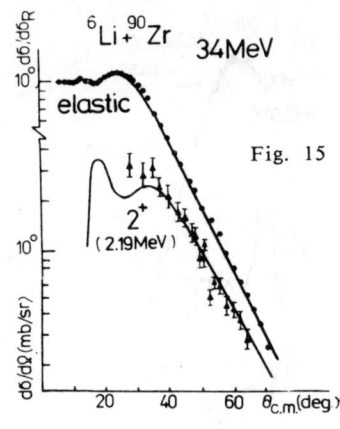

Fig. 14   Cross sections of $^{16}O+^{16}O$ elastic scattering at E_{lab}=49 MeV (---) and 63 MeV (___) for finite-range forces.

Fig. 15 & 16: The calculated cross-sections of ^{6}Li scattering from ^{90}Zr at 34 MeV and ^{58}Ni at 71 MeV.

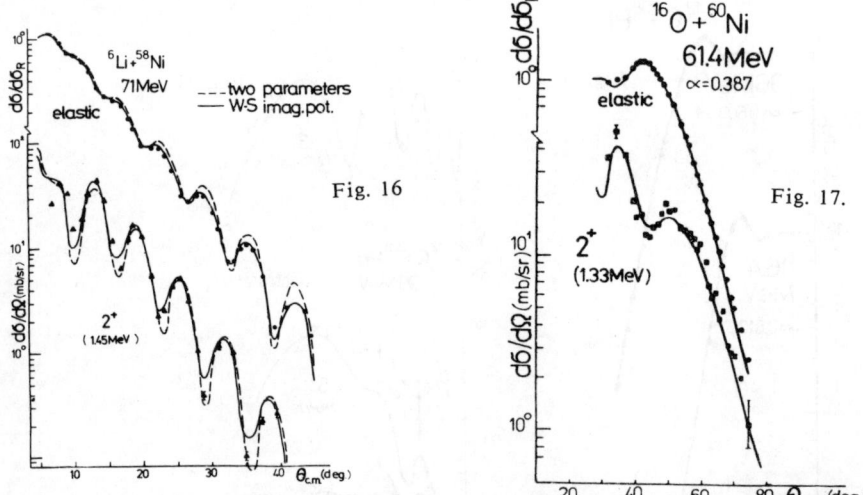

Fig. 17 to 21: The calculated cross-sections of ^{16}O scattering from ^{60}Ni at 61.4 MeV (17); at 114 MeV (18): from ^{208}Pb at 104 MeV (19); [next page:] for ^{12}C + ^{144}Nd at 70.4 MeV calculated with different sets of isoscalar and isovector constants (20): and elastic and inelastic cross-sections of ^{12}C scattering of ^{208}Pb (21).

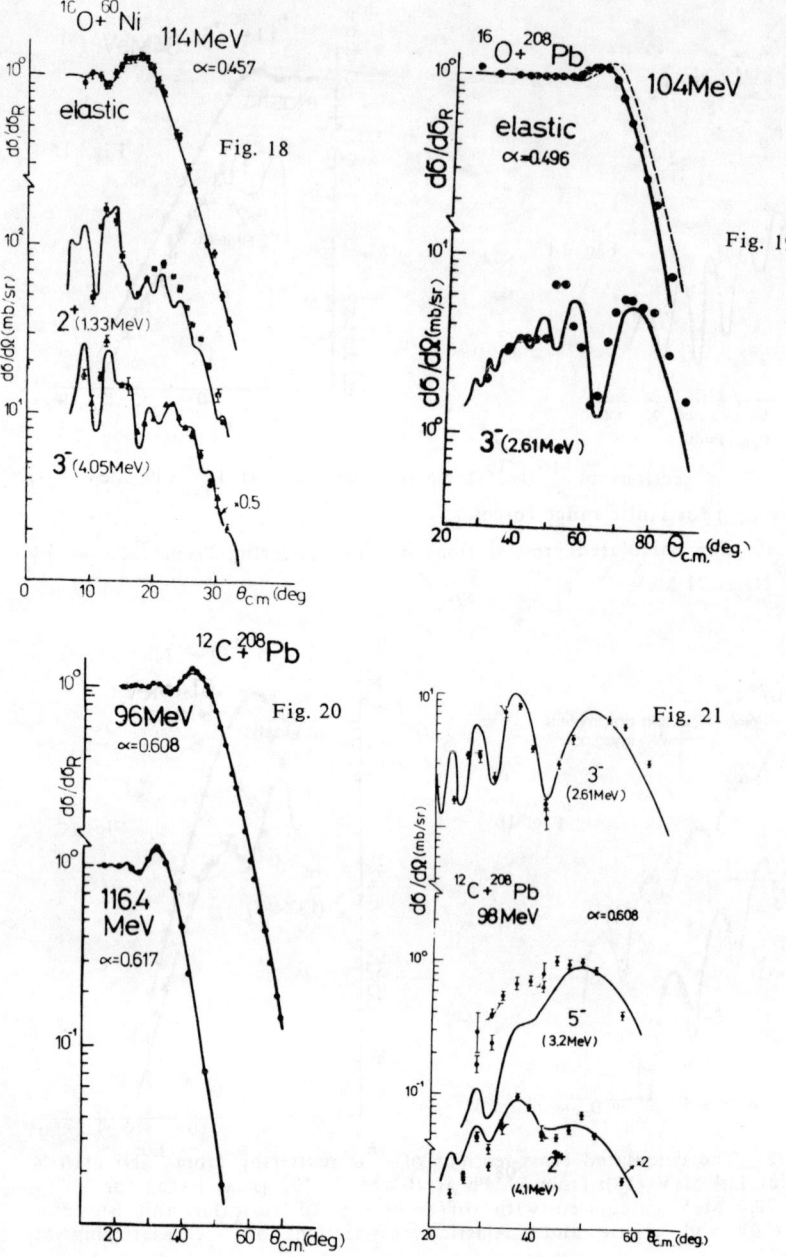

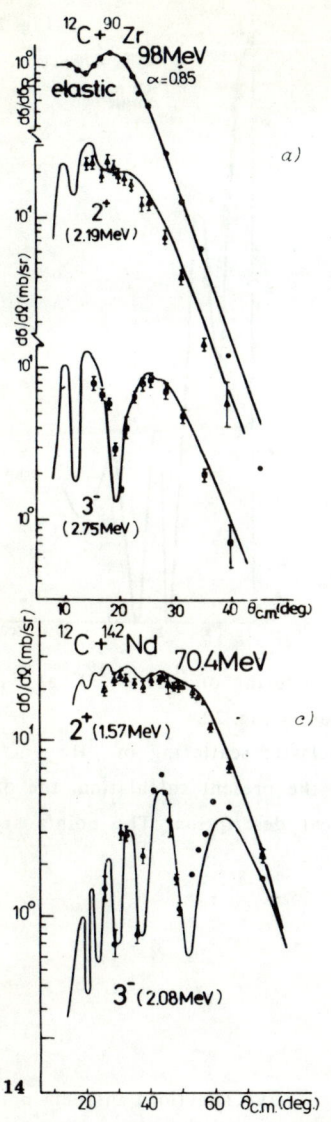
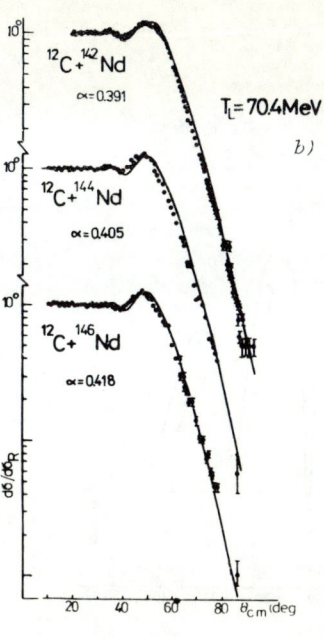

Fig. 22. The calculated cross-sections of ^{12}C scattering on isotopes at 70.4 MeV and ^{90}Zr at 98 MeV.

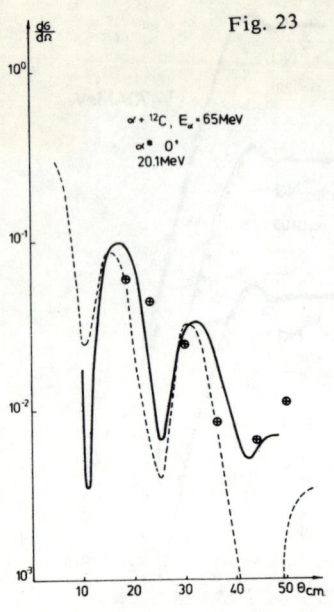
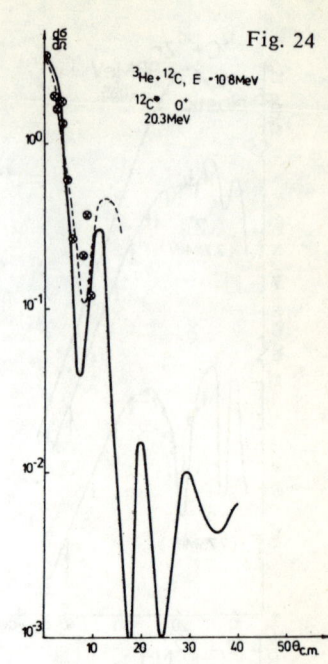

Fig. 23 Cross sections of the inelastic scattering of ^{4}He + ^{12}C at E_{4He} = 65 MeV. The curves have the same meaning as in Fig. 22.

Fig. 24 Angular distribution of the inelastic scattering of ^{3}He + ^{12}C at E_{3He} = 108 MeV. The solid line refers to the present calculation, the dashed line shows the result of a phenomenological description. The points are the experimental data.

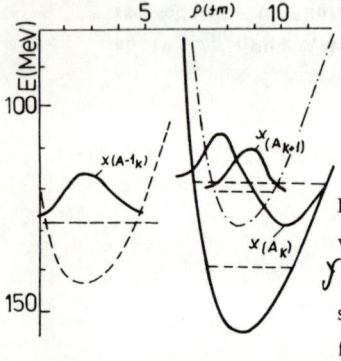

Fig. 25 Three wave functions in their potential wells: the wave function of the monopole $f(^N A_K)$ and dipole $f(^N A_{K+1})$ excited states in A nucleus and the ground state wave function of the A-1 nucleus.